# Maracá

# Maracá

## THE BIODIVERSITY AND ENVIRONMENT OF AN AMAZONIAN RAINFOREST

Edited by

**WILLIAM MILLIKEN**

and

**JAMES A. RATTER**

*Royal Botanic Garden Edinburgh*

## JOHN WILEY & SONS
Chichester • New York • Weinheim • Brisbane • Singapore • Toronto

Copyright © 1998 by John Wiley & Sons Ltd,
Baffins Lane, Chichester,
West Sussex PO19 1UD, England

National 01243 779777
International (+44) 1243 779777
e-mail (for orders and customer service enquiries):
cs-books@wiley.co.uk.
Visit our Home Page on http://www.wiley.co.uk
or http://www.wiley.com

All Rights Reserved. No part of this publication may be reproduced, stored in a retrieval system, or transmitted, in any form or by any means, electronic, mechanical, photocopying, recording, scanning or otherwise, except under the terms of the Copyright, Designs and Patents Act 1988 or under the terms of a licence issued by the Copyright Licensing Agency, 90 Tottenham Court Road, London, UK W1P 9HE, without the permission in writing of the publisher.

*Other Wiley Editorial Offices*

John Wiley & Sons, Inc., 605 Third Avenue,
New York, NY 10158-0012, USA

WILEY-VCH Verlag GmbH, Pappelallee 3,
D-69469 Weinheim, Germany

Jacaranda Wiley Ltd, 33 Park Road, Milton,
Queensland 4064, Australia

John Wiley & Sons (Asia) Pte Ltd, 2 Clementi Loop #02-01,
Jin Xing Distripark, Singapore 129809

John Wiley & Sons (Canada) Ltd, 22 Worcester Road,
Rexdale, Ontario M9W 1L1, Canada

*Library of Congress Cataloging-in-Publication Data*

Maracá : the biodiversity and environment of an Amazonian rainforest /
   edited by William Milliken and James A. Ratter.
      p. cm.
   Includes bibliographical references and index.
   ISBN 0-471-97917-1
   1. Natural history—Brazil—Maracá Island (Roraima)  2. Rain forest ecology—Brazil—Maracá Island (Roraima)  I. Milliken, William.  II. Ratter, J. A.
QH117.M295  1998
508.81'14—dc21                                              97-41832
                                                                CIP

*British Library Cataloguing in Publication Data*

A catalogue record for this book is available from the British Library

ISBN 0-471-97917-1

Typeset in 10/12pt Times from the authors' disks
by Mayhew Typesetting, Rhayader, Powys
Printed and bound in Great Britain by Bookcraft (Bath) Ltd

This book is printed on acid-free paper responsibly manufactured from sustainable forestation, for which at least two trees are planted for each one used for paper production.

# Contents

| | | |
|---|---|---|
| List of Contributors | | ix |
| Foreword by Dr John Hemming, CMG | | xiii |
| Preface | | xv |
| Acknowledgements | | xxi |
| 1 | **The Ilha de Maracá and the Roraima Region**<br>*Michael J. Eden and Duncan F.M. McGregor* | 1 |
| 2 | **Aspects of the Geology of the Ilha de Maracá**<br>*José Mauro Martini* | 13 |
| 3 | **Geomorphology of the Ilha de Maracá**<br>*Duncan F.M. McGregor and Michael J. Eden* | 25 |
| 4 | **The Soils of the Ilha de Maracá**<br>*Stephen Nortcliff and Daniel Robison* | 47 |
| 5 | **The Vegetation of the Ilha de Maracá**<br>*William Milliken and James A. Ratter* | 71 |
| 6 | **Further Contributions to the Botany of the Ilha de Maracá** | |
| | **The pteridophytes of the Ilha de Maracá**<br>*Peter J. Edwards* | 113 |
| | **A small area of young secondary forest on the Ilha de Maracá: structure and floristics**<br>*Robert P. Miller and John Proctor* | 130 |
| | **Phenological observations on a savanna–forest boundary on the Ilha de Maracá**<br>*Robert P. Miller, João Ferraz and José L. dos Santos* | 135 |
| 7 | **Primates of the Ilha de Maracá**<br>*Andrea P. Nunes, José Márcio Ayres, Eduardo S. Martins and José de Sousa e Silva* | 143 |
| 8 | **White-lipped Peccaries and Palms on the Ilha de Maracá**<br>*José M.V. Fragoso* | 151 |
| 9 | **The Bats of the Ilha de Maracá**<br>*Fif Robinson* | 165 |

| | | |
|---|---|---|
| 10 | Small Mammals of the Ilha de Maracá<br>Adrian A. Barnett and Aléxia C. da Cunha | 189 |
| 11 | Birds of the Ilha de Maracá<br>José Maria C. da Silva | 211 |
| 12 | The Reptilian Herpetofauna of the Ilha de Maracá<br>Mark T. O'Shea | 231 |
| 13 | Population and Ecology of the Tortoises *Geochelone carbonaria* and *G. denticulata* on the Ilha de Maracá<br>Debra Moskovits | 263 |
| 14 | The Frogs of the Ilha de Maracá<br>Márcio Martins | 285 |
| 15 | Social Wasps (Hymenoptera, Vespidae) of the Ilha de Maracá<br>Anthony Raw | 307 |
| 16 | Insects of the Ilha de Maracá. Further Contributions I: General Entomology | |
| | Biological data on the Passalidae (Coleoptera) of the Ilha de Maracá<br>Paulo F. Bührnheim and Nair Otaviano Aguiar | 323 |
| | Pollinators, pollen robbers, nectar thieves and ant guards of *Passiflora longiracemosa*<br>Forbes P. Benton | 331 |
| | Field observations on Phoridae (Diptera) associated with ants on the Ilha de Maracá<br>Forbes P. Benton | 338 |
| | An entomological curiosity<br>Forbes P. Benton | 345 |
| | Litter-consuming termites on the Ilha de Maracá<br>Adelmar G. Bandeira | 348 |
| | Butterflies of the Ilha de Maracá<br>Olaf H.H. Mielke and Mirna M. Casagrande | 355 |
| 17 | Insects of the Ilha de Maracá. Further Contributions II: Medical Entomology | |
| | Sandflies (Diptera: Psychodidae) of the Ilha de Maracá<br>Eloy G. Castellón, Nelson A. Araújo Filho, Nelson F. Fé and Josette M.C. Alves | 361 |
| | Triatomine bugs on the Ilha de Maracá<br>Toby V. Barrett | 366 |

|   | *Anopheles* species of the Ilha de Maracá: Incidence and distribution, ecological aspects and the transmission of malaria<br>Iléa Brandão Rodrigues and Wanderli P. Tadei | 369 |
|---|---|---|
| 18 | Arachnids of the Ilha de Maracá | |
|   | Notes on the spiders of the Ilha de Maracá<br>Arno Antônio Lise | 377 |
|   | Pseudoscorpions (Arachnida) of the Ilha de Maracá<br>Nair Otaviano Aguiar and Paulo F. Bührnheim | 381 |
| 19 | Earthworms of the Ilha de Maracá<br>Gilberto Righi | 391 |
| 20 | The Rotifera of Shallow Waters of the Ilha de Maracá<br>Walter Koste and Barbara Robertson | 399 |
| 21 | Biological Indicators in the Aquatic Habitats of the Ilha de Maracá<br>Cecilia Volkmer-Ribeiro, Maria Cristina D. Mansur, Pedro A.S. Mera and Sheila M. Ross | 403 |
| 22 | Soil Properties and Plant Communities Over the Eastern Sector of the Ilha de Maracá<br>Peter A. Furley | 415 |
| 23 | Human Occupation on the Ilha de Maracá: Preliminary Notes<br>John Proctor and Robert P. Miller | 431 |

**Appendices**

| 1 | (Chapter 5) | Plant species mentioned in the account of the vegetation of the Ilha de Maracá | 443 |
|---|---|---|---|
| 2 | (Chapter 10) | Key to small mammals known from the Ilha de Maracá | 447 |
| 3 | (Chapter 10) | Other mammals on the Ilha de Maracá | 449 |
| 4 | (Chapter 11) | Bird species of the Ilha de Maracá | 451 |
| 5 | (Chapter 12) | Reptiles of the Ilha de Maracá | 462 |
| 6 | (Chapter 16) | Butterflies collected on the Ilha de Maracá | 467 |
| 7 | (Chapter 18) | Preliminary list of spiders collected on the Ilha de Maracá | 479 |
| 8 | (Chapter 20) | Species of rotifers collected on the Ilha de Maracá | 482 |
| 9 | (Chapter 22) | Soil profile descriptions from representative vegetation types at the eastern end of the Ilha de Maracá | 486 |

Index   489

# List of Contributors

**Nair Otaviano Aguiar**, Instituto de Ciências Biológicas, Lab. de Zoologia, Universidade do Amazonas, Campus Universitário, 69.077-000 Manaus, AM, Brazil

**Josette M.C. Alves**, Secretaria Estadual de Educação e Cultura, 1984 Av. Perimetral D., 31 de Marco, 69.076-630 Manaus, AM, Brazil

**Nelson A. Araújo Filho**, *deceased*

**José Márcio Ayres**, Museu Paraense Emílio Goeldi, Caixa Postal 399, Campus de Pesquisa, Av. Perimetral Guamá, Belém, PA, Brazil

**Adelmar G. Bandeira**, Instituto Nacional de Pesquisas da Amazônia (INPA), Caixa Postal 478, 69.083 Manaus, AM, Brazil

**Adrian A. Barnett**, Department of Life Sciences, The Roehampton Institute, West Hill, London, SW15 3SN, UK

**Toby V. Barrett**, Instituto Nacional de Pesquisas da Amazônia (INPA), Caixa Postal 478, 69.083 Manaus, AM, Brazil

**Forbes P. Benton**, CEPLAC/CEPEC/SECEN, Caixa Postal 7, 45.600-000 Itabuna, BA, Brazil

**Paulo F. Bührnheim**, Instituto de Ciências Biológicas, Lab. de Zoologia, Universidade do Amazonas, Campus Universitário, 69.077-000 Manaus, AM, Brazil

**Mirna M. Casagrande**, Departamento de Zoologia, Universidade Federal de Paraná, Caixa Postal 19020, 81.504 Curitiba, PA, Brazil

**Eloy G. Castellón**, Instituto Nacional de Pesquisas da Amazônia (INPA), Caixa Postal 478, 69.083 Manaus, AM, Brazil

**Aléxia C. da Cunha**, 110 Hilário de Gouveia, Apto 601, Copacabana, 22.040 Rio de Janeiro, RJ, Brazil

**Michael J. Eden**, Department of Geography, Royal Holloway, University of London, Egham, Surrey, TW20 0EX, UK

**Peter J. Edwards**, Royal Botanic Gardens, Kew, Richmond, Surrey, TW9 3AE, UK

**Nelson F. Fé**, Instituto de Medicina Tropical de Manaus, 23 Av. Pedro Teixeira, Dom Pedro I, 69.040-000 Manaus, AM, Brazil

**João Ferraz**, Instituto Nacional de Pesquisas da Amazônia (INPA), Caixa Postal 478, 69.083 Manaus, AM, Brazil

**José M.V. Fragoso**, Department of Wildlife Ecology and Range Sciences, University of Florida, Gainesville, FL 32611, USA

**Peter A. Furley**, Department of Geography, University of Edinburgh, Drummond Street, Edinburgh, EH8 9XP, UK

**Walter Koste**, Ludwig-Brill-Str. 5, Quakenbrück, Germany

**Arno Antônio Lise**, Instituto de Biociências, Pontifícia Universidade Católica do Rio Grande do Sul, Av. Ipiranga 6681, Caixa Postal 1429, 90.619-900 Porto Alegre, RS, Brazil

**Maria Cristina D. Mansur**, Museu de Ciências Naturais, Fundação Zoobotânica do Rio Grande do Sul, Caixa Postal 188, 90.610 Porto Alegre, RS, Brazil

**José Mauro Martini**, Convênio DNPM/INPA, 8° Distrito, Departamento Nacional da Produção Mineral, a/c INPA, Caixa Postal 478, 69.083 Manaus, AM, Brazil

**Eduardo S. Martins**, Museu Paraense Emílio Goeldi, Caixa Postal 399, Campus de Pesquisa, Av. Perimetral Guamá, Belém, PA, Brazil

**Márcio Martins**, Laboratório de Zoologia, Departamento de Biologia, Instituto de Ciências Biológicas, Universidade de Amazonas, 69.060 Manaus, AM, Brazil

**Duncan F.M. McGregor**, Department of Geography, Royal Holloway, University of London, Egham, Surrey, TW20 0EX, UK

**Pedro A.S. Mera**, Instituto Nacional de Pesquisas da Amazônia (INPA), Caixa Postal 478, 69.083 Manaus, AM, Brazil

**Olaf H.H. Mielke**, Departamento de Zoologia, Universidade Federal de Paraná, Caixa Postal 19020, 81.504 Curitiba, PA, Brazil

**Robert P. Miller**, 118 Newins-Ziegler, School of Forest Resources and Conservation, University of Florida, Gainesville, FL 32611, USA

**William Milliken**, Royal Botanic Garden Edinburgh, Inverleith Row, Edinburgh, EH3 5LR, UK

**Debra Moskovits**, Field Museum of Natural History, Roosevelt Road at Lake Shore Drive, Chicago, IL 60605-2496, USA

**Stephen Nortcliff**, Department of Soil Science, University of Reading, Whiteknights, Reading, RG6 6DW, UK

**Andrea P. Nunes**, Museu Paraense Emílio Goeldi, Caixa Postal 399, Campus de Pesquisa, Av. Perimetral Guamá, Belém, PA, Brazil

**Mark T. O'Shea**, 46 Buckingham Road, Penn, Wolverhampton, WV4 5TS, UK

**John Proctor**, Department of Biological and Molecular Sciences, University of Stirling, Stirling, FK9 4LA, UK

**James A. Ratter**, Royal Botanic Garden Edinburgh, Inverleith Row, Edinburgh, EH3 5LR, UK

LIST OF CONTRIBUTORS

**Anthony Raw**, Departamento de Biologia Animal, Instituto de Ciências Biológicas, Universidade de Brasília, 70.000 Brasília, DF, Brazil

**Gilberto Righi**, Instituto de Biociências, Universidade de São Paulo, Caixa Postal 11.461, 5421-01000 São Paulo, SP, Brazil

**Barbara Robertson**, Instituto Nacional de Pesquisas da Amazônia (INPA), Caixa Postal 478, 69.083 Manaus, AM, Brazil

**Fif Robinson**, The Old Gasworks, The Batts, Whitby, North Yorkshire, YO21 1PG, UK

**Daniel Robison**, Department of Soil Science, University of Reading, Whiteknights, Reading, RG6 6DW, UK

**Iléa Brandão Rodrigues**, Instituto Nacional de Pesquisas da Amazônia (INPA), Caixa Postal 478, 69.083 Manaus, AM, Brazil

**Sheila M. Ross**, Department of Geography, University of Bristol, University Road, Bristol, BS8 1SS, UK

**José L. dos Santos**, Instituto Nacional de Pesquisas da Amazônia (INPA), Caixa Postal 478, 69.083 Manaus, AM, Brazil

**José Maria C. da Silva**, Departamento de Zoologia, Museu Paraense Emílio Goeldi, Caixa Postal 399, Campus de Pesquisa, Av. Perimetral Guamá, Belém, PA, Brazil

**José de Sousa e Silva**, Museu Paraense Emílio Goeldi, Caixa Postal 399, Campus de Pesquisa, Av. Perimetral Guamá, Belém, PA, Brazil

**Wanderli P. Tadei**, Instituto de Biociências, UNESP, Campus São José do Rio Preto, Rua Cristovão Colombo 2265, 15.054-000 São José do Rio Preto, SP, Brazil

**Cecilia Volkmer-Ribeiro**, Museu de Ciências Naturais, Fundação Zoobotânica do Rio Grande do Sul, Caixa Postal 188, 90.610 Porto Alegre, RS, Brazil

# Foreword

The Maracá Rainforest Project started with an invitation from Dr Paulo Nogueira Neto of the Brazilian environment secretariat SEMA to the Royal Geographical Society. In his letter of invitation, Dr Nogueira explained that SEMA had built a research station on Maracá, its riverine island reserve in Roraima, but that it knew very little about the island's natural resources. He wrote: 'Our basic objective is that research should be undertaken to provide a full understanding of the ecosystems that exist there. We want to obtain a general ecological survey of the research station in the following disciplines: geology, geomorphology, soil studies, climate, hydrology, vegetation and fauna.'

The ecological survey was the Maracá Rainforest Project's response to this request. It was by far the largest of the Project's five research programmes. It was led, with wisdom, energy and flair, by Dr James Ratter, one of the editors of this book. It eventually involved 101 scientists and scientific technicians, working for some 3500 person-days. This programme was truly international, with the majority of its scientists coming from Brazil and particularly from the Amazon research institute INPA in Manaus. One of its outstanding researchers was William Milliken, the other editor of this book and a botanist who worked without a break for the full 13 months of the Project's main phase.

As the authors explain in their Preface, the papers in this book represent many months of research, both in the field and in post-project analysis of material. Although the book is only a beginning in recording the seemingly infinite variety of flora, fauna and other properties of Maracá's rich ecosystems, it presents the most thorough study yet made of a forested area in the north of the Amazon basin.

There were many discoveries and surprises on Maracá. It proved to be extraordinarily rich in biodiversity, lying as it does on two biogeographical boundaries: between the Amazon basin and the Guiana Shield and between natural savannas and Hylaean rainforests. Its biological diversity is further boosted by the range of habitats encompassed, which includes riverine and *terra firme* forests, wetlands and low hills.

As the Project's leader, I am really delighted to introduce this book, the last of the many publications that have resulted from it. I congratulate the editors and the authors of the many excellent papers. They have produced a splendid portrait of a remarkable ecosystem.

**Dr John Hemming, CMG**
Royal Geographical Society

# Preface

## THE MARACÁ RAINFOREST PROJECT

This book is the product of a great deal of work by a great many people. For the most part[1] it represents a compilation of the ecological survey results obtained by the INPA/RGS/SEMA Maracá Rainforest Project (Projeto Maracá). This was a major multi-disciplinary international expedition to the Brazilian Amazon, mounted jointly by the Instituto Nacional de Pesquisas da Amazônia and the Royal Geographical Society at the invitation of the Secretaria Especial do Meio Ambiente. At the time (1987–88), SEMA was the Brazilian environmental agency responsible for the management of the Ilha de Maracá, as well as for a large number of the country's other reserves, but it has since been subsumed into the larger Instituto Brasileiro do Meio Ambiente (IBAMA).

The primary purpose of the project was to carry out a comprehensive ecological survey of the Ilha de Maracá, a large and important forest–savanna reserve in Brazil's northernmost Amazon state, Roraima. Between February 1987 and March 1988 a total of approximately 200 scientists, technicians, administrators and ancillaries participated in the project (listed in full by Hemming and Ratter, 1993). An extremely wide range of studies was carried out, contributing both to the ecological survey and to a variety of other scientific sub-projects. The overall result was one of the most complete pictures that exists of the biodiversity and environment of a tropical forest.

## MARACÁ

The Ilha de Maracá is a riverine island in the Rio Uraricoera, one of the principal tributaries of the Rio Branco. At the western end of the island (just above the Cachoeira de Purumame), the river divides into two channels, the northern Furo de Santa Rosa and the southern Furo de Maracá. These rejoin at its easternmost point (Figure Pr.1). Maracá is 60 km long and up to 25 km wide with a total area of rather more than 101 000 ha, making it one of the world's largest riverine islands.

Administratively, Maracá lies within the Municipality of Boa Vista in the State of Roraima, and was established in 1978 as the first of SEMA's *Estações Ecológicas* (ecological reserves). A comfortable, well-equipped research station, referred to here as the Ecological Station, was installed near its eastern end. The few settlers who were living on the island at that time were encouraged to leave (see Chapter 23), and it has since remained completely uninhabited apart from a handful of staff. The island therefore constitutes a large nature reserve virtually free from human interference.

The majority of the Maracá Rainforest Project's research was conducted at the eastern end of the island, in the vicinity of the Ecological Station. This was also true of most of the work conducted before and after the project. Work in this area was facilitated by the immediate logistic support and accommodation provided by the Station, and a substantial system of forest trails originally cut for Debra Moskovits' tortoise research (see Chapter 13). Nevertheless, considerable efforts were made to open up a network of camps and trails into the central and western parts of the island in order to provide a clearer idea of its overall environmental and biological diversity. The names of these trails and camps have been standardized,[2] as have those of the other named features on and around Maracá, and their locations are shown in Figure Pr.1

## PREVIOUS AND SUBSEQUENT RESEARCH ON MARACÁ

Prior to the Maracá Rainforest Project, little scientific research had been conducted on the island and little was known about its environment and biodiversity. Flying visits had been made to Maracá during the massive RADAMBRASIL surveys of the 1970s, when rapid descriptions were made of the soils and vegetation at a couple of sites. However, these data were sketchy and shown by later work to be unrepresentative.

The most detailed and long-term study to have been carried out on the island prior to the project was that made of the *Geochelone* tortoises by Debra Moskovits (see Chapter 13). Her doctoral thesis (1985) constituted a rich source of basic data for the work of the Maracá Rainforest Project, and her careful bird survey provided an important foundation for later ornithological studies (see Chapter 11). In addition, a handful of zoologists had conducted research on the island in varying degrees of detail, including Celso Morato de Carvalho of INPA Roraima (snakes and lizards), Márcio Martins (frogs, see Chapter 14) and Alan Mill (termites). A preliminary analysis of the vegetation had also been carried out as part of Projeto Flora Amazônica by a team from INPA, the New York Botanical Garden, and the Museu Integrado de Roraima, led by David Campbell (NYBG).

It was hoped that the establishment by the Maracá Rainforest Project of such a broad base of environmental and biological information, together with the physical revitalization and re-equipping of the Ecological Station, would open the way for further projects and help to turn the island into a thriving centre for rainforest research. To some extent this has been the case, and there have been some excellent studies conducted which were directly or indirectly inspired by the project. These include phenological research conducted by Robert Miller (see Chapter 6), primatological research by Andrea Nunes (Museu Goeldi) and Antônio Pontes (University of Cambridge), botanical/ecological studies of *Peltogyne* forests by Marcelo Nascimento and Dora Villela (Universidade Estadual do Norte Fluminense), and José Fragoso's studies of the white-lipped peccaries (see Chapter 8).

## PUBLICATIONS

Several years have now passed since the completion of the fieldwork of the Maracá Rainforest Project. During that time much has been written about the project and

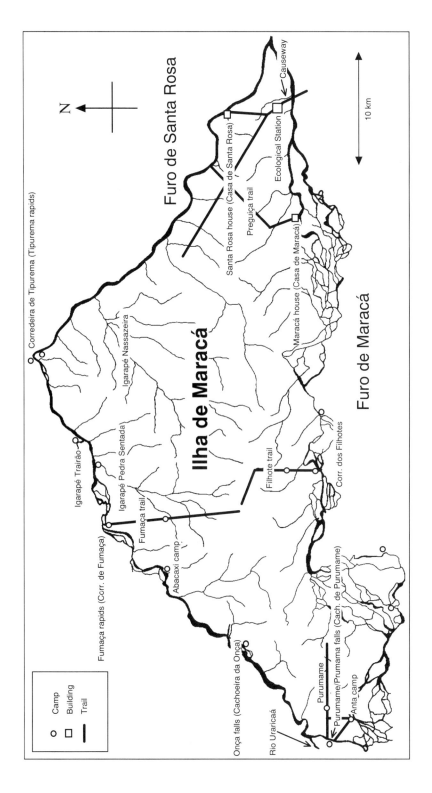

**Figure Pr.1.** Map of the Ilha de Maracá showing principal trails, camps and other sites referred to in the text

its results, and it has received much acclaim both within Brazil and abroad. From the outset a deliberate policy was put in place encouraging the initial dissemination and publication of the scientific results in Brazil (and ideally in Portuguese), and to a large extent this has been successful. In March 1988 a workshop was held in Boa Vista to present and discuss the initial results of the project, and this was followed by a full conference in Manaus in April 1989 (and another in London later that year). One issue of the INPA journal *Acta Amazonica* (Vol. 21, 1991) was devoted entirely to the results of the project, and many learned articles have since been published in other Brazilian journals. In addition, a series of bound reports, covering all aspects of the ecological survey, was submitted to SEMA in 1989 as a basis for management planning.

A substantial body of literature has also appeared in English. The amount of follow-up work put into the results of the ecological survey of Maracá has probably far outstripped the time originally spent in the field, but nevertheless in certain cases, e.g. the identification of insect and spider specimens, work is still proceeding and it will in fact be many years before all the *t*s have finally been crossed and all the *i*s dotted. To wait until all of the results are available before publishing an overall account of the biodiversity and environment of Maracá would be unrealistic, and this book is intended to represent the last major contribution to this field by the Maracá Rainforest Project. It complements the volume edited by Hemming (1994), which covers the other studies conducted by the project on and around Maracá (forest regeneration, litter and nutrient cycling, savanna–forest boundary dynamics, remote sensing and land development). Popular, illustrated accounts of the expedition have also been produced (Hemming *et al.*, 1989; Hemming and Ratter, 1993), as has an education pack for schools.

## CONSERVATION OF THE ILHA DE MARACÁ

The rationale underlying the Maracá Rainforest Project was fundamentally one of conservation. The Ilha de Maracá is an important example of the fringe forest–savanna environments of the northern rim of the Amazon basin, and yet without an understanding of its environment, biodiversity and ecology, SEMA was not in a position to formulate a management plan which would ensure its long-term survival. During the time that has elapsed since the termination of the field phase of the Maracá Rainforest Project in 1988, dramatic events have taken place in Roraima which have served to highlight the extreme importance of the establishment of an effective and informed strategy for conservation enforcement in the region.

As the Maracá Rainforest Project was finishing its field phase in early 1988, the stage was being set for one of the most serious human and environmental disasters ever to have taken place in Roraima. Gold and diamonds were discovered in large quantities in the headwaters of the tributaries of the Rio Branco, and when this became common knowledge through the media it triggered a massive rush of prospectors. Tens of thousands of illegal wildcat miners (*garimpeiros*), many of them desperate people from the poverty-stricken Brazilian Northeast or the dwindling mines of the eastern Amazon, invaded the lands of the Yanomami Indians and the other indigenous communities strung along the frontier zone (see MacMillan, 1995). Access to these remote areas was established through the building of over 100 illegal airstrips, and the *garimpeiros* set about causing such levels of havoc that the repercussions are still being felt by the environment and the indigenous inhabitants today, several years after the majority of the *garimpeiros* have left.

By the second half of 1988, the scientists of the Maracá Rainforest Project were back in their laboratories and libraries, analysing their specimens and data and drawing their conclusions about the island's virtually pristine environment. Meanwhile, a short way up the Uraricoera from Maracá, forests were being felled, rivers were being polluted with mercury and silt, game animals were being hunted out or chased away, and indigenous populations were literally being decimated by disease and violence. The cultures and lifestyles of people who had previously experienced little or no exposure to the outside world were being shaken to their roots by an uninvited encounter with some of the least desirable elements and aspects of Brazilian society. The government had effectively relinquished control of the region, and the *garimpeiros* had taken the law into their own hands.

As it happened, no significant deposits of gold were found on the Ilha de Maracá, and it appears that *garimpeiros* never established any large-scale mining operations on the island itself. If, however, gold had been found, untold damage would undoubtedly have been caused and it is almost certain that SEMA, with its meagre resources, would have been unable to do anything to prevent it. Maracá was not entirely unaffected by the gold rush, however. The Furo de Santa Rosa and (to a lesser extent) the Furo de Maracá became much-travelled routes for *garimpeiros* on their way up to the mines on the upper Uraricoera and Uraricaá rivers, and temporary transit camps were established along their lengths. For a period, also, mining barges were reported to be in operation towards the western end of the island, dredging gold-rich sediments from the bed of the Uraricoera. Inevitably the people inhabiting these camps hunted deep into the forests of the Ilha de Maracá, enjoying the abundance and relative fearlessness of its large mammal and bird populations. The damage which was done to wildlife during that period may have been significant.

The population of Roraima is expanding rapidly, and its frontiers are being pushed back at an alarming rate. In 1987 and 1988 the Ilha de Maracá was still in a state of relative isolation, being surrounded largely by untouched forest or open savannas, but in recent years a pincer movement of development has been closing on the island. To the north, the expanding colony of Trairão has been moving down towards the Furo de Santa Rosa, where it threatens to spread along the banks of the river. When this occurs, the island will be little more than a stone's throw from the homes of poor settlers who are unlikely to see the conservation status of Maracá as a serious obstacle to their exploitation of its resources. On the southern side, the *fazendas* (farms) which lie alongside the Furo de Maracá are spreading inexorably upstream as developers clear new tracts of forest and sell them off as farmland.

One day, in the not so very distant future, it is likely that the Ilha de Maracá will be an island cut off not only by a river but by a wasteland of degraded pasture.

When this has happened, the issues surrounding its conservation will have shifted from the largely theoretical to the intensely practical. An understanding of its ecology, its biodiversity and its environment, to which it is hoped that this book will have made a significant contribution, will then be of paramount importance.

**William Milliken and James A. Ratter**
Royal Botanic Garden Edinburgh

## NOTES

1. The research described in Chapters 8, 13 and 14 was conducted separately from the Maracá Rainforest Project.
2. In a few cases (e.g. the Cachoeira de Purumame/Prumama), where a place name has been transcribed from an indigenous language, there are alternative spellings and these have been left to the personal preference of the authors.

## REFERENCES

Hemming, J.H. (1994). *The rainforest edge. Plant and soil ecology on Maracá Island, Brazil.* Manchester University Press.

Hemming, J.H. and Ratter, J.A. (1993). *Maracá. Rainforest island.* Macmillan, London.

Hemming, J.H., Ratter, J.A. and Santos, A. dos (1988). *Maracá.* Empressa das Artes, São Paulo.

MacMillan, G.J. (1995). *At the end of the rainbow? Gold, land and people in the Brazilian Amazon.* Earthscan, London.

# Acknowledgements

To avoid duplication, the acknowledgements in the individual chapters of this book refer only to people and organizations whose assistance was specific to the sub-projects concerned, rather than general to the Maracá Rainforest Project (Projeto Maracá). However, it should be emphasized that almost all of these studies owe a part of their success to the people and organizations acknowledged below for their contributions to the overall project.

This book owes an acknowledgement to everybody who supported or participated in the Maracá Rainforest Project and contributed towards its success, and several others besides. To name them all here would not, unfortunately, be feasible.

Financial support for the project was received from a great many sources, which are listed in full by Hemming and Ratter (1993). Of these, particular thanks should be given to the Ford Foundation, ICI, Souza Cruz, the Headley Trust, the Baring Foundation, the Dulverton Trust and Mercedes Benz do Brasil.

The original invitation to undertake an ecological survey on the Ilha de Maracá was given by Dr Paulo Nogueira Neto of SEMA, and without this invitation the expedition would not have taken place.

We particularly wish to thank Dr John Hemming, leader of the Maracá Rainforest Project and then Director and Secretary of the Royal Geographical Society, for his persevering support and encouragement both in the UK and in Brazil. The success of this project was dependent on the efforts of the RGS and its tireless and ever cheerful administrative and logistic support team, of whom the following should be acknowledged: Steve Bowles, Fiona Watson, Sarah Latham, Daphne Hanbury and Sue Whinney.

Dr Ângelo dos Santos, then Chief of INPA's ecology department, was Brazilian leader of the Ecological Survey and the *responsável* (official liaison) appointed for the expedition by CNPq. We much appreciate his hard work on our behalf and are extremely grateful for his efforts to ensure the smooth running of the expedition. José Antônio Alves Gomes, who acted as liaison and coordinator for INPA's scientific participants in the field, also played a vital role in this capacity.

Were it not for Guttemberg Moreno de Oliveira's genuine dedication to the conservation of the Ilha de Maracá, of which he was and still is the administrator, the research facilities on the island would probably not have existed by the start of 1987 and the project would never have happened. Throughout the project he offered invaluable support, and the contributions of both he and his team of employees (Valquimar Felix de Souza or 'Amazonas', his wife Francinete, Antônio Alves da Silva, Eronias Silva dos Santos and his wife Patricia) are gratefully acknowledged.

It would be easy but unjust to overlook the contributions of many of the people whose roles in the Maracá Rainforest Project were perhaps less prominent than those of the scientists and leaders, but equally vital. Numerous *técnicos* or technicians took part in the research, mainly from the Instituto Nacional de Pesquisas da Amazônia (INPA) but also from the Museu Integrado de Roraima (MIRR) and other participating organizations. Many scientists are far more reliant upon the knowledge and technical abilities of these people than they would like to admit, and their contributions to research are rarely properly acknowledged. Luiz Pestana (driver and boatman), the Maiongong Indians of Uaicas and the Macuxi Indians of Mangueira also made invaluable contributions to the expedition.

The editing of this book was largely funded by the Royal Botanic Garden Edinburgh (RBGE), by the Royal Geographical Society and by the Overseas Development Administration (ODA) from whom William Milliken received a grant in 1991. Mary Mendum and Maureen Warwick are thanked for their assistance with illustrations, and our appreciation goes to Helen Foster for her tireless typing of manuscripts during the earlier stages of the work, and to Arnildo Pott for his help in the translation of the summaries.

# REFERENCE

Hemming, J. and Ratter, J.A. (1993). *Maracá. Rainforest island*. Macmillan, London.

# 1 The Ilha de Maracá and the Roraima Region

**MICHAEL J. EDEN AND DUNCAN F.M. McGREGOR**
*Royal Holloway, University of London, Egham, UK*

## SUMMARY

An introductory account is given of the State of Roraima and of the Ilha de Maracá which lies in its northern part. The Ilha de Maracá covers only a small portion of the State, but is of marked scientific interest and should continue as a designated conservation area. Mean annual rainfall in the Maracá area is estimated at 1900–2000 mm. The outline of the island, as well as elements of its macro-relief, appears substantially to be tectonically controlled, while dissected terrain surfaces at 250–330 m and at about 135 m above sea level may provide evidence of phases of etchplanation during the late-Tertiary and end-Tertiary respectively. The most typical soils of Ilha de Maracá are red–yellow podzols and latosols. The island is almost entirely forested, with a well-developed, closed canopy formation of semi-evergreen to semi-deciduous character. Locally, forest cover is displaced by small areas of poorly drained savanna and herbaceous swamp.

## RESUMO

Apresenta-se um resumo introdutório das características do Estado de Roraima, e em particular da Ilha de Maracá que encontra-se na parte setentrional desta região. A ilha constitui apenas uma porção pequena da área do Estado, mas tem grande interesse científico e deveria continuar designada como uma área de conservação. Estima-se 1900–2000 mm como a precipitaço anual média. O contorno da ilha, bem como os elementos do macrorelevo, aparentemente são de origem tectônica, enquanto extensões de terrenos dissecados, localizando-se a aproximadamente 250–333 m e 135 m acima do nivel do mar, possivelmente dêem evidências de fases de 'etchplanation' durante a parte última e no final do Terciário, respectivamente. Os solos mais típicos da ilha são podzólicos vermelho-amarelos e latosolos. Reveste-se a ilha quase completamente com mata semi–sempreverde ou semi-decídua de dossel fechado e estrutura bem desenvolvida. Mas, em alguns lugares a cobertura florestal é quebrada por manchas de savana mal-drenada e trechos de brejo com vegetação herbácea.

## INTRODUCTION

The Ilha de Maracá is an island by virtue of the bifurcation of the Rio Uraricoera around a triangular tract of land, covering some 112 000 ha, in the north of the State of Roraima. The State, which has an area of 23 000 000 ha, is broadly coincident with the basin of the Rio Branco, of which the Uraricoera is a major

*Maracá: The Biodiversity and Environment of an Amazonian Rainforest.*
Edited by William Milliken and James A. Ratter. © 1998 John Wiley & Sons Ltd.

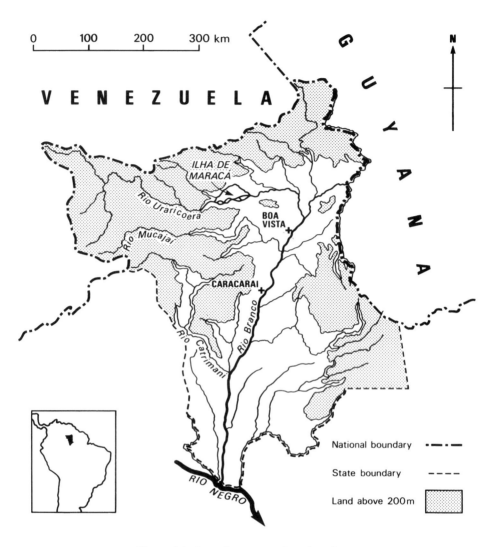

**Figure 1.1.** Location map – Roraima State

tributary. This most northern part of Brazil, flanked by Venezuela to the north and west and by Guyana to the east, culminates in the high summit of Mount Roraima (2810 m), which marks the conjunction of the three nation states (Figure 1.1).

Much of Roraima State is of rather low elevation and relief. Its capital of Boa Vista, which lies more than 2000 km by river from the ocean but has an elevation of only 120 m above sea level, lies within a broad plain that covers much of the middle Rio Branco basin. Further south, in the vicinity of Caracaraí, the plain is locally obstructed by mountains, but re-establishes itself subsequently and extends to the Rio Negro and beyond. Only in the north and west of the State, where the frontier with Venezuela coincides with the Serra Pacaraima and Serra Parima, is there more continuous, elevated terrain. In these remote areas, drained by the Mucajaí, Uraricoera and Surumú rivers, there exist recurrent high peaks, entrenched valleys and rivers obstructed by continuous rapids. This ancient landscape of the Guiana Shield, overlain in places by the Roraima sandstone, is both of striking appearance and of considerable attraction for its gold and diamonds.

Much of Roraima State is covered by tropical moist forest, which ranges from lowland to montane in character. As a function of climate and soil conditions, extensive areas also exist where the forest is replaced by open wooded or herbaceous communities. These include savanna, steppe savanna, and caatinga-type vegetation similar to that of the upper Negro basin (RADAMBRASIL, 1975a).

Very little information exists on pre-Columbian occupancy of Roraima. Indian groups were certainly present in earlier times, although low population densities seem to have been characteristic of both forest and non-forest areas. Indigenous groups are still present in many parts of the territory; among them are several thousand Yanomama in the forests of northwest Roraima, as well as groups such as the Wapixana and Makuxi who live in the savannas to the northeast. According to Migliazza (1978), the indigenous population of the State totals approximately 12 000, of whom about a third are integrated.

Modern settlement and exploitation in Roraima effectively dates from the late 18th century, when cattle were introduced to the northeast savannas and ranching activities commenced. The cattle population, which currently numbers some 300 000 (IBGE, 1987), is widely distributed within the savanna, but overall density (6–8 animals per $km^2$) is low and productivity levels are modest. The isolation and inaccessibility of the region has long hindered development, and the traditional cattle-based economy has aptly been described by Rivière (1972) as 'subsistence ranching'. By 1943, northern Roraima had a reported population of only 18 000 people, and although construction of a roadlink between Manaus and Boa Vista in the late 1970s has begun to open up the area, it is still relatively undeveloped even by Amazonian standards. In recent decades, some agricultural colonization has occurred in forest areas of central and northern Roraima, notably at Alto Alegre and east of Caracaraí, while in the last few years numerous *garimpeiros* (mineral prospectors) have established themselves on reserved Yanomama land in the extreme northwest of the state. In 1985 the population of Roraima as a whole was reported to be 102 491, of whom 66 028 were resident in Boa Vista (IBGE, 1987).

## CLIMATE

Scanty climatic data are available for the State, although there is an old-established meteorological station at Boa Vista. Mean annual temperatures in lowland areas are 26–27°C, with a mean monthly range of 2–3°C. Rainfall levels vary somewhat across the region (Figure 1.2). Boa Vista in the north is reported to have an annual total of 1751 mm, with seven dry months (<100 mm) in the year (RADAMBRASIL, 1975a). A similar regime prevails at St Ignatius in nearby Guyana. Rainfall totals generally increase to the south and west of Boa Vista, with most forested areas of the State having annual rainfall in the range 1800–2200 mm. With

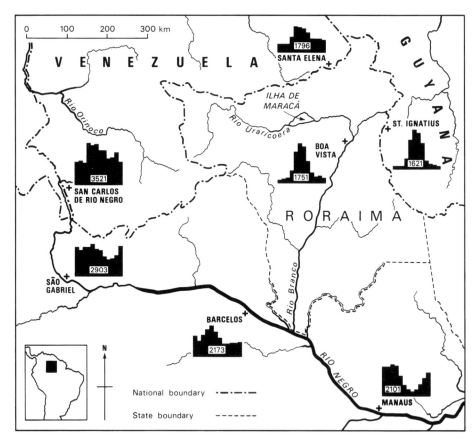

**Figure 1.2.** Annual rainfall totals and monthly distribution for the Roraima region

increasing rainfall total, the number of wet months in the year also increases. Manaus (2101 mm), to the southeast of the State, has only four dry months (<100 mm) each year, while Barcelos (2173 mm) experiences perennially wet conditions (Figure 1.2). Rainfall totals also vary significantly from year to year, especially in the drier northeast, where the onset and magnitude of the rainy season are highly variable (Frost, 1968).

No long-term meteorological records are available for the Ilha de Maracá. As at Boa Vista, marked rainfall seasonality is evident, with drier conditions commonly extending from September to March. Mean annual rainfall in the area is estimated at 1900–2000 mm.

## GEOLOGY

The State of Roraima lies on the southern flank of the Guiana Shield, which comprises an ancient complex of crystalline rocks extending across northeast South

America. The rocks are mainly granites, gneisses and mica schists of Pre-Cambrian age, although they are locally overlain by Proterozoic sandstones of the Roraima Formation (McConnell, 1968). The sandstones are most extensively developed in southeast Venezuela and adjacent areas of Guyana, and exist only locally in the State of Roraima. The main occurrence in the State is in the extreme north, flanking the sandstone mesa of Mount Roraima itself. Subsidiary outliers exist at Serra Tepequém and Serra Uafaranda to the west.

Elsewhere in the State, Shield rocks are commonly encountered. The northern frontier zone with Venezuela comprises mainly dacites, andesites and rhyolites of the Surumu Formation, while Serra Parima to the west consists largely of the alkaline Surucucu granite. Much of the remainder of the territory is made up of rocks belonging to the Guiana Complex, which is of earlier Pre-Cambrian age and comprises mainly amphiboles, diorites, granites and gneisses. In the northeast of the territory, Shield rocks at low elevation are locally overlain by Quaternary alluvium of the Boa Vista Formation, while in the extreme south there is a similar overlay of Tertiary and Quaternary alluvium known as the Solimões Formation; these sediments are commonly sandy in nature (RADAMBRASIL, 1975a, 1978).

The Ilha de Maracá lies within the area of the Guiana Complex, and consists mainly of quartz–biotite schists, quartz–feldspar gneisses, and associated tonalitic granites (Martini, Chapter 2). As elsewhere on the Guiana Shield, there is recurrent local evidence of faulting and related tectonic effects. Fault patterns in the area strongly influence drainage alignments, as in the case of the Furo de Santa Rosa; indeed, the outline of the island itself as well as elements of its macro-relief seem substantially to be tectonically controlled.

## GEOMORPHOLOGY

The landscapes of the Guiana Shield have undergone polycyclic development, a process well exemplified in the State of Roraima. Planation surfaces of gently undulating to dissected character, in places separated by pronounced scarps, have developed as a result of periodic landscape rejuvenation. The remnants of older, higher surfaces are rather fragmentary, although distinct levels of 500–600 m above sea level occur locally in northern parts of the State as well as on isolated mountain blocks further south such as the Serra do Demini and Serra da Lua. These planation remnants lie within morphostructural units described by Projeto Radambrasil as the *Planalto do Interflúvio Amazonas–Orenoco* and the *Planalto Residuais de Roraima* (Figure 1.3). A more extensive planar surface exists at some 250–400 m above sea level; the surface is well developed in the area immediately west of the Ilha de Maracá, while smaller and more isolated outliers occur in central and southern parts of the territory. The surface corresponds to the unit described as the *Planalto Dissecado Norte da Amazonia* (RADAMBRASIL, 1975a). The most extensive level in the region lies at 80–160 m above sea level, occupying the majority of the eastern and southern parts of the territory; this also is a planar surface, although it has locally been overlain by Tertiary and Quaternary deposits.

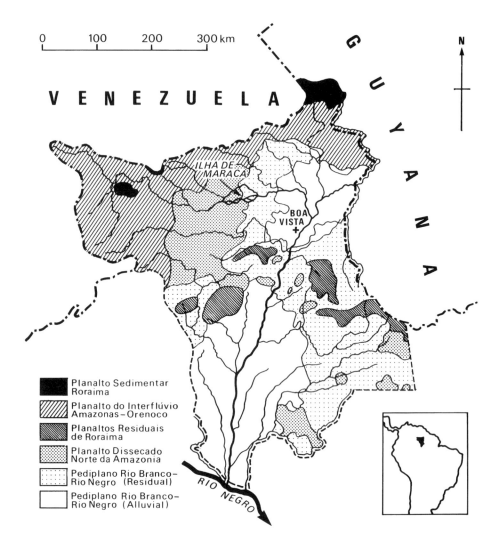

**Figure 1.3.** Morphostructural units of Roraima (after RADAMBRASIL, 1975a, 1975b, 1978)

It corresponds to the unit described as the *Pediplano Rio Branco–Rio Negro* (RADAMBRASIL, 1975a). Local incision into this surface exists along major drainage lines, forming contemporary floodplains.

The planation levels evident in the State of Roraima are widely reproduced across the Guiana Shield (Choubert, 1957; Short and Steenken, 1962; McConnell, 1968); they correspond to the following levels identified generally for Brazilian Amazonia (Bigarella and Ferreira, 1985): Pediplane $Pd_3$ – a mid-Tertiary (Oligocene) surface at an elevation of 400–600 m; Pediplane $Pd_2$ – a late-Tertiary (upper Miocene–lower Pliocene) surface at an elevation of 250–400 m; Pediplane $Pd_1$ – a Plio-Pleistocene surface at an elevation of 50–200 m.

A notable feature of the regional morphology of Roraima is the separation of the 80–160 m surface into northern and southern parts by remnants of older surfaces extending across the central part of the territory (Figure 1.3). This reflects an earlier drainage alignment; originally the eastward-flowing Uraricoera and Mucajaí rivers continued their courses, via the Tacutu, into Guyana and thence flowed northeastwards to the Atlantic. The Proto-Berbice, as this earlier river was called (McConnell, 1968), suffered capture of its original Uraricoera and Mucajaí headwaters by the Rio Branco in late-Tertiary times. The remnants of older surfaces in central Roraima are part of the original watershed of the Amazon and Proto-Berbice basins.

Whether the main surfaces in Roraima are appropriately described as pediplanes (McConnell, 1968; Bigarella and Ferreira, 1985) is open to question. Evidence of deep weathering processes and the existence of residual, domed inselbergs in many parts of the Guiana Shield are arguably more suggestive of etchplanation, associated with deep weathering under humid conditions and stripping of saprolite under subsequent semi-arid conditions, than of pediplanation (Eden, 1971; Kroonenberg and Melitz, 1983). The age of the older surfaces in Roraima is also uncertain. According to McConnell (1968), older levels in the Shield landscape date from at least late Cretaceous to early Tertiary times, while Bigarella and Ferreira (1985) claim, for Brazilian Amazonia at least, that there are no remains in the present-day relief of features developed before mid-Tertiary times. As yet, only preliminary landform studies have been undertaken in the territory, and there is scope for more detailed investigation.

As Figure 1.3 shows, the Ilha de Maracá is located at the transition between the 80–160 m surface to the east (*Pediplano Rio Branco–Rio Negro*) and the 250–400 m surface to the west (*Planalto Dissecado Norte da Amazonia*). The former level comprises lightly to moderately dissected terrain at about 135 m above sea level in eastern and central Maracá, while higher and more dissected areas at 250–330 m in the island interior are remnants of the older surface.

## SOILS

The soils of Roraima exhibit significant variability at local and regional levels. As yet, only reconnaissance survey has been undertaken over most of the territory, and, particularly in more remote areas, no more than preliminary soil data are available. As expected, emerging soil patterns reflect the variety of climatic, geologic and geomorphic conditions in the territory.

Numerous soil types have been identified in the Projeto Radambrasil surveys that cover Roraima (RADAMBRASIL, 1975a, 1975b, 1978). Among the soils most commonly reported are latosols (cf. Oxisols, US taxonomy), red–yellow podzols (cf. Ultisols, US taxonomy), lithosols (cf. Lithic subgroups, US taxonomy) and a variety of hydromorphic soils, including gleyed, lateritic, and podzolic types. The soils are generally variable in texture, commonly as a function of parent materials, but are mostly dystrophic and of low chemical fertility. A few eutrophic soils exist, associated for example with basic intrusive parent materials that give rise to *Terra Roxa Estruturada* (cf. Alfisol, US taxonomy).

The most common soils in the State are latosols and red–yellow podzols, which are widely encountered on planar surfaces of low to moderate relief. The soils are mostly dystrophic, fine to medium in texture, and often plinthitic. The older 250–400 m surface (*Planalto Dissecado Norte da Amazonia*) carries mainly fine-textured red–yellow podzols, while similar residual areas of the 80–160 m surface (*Pediplano Rio Branco–Rio Negro*) carry red–yellow podzols and red–yellow latosols, again mainly fine-textured in character; alluvial areas of the latter surface have mainly medium-textured yellow latosols, at least where free drainage prevails.

There are extensive areas of lithosols in Roraima. They mostly occur in more elevated and mountainous areas to the north and northwest. They are present over Roraima sandstone and in areas of dissected Shield rocks. Lithosols also occur in more elevated areas of central Roraima, such as the Serra da Mocidade and Serra do Mucajaí.

Hydromorphic soils of various kinds are also present in the region. They are widespread in low-lying, depositional areas of the *Pediplano Rio Branco–Rio Negro*. In the savanna zone of northeast Roraima there are extensive areas of hydromorphic quartz sands and hydromorphic laterites, while alluvial areas in the southern territory carry hydromorphic podzols and some hydromorphic quartz sands. River floodplains mostly have hydromorphic gley or alluvial soils. The hydromorphic soils are mainly dystrophic and of low chemical fertility.

The Ilha de Maracá has soils that are mostly developed from residual Shield parent materials and are mainly fine-textured and dystrophic in character. According to Projeto RADAMBRASIL, they comprise both red–yellow podzols and red–yellow latosols, although, as elsewhere in the State, recurrent toposequences exist that contain hydromorphic downslope members. Marginal areas of the island have some more continuous areas of hydromorphic soils, including alluvial soils (RADAMBRASIL, 1975a).

## VEGETATION

Much of Roraima State is covered by forest vegetation, which includes montane, sub-montane and lowland formations. Other vegetation types identified by Projeto Radambrasil are savanna (*cerrado*), steppe savanna (*savana estépica*), and pioneer formations (*formações pioneiras*) (Figure 1.4). These various types, which are largely a response to contrasting climate and soil conditions, create a highly complex regional pattern of vegetation.

The forest vegetation of Roraima is extensively developed in the west and southeast parts of the state. In western Roraima there are extensive tracts of montane and sub-montane evergreen forest; in more seasonal areas, such as parts of the lower Mucajaí basin, the evergreen cover is replaced by semi-evergreen to semi-deciduous forest. In southeast Roraima, there are further extensive tracts of evergreen forest, again mainly sub-montane in character; areas of more open forest cover also exist (RADAMBRASIL, 1975a, 1975b).

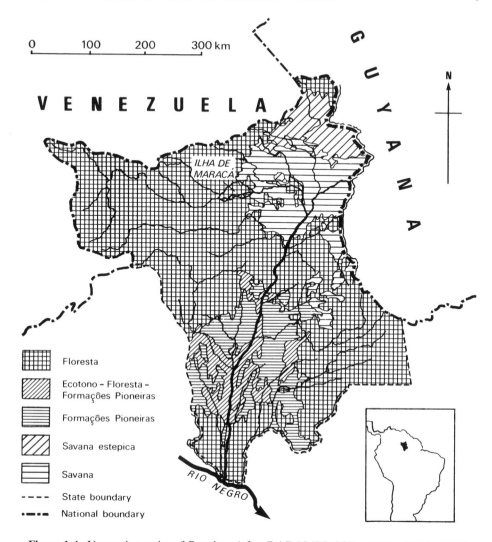

**Figure 1.4.** Vegetation units of Roraima (after RADAMBRASIL, 1975a, 1975b, 1978)

The savanna or *cerrado* vegetation of northeast Roraima is continuous with the Rupununi savanna of Guyana. The formation is xeromorphic in character, reflecting pronounced moisture seasonality. Floristically, the formation is related to the *llanos* and to other northern Amazonian savannas (Goodland, 1966; Eden, 1974). Some species diversity is apparent, but individual savanna communities are mostly dominated by a few, widespread species. In free-draining sites, open wooded savanna commonly contains *Curatella americana* L. and *Byrsonima crassifolia* (L.) Kunth, with an understorey of bunch grasses such as *Trachypogon spicatus* (L.f.) Kuntze and *Andropogon angustatus* (J.L. Presl) Steud. Poorly drained areas usually support herbaceous communities of sedges or mixed sedges and grasses; the palm *Mauritia flexuosa* L.f. is commonly present along drainage lines (see Plate 1a). Riparian forest and scattered 'bush islands' also occur within the savanna. The transition between savanna and forest is generally abrupt, possibly as a function of savanna fires.

More dissected areas to the north carry a steppe savanna formation, which includes stands of more or less woody character. The formation is widespread in the upper Surumu basin, which experiences similar moisture seasonality to the savanna. The steppe savanna is again xeromorphic in character, but its physiognomy differs from that of the savanna. This is especially the case with woody species, which are spiny and mostly deciduous; herbaceous species are commonly those prevalent in the savanna, notably *Trachypogon spicatus* and *Andropogon* spp. (RADAMBRASIL, 1975a).

The pioneer formations of Roraima are associated with low-lying alluvial areas in southern parts of the State. Formations range from open herbaceous and shrub communities to low forests, and are perceived as early successional in character. They are also an edaphic response, being associated with sandy parent materials subject to periodic inundation. Floristically, the vegetation is related to the *caatinga* of the upper Rio Negro basin; characteristic woody species include *Eperua leucantha* Benth. and *Humiria balsamifera* (Aubl.) St Hil.

The Ilha de Maracá, which is transitional in terms of climate and morphology, is somewhat similar in respect of its vegetation cover. The island itself is almost entirely forested, although immediately to the north and east there are extensive tracts of *Curatella–Trachypogon* savanna. The forest on the island comprises a well-developed, closed canopy formation of semi-evergreen to semi-deciduous character. Taller trees reach 30–35 m in height, with occasional emergents to 40 m or so; the majority of species are evergreen but some, such as *Tabebuia uleana* (Kranzl.) A. Gentry, *Lueheopsis duckeana* Bussett and *Peltogyne gracilipes* Ducke, lose their leaves during the dry season (Milliken and Ratter, Chapter 5). Locally, the forest cover on the island is displaced by small tracts of poorly drained savanna and herbaceous swamp (see Plates 1b and 4).

## CONCLUSION

The Ilha de Maracá occupies a fraction of the area of Roraima and contains only a sample of its diverse habitats. Yet, the transitional bio-physical status of the island gives it a distinctive character and generates marked scientific interest, notably in relation to the renewability of forest systems and to the dynamics of savanna–forest boundaries. In earlier times, the island was no doubt settled by Indians (see Proctor and Miller, Chapter 23), but of late there has been minimal occupancy by either Indians or neo-Brazilians, and consequently little disturbance of the natural vegetation and fauna. A notable abundance of larger animals exists, especially peccary and caiman, while adjacent savannas carry deer and anteater as well as introduced cattle. Elsewhere in Roraima, resource exploitation is increasing and the natural landscape is in places being disturbed, particularly by agricultural colonization and mining activities. The Ilha de Maracá, with its discrete and protectable boundaries and its established status as an Ecological Station, is a natural resource of a rather different kind and a valuable asset to Roraima and to the larger Amazonian region.

# REFERENCES

Bigarella, J.J. and Ferreira, A.M.M. (1985). Amazonian geology and the Pleistocene and the Cenozoic environments and paleoclimates. In: *Key environments: Amazonia*, eds G.T. Prance and T.E. Lovejoy, pp. 49–71. Pergamon Press, Oxford.
Choubert, B. (1957). *Essai sur la Morphologie de la Guyane*. Mém. Carte Géol. Dét. France. Dép. de la Guyane Francaise, Paris.
Eden, M.J. (1971). Some aspects of weathering and landforms in Guyana (British Guiana). *Z. Geomorph.*, N.F. 15, 181–198.
Eden, M.J. (1974). Paleoclimatic influences and the development of savanna in southern Venezuela. *J. Biogeog.*, 1, 95–109.
Frost, D.B. (1968). *The climate of the Rupununi savannas*. McGill Univ. Savanna Research Series, Montreal, 12, 1–92.
Goodland, R. (1966). *South American savannas. Comparative studies. Llanos and Guyana*. McGill Univ. Savanna Research Series, Montreal, 5, 1–52.
IBGE (1987). *Anuário Estatístico do Brasil – 1986*. Fundação Instituto Brasileiro de Geografia e Estatística, Rio de Janeiro.
Kroonenberg, S.B. and Melitz, P.J. (1983). Summit levels, bedrock control and the etchplain concept in the basement of Suriname. *Geol. Mijnbouw*, 62, 389–399.
McConnell, R.B. (1968). Planation surfaces in Guyana. *Geog. J.*, 134, 506–520.
Migliazza, E.C. (1978). The integration of the indigenous peoples of the Territory of Roraima, Brazil. *International Work Group for Indigenous Affairs, Document* 32, 1–29.
RADAMBRASIL (1975a). *Projeto Radambrasil. Folha NA.20 Boa Vista e Parte das Folhas NA.21 Tumucumaque, NB.20 Roraima e NB.21. Levantamento de Recursos Naturais*, Volume 8. Ministério das Minas e Energia, Rio de Janeiro.
RADAMBRASIL (1975b). *Projeto Radambrasil. Folha NA.21 Tumucumaque e Parte da Folha NB.21. Levantamento de Recursos Naturais*, Volume 9. Ministério das Minas e Energia, Rio de Janeiro.
RADAMBRASIL (1978). *Projeto Radambrasil. Folha SA.20 Manaus. Levantamento de Recursos Naturais*, Volume 18. Ministério das Minas e Energia, Rio de Janeiro.
Rivière, P. (1972). *The forgotten frontier. Ranchers of North Brazil*. Holt, Rinehart & Winston, New York.
Short, K.C. and Steenken, W.F. (1962). A reconnaissance of the Guyana Shield from Guasipati to the Rio Aro, Venezuela. *Bol. Inform. Venezuela*, 5, 189–221.

# 2 Aspects of the Geology of the Ilha de Maracá

**JOSÉ MAURO MARTINI**
*Departamento Nacional da Produção Mineral, Manaus, Brazil*

## SUMMARY

This chapter presents the results of semi-detailed geological field mapping in the area of the Ilha de Maracá. During the mapping operation, seven distinct litho-stratigraphic units were defined. The stratigraphy and petrography of these units is discussed.

## RESUMO

Este capítulo apresenta os resultados obtidos nos trabalhos de mapeamento geológico em semi-detalhe na área que compreende a Ilha de Maracá. O mapeamento permitiu melhor conhecimento da área. Foram individualizadas sete unidades petro/estratigráficas com características distintas. São discutidas a estratigrafia e petrografia destas unidades.

## METHODS

Prior to the initiation of fieldwork on the Ilha de Maracá, efforts were made to identify the principal stratigraphic units of the island from aerial photographs. The fieldwork itself entailed the collection of rock samples, observations on the geological/geomorphological configuration of the area (structures, tectonic processes, etc.) and alluvial prospecting in sites previously chosen on the basis of photo-interpretation. The results of the latter study are not presented here, but are detailed in Martini (1988).

Mapping expeditions were mounted on both branches of the Uraricoera river (the Furos de Santa Rosa and Maracá), as well as along certain of their tributary streams and on forest trails within the island itself. The first field study phase took place between March and April 1987, during which the Furo de Santa Rosa was surveyed as far west as the Fumaça rapids (see Figure Pr.1, p. xvii), and the lower reaches of the Furo de Maracá were also investigated. In the second phase, which took place in October of the same year, the entire length of the Furo de Maracá (from its western extremity at the Purumame falls to its confluence with the Furo de Santa Rosa) was geologically mapped. During this period further exploratory trips were made into the island's interior.

## INTRODUCTION TO THE GEOLOGY AND GEOMORPHOLOGY

The Ilha de Maracá lies on the crystalline basement of the Guiana Shield, a formation principally composed of granites and gneisses. To the north (20 km) lies the flat-topped Serra de Tepequém, an outlier of the Roraima Formation of southeast Venezuela and Guyana.

The examination of aerial photographs revealed two general surfaces, apparently reflecting the litho-stratigraphic units observed on the island. The first of these surfaces (Surface 1), which is relatively hilly and dissected and constitutes the higher regions of the island (c. 10%), coincides with the predominance of a band of metamorphic rocks enclosing granitic cataclastic rocks. The second of these surfaces (Surface 2), which is little dissected and heavily denuded (and makes up approximately 90% of the island), is associated with a variety of lithotypes including quartz-feldspathic gneisses and quartz–biotite schists. Its otherwise monotonous topography is interrupted in places by low granitic hills.

The criteria which were used to guide interpretation of the field observations were, principally: tonality, topographic form, drainage and tectonics. The tonalitic granite intrusions mentioned above, for example, generally exhibited distinct linkage to the topography, forming elongated or somewhat semi-circular hills with convex crests, smooth sides and gentle slopes (as seen at the centre of Surface 2). The secondary and tertiary drainage (in relation to the Furos de Santa Rosa and Maracá) is characterized on Surface 1 by converging series of water-courses accompanied by deep valleys, as was observed on the Purumame trail at the western end of the island (Figure Pr.1, p. xvii). The drainage of Surface 2 takes the form of parallel courses, possibly controlled by the large-scale regional faulting (see Figure 2.1). A more detailed aerial photographic interpretation of topographic and drainage patterns is given in McGregor and Eden (Chapter 3).

## STRATIGRAPHY AND PETROGRAPHY

### GENERAL

Semi-detailed geological mapping, supported by photo-geological interpretation, permitted the identification and localization of several distinct lithological units: tonalitic gneisses, mylonitized tonalitic gneisses, mylonite gneisses, quartz–biotite schists, hornblende granodiorites, tonalites and dolerites.

Using the data obtained during the fieldwork phase, a study was then set up to characterize the identified lithotypes, with the object of correlating them and establishing the stratigraphic position of the area.

Bearing in mind that the fieldwork was carried out in a relatively small area in terms of the scale of the maps being used (1:100 000), an attempt was made to establish a correlation between the geological units of Maracá and those identified during the Catrimani–Uraricoera and Roraima projects of DNPM/CPRM (Ramgrab, 1971; Bomfim, 1974; Braun and Ramgrab, 1976). The gneisses with

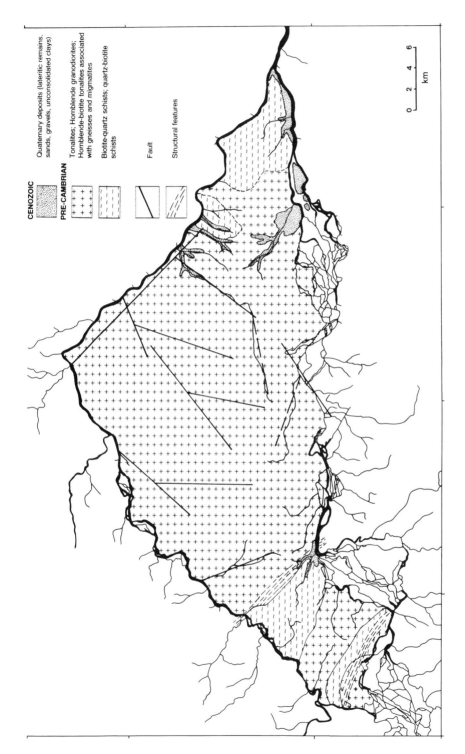

**Figure 2.1.** Preliminary geological map of the Ilha de Maracá

migmatitic features observed at P-56 (Figure 2.2), for example, could be correlated with the 'Uraricoera' metamorphic suite of the previous studies, and similarly the quartz–biotite schists with the 'Parima' metamorphic suite. The tonalitic granitic rocks observed near the Fumaça rapids can, at least on textural/petrographic grounds, be correlated with the 'Mucajaí' granites of those surveys.

From the fieldwork carried out, it became clear first that the study area represents an igneous/metamorphic basement characterized by schists, gneisses with migmatitic features, and general occurrence of granitic rocks and metamorphic rocks with cataclastic features, and second that in the northern part of the island (P-56) there is a small dolerite dyke. Contact between the differing lithotypes can be observed at a number of locations, both on the Furo de Santa Rosa and in the interior of the island. At P-55, for example, the hornblende–biotite–tonalite encloses xenoliths of up to 40 cm of fine-grained, distinctly banded quartz-feldspathic gneisses, suggesting that an igneous body has enveloped an older intrusive rock. At another site a dolerite dyke (orientation [N60E-vert.], corresponding to the fault plane of the older rock) can be seen cutting through a gneissic formation. Zones of metamorphic rocks were observed with a general [SE–NW] orientation, including some distinctly cataclastic rocks and some of more variable composition. The majority of these were sited on the western end of the island, a notable example being the outcrops of cataclastic granites at the Purumame waterfall.

## LITHOTYPES OBSERVED ON THE ILHA DE MARACÁ

### I – Tonalitic gneisses

These rocks are well represented in the area, and were observed at several outcrops including four along the Furo de Santa Rosa. They were not observed on the Furo de Maracá, however. These units generally take the form of extensive pavements alongside the river (to 20 m or more), or small outcrops arising from the bed of the river itself. They are medium- to coarse-grained rocks, crystalline, phaneritic and displaying characteristic banding of quartz–feldspars interleaved with biotite. In hand specimen (as observed at P-56), these rocks present a folded appearance, presenting a single surface of millimetric proportions with centimetric folds. The planes are separated by marked mineralogical differences, with layers of quartz–feldspars and micas sometimes exhibiting axial plane foliation. In P-81 and P-67 (Nassazeiro), rocks of this lithotype are found with parallel plane banding in the direction [N60E-sub-vert.].

This lithotype is generally associated with faulting of varying orientation, but primarily in the direction [N60E-vert.] which is exactly parallel to the plane of the rock. Secondary faulting occurs in the direction [N60W-vert.]. At an outcrop at the Topé rapids, one of these primary faults [N60E-vert.] has been filled by a dolerite dyke.

This unit appears to exhibit migmatitic features (principally at P-56), but for positive identification a more detailed study, with proper characterization of the paleosome and neosome, would be necessary.

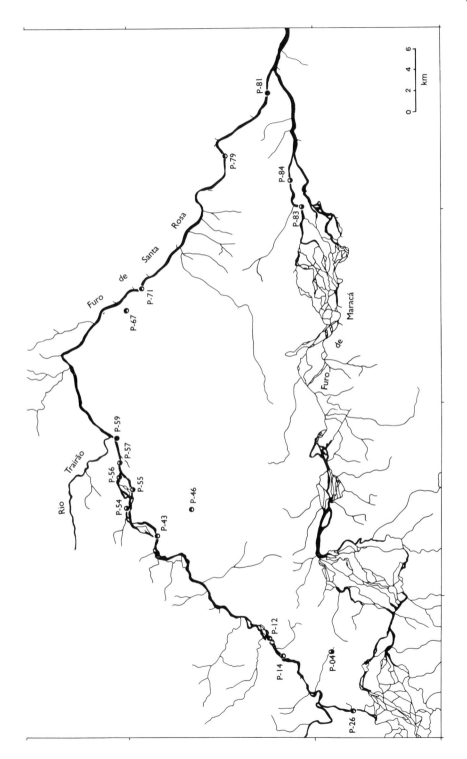

**Figure 2.2.** Locations of geological collecting sites referred to in the text

Using thin-section analysis, the petrographic characteristics of the rock were defined as: gneissic with tonalitic composition, banded, predominantly granoblastic in texture (over lepidoblastic), heavily altered by deformational action, and developing axial plane foliation of the planar minerals. Cataclastic features such as microfissures and sutured and interpenetrated contacts are exhibited. The quartz grains exhibit strongly undulatory extinction, with evidence of recrystallization of the margins around larger nuclei which display deformation laminae, and make irregular contacts resulting in stretching of the weaker zones. This process is indicative of conditions of high temperature and pressure, and intense deformation.

The plagioclase forms partially sericitized hypidioblastic crystals containing deformation lamellae, with microfissures in which muscovite has developed. The phyllosilicates, principally muscovite, are folded and contain crenulation cleavage and kink-bands where their lamellae have been shortened. Biotite occurs in smaller quantities, in the form of aggregates which define the foliation.

## II – Tonalites

This unit is composed of granitic rocks of tonalitic composition, which comprise the eastern end of the Ilha de Maracá and take the form of pavements and blocks of varying size (e.g. P-84). At the latter location the rock is fine- to medium-grained, holocrystalline and lacking in evidence of deformational action. Further to the east (P-81 and P-83) the rock tends to exhibit a coarser grain, which suggests that we are dealing with a single igneous body across which exist gradations in granularity. At P-83 these rocks demonstrate slight foliation, which may be an artefact of the intrusion of the igneous body rather than a regional metamorphic effect.

This lithotype was also observed at P-26 where, as is also the case at P-46, foliated facies appear to be absent in spite of their proximity to the shear zone.

Structurally these rocks appear faulted in a variety of patterns, but principally in the direction [N40W-vert.] (faulting in this orientation was observed at all outcrops of this unit). In hand specimen they are phaneritic and holocrystalline, fine- to medium-grained, isotropic, compact, and composed of quartz, feldspar and mica. In thin section they show a granular–hypidiomorphic texture, and the presence of biotite is evident within the grains of microcline, illustrating their poikilitic nature. The grains of alkaline feldspar themselves display a graphic texture.

Plagioclase is the predominant mineral, with a cloudy, clayey appearance and twinning of the albite–albite variety. Quartz appears in the form of anhedric grains, with undulatory extinction and often with fracturing possibly attributable to the forces related to intrusion and not to cataclastic phenomena. The biotite consists of platelets, sometimes chloritic and sometimes associated with crystals of sphene or with opaque minerals. Muscovite, titanite, apatite and opaques are accessory minerals present within this unit.

## III – Quartz–biotite schists

This unit incorporates metamorphic, schistose, fine-grained rocks with lamination defined by foliation. Seven of the described outcrops were included in this category.

In the field these rocks were observed as extensive pavements, as for example at P-54 where the rocks are cut by veins of fine- to medium-grained granitic material in the directions [NS-60W] and [N50W-60SW]. They were also found at both the western and eastern ends of the island in the form of xenoliths or inclusions within other geological units.

Structurally, the unit exhibits foliation in the direction [N10W], and faulting in a variety of patterns but predominantly [N35E-vert.] and [N60W-70NE]. At P-54, faults of [N60W] were observed displacing similar faults of [N15E].

In hand specimen the rocks are whitish-yellow in colour and divided by quartz bands of millimetric dimensions. They are strongly isotropic, and occasionally contain thicker bands with open, isoclinal, syntectonic deformational folding.

In thin section they are seen to be fine- to medium-grained rocks with an absence of porphyroclasts, presenting a granuloblastic texture and primarily composed of biotite, quartz, muscovite and plagioclase. The biotite is yellowish in coloration, shows a rudimentary alignment possibly resulting from metamorphic differentiation phenomena, and sometimes displays twisted cleavage. The quartz occurs with a varied granulometry, and can clearly be divided into two generations. The first is irregular in outline with corroded borders, undulatory extinction and no fracturing. The second is limpid, rounded and with homogeneous extinction.

The muscovite is colourless, occurs in larger crystals than the biotite (to 2 mm), and is sometimes limpid and sometimes partially sericitic in the nucleus.

At P-14 there is a predominance of muscovite, exhibiting fine lepidoblastic textural features and with a tendency to develop mylonitic features, and strongly deformed by the formation of kink-bands in the muscovite and biotite twins. Extinction of the quartz crystals is strongly undulatory.

The most striking features are the development of syntectonic foliation, and of a mineral streaking perpendicular to the foliation (whose orientation is defined by the muscovite lamellae). Fibrous sillimanite, in the form of well-developed crystals, is sometimes present in conjunction with the granuloblastic quartz crystals. The quartz crystals have pulverized borders resulting from a mylonitic flux process, which represents deformation in the plastic phase. The planar minerals have enveloped the more resistant minerals, although with noticeable deformation.

Mineralogically P-14 corresponds to a muscovite–biotite schist created under conditions of moderate pressure, and due to the presence of sillimanite it can be classified as medium-grade facies of pelitic derivation. However, its textural and petrographic properties permit its inclusion in unit III.

## IV – Hornblende–biotite tonalites

This lithotype was mainly observed in the northern part of the Ilha de Maracá, close to the Tipurema rapids and the Trairão river. They are coarse-grained rocks with a high proportion of mafic minerals (visible in thin section), and occur as pavements up to 50 m across. At P-55, where the Furo de Santa Rosa divides into two distinct channels, an outcrop of this unit with a lateral exposure of over 100 m was examined. At P-55 tonalitic gneisses with xenoliths of these rocks within them were also observed, characterizing two stratigraphically distinct units.

At P-59 an outcrop of this unit is faulted in the direction [N15E-80SE] (as are the rocks at P-55), and the cracks have been filled with a fine- to medium-grained quartz-feldspathic material. In hand specimen these rocks show a whitish–pinkish coloration, with conspicuous black spots due to the presence of biotite. There is no visible evidence of foliation, and they are generally smooth and holocrystalline.

In thin section they exhibit a granular–hypidiomorphic and poikilitic texture, with the inclusion of biotite crystals within the plagioclase. The plagioclase sometimes exhibits plastically twisted twinning, is frequently fractured and shows undulatory extinction. The crystals are up to 5 mm in size with $An_{40}$ (andesine) composition, and were observed to form triple contacts which may be a result of stretching (cataclasis?).

Quartz occurs as fractured crystals up to 3 mm in size, with undulatory extinction and corroded outlines, or as rounded limpid crystals of up to 0.1 mm with homogeneous extinction. These are probably two distinct generations. The biotite occurs in elevated quantities, is greenish pleochroic, and sometimes contains chlorite and opaque minerals in its margins. Hornblende with greenish pleochroism is the second most frequent mafic mineral in these rocks; its occurrence is associated with biotite and titanite. Opaques, chlorite and zircon are accessory minerals in this unit.

### V – Hornblende granodiorites

This lithotype occurs at five of the described outcrops, predominantly along the Furo de Santa Rosa, close to the Fumaça rapids and in isolation at P-79. They are faulted in the direction [N10W-vert.], and occur as small nuclei. Their contacts with other lithotypes were not properly delimited, as a result of the scarcity of their outcrops.

Mesoscopically these are dark grey, holocrystalline, isotropic rocks of medium to coarse granulation. In thin section they exhibit a granular–hypidiomorphic texture, with plagioclase occurring as elongate crystals up to 3 mm in length whose nuclei are more altered than their margins. They are greyish, clayey and fractured, display non-homogeneous extinction and are andesine in composition ($An_{45}$).

Hornblende is present at high levels, is green–yellow pleochroic (altered with rusty tinges), and is up to 2 mm in size. The biotite, viewed in natural light, also displays strong pleochroism and is up to 2 mm. The quartz consistently displays undulatory extinction and is generally fractured.

This unit can be regarded as related to unit IV, its compositional difference being the result of assimilation phenomena, or of contamination with the rocks which lie alongside it.

### VI – Dolerites

This lithotype was observed at P-56, in the form of a dyke approximately 1 m wide filling a [N60E-vert.] fault. This was the only observed occurrence of this unit in the study area.

In hand specimen the rock is compact, fine-grained and dark in coloration; it is strongly isotropic and has an aphanitic texture. In thin section the texture is granular–ophitic with a maximum grain size of 0.5 mm, and incorporates grains of labradoritic plagioclase (composition $An_{60}$) enclosed in mafic minerals such as uralitized hornblende, hedenbergite and chlorite. Plagioclase is the dominant constituent, and takes the form of elongated hypidiomorphic crystals. The hornblende is also present in high concentrations, its form being predominantly prismatic. Alongside these crystals opaque minerals occur, possibly providing evidence for the secondary metamorphism of these rocks through hydrothermal metamorphism of pyroxenes. Chlorite always occurs in association with these crystals.

## VII – Mylonitized tonalitic gneisses

This unit, which was found principally in the western part of the island, includes metamorphites in which one can distinctly observe foliation resulting from dynamic effects. It occurs as elongate ridges running in a northwesterly direction, as observed along stretches of the Furo de Santa Rosa at P-71, P-43 and P-12. In the outcrops these rocks display clear foliation, generally in the directions [N55W-85NE], [N64W-80SW] and [N80W-85NE], with associated fracturing in various directions but principally [N72W-77SW] and [N60W-vert.].

They vary from cataclastic rocks to true cataclasites, depending upon the varying degree of intensity of cataclastic activity which they have undergone. Those which have undergone the most intense deformation tend towards mylonites, whose most characteristic features are banding and striping as observed at P-04. The three great steps of the Purumame waterfall (where this unit can clearly be observed), and the alignment of stretches of the Furo de Santa Rosa with the regional foliation, could perhaps be taken to be direct effects of the cataclasis which took place in the region.

In terms of composition these metamorphites are similar to their parent rocks, differing principally in their texture which is dependent upon the intensity of the action that they have undergone. It seems probable that alongside the cataclastic processes, textural modifications were brought about by hydrothermal solution, as can be observed at P-71 where quartz and feldspar veins have been injected into the rock. These veins cut across the existing foliation, indicating a syntectonic/post-tectonic magmatic event.

It is interesting to point out that the further away that one goes from the cataclastic zones, the more these rocks are seen to preserve (at least in part) their igneous character.

In hand specimen the rocks are medium- to fine-grained, pale grey, phaneritic, holocrystalline and strongly isotropic, showing foliation along an S-plane. In composition they represent quartz monzodiorites, but take the form of protomylonites composed of 50% plagioclase ($An_{20}$), 10% microcline, 10% quartz, 10% chlorite and 5% amphibolite or, as was observed at P-43, tonalites composed of quartz, plagioclase and biotite.

At P-04 the rocks show a fine banding, which has undergone ductile deformation during a period of high plasticity (i.e. in the lower crustal layer).

Texturally they exhibit a granular–hypidiomorphic character, within a finely foliated matrix associated with the cataclastic effects superimposed on the rock. These have resulted in an intensely deformed porphyroclastic texture. At P-43 the rocks are granoblastic–hypidiomorphic, with development of phase differentiation relative to ductile deformation. These phases represent stress-shadow, crystal comminution, primary phase recrystallization and crystal rotation, with the preservation of a mylonitic texture.

At P-04 one also encounters mylonitic flux texture, with the occurrence of intensely rotated plagioclase porphyroclasts. The matrix of these rocks shows a high level of recrystallization, being made up of microgranular quartz crystals in syncinematic ribbons accompanying ductile shearing. This rock shows the highest level of deformation encountered on the island.

Mineralogically these rocks demonstrate the effects of deformation, with normal zoning of the plagioclase (with calcium levels becoming progressively higher towards the centre), and with almandine garnets surrounded by chlorite whose pleochroism varies from green to colourless. This is indicative of moderate pressure conditions in the amphibolite facies. The amphibole is of the green hornblende variety. The chlorite and biotite indicate the presence of a second metamorphic event, associated with retrometamorphism.

At P-57 the rock shows excellent foliation, differing considerably from the other rocks of this unit. It can be characterized as a biotite–quartz schist, with a predominantly granolepidoblastic texture. This rock exhibits clear schistosity with evidence of comminution (trituration and pulverization) of the crystals (the beginning of deformation). The alternation of bands separated by arched or kink-band structures of chloritized biotite characterizes a mylonitic texture which has passed through a phase of plasticity.

## ECONOMIC GEOLOGY

The Ilha de Maracá lies in the basin of the Rio Uraricoera, in which the presence of economically important minerals was established by surveys prior to this study. In studies conducted by CPRM/DNPM, the north of Maracá is mentioned together with the Serra Tepequém, an area delimited by federal decree for gold and diamond prospecting. As a consequence of the presence of these minerals, the Uraricoera basin was the scene of intensive prospecting during the Roraima gold rush of the late 1980s and the early 1990s (see MacMillan, 1995). Prospecting had already been taking place in the Maracá region prior to the rush – abandoned prospecting camps were found along the Furo de Santa Rosa in 1987 at the mouth of the Trairão and near the Purumame falls (see Figure 2.2) – and during the gold rush dredging barges were operating near the western end of Maracá.

Prospecting for economically important minerals was carried out on the island during the present study, concentrating on alluvial materials, and the presence of gold, wolfram and columbite was recorded. The results of this survey are discussed in more detail by Martini (1988).

## CONCLUSION

Two geomorphological surfaces were identified on the Ilha de Maracá, the first corresponding to the areas of higher elevation in the west of the island, constituting crests or morphological ridges aligned in a northwesterly direction. The second is a denuded and relatively flat surface with isolated hills, making up the greater part of the island. Parallel drainage predominates in the study area, demonstrating the presence of structural control in the region.

The area mapped in this study was made up of an association of rocks represented by granites of varying composition, quartz-feldspathic gneisses, quartz-feldspathic gneisses with migmatitic features, schists and diabases. All of these units, with the exception of a dolerite dyke [P-56], show the effects of tectonic activity.

The dolerite dyke may be related to the large basic non-metamorphic intrusions widely distributed through the northern part of the State of Roraima, where intrusions are found in Pre-Cambrian formations. On the tectonic side, the area has been affected by various dynamic/metamorphic episodes, the last of which is suggested by the shear zones in which mylonites are found at the western end of the island. The rigid tectonics manifest themselves as NW-orientated faults, such as those of the Rio Uraricaá, the Rio Trairão and the Furo de Santa Rosa, as well as other associated features.

## REFERENCES

Bomfim, L.F.C. (1974). *Projeto Roraima; Relatório final.* Ministério das Minas e Energia, Departamento Nacional da Produção Mineral, Manaus. Convênio DNPM/CPRM.

Braun, O.P.G. and Ramgrab, G.E. (1976). *Geologia da área do Projeto Roraima.* Ministério das Minas e Energia, Departamento Nacional da Produção Mineral, Manaus. Convênio DNPM/CPRM.

MacMillan, G.J. (1995). *At the end of the rainbow? Gold, land and people in the Brazilian Amazon.* Earthscan, London.

Martini, J.M. (1988). *Projeto Geologia – Ilha de Maracá. Relatório Final.* Ministério das Minas e Energia, Manaus. Convênio INPA/DNPM.

Ramgrab, G.E. (1971). *Projeto Roraima: Relatório Progressivo (1971 No.1). Mapeamento Geológico da Area Divisor.* Ministério das Minas e Energia, Manaus. Convênio DNPM/CPRM.

# 3 Geomorphology of the Ilha de Maracá

DUNCAN F.M. McGREGOR AND MICHAEL J. EDEN
*Royal Holloway, University of London, Egham, UK*

## SUMMARY

The nature and significance of geomorphological processes observed on and around the Ilha de Maracá are discussed. A brief introduction to regional geology and geomorphology is followed by an examination of fluvial patterns. Broad patterns of relief on the island are then examined through interpretation of aerial photographs; and slope profiles representative of conditions observed on and off the island are discussed. Analysis of surface materials indicates the widespread presence of saprolite. Evidence of laterization is relatively slight under forest, but more common in savanna areas. In terms of regional geomorphology and paleoclimates, the presence of extensive dune systems along the left bank of the Tacutu suggests that northern Roraima was significantly drier at times in the past than it is at present. However, in the Maracá area, clearly contrasting pedogeomorphic processes are in evidence on either side of the present forest–savanna boundary. This does not accord with a hypothesis of past fluctuations in the forest–savanna boundary, and requires further investigation.

## RESUMO

São discutidos o cárater e a importância dos processos geomorfológicos observados na Ilha de Maracá e nas suas proximidades. Uma breve introdução à geologia e geomorfologia regional é seguida por uma análise de padrões fluviais. Padrões de relevo em grande escala são examinados pela interpretação de fotos aéreas; e perfis de declive representativos de condições observadas dentro e fora da Ilha são discutidos. As análises de material superficial indicam a presença extensiva de saprólitos. Evidência de laterização sob floresta é relativamente pouca, mas é bem mais comum nas áreas de savana. No sentido de geomorfologia regional e paleoclimas, a presença de sistemas extensivos de dunas ao longo da margem esquerda do rio Tacutu sugere que o norte de Roraima era significativamente mais seco no passado do que no presente. Porém, em cada lado da transição mata–savana na área de Maracá se mostram processos geomorfológicos nitidamente constante. Isto não concorda com uma hipótese de flutuações da divisa mata–savana no passado, e o assunto precisa de mais investigação.

## INTRODUCTION

This chapter is concerned with the nature and significance of geomorphological processes in the Maracá area. Due to constraints of time and access, fieldwork undertaken by the authors on the Ilha de Maracá itself was limited to the eastern end of the island and to short transects from the Furo de Santa Rosa and the Furo de Maracá. The authors' main investigations were concentrated on the area immediately to the south of the island (Eden *et al.*, 1990) and on the wider area of northern Roraima. Interpretation of the geomorphology of the Ilha de Maracá is thus partly reliant on secondary information, particularly that gleaned from examination of remotely sensed imagery, as well as on field transects and laboratory examination of materials. Supporting information is provided from the immediate vicinity, while the place of Maracá within the wider context of Roraima and northern Amazonia is also discussed.

Geologically, the Territory of Roraima lies on the southern flank of the Guiana Shield, which comprises an ancient complex of crystalline rocks. These are mainly granites, gneisses and mica schists of Pre-Cambrian age, and are locally overlain by Proterozoic sediments of the Roraima Formation. The Ilha de Maracá is shown on the RADAMBRASIL geological map (RADAMBRASIL, 1975) as lying within the area of the Guiana Complex. Exposures of bedrock within or adjacent to the main streams consist mainly of quartz–biotite schists, quartz–feldspar gneisses, and associated tonalitic granites (Martini, 1988, and Chapter 2). There is recurrent evidence of faulting and other tectonic effects in the vicinity of Maracá. Fault patterns in the area strongly influence local drainage alignments, as in the case of the Furo de Santa Rosa.

The geomorphology of the State of Roraima has been reviewed generally at various times (Guerra, 1956; Ruellan, 1957; Barbosa and Ramos, 1959; Beigbeder, 1959; RADAMBRASIL, 1975; Schaefer and Dalrymple, 1995). These studies have been concerned mainly with general topographic characteristics and long-term landform evolution.

Landscapes of the Guiana Shield have undergone polycyclic development, with gently undulating to dissected planation surfaces produced. The general distribution of planation surfaces in the Territory of Roraima is outlined in Chapter 1 (Figure 1.3, p. 6). These correspond, as shown in Table 3.1, to levels identified generally for Brazilian Amazonia (Bigarella and Ferreira, 1985), and widely across the Guiana Shield (Choubert, 1957; Short and Steenken, 1962; McConnell, 1968; Berrangé, 1977).

The most extensive surface to the east of the Maracá area, towards Boa Vista, is at elevations of about 100–130 m. An extensive planar surface occurs at approximately 250–400 m in the area to the west of the island. These units are shown on the morphostructural map of Roraima produced by Projeto Radambrasil (Figure 1.3). This map shows the Ilha de Maracá as split by the relief boundary between the lower, end-Tertiary (Plio-Pleistocene) surface, denoted by Bigarella and Ferreira (1985) as *Pediplane $Pd_1$*, to the east, and the higher, late-Tertiary (upper Miocene–lower Pliocene) surface, denoted by Bigarella and Ferreira (1985) as *Pediplane $Pd_2$* to the west.

The eastern end of the Ilha de Maracá lies at a height of approximately 130–140 m on the margin of the $Pd_1$ surface, whereas much of the western end of the island lies at over 250 m. Owing to the forest cover and the dissected nature of the terrain, no clear scarp feature has been traced on the island, nor do east–west cross-sections through the island taken from the 1:100 000 IBGE topographic maps

Table 3.1. Correlation of erosion surfaces in Roraima (Bigarella and Ferreira, 1985) and southern Guyana (McConnell, 1968)

| Age | Approx. elevation (m) | Roraima | Southern Guyana | Approx. elevation (m) | Age |
|---|---|---|---|---|---|
| Mid-Tertiary (Oligocene) | 500–600 | $Pd_3$ | Kopinang | 650–700 | Late-Cretaceous– Early-Tertiary |
| Late-Tertiary (Miocene–Pliocene) | 250–400 | $Pd_2$ | Kaieteur | 400–460 | Late-Tertiary |
| End-Tertiary (Plio–Pleistocene) | 50–200 | $Pd_1$ | Rupununi | 100–135 rising southwards to 215 | End-Tertiary |

indicate any consistent step in the landscape. However, it appears that the higher ground towards the western end of the island represents dissected remnants of the $Pd_2$ surface.

Examination of the available maps confirms the existence of broad regional levels as indicated by Bigarella and Ferreira (1985) and Schaefer and Dalrymple (1995). A generalized east–west cross-section confirms a regional, though gradual, increase in height from about 130 m to about 250 m. A north–south cross-section through the Ilha de Maracá shows a regional concordance of isolated summit areas at around 300 m, while to the northwest of Maracá generalized summit levels exist at around 450 m and at around 600 m.

## GEOMORPHOLOGY OF THE ISLAND

### DRAINAGE

The Ilha de Maracá owes its existence as an island to a combination of geological control and river capture. Major fault directions within the Guiana Complex characteristically trend southwest to northeast and northwest to southeast. These trends are clearly seen in the network of faults and fractures which characterizes the basement geology of the Maracá area. The sharp change of direction encountered at the northernmost point of the Furo de Santa Rosa is the most obvious example of the intersection of two major faults, but similar fault alignments are visible on aerial photographs along several reaches of both the Furo de Santa Rosa and the Furo de Maracá.

The existence of numerous rapids also affects river behaviour. A series of north–south topographic sections was taken from the 1:100 000 IBGE topographic map sheets MI-26, MI-38 and MI-39 (Figure 3.1). Only gross generalizations may be made, as contours are at 50 m intervals and subject to errors of estimation in areas under forest. However, the sections show clearly (Figure 3.2) that the distribution

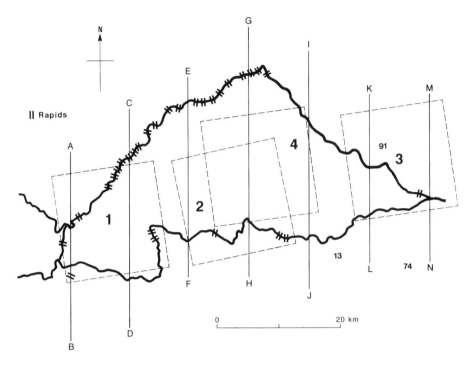

**Figure 3.1.** Location of topographic sections (Figure 3.2), photographs (Figures 3.5–3.8), profiles (Figure 3.9) and rapids shown on the IBGE 1:100 000 topographic maps. Schematic outline of river system shown. For detailed channel pattern, see Figure 3.3

of height loss is markedly different in the two Furos. Channel pattern is also significantly different.

At the western end of the island, the Rio Uraricoera splits into a network of distributaries, of which the main one swings northwards to join the Rio Uraricaá in the Furo de Santa Rosa. At this point both Furos are at the same height, approximately 200 m above sea level. The Furo de Santa Rosa descends about 100 m through a long series of rapids to its apex at the Corredeira de Tipurema, whereafter it loses relatively little further height to the eastern end of the Ilha de Maracá. It generally meanders over its entire course, with occasional relatively small islands and one section of about 10 km of distributaries. In contrast, the channel pattern of the Furo de Maracá (see Plate 2) has an anastomosing form with numerous distributaries occurring over much of its course (Figure 3.3).

The difference in channel patterns is explicable in terms of geological control, namely the outcropping of rapids in the stream bed and their influence on the distribution of stream energy. Both Furos appear to have similar discharge and load. Sand bars protrude at low water, but the majority of the load appears to be carried in suspension.

The presence of rapids in the stream bed reflects the outcrop of unweathered bedrock, the outcrop itself representing the exhumation of the weathering front. In

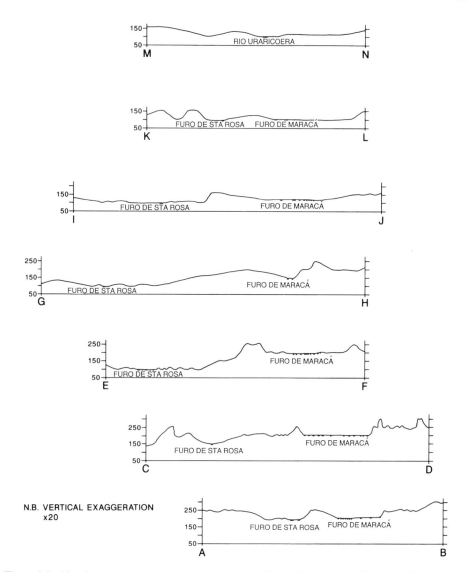

**Figure 3.2.** North–south topographic sections of the Ilha de Maracá. See Figure 3.1 for locations

the absence of detailed geological maps, it is not possible to ascertain whether or not the incidence of rapids is related to the outcrop of particular rock types. However, the preliminary investigations of Martini (1988 and Chapter 2) and Nortcliff and Robison (1989) suggest that granitic rocks may be preferentially exposed at river level, and that much of the higher ground of the Ilha de Maracá may be metamorphic, comprising schists, gneisses and quartzites.

Height loss is more marked in the southwest to northeast stretch of the Furo de Santa Rosa, and takes place over a series of rapids, yet a broadly meandering habit persists. This indicates that stream energy does not vary significantly over that

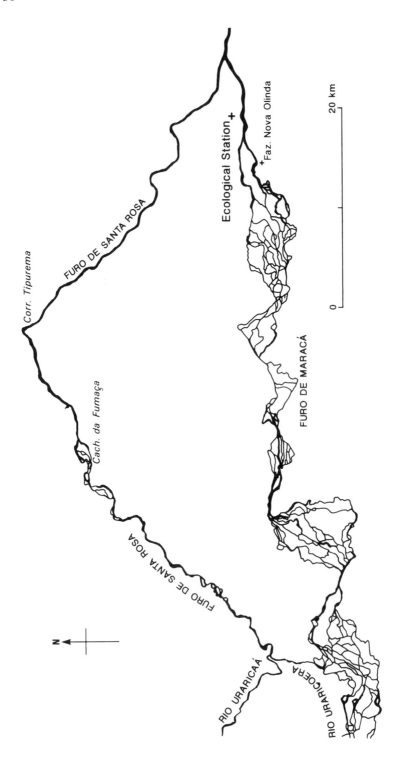

**Figure 3.3.** Channel patterns, Furo de Santa Rosa and Furo de Maracá

stretch. Fewer rapids are shown on the IBGE map over the length of the Furo de Maracá. However, observation, particularly in the dry season, suggests that the rapids present in the Furo de Maracá are steeper than those in the Furo de Santa Rosa. As the drop in river level is the same in both channels, it follows that stretches between major rapids in the Furo de Maracá have lower river gradients. Energy levels will, correspondingly, be more variable, being higher over the short stretches of rapids and lower in the intervening reaches. The strongly anastomosing pattern may reflect this.

Local river capture by the Rio Uraricoera may have occurred by partial abstraction of discharge by the Rio Uraricoera from a formerly separate Rio Uraricaá at the western end of the Ilha de Maracá. With respect to its course beyond the Corredeira de Tipurema, the presence of a wide band of low ground suggests that it is possible that the Rio Uraricaá may have formerly continued northeastwards to join the Rio Amajari. Local river capture, if such took place, may have been associated with the regional-scale capture of major rivers of this part of the territory, including the Uraricoera, the Mucajaí and the Tacutu–Ireng system, by the Rio Branco (Schaefer and Dalrymple, 1995). Regional tilting of the southern flank of the Guiana Shield may explain the fact that the present watershed on the Ilha de Maracá lies relatively close to the southern margin of the island.

# MORPHOLOGY

## RELIEF PATTERNS

Considerable local dissection has occurred in the Maracá area, giving rise to undulating relief with an amplitude of dissection of 25–30 m. On the Ilha de Maracá, local changes in dissection patterns are identifiable on aerial photographs at a scale of *c.* 1:70 000. The changes have been mapped (Figure 3.4), are partly structural in origin, and are only partially related to the $Pd_1/Pd_2$ boundary.

Five terrain classes are identified. They comprise two classes of upland terrain, two classes of lowland terrain, and the contemporary floodplain (Figure 3.4).

*Class 1: Upland terrain (summit surface)* This comprises the highest terrain on the Ilha de Maracá, and is mainly encountered in the western half of the island. Summit areas reach over 250 m, with a maximum height of 331 m according to the IBGE topographic map. The terrain is characterized on the aerial photographs by a relatively high frequency of deep dissection and by rounded summits (Figure 3.5), and may represent remnants of the $Pd_2$ surface of Bigarella and Ferreira (1985). Fault lines are visible on the photographs, with the margins of the terrain appearing to be largely fault-controlled. The terrain covers approximately 9.2% of the island.

*Class 2: Upland terrain (sub-summit surface)* This unit comprises terrain at about 170–250 m above datum. It is characterized on the aerial photographs by rounded summits, but appears to be less heavily dissected and has a lower frequency of

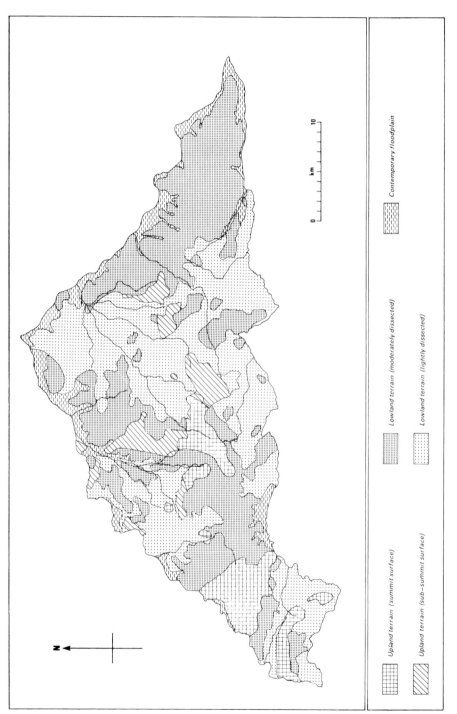

**Figure 3.4.** Terrain classification of the Ilha de Maracá

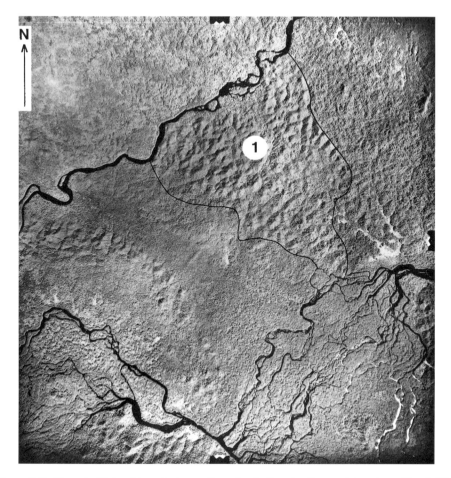

**Figure 3.5.** Class 1: Upland terrain (summit surface). Deeply dissected terrain outlined. This unit is largely bounded by fault lines, as, for example, along its northern edge. See Figure 3.1, Box 1 for location

dissection compared with the Class 1 terrain (Figure 3.6). It fringes Class 1 in part, and also extends in fragmented form across the eastern half of the island, where it gives the appearance on photographs of separate, domed summits. The terrain appears to comprise degraded remnants of the $Pd_2$ surface. It covers approximately 5.3% of the island.

*Class 3: Lowland terrain (moderately dissected)*   This terrain occurs throughout the island, but comprises almost all the eastern third of the Ilha de Maracá. It is characterized by relatively broad interfluves, at heights of 110–150 m. The degree of dissection is generally less than that of Classes 1 and 2 (Figure 3.7), but increases locally, as for example at the relatively sharp topographic junction with the contemporary floodplain in the extreme eastern tip of the island. Together with Class 4,

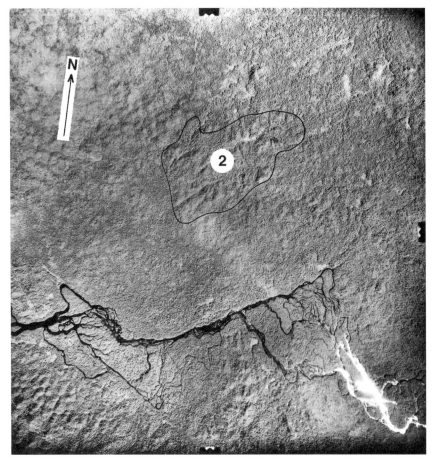

**Figure 3.6.** Class 2: Upland terrain (sub-summit surface). Rounded summits outlined in centre of photo. See Figure 3.1, Box 2 for location

this terrain appears to represent the margins of the $Pd_1$ surface of Bigarella and Ferreira (1985). The unit covers about 39.3% of the island.

*Class 4: Lowland terrain (lightly dissected)* This terrain mainly exists in a north–south band through the middle of the island. It rises gently in height from about 100 m above datum in the north to about 180 m in the south of the island. There is thus a height overlap with Class 2, though the appearance of the two units is markedly different on the aerial photographs. Class 4 terrain is characterized by a low degree of dissection, giving an almost uniform appearance on the photographs (Figure 3.8). It is the largest terrain unit, covering about 40% of the island.

*Class 5: Contemporary floodplain* Areas of contemporary floodplain are identifiable on the photographs by the presence of fringing bluffs (as, for example,

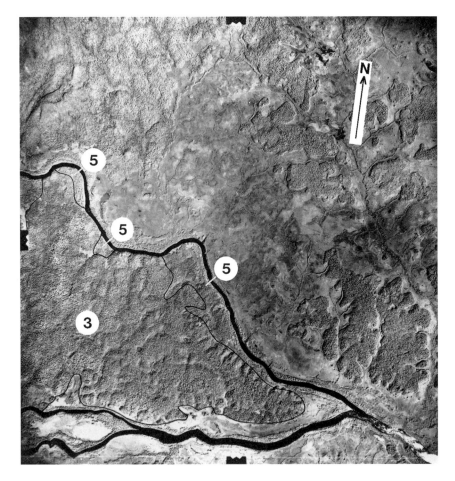

**Figure 3.7.** Class 3: Lowland terrain (moderately dissected). Area outlined in interior of island. Class 5: Contemporary floodplain. Linear areas outlined between rivers and Class 3 terrain. See Figure 3.1, Box 3 for location

already noted at the extreme eastern tip of the island). Cutoff morphology is visible in places, together with areas of seasonally flooded terrain like that immediately south of the Maracá Ecological Station. The unit, covering approximately 6% of the island, is illustrated in Figure 3.7.

Floodplain terrain occurs at multiple elevations, but is presumably consistent with adjacent river level. Areas of floodplain visible on the aerial photographs are mainly located on the right bank of the Furo de Santa Rosa, and are much less frequent on the left bank of the Furo de Maracá. This is consistent with regional tilting in this area of Roraima, with rivers subsequently favouring the northern margins of their floodplain. The areas of floodplain are associated preferentially with meandering, rather than anastomosing, courses, which confirms the longer-term permanence of the anastomosing reaches.

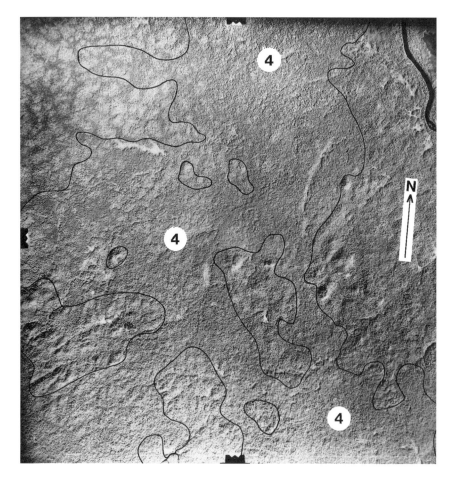

**Figure 3.8.** Class 4: Lowland terrain (lightly dissected). Comprises the majority of the centre of the photo, with isolated remnants of Class 2 and Class 3 terrain outlined. See Figure 3.1, Box 4 for location

Quantitative assessment of dissection differences between areas presently under forest and adjacent off-island areas under savanna proved impossible, due to the masking effects of the forest cover. However, a qualitative assessment of pattern under forest and savanna at the margin of the $Pd_1$ surface suggests that more ridged and narrower summits under forest contrast with flatter and wider summits in savanna areas. Summits in forest areas show a slightly more linear plan, compared with a slightly more circular plan in savanna areas.

Only preliminary field checking of the terrain units has been carried out, and their distribution and geomorphological significance require further investigation. The units proposed here agree in broad terms with those identified by Nortcliff and Robison (1989) and used by them to characterize the soils of the Ilha de Maracá.

Relief contrasts are also visible on radar imagery (RADAMBRASIL, 1975). The texture of tonal change on such imagery relates to the degree of dissection. Thus, a higher frequency of tonal change corresponds, on surfaces of the same age, to lesser but more frequent vertical incision, while a lower frequency of tonal change corresponds to greater and presumably more 'mature' incision (Eden et al., 1984). In this area, where surfaces of different ages are present, the higher $Pd_2$ surface is characterized by higher-frequency dissection patterns predominating to the west and southwest of the Ilha de Maracá, in areas presently under continuous forest. This reflects a longer period of incision of the $Pd_2$ surface, and may be related to the persistence and rejuvenation of drainage lines under a protective forest cover, as postulated by Löffler (1977) for the northern part of the Fly Platform in Papua New Guinea.

The margin of the lower $Pd_1$ surface, partly under forest and partly in savanna, is characterized by a more open texture, with fewer tonal changes. This is well illustrated in the forest zone running approximately north–south through the eastern third of the Ilha de Maracá. This reflects not only a lower amplitude of relief on the lower surface, but also the dominance of sheet-flow in savanna areas (Löffler, 1977). Further to the east, the $Pd_1$ surface is predominantly in savanna, and is characterized on the imagery by a progressively more open texture, corresponding to colluvial/alluvial plains of low relief.

## SLOPE PROFILES

Slope profiles are generally smooth throughout the field area, with the exception of profiles in which rock outcrops are present. Representative surveyed profiles are shown in Figure 3.9 (see Figure 3.1 for locations).

Slope profiles are similar to those reported elsewhere in Amazonia (e.g. Young, 1970; Journaux, 1975; Eden et al., 1982), more angular savanna profiles contrasting with more rounded profiles under forest. However, it is notable that in many savanna profiles, the lower slope is markedly steeper than that under forest (Figure 3.9). Maximum slope angles in savanna and under forest recorded in this survey are not conspicuously dissimilar (28°20' in savanna compared with 26°30' under forest), but angles exceeding 20° are relatively frequent in savanna and uncommon under forest.

Elsewhere in Amazonia, steep lower slopes have been noted under forest (Journaux, 1975). These are ascribed to a combination of periodic fluvial undercutting and remobilization of colluvial material during drier climatic conditions in the Pleistocene. While forms identical to the 'half-oranges' described by Journaux (1975) have not been identified in the Maracá area, field data point to the existence of slope undercutting (either by stream erosion or basal sapping) as a contributory factor in steepening of profiles in savanna areas. Wash features are common in off-island savanna areas, and thicken towards the foot of the slope. Minor gullying and mass movement (terracettes) are commonly present. The sharp break of slope frequently noted just off the summit edge may relate to the armouring of many off-island savanna summits by lateritic gravels.

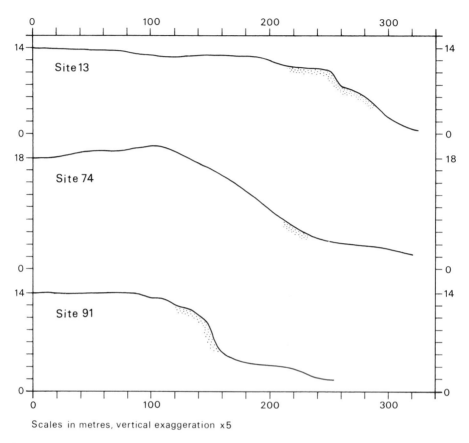

**Figure 3.9.** Representative surveyed profiles in forest (Site 74) and savanna (Sites 13 and 91). See Figure 3.1 for location. Stipple indicates outcrop of saprolite

## MATERIALS

The marginal area of the $Pd_1$ surface in the vicinity of Maracá is a partially stripped landscape. Unweathered rock outcrops are localized, and are mainly micaceous schist with some granitic material. Tors, with rounded blocks, occupy some off-summit slopes, though large-scale rock residuals are less frequent in this area than is often the case elsewhere on this surface (Eden, 1971). Such features are, of course, more readily observed in savanna areas, but have been encountered under forest on the Ilha de Maracá, notably along the trail leading inland from the Cachoeira de Fumaça (see Figure Pr.1).

Exposures of *in situ* weathered bedrock (saprolite) are encountered on many summit flanks in both savanna and forest zones. Preliminary analysis of representative slope profiles suggests that such exposures tend to occur nearer to summits in savanna areas close to the Ilha de Maracá than under forest. Figure 3.9 shows some typical slope profiles measured off-island (see Figure 3.1 for location).

**Table 3.2.** Chemical analysis of *in situ* weathered mica schist, Boqueirão; XRF determination (percentage composition)

| Compound | Sample depth (cm) | | | |
|---|---|---|---|---|
| | 170 | 270 | 320 | 420 |
| $SiO_2$ | 69.30 | 58.64 | 69.21 | 70.84 |
| $TiO_2$ | 0.77 | 0.67 | 0.66 | 0.52 |
| $Al_2O_3$ | 19.58 | 17.33 | 15.94 | 15.00 |
| $Fe_2O_3$ | 2.33 | 13.77 | 5.81 | 4.36 |
| MnO | 0.01 | 0.01 | 0.01 | 0.03 |
| MgO | 0.17 | 0.18 | 0.94 | 1.17 |
| CaO | 0.03 | 0.03 | 0.06 | 0.61 |
| $Na_2O$ | 0.19 | 0.17 | 0.32 | 1.59 |
| $K_2O$ | 1.42 | 1.75 | 2.34 | 2.16 |
| $P_2O_5$ | 0.02 | 0.03 | 0.03 | 0.05 |
| Loss on ignition | 6.50 | 7.27 | 4.79 | 3.56 |

The saprolite also appears to be somewhat thicker where exposed in savanna (on average about 4 m thick, compared with about 2 m thick in forest profiles), though this is speculative owing to the relatively small numbers of exposures encountered. It seems likely that these materials are associated with a process of etchplanation rather than pediplanation (Eden, 1971). The resultant clays are mainly kaolinitic, with gibbsite and some traces of illite (RADAMBRASIL, 1975).

Chemical analysis of *in situ* weathered mica schist by XRF determination at Boqueirão, approximately 10 km southeast of the Ilha de Maracá, shows a relative increase of aluminium towards the top of the profile and a zone of iron accumulation at 270 cm (Table 3.2). The profile was in a roadside gully section on a mid-slope savanna site, and was overlain by sandy colluvium.

In general, evidence of laterization is comparatively slight in forest profiles in the study area. Incipient iron concretions are present in most forest profiles, but hardened concretions are few, and often limited to occasional small pisoliths in upper horizons. Nortcliff and Robison (1989) note the presence of plinthite under rainforest on the Ilha de Maracá, although they do not detail the nature of the plinthite.

Occasionally, thin (<1 m) layers of conglomeritic material are found, with rounded pisolithic and quartzitic pebbles (commonly *c.* 1 cm in diameter) in an iron-rich sandy matrix. These materials were noted in the vicinity of the Ecological Station, where they form a flat local summit. The wider distribution of this material is not known, and its origin remains speculative. It closely resembles, however, material encountered at approximately the same height on a local summit near Fazenda Nova Olinda on the southern flank of the Furo de Maracá, and may represent the remnants of former alluvial deposits of contrasting calibre to the load carried by the present Uraricoera.

Topsoil and subsoil samples from the Maracá area in general show relatively little textural range, despite observed differences in bedrock. Nor is there any significant difference between forest and savanna sites (Figure 3.10). What does

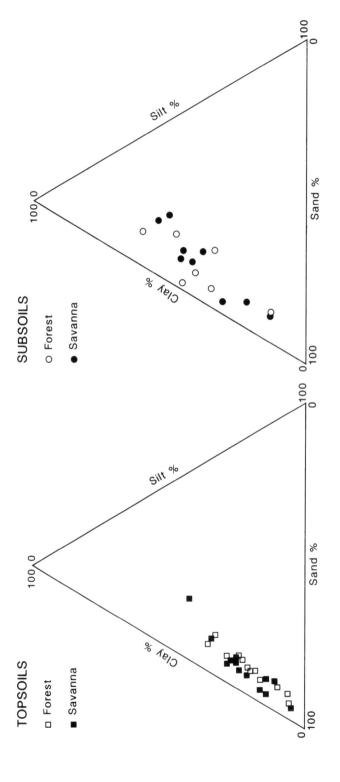

**Figure 3.10.** Texture analyses of topsoil and subsoil samples from the Maracá area

differ markedly is the degree of hardening of lateritic materials, which is much greater under savanna than in the adjacent forest.

# REGIONAL GEOMORPHOLOGY

## PRE-QUATERNARY

Bigarella and Ferreira (1985), in a review of the geomorphology of Brazilian Amazonia, contend that there are no remains in the present landscape of geomorphological features developed before mid-Tertiary times. They argue that towards the end of the Oligocene, the landscape of much of the continent was planated and topographically low. Thereafter, epeirogenic movements accompanied by fault tectonics rejuvenated the landscape, and initiated the polycyclic development of planation surfaces.

Such surfaces, deduced in Roraima by analysis of dissection patterns on radar imagery (RADAMBRASIL, 1975), have been recognized in the wider area of the Guiana Shield (Choubert, 1957; Short and Steenken, 1962; McConnell, 1968; Eden, 1971; Berrangé, 1977) (Table 3.1). Most authors have identified the surfaces as pediplains, but Eden (1971) and Kroonenberg and Melitz (1983) have argued, on the basis of evidence of deep weathering profiles and the presence in the landscape of domed inselbergs, that the surfaces should be regarded as products of etchplanation.

As yet, no systematic field investigations have been undertaken of surface forms or materials which would confirm, or deny, the reality of the $Pd_1$, $Pd_2$ and $Pd_3$ levels of Bigarella and Ferreira (1985). The present authors are currently examining materials collected from the $Pd_1$ surface to the east of Maracá towards Boa Vista. Only generalized comments on the materials of the $Pd_2$ and $Pd_3$ surfaces are found in the RADAMBRASIL volumes of the area (RADAMBRASIL, 1975, 1976); and a general description of the surface features towards Boa Vista is given by Schaefer and Dalrymple (1995). This topic requires additional field study.

As noted in Chapter 1, the $Pd_1$ surface in Roraima is separated into two parts by remnants of the higher $Pd_3$ surface (Figure 1.3, p. 6). The remnants represent the original watershed between the Amazon basin to the south and the Proto-Berbice basin of McConnell (1968) to the north. Until the late Tertiary, the major rivers of northern Roraima, principally the Uraricoera and the Mucajaí, flowed northeastwards, via the Tacutu, and formed part of the headwaters of the Proto-Berbice which itself flowed northeastwards through Guyana and into the Atlantic.

The Proto-Berbice river suffered capture of its original Uraricoera and Mucajaí headwaters by the Rio Branco, probably at some time during the late Tertiary, thus radically realigning the drainage of northern Roraima. The field evidence for this realignment in the northern Rupununi area is discussed by Sinha (1968), who illustrates how the capture of the Proto-Berbice headwaters by the Rio Branco led firstly to the northwestwards deflection of the Tacutu, and secondly to the development of a southwesterly course for the Ireng which utilized, in reverse direction, part of the former northeastwards course of the Proto-Berbice. Sinha ascribes these events to the last 'interglacial/glacial' cycle of the Pleistocene, but it is more likely that the requisite erosion took much longer, particularly with respect to the major realignment of the Ireng. In any case, as is discussed below, the last 'glacial' phase of the Pleistocene was probably associated with a degree of desiccation in the area, which would have reduced fluvial activity.

## QUATERNARY

It is now recognized that significant climatic changes affected Amazonia during the Pleistocene. Tricart (1985) summarizes the evidence for alternations, during the Pleistocene, of wetter and drier climates than those of the present. Haffer (1969) argued that the associated effects on vegetation types and distributions led to marked faunal speciation. It has been argued (e.g. Prance, 1982) that as much as 80% of existing forest may have been replaced by savanna during drier periods, though Prance has latterly amended this view, and now envisages widespread development of more open or lower types of forest during drier periods (Whitmore and Prance, 1987).

The presence of extensive fossil linear dune systems along the left bank of the Tacutu downstream of its confluence with the Ireng suggests that northern Roraima was significantly drier at some time in the past. The dune systems are aligned southwest–northeast, a trend similar to that of the fossil dune fields present on the left bank of the Orinoco in the 'bajo llano' of Venezuela (Tricart, 1974). In addition, Sinha recorded the presence of dunes in adjacent areas of Guyana (Sinha, 1968).

The Tacutu dunes are clearly visible on LANDSAT imagery over a distance of at least 30 km (Figure 3.11), and occupy a stretch of relatively lightly dissected terrain approximately 5 km wide, at an elevation of approximately 100 m. It is likely that this terrain represents a former floodplain of the Uraricoera before its capture by the Rio Branco, which would provide suitable dune-forming materials at a time of increased desiccation in northern Roraima. That at least some of this material has been derived from Roraima sandstone carried by the Proto-Berbice river is suggested by Sinha (1968).

Such dune systems indicate either drier conditions or stronger and unidirectional winds sufficient to mobilize fine-grained materials. Further investigation of the satellite imagery is being undertaken with a view to tracing the extent of such materials, and in particular the extent of recognizable dune forms in this area. The presence of rectangular drainage patterns in the area is further evidence of dune ridge formation.

The timing of the dune-forming event cannot yet be confirmed, but the presence of an extensive fossil dune field some 130 km east of the present forest–savanna boundary does suggest radical, and perhaps repeated, changes of climate or atmospheric circulation patterns in northern Roraima in the past.

However, the extreme degree of Pleistocene aridity envisaged by Haffer (1969) has yet to be substantiated, and the conflicting view of Endler (1982) and Colinvaux (1987) is that the endemism observed by Haffer can be explained by species changes along environmental gradients, without invoking massive shifts in the forest–savanna boundary. Nelson *et al.* (1990) suggest that some proposed centres of endemism may be sampling artefacts. Further, Salo (1987) suggests that recent

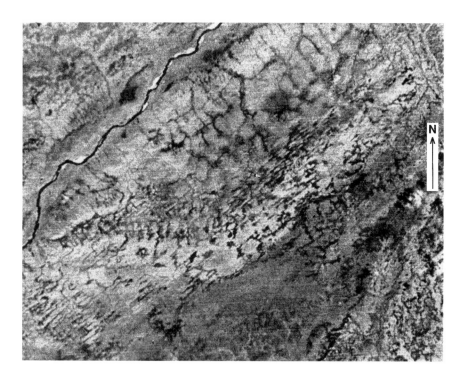

**Figure 3.11.** Fossil sand dunes of the Tacutu. Dunes aligned southwest–northeast, on left bank of the present Tacutu river

research into the fluvial geomorphology of the Amazon basin reveals geomorphological barriers which could explain the observed biological distribution patterns.

The effects of Pleistocene climatic change – as evidenced by the Tacutu dune field – on the forest–savanna boundary in the Maracá area must be considered, and the field evidence for any such change examined. In general, in areas of relatively uniform bedrock geology such as the Maracá area, contrasting pedogeomorphic processes may be expected either side of a stable forest–savanna boundary.

Areas of Amazonia in which savanna appears to have predominated throughout the Quaternary commonly exhibit stripped profiles, with hardened plinthite at or close to the surface in many places. Flat summits (often with associated gravel deposits) and areas of sheetwash also occur, giving parts of the savanna a rather angular appearance. This is broadly the case in savanna areas adjacent to the Ilha de Maracá.

In contrast, areas which appear to have remained under forest commonly exhibit a lower degree of plinthite hardening, reflecting less exposure to stripping. Incipient concretions are found within soil profiles, but indurated layers infrequently occur. Summits are generally rounded rather than flat, while slope forms are generally convexo-concave. Sheetwash processes occur, but are less efficient than in the savanna. Again, this typifies the areas of forest investigated on and adjacent to the Ilha de Maracá.

Where forest has been temporarily displaced by savanna as a result of climatic change, savanna processes would extend into the area formerly under forest. It would be expected, therefore, that relict features of savanna processes, such as lateritic gravels and sheetwash deposits, would persist under the restored forest. In the Maracá area, however, pedogeomorphic contrasts are encountered across the present forest–savanna boundary, despite a restricted range of bedrock lithology and relatively uniform geochemical and textural weathering products. This is not in accord with the hypothesis of regional shifts in forest–savanna boundaries, and is surprising in view of the evidence of aridity adjacent to the Maracá area. It may be that such shifts did not occur in the Maracá area. This requires further investigation, and field evidence from a wider area.

## CONCLUSION

Preliminary observations of the geomorphology of the Ilha de Maracá indicate topics for further investigation. Undoubtedly, complex patterns of landforms and materials exist on the island and in its vicinity, which reflect a complex geomorphological history. The geomorphology of the interior of the island remains largely unexplored, although examination of aerial photographs and radar images enables broad generalizations to be drawn.

In respect of long-term landform evolution in the Maracá area, mapping of distinctive patterns of dissection on the Ilha de Maracá indicates a clear demarcation of the terrain types related by Radambrasil (1975) to pre-Quaternary surfaces elsewhere in Amazonia. This would undoubtedly repay further field investigation. In particular, the critical threshold area between the $Pd_2$ and $Pd_1$ surfaces of Bigarella and Ferreira (1985), both on the island and to the west, deserves examination.

Where more detailed field investigations have been made, pedogeomorphic contrasts have been encountered between forest and savanna areas. The contrasts are contrary to the view that massive ecosystem shifts, postulated for Amazonia during the Pleistocene, occurred in the Maracá area. Paradoxically, the presence of extensive dune fields some 130 km east of the Ilha de Maracá suggests that climatic change affected northern Roraima. Further examination of surface forms and materials over a wider area is necessary to clarify whether or not significant vegetational changes occurred.

Further research is thus needed to examine the complexities of the geomorphology of the Ilha de Maracá, and to place these findings in the context of Tertiary and Quaternary environmental change in northern Roraima.

## ACKNOWLEDGEMENTS

Grant assistance was received from the Ford Foundation/Royal Geographical Society (MJE/DFMM), the Carnegie Trust for the Universities of Scotland (DFMM), and the Central Research Fund of London University (MJE). Laboratory analyses were undertaken by M. Onwu and G. Marriner, and cartographic work by R.C. Halfhide and J. Jacyno.

# REFERENCES

Barbosa, O. and Ramos, J.R. de A. (1959). *Território do Rio Branco: aspectos principais da geomorfologia, da geologia e das possibilidades minerais de sua zona setentrional.* Departamento Nacional da Produção Mineral, Rio de Janeiro.

Beigbeder, Y. (1959). *La région moyenne du haut Rio Branco (Brésil): étude géomorphologique.* Institut des hautes études de l'Amérique Latine, Paris.

Berrangé, J.P. (1977). *The geology of southern Guyana, South America.* HMSO, London.

Bigarella, J.J. and Ferreira, A.M.M. (1985). Amazonian geology and the Pleistocene and the Cenozoic environments and paleoclimates. In: *Amazonia*, eds G.T. Prance and T.E. Lovejoy, pp. 49–71. Pergamon Press, Oxford.

Choubert, B. (1957). *Essai sur la morphologie de la Guyane.* Mém. Carte Géol. Dét. France. Dép. de la Guyane Francaise, Paris.

Colinvaux, P.A. (1987). Amazon diversity in the light of the paleoecological record. *Quaternary Science Reviews*, 6, 93–114.

Eden, M.J. (1971). Some aspects of weathering and landforms in Guyana (British Guiana). *Z. Geomorph N.F.*, 15, 181–198.

Eden, M.J., McGregor, D.F.M. and Morelo, J.A. (1982). Geomorphology of the middle Caquetá basin of eastern Colombia. *Z. Geomorph. N.F.*, 26, 343–364.

Eden, M.J., McGregor, D.F.M. and Morelo, J.A. (1984). Semiquantitative classification of rainforest terrain in Colombian Amazonia using radar imagery. *Int. J. Remote Sensing*, 5, 423–431.

Eden, M.J., McGregor, D.F.M. and Vieira, N.A.Q. (1990). Pasture development on cleared forest land in northern Amazonia. *Geog. J.*, 156, 283–296.

Endler, J.A. (1982). Pleistocene forest refuges: fact or fancy? In: *Biological diversification in the Tropics*, ed. G.T. Prance, pp. 641–657. Columbia University Press.

Guerra, A.T. (1956). Aspectos geográficos do território do Rio Branco. *Revista Brasileira de Geografia*, 18, 117–128.

Haffer, J. (1969). Speciation in Amazonian forest birds. *Science*, 165, 131–137.

Journaux, A. (1975). Recherches géomorphologiques en Amazonie brésilienne. *Bull. Centre Géom. CNRS, Caen*, 20, 1–68.

Kroonenberg, S.B. and Melitz, P.J. (1983). Summit levels, bedrock control and the etchplain concept in the basement of Suriname. *Geol. Mijnbouw*, 62, 389–399.

Löffler, E. (1977). Tropical rainforest and morphogenic stability. *Z. Geomorph. N.F.*, 21, 251–261.

Martini, J.M. (1988). *Projeto Geologia – Ilha de Maracá.* Convênio DNPM/INPA, Manaus.

McConnell, R.B. (1968). Planation surfaces in Guyana. *Geog. J.*, 134, 506–520.

Nelson, B.W., Ferreira, C.A.C., Silva, M.F. da and Kawasaki, M.L. (1990). Endemism centres, refugia and botanical collection density in Brazilian Amazonia. *Nature*, 345, 714–716.

Nortcliff, S. and Robison, D. (1989). The soils and geomorphology of the Ilha de Maracá, Roraima: the second approximation. Maracá Rainforest Project First Report, Soils Part. Royal Botanic Garden, Edinburgh.

Prance, G.T. (1982). Forest refuges: evidence from woody angiosperms. In *Biological diversification in the Tropics*, ed. G.T. Prance, pp. 137–158. Columbia University Press.

RADAMBRASIL (1975). *Folha NA.20 Boa Vista e Parte das Folhas NA.21 Tumucumaque, NB.20 Roraima e NB.21*, Vol. 8. Ministério das Minas e Energia, Rio de Janeiro.

RADAMBRASIL (1976). *Folha NA.19 Pico da Neblina*, Vol. 19. Ministério das Minas e Energia, Rio de Janeiro.

Ruellan, F. (1957). *Expedições geomorfológicas no território do Rio Branco.* Instituto Nacional de Pesquisas da Amazônia, Rio de Janeiro.

Salo, J. (1987). Pleistocene forest refuges in the Amazon: evaluation of the biostratigraphical, lithostratigraphical and geomorphological data. *Ann. Zool. Fennici*, 24, 203–211.

Schaefer, C.E.G.R. and Dalrymple, J.B. (1995). Landscape evolution in Roraima, North Amazonia: planation, paleosols and paleoclimates. *Z. Geomorph. N.F.*, 39, 1–28.

Short, K.C. and Steenken, W.F. (1962). A reconnaissance of the Guayana Shield from Guaspati to the Rio Aro, Venezuela. *Bol. Inform. Venezuela*, 5, 189–221.

Sinha, N.K.P. (1968). *Geomorphic evolution of the northern Rupununi Basin, Guyana*. McGill University Savanna Research Series, No. 11, Montreal.

Tricart, J. (1974). Existence de périodes sèches au Quaternaire en Amazonie et dans les régions voisines. *Rev. Géomorph. Dynamique*, 23, 145–158.

Tricart, J. (1985). Evidence of Upper Pleistocene dry climates in northern South America. In: *Environmental change and tropical geomorphology*, eds I. Douglas and T. Spencer, pp. 197–217. Allen & Unwin, London.

Whitmore, T.C. and Prance, G.T. (1987). *Biogeography and Quaternary history in Tropical America*. Clarendon Press, Oxford.

Young, A. (1970). Slope form in the Xavantina–Cachimbo area. *Geog. J.*, 136, 383–406.

**Plate 1a** The approach to the Ilha de Maracá from the south, showing its position on the natural boundary between savanna (foreground) and forest (background). The characteristic lines of *buriti* palms *(Mauritia flexuosa)* crossing the savanna follow the lines of seasonal or permanent watercourses. The grassy savanna in the foreground is periodically burned to produce new growth for cattle. Photograph by J.A. Ratter

**Plate 1b** The Santa Rosa savanna on the Ilha de Maracá, showing the natural transition from savanna *(campo)* to forest. Note the dense tussock grass, the typical bushy *Curatella americana* trees on the left, the lone *buriti* palm in a damp depression, and the yellow flowers of *Cochlospermum orinocense* (a common species on the forest margin). A substantial number of leafless crowns are visible in the background, illustrating the semi-deciduous nature of the forests in the eastern part of the island. Photograph by J.A. Ratter

**Plate 2a** Banks of the Furo de Maracá (close to the Maracá house) during the dry season. The exposed boulders are fully submerged during the rainy season. On the banks there is a dense riverine thicket, behind which lies forest dominated by *Peltogyne gracilipes* (Leguminosae), recognizable at this season by its sparsely leaved crowns and pale trunks. Photograph by W. Milliken

**Plate 2b** Islands of the Furo de Maracá, upstream of the Maracá house. This combination of small islands, boulders and sandbanks among a maze of narrow channels and rapids is typical of much of the Furo de Maracá and of some stretches of the Furo de Santa Rosa. The photograph was taken during the rainy season, as evidenced by the turbidity of the water. The branches in the foreground are of the leguminous *Macrolobium acaciifolium*, a common species on the sandy river banks. Photograph by W. Milliken

**Plate 3a** Mixed *terra firme* forest at the northern end of the Preguiça trail. The collector, in the crown of an understorey tree, is cutting specimens with a long-arm pruner from *Zollernia grandifolia* (Leguminosae), a canopy species. Note the large perforated leaves of the aroid *Monstera dubia* on the *Zollernia* trunk. Photograph by J.A. Ratter

**Plate 3b** Forest dominated by *Peltogyne gracilipes* on the Preguiça trail. The rather sparse ground layer, the abundant regeneration of *Peltogyne* seedlings (with their characteristic double leaflets), and the smooth pale trunks are typical of this forest type, which covers large tracts of the Ilha de Maracá. Photograph by W. Milliken

**Plate 4a** Aerial view of a *vazante* drainage line at the eastern end of the Ilha de Maracá, with its typical shrubby/herbaceous vegetation and its abrupt transition to adjacent forest. In the background lies the strip of seasonally flooded savanna below the Ecological Station, crossed by the artificial causeway. Seasonal lagoons are also visible on the savanna, and the fronds of the *inajá* palm *(Maximiliana maripa)*, a common species at the eastern end of the island, are abundant in the forest canopy. Photograph by W. Milliken

**Plate 4b** Aquatic vegetation in a seasonal lagoon close to the Ecological Station. Note the *Elaeocharis* reeds (Cyperaceae) in the foreground and the fleshy water hyacinths *(Eichhornia azurea,* Pontederiaceae) behind. In the distance lies a stretch of open grassy savanna, also partially flooded, giving way to the narrow strip of forest which covers the levee beside the river. Photograph by W. Milliken

# 4 The Soils of the Ilha de Maracá

**STEPHEN NORTCLIFF AND DANIEL ROBISON**
*University of Reading, Reading, UK*

## SUMMARY

A number of researchers have collected detailed information on various aspects of the soils of a restricted area in the east of the Ilha de Maracá, the results of which are published elsewhere (e.g. Ross *et al.*, 1990; Furley and Ratter, 1994; Furley, Chapter 22). This study is an attempt to bring together observations on and analyses of the soils collected during a number of expeditions to various parts of the island. It is intended that the material discussed here may serve as an information base for other studies in related and other disciplines. The information and discussion describe the nature and distribution of the soils and give a partial description of the soil properties based on around 100 soil profiles unequally distributed over the island. Most of the observations are based on the 30 profiles described and sampled by the authors. Data available from other studies on the island have been incorporated where possible.

## RESUMO

Descrições detalhadas dos solos da parte oriental da Ilha de Maracá já foram feitas por vários pesquisadores (e.g. Ross *et al.*, 1990; Furley e Ratter, 1994; Furley, Chapter 22). O objetivo deste capítulo é dar um resumo de observações e análises dos solos coletados durante uma série de viagens pelo interior da ilha. A maioria das observações é baseada nos 30 perfis descritos pelos autores, mas, quando era possível, dados provenientes de outros autores foram incorporados. As informações e a discussão descrevem o caráter e distribuição dos solos, e uma descrição parcial das propriedades dos solos, baseada num levantamento de aproximadamente 100 perfis distribuídos por toda a ilha. Nossa intenção é que os dados aqui apresentados sirvam como uma base de informação para outros estudos.

## GEOMORPHOLOGY AND PARENT MATERIAL

### RELIEF

Detailed investigations of the soils and associated landforms have been undertaken along a series of transects at different locations across the island (Figure 4.1). Several subdivisions (in terms of relief) have been used to categorize the soils of the island. The subdivisions identified, which correspond closely to those identified by McGregor and Eden (Chapter 3: Figure 3.4), are:

*Maracá: The Biodiversity and Environment of an Amazonian Rainforest.*
Edited by William Milliken and James A. Ratter. © 1998 John Wiley & Sons Ltd.

1. *The Eastern Shield* This is relatively flat, but generally lies about 5 m above the level of the seasonally flooded river edges. The Ecological Station is situated on the southern edge of the shield. From aerial photographs, this area resembles terrain on the mainland on the northeast side of the river more than it does the rest of the island.

2. *The Nassasseira Catchment Area* This extends from the Furo de Santa Rosa to the Central Highlands. Ground observation as well as aerial photographs suggest this area to be extensively flooded during the rainy season.

3. *The Pedra Sentada Catchment Area* This covers the northwestern part of Maracá to the south of the Fumaça falls (see Figure Pr.1, p. xvii). Whilst much of the lower-lying area shows signs of regular seasonal flooding, a significant part of the catchment is well-drained land. Much of this well-drained land appears to be related to ridges of granite that are generally not exposed, but raise the ground surface by 5–10 m.

4. *The South Central Highlands* Much of the drainage of the island originates in this area, to which the Brazilian topographic maps attribute the highest points. It remains largely unexplored, but the variability in vegetation, parent material and soils that was encountered suggests it as a promising place for future work.

5. *The Southwestern Highlands* These end at the Purumame falls, and are composed of two separate ranges of hills with a seasonally flooded valley in between. The local relief is on the average steeper than that of the Central Highlands, but the summits do not appear as high.

## RELATIVE DISTRIBUTION OF DRIFT VERSUS RESIDUA

The rocks beneath the island are of Pre-Cambrian origin and relatively unaltered through the ages. This raises two important questions:

- How much of the island surface is comprised of this Pre-Cambrian parent rock or weathered deposits derived from it?
- How much is drift?

Drift in this case is used to denote material that has been through at least one cycle of alteration, erosion and deposition prior to soil development in its current position. It is suggested (IBGE, 1981) that at some point in geological history there was sedimentary material overlying the present area of the island to a depth similar to the present height of Tepequém (i.e. approx. 800 m) – a table-topped mountain (*tepui*) some 30 km to the north of Maracá.

The present nature and distribution of these deposits of altered material throughout the island are complicated, being the result of a complex interaction of geological

# SOILS

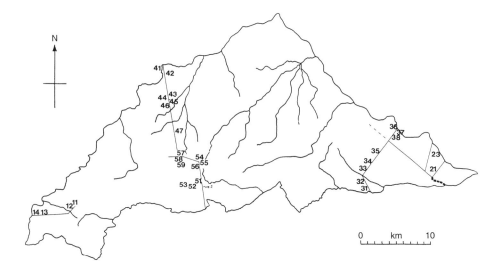

**Figure 4.1.** Location of sample transects and soil pit sites

and geomorphological processes which lowered the island's surface to its present level. These deposits are in places superimposed upon the parent rocks and are variously sandy, gravelly or even clayey. Contemporary soil development is taking place within these deposits rather than directly upon the weathering residue of the underlying parent rocks.

## DRIFT

During the survey a number of different types of drift were encountered. Part of the Eastern Shield is underlain by a layer of conglomerate-type material with rounded pebbles cemented together by quartz and iron. It is unknown to what degree the layer of conglomerate represents a geological alluvial deposit which was cemented in the past on a geological scale, or the degree to which the iron enrichment and induration is a current process. The Ecological Station is underlain by such material, which in turn is capped by a layer of very sandy drift (e.g. profile 21). Parent material samples that were retrieved included rounded pieces of white sandstone that may have come from the Roraima geological formation.

Another important category of drift on the island is alluvial and colluvial clay deposits. These and related soils are common around the edge of the island. They also occur in localized areas of poor drainage in the interior, frequently with dense occurrence of low hummocks. Sandy deposits of limited area are also common around the island; near the Ecological Station and on the Preguiça trail they tend to occur in narrow east–west bands. In the Pedra Sentada catchment (Fumaça trail), deposits of coarse sand in broad valley bottoms alternate with the granite ridges.

At the northern end of the Fumaça trail and along the eastern fringes of the Western Highlands, there are deposits of drift in which deep, red, relatively uniform soils are developing (profile 41). In the Central Highlands themselves there is a different clay drift deposit, which is lighter red than the previous and lower in weatherable minerals. This layer appears to overlie a horizontal deposit of ironstone.

## RESIDUA

Soils developing directly from the *in situ* weathering of fresh rock (residua) have to date been observed in a limited number of locations, chiefly on the granite ridges on the Fumaça trail and on hills on the Filhote and Purumame trails, and on the steeply sloping hills. Granite is the most commonly occurring outcrop in the river, but one must be cautious in extrapolating for the interior of the island. The following generalizations, however, appear to hold:

- The granite is limited as a primary parent material to the low ridges.
- All of the steeper hills encountered (bar the Central Highlands) are of metamorphic rocks (schist, gneiss and quartzite), with two examples of amphibolite.
- The Central Highlands are underlain by the dissected drift mentioned above.

## MINERALOGY OF SOILS (WHOLE SOIL)

X-ray diffraction was conducted on selected horizons (principally the C or lower B horizons) of 32 profiles sampled during the study. The method is described by Brindley and Brown (1980), but briefly involves grinding the soil very finely in a TEMA mill and presenting this material to an X-ray beam. The resulting diffractogram gives an indication of the types of crystalline minerals in the whole-soil sample. To a limited degree, the relative intensity of the signals for certain minerals indicates the relative amounts of the minerals present.

The intensity of the signal was measured at 10 different angles that correspond to the more important soil minerals. The information was summarized by cluster analysis using Ward's minimum variance method (Ward, 1963). This procedure clusters together the samples that are most similar at progressively lower levels of similarity until there is a single cluster. The results of this analysis are shown in Figure 4.2. The semi-partial $R^2$ values indicate the degree of similarity at which groups (clusters) are formed. The samples are referenced by three digits; the first two digits refer to the profile number (which is located on Figure 4.1) and the third digit refers to the horizon number. Hence 515 is horizon 5 of profile 51.

The analysis shows a distinctive pattern of clustering with most of the sites forming distinct clusters at a high level of similarity, with just three samples distinctly separated. Four major groups were distinguished in the analysis.

### GROUP 1

This group (Figure 4.2) is predominantly composed of soils from the lowlands. There are two subgroups: the first (profiles 11, 31, 32, 45, 46 and 47) are from

# SOILS

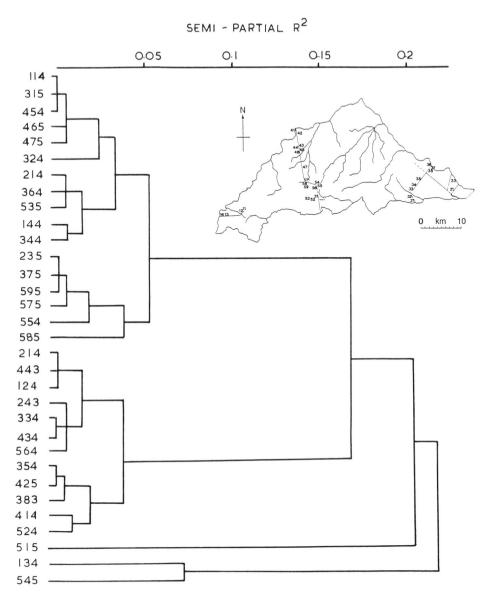

**Figure 4.2.** Dendrogram showing groupings of soil samples based on mineralogical data. The first two groupings refer to profile number, the third refers to horizon number

lightly dissected landscapes, and the second (profiles 14, 21, 34, 36 and 53) are from low- to mid-slope positions. All the soils in this group have quartz as a major component, with marked peaks for kaolinite and halloysite (a hydrated kaolin mineral) in many of the samples. Samples in the second subgroup also show evidence for haematite and goethite. Feldspars are absent in samples from this group. These soils are predominantly developed on drift materials.

GROUP 2

The majority of the soils in this group (profiles 23, 37, 55, 57, 58 and 59) are from upland locations and, with the exception of profile 23 from the Santa Rosa Savanna Transect, are well drained. The mineralogy of these soils consists of quartz and kaolinite, with distinct peaks of goethite and haematite.

GROUP 3

The soils in this group (profiles 12, 21, 33, 43, 44 and 56, and sample 24 from the Santa Rosa Savanna transect which was not included in other analyses) are all found in mid- to lower-slope positions. All soils in this group have very strong peaks for quartz, which indicates soils with very sandy subsoil textures. Some of the soils have significant proportions of halloysite and feldspar. These soils are probably of low fertility.

GROUP 4

This group (profiles 35, 38, 41, 42 and 52) consists of soils with varied mineralogy from a wide range of landscape locations. It is probable that these soils are grouped by the algorithm because of their lack of close similarity with members of other groups. All the soils have distinct peaks for quartz and kaolinite, but exhibit considerable variability in the nature of the less important minerals.

Three soils (profiles 13, 51 and 54) do not join any of the four groups defined above. These profiles are all located at crest locations and have varied mineralogy with strong evidence of the presence of chlorite in profiles 13 and 51. The nature of the mineralogy of these three soils would appear to show direct influence of the underlying solid geological materials, rather than reflecting the varying degrees of influence of drift and residual deposits as is evident in many of the other sites.

## SOIL CHEMICAL PROPERTIES

The analysis to date of profiles from around the island shows that small-scale maps of chemical properties are of limited value. That is, the range of values for almost any property is very wide relative to tropical soils in general. For example, two hillslopes in the Central Highlands represent the extremes of the island soils: Filhote Ravine Transect soils (profiles 55 and 56) are among the most weathered and leached, while soils on the two nearby transects, 'Filhote Soldier Ant' (profiles 57, 58 and 59) and 'Filhote Sweat Bee' (profiles 51, 52, 53 and 54), are the most base-rich of the island and may be considered very high for any tropical soils. These points notwithstanding, included below are points that result from visual and numerical analysis of the soils and help to summarize the range and distribution of the properties. A summary of the selected chemical data is given in Table 4.1, with the soils grouped according to broad classes of the Brazilian Classification (EMBRAPA-SNLCS, 1981).

Table 4.1 Summary of selected soil chemical data from the Ilha de Maracá, for soil groupings based on Brazilian Classification of soils

| Soil type by Brazilian Classification (number of profiles represented) | Surface (0–20 cm) | | | | | Subsurface | | | | | |
|---|---|---|---|---|---|---|---|---|---|---|---|
| | pH | Sum of bases* | Total P (ppm) | Ca:Mg | pH | Sum of bases* | ECEC* | Exch. Al* | Fe oxide (%) | Al oxide (%) | Total P (ppm) |
| Dark red podzolic (7) | 5.40 | 4.42 | 155 | 2.90 | 5.56 | 1.61 | 3.59 | 1.99 | 4.23 | 4.01 | 69 |
| Dystrophic yellow podzolic (7) | 4.47 | 0.62 | 103 | 1.04 | 5.15 | 0.40 | 2.08 | 1.67 | 1.04 | 2.69 | 1 |
| Red–yellow podzolic (4) | 5.52 | 3.70 | 96 | 1.87 | 5.58 | 4.71 | 6.50 | 1.78 | 2.03 | 3.03 | 104 |
| Brown–grey podzolic (1) | 6.32 | 17.08 | 336 | 5.89 | 6.03 | 25.72 | 26.41 | 0.69 | 8.18 | 6.43 | 112 |
| Eutrophic yellow podzolic (1) | 5.89 | 12.70 | 229 | 4.78 | 6.40 | 32.39 | 33.53 | 1.14 | 6.15 | 1.90 | 59 |
| Dystrophic red–yellow latosol (2) | 4.16 | 0.49 | 137 | 2.44 | 5.08 | 0.19 | 1.08 | 0.89 | 2.31 | 5.60 | 174 |
| Dystrophic yellow latosol (1) | 4.30 | 2.68 | 384 | 1.28 | 5.25 | 0.74 | 3.12 | 2.37 | 2.03 | 3.45 | 321 |
| Dystrophic dark red latosol (1) | 3.93 | 0.31 | 145 | 1.18 | 5.07 | 0.32 | 4.37 | 4.05 | 3.55 | 5.96 | 116 |
| Greyish eutropic hydromorphic (2) | 4.52 | 12.21 | 542 | 2.10 | 6.31 | 10.79 | 11.09 | 0.30 | 3.76 | 4.87 | 109 |
| Dystrophic cambisol (2) | 4.19 | 1.39 | 106 | 2.3 | 4.97 | 0.55 | 1.24 | 0.74 | 2.21 | 2.09 | 170 |
| Dystrophic hydromorphic sands (1) | 4.78 | 1.85 | 50 | 4.75 | 5.25 | 0.75 | 1.69 | 0.94 | 0.04 | 0.24 | 30 |
| Structured brown earth (1) | 4.22 | 1.28 | 386 | 1.86 | 5.91 | 0.14 | 1.52 | 1.38 | 14.28 | 9.43 | 268 |

* $cmol_c\ kg^{-1}$.

## pH TRENDS AND DISTRIBUTION ON MARACÁ

The pH of the samples obtained on the island range from 3.64 to 6.37 on surface horizons and 4.71 to 6.97 in the subsoil. In short, the range for the island is the same as the range for most humid tropical soils. The lowest pH values were encountered in the hills on the Purumame trail and on the Filhote Ravine Transect. The highest pH values were on the amphibolite-related soils in the Central Highlands and in the dark, clayey soils of some of the *brejos* (boggy drainage areas). Most other soils have pH values ranging between 4.5 and 5.5, which would be considered acid but not extremely acid.

## BASE STATUS AND EFFECTIVE CATION EXCHANGE CAPACITY (ECEC)

Other indices important to plant growth and more quantitative than pH are exchangeable cations, exchangeable acidity and the effective cation exchange capacity (ECEC). The exchangeable bases discussed here were extracted in 1 M KCl following recommended Brazilian practice. Dr Furley's samples (Furley, 1989) were extracted with 1 M ammonium acetate, and consequently may not be directly comparable with the results presented in this report. Once again, almost the full range of levels of extractable bases for tropical soils was observed on Maracá. The sum of bases (Ca+Mg+K) ranges from 0.14 to 17.08 $cmol_c$ $kg^{-1}$ at the soil surface, and from 0.04 to 32.39 $cmol_c$ $kg^{-1}$ in the subsoil. There seems to be a broad correlation between sum of bases and pH. Exchangeable acidity (KCl-extractable H and Al) in the subsoils ranges from 0.2 to 4.05; the lowest values are in the darker alluvial soils while the highest are in soils rich in clay and low in nutrient bases (profiles 55 from the Filhote trail and 11 and 12 from the end of the Purumame trail).

Subsoil ECEC (Ca+Mg+K+Al+H) ranged from 0.92 in profile 34 on the Preguiça trail to 33.52 at the top of Sweat Bee Transect (profile 51). These compare to a range of 0.23 for a Regosol and 32.44 for a hydromorphic Podzol in the soils described and analysed around Manaus and Itacoatiara (IPEAN, 1969).

Cluster analysis was conducted using the following variables: the subsoil sum of bases and ECEC together with exchangeable acidity, the sum of bases and pH at the surface (Figure 4.3). Two profiles (51 and 58) are clearly identified as separate from the rest of the profiles sampled. The remaining profiles show distinct, reasonably compact groupings. Both these separate sites were located on the Filhote transect and field evidence suggests they may be related to possible amphibolite rock intrusions (it should be noted that profile 51 was identified as a site with only limited similarity to other sites in the cluster analysis based on mineralogy). Profiles 32 and 47 – two sites with soils described as dark alluvial soils – are grouped together. These exceptions apart, there are two broad groupings: Group 1 (profiles 11, 12, 36, 37, 52, 53, 55 and 59), consisting of soils with medium base saturation, and the remaining sites (Group 2) characterized by very low base saturation.

The difficulty in making small-scale maps is illustrated by soils on the Preguiça trail (all beginning with the number 3), which are scattered across the groups. However, it can be said that the profiles that probably represent the majority of the

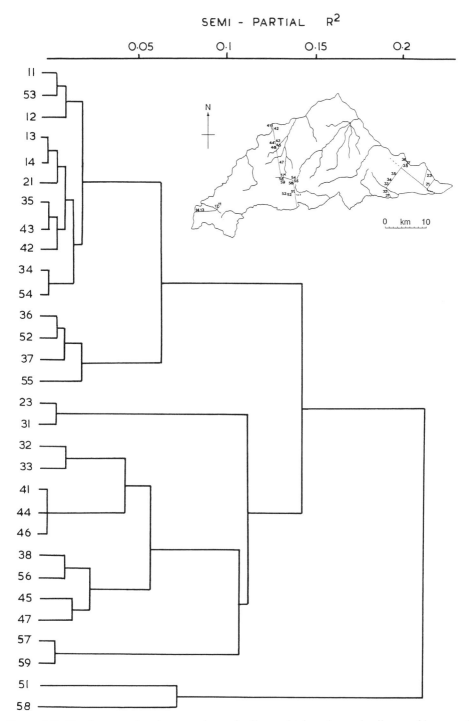

**Figure 4.3**. Dendrogram showing groupings of soil samples based on subsoil sum of bases and ECEC, and surface sum of bases and pH

island's area fall into the low base saturation, relatively low pH category. It is also possible to suggest that the internal alluvial soils and soils at the bases of hills are on average higher in exchangeable bases and ECEC. The Uraricoera alluvial soils, put into this context, are by comparison very low in base saturation.

It has been observed that a large number of the soils sampled have very high Mg:Ca ratios. Most plants need Ca in larger amounts than Mg, and the imbalance may actually inhibit plant growth. The areas of the island that are predominantly under *Peltogyne* forests (see Milliken and Ratter, Chapter 5) tend to coincide with these soils with abnormally high Mg:Ca ratios. Reasons for the imbalance are likely to be a combination of the nature of the parent material and the poor drainage found in many of the *Peltogyne* areas.

## LATERIZATION COMPLEX

Though a significant part of the Brazilian Classification system for latosols is based on the 'laterization complex', there was not time to conduct laboratory analyses specifically for these properties. As an alternative, Fe and Al were measured in the Kjeldahl digest produced primarily for total nitrogen. Stereotypically, high iron and aluminium oxide are generally associated with latosols or oxisols and therefore with low fertility. However, within the context of the island, iron and aluminium are only weakly correlated with base saturation ($R^2 = 0.5$). The 'poorest' soils on the island are the sandier, silica-enriched soils on the Fumaça and Preguiça trails, and those surrounding the Ecological Station. The soils highest in both aluminium and iron are the amphibolite-related soils, followed by the dark alluvial soils. In this framework, it would seem that high aluminium and iron are less a sign of excessive weathering, but more specifically reflect soils developed from parent material high in these elements.

## **PROFILE CLASSIFICATION**

### GENERAL CLASSIFICATION OVERVIEW

Following the field observation of soil profiles as a component of soil survey, and the laboratory analysis of soil samples, it is customarily the next step to endeavour to classify the soils into groups. This classification process may be undertaken independently of any previously established system, and similarities between profiles assessed solely in the context of the profiles examined during the survey. This independent approach often involves a comparison of soils using all the information available with no prior weighting of particular soil properties. Such an approach maximizes the use of the limited data set under consideration, and has in the past been referred to by a number of authors as 'natural' classification. An alternative approach is to place the observed soils in a previously defined soil classification system. Allocation to classes in an established system enables comparisons to be made with similar soils at different locations, hence facilitating the transfer of knowledge and technology. Additionally this frame of reference enables information from the sites under consideration to be placed in a broader context. In this study a preliminary attempt has been made to place soils in two soil classification systems: USDA's *Soil Taxonomy* (Soil Survey Staff, 1975; Soil Management Support Services, 1983), and the provisional *Brazilian Classification* (Camargo et al., 1987). Of these, *Soil Taxonomy* is established upon worldwide experience over a number of years, but whilst there have been considerable improvements since the introduction of the *Seventh approximation* (Soil Survey Staff, 1960), the precursor to *Soil Taxonomy*, it is still considered by many workers to be a relatively poor differentiator of tropical soils. The Brazilian system, whilst it has not yet been formally presented, has appeared in outline as a second approximation (EMBRAPA-SNLCS, 1981), and some of the major features are presented by Camargo et al. (1987).

In this study an attempt has been made to group soils using the following approaches:

- Natural Classification – using Ward's Grouping Algorithm (Ward, 1963) with data from laboratory analysis of soil samples.
- USDA *Soil Taxonomy*.
- The *Brazilian Classification*.

## NATURAL CLASSIFICATION

The use of the term 'natural' is perhaps rather misleading; in the context used here it indicates that the soils have been grouped using the data available in various combinations, with no *a priori* decision on key or diagnostic properties. All properties used in the analysis are allocated equal weight. Groupings of soils occur because there is a degree of similarity based upon the interrelationships of all soil properties incorporated in the analysis. Three 'natural' groupings were undertaken using different data sets:

- All analysis results for top (0–20 cm) and bottom (C) soil horizons.
- All analysis results for top horizons.
- All analysis results for bottom horizons.

The results are presented in Figures 4.4–4.6.

### Top and bottom horizons

The results from this analysis are shown in Figure 4.4. It is clear that there are two broad groups, the first with 23 profiles and the second with 6 (profiles 33 and 34 are not included because of incompatible data sets). On further analysis of these results, no clear trends emerge. The second, smaller, group contains the three soils from the Filhote Soldier Ant Transect (profiles 57, 58 and 59), but also has two examples of gleyed soils from the Preguiça and Fumaça Transects.

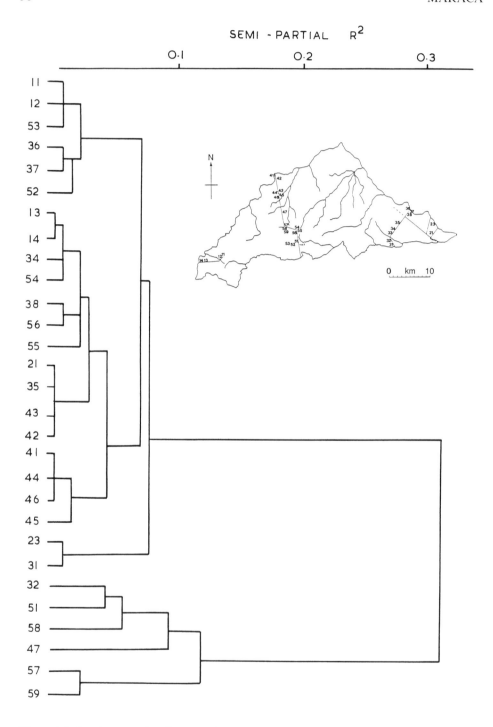

**Figure 4.4**. Dendrogram showing groupings of soil samples based on all measured properties for top and bottom horizons

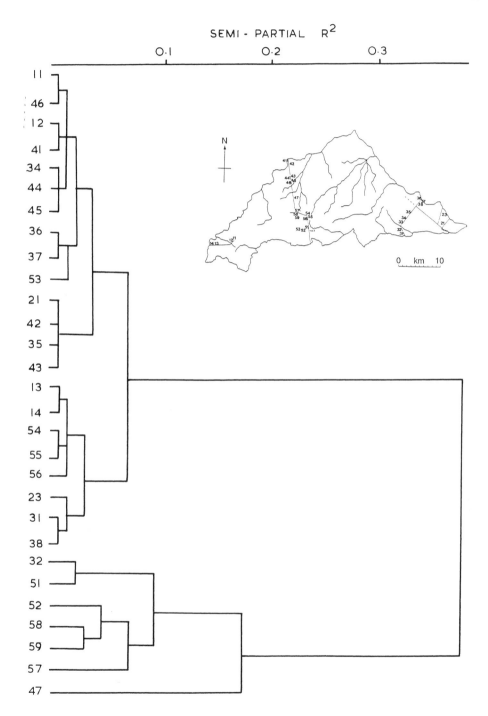

**Figure 4.5**. Dendrogram showing groupings of soil samples based on all measured properties for top horizons

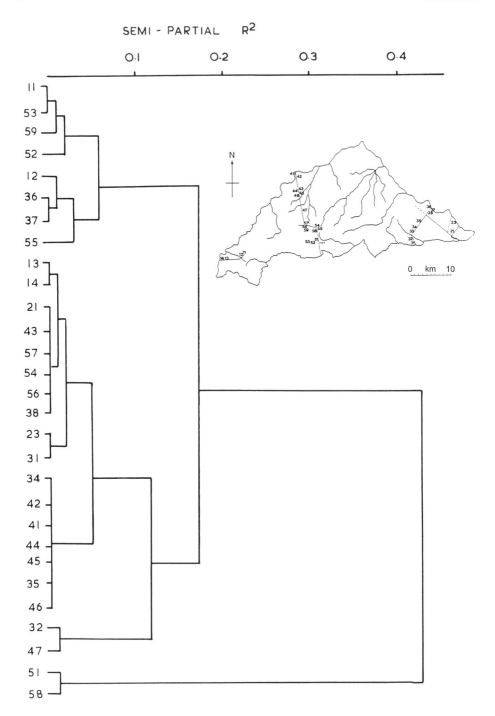

**Figure 4.6**. Dendrogram showing groupings of soil samples based on all measured properties for bottom horizons

## Top horizons

In broad terms the results of the grouping based on topsoil data (Figure 4.5) show a broad similarity with the analysis for top and bottom horizons combined. There are two broad subdivisions, with similar profile members. The smaller of the two groups has seven rather than six members (profile 52 has joined this group). Within the larger group the relationships and composition of the subgroups have some similarities with the groupings for top and bottom horizons combined, but there are sufficient differences to suggest an exceedingly complex set of interrelationships.

## Bottom horizons

The groupings from this set of analyses (Figure 4.6) show marked differences from the previous groupings. Three distinct groups are present, with 15, 13 and 2 soils respectively.

## USDA CLASSIFICATION METHOD

*Soil Taxonomy* is a hierarchical classification with six categorical levels. There are large numbers of classes at the lowest level, which are needed in order to consider all the soil properties that are important for soil use. The hierarchical nature of the system allows classes to be grouped on a rational basis into progressively smaller numbers of classes in the higher categories. In general the *differentiae* used are soil properties that can either be observed or inferred in the field. Limits are set which enable the establishment of consistent groups. The soil properties selected are those that are believed to result from soil genesis or those that affect soil formation, because they normally carry with them a set of accessory properties. Diagnostic horizons and properties have been defined to encompass many of these sets of accessory properties, and these are used as *differentiae* for classes, especially at the higher categorical levels. The *differentiae* selected are normally stable subsoil properties that are not subject to change as a result of normal management practices. In this study the diagnostic horizons and properties of major importance are the presence or absence of an argillic B horizon (identified by between-horizon textural contrast), the presence or absence of an oxic B horizon, the overall activity of the clay, the percentage base saturation and the presence of waterlogging within the profile. Details of the classification of the selected soil profiles with reference to *Soil Taxonomy* are given in Table 4.2.

## THE BRAZILIAN CLASSIFICATION

The soil classification system currently under development in Brazil has many similarities to *Soil Taxonomy*, but because of the different range of soils and soil conditions prevailing, the classes and limits for a number of key properties are different. As with *Soil Taxonomy* the system is multicategorical and essentially morphogenetic. Diagnostic horizons and properties form the basis of differentiation, and these have been drawn from the *Seventh approximation* (Soil Survey Staff,

Table 4.2  Classification of soil pits on the Ilha de Maracá. Pits classified according to the Brazilian Soil Classification and the USDA Classification

| Pit | Brazilian Classification | USDA Classification |
|---|---|---|
| 11 | Dystrophic dark red podzolic soil, clay-textured | Fragiustult |
| 12 | Dark red podzolic soil, clay-textured, fragic | Fragiustult |
| 13 | Dystrophic podzolic cambisol, medium-textured | Ustic Dystropept |
| 14 | Dystrophic yellow podzolic, clay-textured | Oxic Haplustult |
| 21 | Dystrophic yellow podzolic, plinthitic, sandy-textured | Grossarenic Plinthic Paleudult |
| 23 | Dystrophic yellow latosol, medium-textured | Aquic Haplorthox |
| 31 | Dystrophic yellow podzolic, clay-textured | Aquic Paleudult |
| 32 | Greyish eutrophic hydromorphic soil, clay-textured with vertic features | Oxic Haplustult |
| 33 | Dystrophic dark red podzolic, clay-textured | Typic Tropaquept |
| 34 | Dystrophic red–yellow podzolic, clay-textured | Oxic Haplustult |
| 35 | Dystrophic yellow podzolic, sandy-textured | Arenic Haplustult |
| 36 | Eutrophic red–yellow podzolic, sandy-textured | Typic Tropudalf |
| 37 | Dark red dystrophic podzolic, clay-textured | Ustic Paleustalf |
| 38 | Dystrophic cambisol | Oxic Dystropept |
| 41 | Dystrophic dark red podzolic, clay-textured | Ustoxic Dystropept |
| 42 | Dystrophic hydromorphic quartzose sands | Dystropept |
| 43 | Dystrophic yellow podzolic, sandy-textured (hydromorphic?) | Aeric Tropaquept |
| 44 | Dystrophic yellow podzolic, medium-textured | Orthoxic Tropudult |
| 45 | Dystrophic yellow podzolic, medium-textured | Lithic Dystropept |
| 46 | Dystrophic red–yellow podzolic, clay-textured | Oxic Haplustult |
| 47 | Eutrophic grey hydromorphic soil with vertic features | Typic Tropaquept |
| 51 | Eutrophic yellow podzolic, clay-textured | Typic Haplustalf |
| 52 | Eutrophic red–yellow podzolic, medium-textured | Typic Haplustalf |
| 53 | Dystrophic dark red podzolic, clay-textured | Typic Haplustult |
| 54 | Dystrophic red–yellow latosol, medium-textured | Ultic Haplustult |
| 55 | Dystrophic dark red latosol, clay-textured | Typic Haplustox |
| 56 | Dystrophic red–yellow latosol, clay-textured | Typic Haplustox |
| 57 | Structured leached brown earth, clay-textured | Ustic Dystropept |
| 58 | Eutrophic brownish grey podzolic, medium-textured | Udic Haplustalf |
| 59 | Dystrophic dark red podzolic, clay-textured | Ustic Dystropept |

1960), from the legend of the *Soil map of the world* (FAO-UNESCO, 1974) and *Soil Taxonomy* (Soil Survey Staff, 1975). In the context of the soils encountered on Maracá, the important properties are the nature of the B horizon, the activity of the clay, and the base saturation. Classification of the same selected profiles as used for the USDA system are given in Table 4.2.

# SOIL MAPS OF MARACÁ

### THE RADAM MAP

The exploratory soil map of RADAMBRASIL (1975), based on interpretation of RADAM imagery with only very limited ground truth observation, identifies three broad groupings of soils on the Ilha de Maracá. Approximately 50% of the island is identified as Unit LV14, consisting of red–yellow latosols with less important occurrences of red–yellow podzolic soils. The other two units are PB12 consisting of red–yellow podzolic soils with latosols in subdominance, and PB19 of loamy-textured red–yellow podzolic soils with quartzose sands and clay-textured podzolic soils in subdominance. This map is essentially a reconnaissance survey and the boundaries appear to have been drawn from an interpretation of the gross relief features, with just two sampling sites indicated on the northern tip of the island.

## PROVISIONAL SOIL MAP BY NORTCLIFF AND ROBISON

It has been possible to attempt an approximation of a soil map for the whole island (Figure 4.7), based on field observations of soils, and where possible vegetation, along three major transects (Preguiça, Fumaça and Filhote trails). This was combined with a number of other less systematic observations around the island, and with observations and interpretations from the air photo cover of the island and RADAM imagery. Units have been tentatively drawn, and dominant soils within each unit identified together with the extensive subdominants. Nine major soil units are provisionally identified, with a tenth unit associated with the *vazante* areas not portrayed on the map. *Vazante* is of islandwide importance (see Milliken and Ratter, Chapter 5), but occurs in relatively small areas which are not convenient to include.

# CONCLUSIONS FROM THE PROVISIONAL SOIL MAP

All evidence suggests that the RADAM reconnaissance map has overestimated the extent of latosols or oxisols on the island. Whereas on the RADAM map approximately half of the island is attributed dominant latosols, only four of the 31 profiles characterized in detail in this study (Table 4.2) were clearly of this class. Furthermore, this study suggests that there is an overall dominance of soils with argillic horizons, with dystrophic red–yellow podzolic soils the most extensive. Within this group, the sandy variants are quite different from those with finer textures. Additionally, there are significant areas with dystrophic yellow podzolic soils and dystrophic dark red podzolic soils. There are also occurrences of eutrophic variants (high base saturation) of each of these, though it was not possible to make an estimate of their spatial extent. Finally, the soils described as various forms of 'podzolics' should not be confused with the current European concepts of podzols, but rather they are similar to the red–yellow podzolics of pre-1960 USDA soil classification and similar soil classifications presented on a broad world scale. The podzolic soils discussed here correspond with a few soils currently described as alfisols and most described as ultisols in *Soil Taxonomy*.

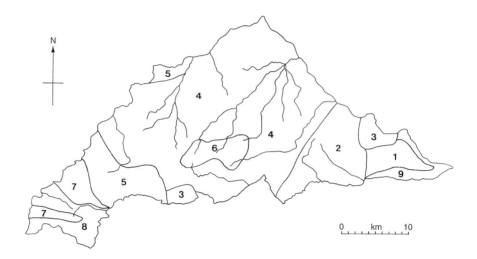

**Figure 4.7**. Provisional soil map of the Ilha de Maracá:

- *Unit 1* – Ecological Station Area (considered a dissected plinthite landscape). Dominant soils are red–yellow podzolics, plinthic and sandy-textured, together with quartzose sands and localized grey hydromorphic soils.
- *Unit 2* – Preguiça Trail Area. Dominant soils are red–yellow podzolics both clay and loamy-textured, with occurrences of dystrophic yellow podzolics and eutrophic clay-textured grey hydromorphic (no plinthite found near surface – a rolling landscape).
- *Unit 3* – Quartzite Hill and Filhote Trail Area (metamorphic rocks, minor metamorphic intrusions). Dystrophic red–yellow podzolic soils on the upper slopes with eutrophic red–yellow podzolic soils on the lower slopes, occurrences of dystrophic clay-textured dark red podzolic soils, and dystrophic cambisols.
- *Unit 4* – Fumaça 4–5 Transect Area (low gradient hills, with granite ridges and infilled valleys). Dystrophic sandy-textured yellow podzolic soils (hydromorphic in places) with some loamy-textured variants. Dystrophic red podzolic soils on local ridges and grey hydromorphic soils locally in valley bottoms.
- *Unit 5* – Flat crests with deep dissection by rivers. Dystrophic dark red podzolic soils and locally grey hydromorphic soils.
- *Unit 6* – Central Highlands. Eutrophic and dystrophic clay- and loamy-textured yellow podzolic soils dominant, with dark red podzolic soils with limited but locally dominant dystrophic red-yellow latosols.
- *Unit 7* – Purumame Area (steep-sided hills). Clay-textured dystrophic yellow podzolic soils, with minor occurrences of eutrophic yellow podzolic soils (slope-foot sites) and dystrophic cambisols.
- *Unit 8* – Lowland with low ridges (similar to Unit 4). Fragic dystrophic dark red podzolic soils in well-drained sites with occurrences of quartzose sands (associated with poorly drained positions), and loamy-textured red–yellow podzolic soils.
- *Unit 9* – Contemporary Floodplain. Undetermined combination of loamy-textured dystrophic yellow latosols and clay-textured dystrophic yellow podzolic soils.

NB There are also localized occurrences of *vazante* (Fumaça Transect Vegetation Unit 9) with grey hydromorphic soils (see above).

# RELATIONSHIP OF VEGETATION TO SOIL AND GEOMORPHOLOGY

At least three obviously different types of tall forest are found on the Ilha de Maracá. The first, referred to here as $F_1$, covers the raised shield immediately around the Station and extends at least 5 km westwards. Within the forest the large tree species composition varies considerably. The soils in this area are predominantly dystrophic plinthitic sandy-textured yellow podzolic soils with quartzose sands and locally grey hydromorphic soils.

The second forest type, which covers the steep hills at the western end of Maracá with heavily dissected relief, is referred to as $F_2$. The soils are predominantly clay-textured dystrophic yellow podzolic soils, with occurrences of eutrophic yellow podzolic soils (particularly in slope-foot positions) and dystrophic cambisols.

These two forest types are not mutually exclusive in terms of species, but each contains a considerable number of species that are not found in the other (see Milliken and Ratter, Chapter 5). Some of the hills in the Central Highlands (also with heavily dissected relief) support forest extremely similar to type $F_2$ described in the west. The soils in this area are eutrophic and dystrophic, clay and loam-textured red–yellow podzolic soils, with locally dominant dystrophic red–yellow latosols.

The rest of Maracá is covered by a number of other forest types of which the most prominent is *Peltogyne* forest, $F_P$ (Plate 3b). This is possibly the dominant vegetational unit on the island, and frequently corresponds to the lower undulating regions with broad, poorly drained, valley bottoms. The soils are predominantly dystrophic yellow podzolics, dominantly sandy-textured with some loamy-textured, locally hydromorphic, with dystrophic red podzolic soils on the low relief (granite) ridges and grey hydromorphic soils in the wetter areas. Many of the soils underlying the *Peltogyne* are characterized by some of the highest levels of exchangeable Mg recorded on the island.

On higher ground on metamorphic rocks corresponding to minor metamorphic intrusions, there are forests intermediate in composition between types $F_1$ and $F_2$, as they are in their geographical position (see Milliken and Ratter, Chapter 5). The soils show a distinct pattern with slope, with predominantly dystrophic red–yellow podzolic soils on the upper slopes and the eutrophic version of the same soils on lower slopes. Locally there are occurrences of eutrophic dark red podzolic soils and dystrophic cambisols.

It is important to note that these described forms of forest are not definitive units, but grade into each other to produce intergrades, as do the soil types on which they are found. They may also take lower forms and undergo significant compositional changes resulting from the combined local influences of hydrological, pedological and topographical features, for example stream beds, patches of white sands (probably associated with previous river channels), etc.

A small proportion of the Ilha de Maracá is covered by non-forest vegetation types, of which the most important are *campo* (savanna) and *vazante* (see Plates 1 and 4). *Campo*, although the least important in terms of spatial extent, is probably most significant because of Maracá's position on the natural boundary between savanna and evergreen forest. These savannas correspond to seasonally high water-tables (often at or very close to the soil surface during the wet season) and lie on soils with a substantial micaceous composition. *Vazante* tends to occur in areas of seasonal inundation and poor drainage, the soil surface generally showing hummocks. The soils are grey and hydromorphic, with sharp seasonal contrasts; in the dry season they are split by deep (to 50 cm+) cracks and in the wet season they are inundated.

## AN EXAMPLE: THE FUMAÇA TRAIL

The majority of the soils investigated along this trail have an exchangeable Mg:Ca >1.0. Soils on units 1, 4 and 5 have relatively high levels of exchangeable K. Percentage base saturation is defined as follows:

Low base saturation <15%
Medium base saturation 15–35%
High base saturation >35%

### Fumaça Unit 1

*Vegetation*  Mixed *terra firme* forest similar to $F_1$ but with presence of *Peltogyne gracilipes* Ducke (Leguminosae).

*Soils*  Deep well-drained soils, clayey-textured, pH in the region of 5.5, medium base saturation, medium cation exchange capacity (8–14 $cmol_c$ $kg^{-1}$ soil). Brazilian Classification – Dystrophic dark red podzolic, clay-textured. USDA Classification – Ustoxic Dystropept.

### Fumaça Unit 3

*Vegetation*  An intergrade between tall mixed *terra firme* forest and *Peltogyne*-rich low-diversity *terra firme* forest.

*Soils*  Deep light-coloured, well-drained sandy-textured soils, pH 5.2, medium base saturation (high relative Mg), very low cation exchange capacity (1.9 $cmol_c$ $kg^{-1}$ soil). Brazilian Classification – Quartzose sands. USDA Classification – Dystropept.

### Fumaça Unit 4

*Vegetation*  Low forest with a high incidence of *Peltogyne gracilipes* and emergent specimens of the same.

*Soils*  Deep light-coloured sandy over sandy-clay-loam-textured soils, showing evidence of poorer drainage below 25 cm, pH 4.9, medium base saturation (*c*.16%), low cation exchange capacity (3.9 $cmol_c$ $kg^{-1}$ soil), high exchangeable acidity.

Brazilian Classification – Dystrophic yellow podzolic, medium-textured. USDA Classification – Orthoxic Tropudult.

**Fumaça Unit 5**

*Vegetation* Mixed *terra firme* with some *Peltogyne gracilipes* and a heavy ground cover of *Calathea* spp. (Marantaceae) and *Phenakospermum guyannense* (L.C. Rich.) Endl. ex Miq. (Musaceae).

*Soils* Deep red to yellowish red well-drained sandy loam over clay soils, with granite boulders on surface and weathered granite boulders within the soil, pH 5.5, low base saturation, medium cation exchange capacity (8.5 $cmol_c$ $kg^{-1}$ soil), no labile phosphorus. Brazilian Classification – Dystrophic red–yellow podzolic, clay-textured. USDA Classification – Oxic Haplustult.

**Fumaça Unit 9**

*Vegetation* *Vazante*, i.e. seasonally flooded vegetation made up of herbs and low shrubs with abundant cover of tangled herbaceous climbers. Typical species: *Thalia geniculata* L., *Canna glauca* L., *Ipomoea tiliacea* (Willd.) Choisy, etc.

*Soils* Hummocky soil surface, with deep vertical cracks when dry, excellent structure, grey and greyish brown clay to silty clay, high base saturation, high cation exchange capacity (21 $cmol_c$ $kg^{-1}$ soil). Brazilian Classification – Eutrophic grey hydromorphic soil with some vertic features. USDA Classification – Typic Tropaquept.

# CONCLUSIONS

The conclusions from this study suggest that the pattern of soils within the landscape, and the distribution of landscapes and soils across the island is of far greater complexity than has been anticipated by earlier studies, albeit reconnaissance studies of an exceptionally general nature. This pattern of soils is in part explained by the complexity of soil parent materials, an important feature of which is the nature and distribution of drift materials, possibly of considerable age. Many soils are, however, developed in the weathering residua derived directly from the underlying rocks. From these preliminary observations there appears to be a far greater complexity of geological materials across the island than has been identified to date in the reconnaissance mapping of the solid geology, or than might be inferred from observations of river bed geology in either of the two major river channels to the north and south of the island.

Superimposed upon the patterns of soils related to the parent materials from which they are derived, there is strong evidence for substantial variation in relation to landscape position, with marked contrasts from crest to valley bottom. In some locations this may also reflect variations in the soil parent material, with soils in parts of the slope derived from *in situ* residua and parts from drift material. This problem is currently the subject of further investigation by the authors of this paper, and the findings will be published later.

In contrast to the previous reconnaissance studies of the island, it is apparent that the dominant soils on an island-wide basis are dystrophic red–yellow podzolics. Latosols are present, but far less widely distributed than the RADAM map would suggest. Similarly there is evidence to suggest that eutrophic soils are more widespread than indicated on the RADAM map. The patterns of the soils in relation to relief units and landscape are complex, and whilst it may be possible to draw a broad outline framework of soil distribution from topographic base maps, the characterization of the units is not possible without detailed analysis on the ground.

## SOILS OF MARACÁ IN THE REGIONAL CONTEXT

Whilst most of the previous attempts to map the distribution of soils of northern Brazil are associated with broad-scale reconnaissance interpretations often from remotely sensed imagery such as Projeto RADAMBRASIL, there have been a number of wide-scale studies which have attempted to identify the range of soils encountered, or likely to be encountered, in this region. For example Bennema (1963) discussed the range of soils found in Brazil, placing particular emphasis on the occurrence and nature of the latosols. He suggested that the differences between red and yellow latosols occurred because of variations in the nature of the parent materials, in particular the levels of iron oxides and hydrous oxides in the parent rocks. Similarly Costa de Lemos (1968) distinguished a major division of Brazilian soils between those with latosolic B horizons (broadly equivalent to the Oxic horizon of *Soil Taxonomy*) and soils with a textural B horizon (broadly equivalent to the Argillic horizon of *Soil Taxonomy*). Both the influence of the nature of the parent material and the nature of the B horizon have been found to be major differentiators in the soils across Maracá, together with localized influences of wetter soil conditions producing hydromorphic soils. In a general survey of soils of tropical America, Cochrane *et al.* (1985) suggested that the area of Maracá was totally covered with ultisols (broadly equivalent to dystrophic podzolic soils) and did not indicate the presence of eutrophic soils.

A team from Hokkaido University in Japan have undertaken reconnaissance studies of the soils and agroecological zonation of the land in the Amazon basin (Tanaka *et al.*, 1984), in which they similarly encountered a range of soils including latosols, red and yellow podzolic soils, hydromorphic soils and soils where the development is dominated by sands. In a more detailed study of the Mato Grosso region (Tanaka *et al.*, 1989) they identified a similarly wide range of soils, but drew only broad soil–landscape relationships. These authors do, however, draw attention to the major influence on soil development exerted by the nature of the parent materials, particularly with respect to the broad soil fertility status. These differences are also highlighted in this present study by the differentiation of dystrophic (broadly base-deficient, low fertility) and eutrophic (broadly less base-deficient, moderate fertility), again distinguishing differences in parent materials.

## ACKNOWLEDGEMENTS

This work was undertaken with the financial support of the Royal Society in the form of an Overseas Field Research Grant (SN), and the Royal Geographical Society (DR). The authors are extremely grateful for this support, which provided the opportunity to investigate a wide range of soils within their natural environment, and additionally to have the opportunity to visit remote locations and gain a new insight into the functioning of the natural ecosystem in the region of Ilha de Maracá.

## REFERENCES

Bennema, J. (1963). The red and yellow soils of the tropical and sub-tropical uplands. *Soil Science*, 95, 250–257.

Brindley, G.W. and Brown, G. (eds) (1980). *Crystal structures of clay minerals and their x-ray identification*. Mineralogical Society, London.

Camargo, M.N., Klamt, E.E. and Kaufman, J.H. (1987). Sistema brasileiro de classificação de solos. *Separata do B. Inf., Sociedade brasileiro de Ciência de Solo, Campinas*, 12 (1), 11–33.

Cochrane, T.T., Sanchez, L.G., Azevedo, L.G., Porras, J.A. de and Garver, C.L. (1985). *Terra na America tropical*. Centro Internacional de Agricultura Tropical, Cali, Colombia.

Costa de Lemos, R. (1968). *The main tropical soils of Brazil*. World Soil Resource Report no. 32, FAO, Rome.

EMBRAPA-SNLCS (1981). *Manual de métodos de análise de solo*. Rio de Janeiro.

FAO/UNESCO (1974). *Soil map of the world*. UNESCO, Paris.

Furley, P.A. (1989). The soils and soil–plant relationships of the eastern sector of Maracá Island. In: *The vegetation of the Ilha de Maracá*, eds W. Milliken and J.A. Ratter, pp. 229–276. Royal Botanic Garden Edinburgh.

Furley, P.A. and Ratter, J.A. (1994). Soil and plant changes at the forest–savanna boundary on Maracá Island. In: *The rainforest edge: plant and soil ecology of Maracá Island, Brazil*, ed. J. Hemming, pp. 92–114. Manchester University Press.

IBGE (1981). *Atlas de Roraima*. IBGE, Rio de Janeiro.

IPEAN (1969). *Os solos da área Manaus–Itacoatiara*. IPEAN, Rio de Janeiro.

RADAMBRASIL (1975). *Folha NA.20 Boa Vista e Parte das Folhas NA.21 Tumucumaque, NB.20 Roraima e NB.21*, Volume 8. Ministério das Minas e Energia, Rio de Janeiro.

Ross, S.M., Thornes, J.B. and Nortcliff, S. (1990). Soil hydrology, nutrient and erosional response to the clearance of terra firme forest, Maracá Island, Roraima, Northern Brazil. *Geographical Journal*, 156, 267–282.

Soil Management Support Services (1983). *Keys to Soil Taxonomy*. SMSS Technical Monograph No. 6, Washington.

Soil Survey Staff (1960). *Soil classification, a comprehensive system; seventh approximation*. USDA, Washington.

Soil Survey Staff (1975). *Soil taxonomy*. USDA Agriculture Handbook No. 436, Washington.

Tanaka, A., Sakuma, T., Okagawa, M., Imai, H. and Ogata, S. (1984). *Agro-ecological conditions of the Amazon basin: report of a preliminary survey*. Faculty of Agriculture, Hokkaido University, Sapporo.

Tanaka, A., Sakuma, T., Okagawa, M., Imai, H., Sato, T., Yamaguchi, J. and Fujita, K. (1989). *Agro-ecological conditions of the oxisol–ultisol area of the Amazon river system*. Faculty of Agriculture, Hokkaido University, Sapporo.

Ward, J.H. (1963). Hierarchical grouping to optimize an objective function. *Journal of the American Statistical Association*, 58, 236–244.

# 5 The Vegetation of the Ilha de Maracá

**WILLIAM MILLIKEN AND JAMES A. RATTER**
*Royal Botanic Garden Edinburgh, Edinburgh, UK*

## SUMMARY

This chapter reports the results of the vegetation survey of the Ilha de Maracá, which is situated at the junction of the Amazonian forest (Hylaea) and the great areas of dry savanna which lie to the south and east on the mainland, running through into the Rupununi savannas of Guyana. About 90% of the island is forested and dry savanna only occurs as a few small patches at the eastern end of the island. The forests were intensively studied by transect work, general observation ('wide patrolling') and remote sensing, and a number of different types were recognized.

The *terra firme* forests in the east and centre of the island have a canopy height of c. 25–40 m with emergents reaching 50 m. Some areas qualify as 'exceptionally large forests' with more than 40 $m^2$/ha basal area of trees $\geq$10 cm dbh (diameter at breast height). Common trees of the canopy layer are *Pradosia surinamensis*, *Licania kunthiana*, *Tetragastris panamensis*, etc., while the palms *Maximiliana maripa* and *Astrocaryum aculeatum* are abundant in the upper understorey. In this area Sapotaceae are the most important family. The *terra firme* forest of the western hills is floristically considerably different from that of the east end, and includes species such as *Hevea guianensis* and *Alexa canaracunensis*. The overall species diversity is not high by Amazonian standards, particularly at the (relatively species-poor) eastern end of the island where only 80 spp. $\geq$10 cm dbh were recorded in 1.5 ha of forest. However, these figures are more or less comparable with those occurring in fringe Amazonian forest in Mato Grosso and Tocantins.

An interesting floristic community which we term *Peltogyne gracilipes* forest is common throughout the island, except in the extreme east. It is tall forest dominated by trees of *P. gracilipes* to 40 m in height, and often has remarkably low species diversity. A fairly high number of the top-storey trees of the *terra firme* forests are deciduous, and in places the proportion exceeds one-third. Following the criteria of Beard (1944), these areas should be classified as semi-evergreen seasonal forest or even, where the proportion of deciduous trees exceeds two-thirds as it does in some areas of *Peltogyne gracilipes* forest, as deciduous seasonal forest. However, in large areas less than one-third of the top-storey trees lose their leaves, and here, according to Beard (1944), the vegetation is evergreen seasonal forest.

Various categories of low forest have been recognized, such as *campina* forest, thicket (*carrasco*), etc. They generally occur on sandy soils and have characteristic floristic compositions. Vegetation dominated by tall *Mauritia flexuosa* palms (*buritizal*) occurs along streams and in some swamps, while low woodland dominated by *Clusia renggerioides* (*manguezal*) occurs in other damp areas. In many places the Rio Uraricoera has steep banks and the *terra firme* forest comes right to the river edge, but where banks are shallower there is often a lower community with trees of *Triplaris surinamensis* and/or *Cecropia* spp. If the banks are very shelving there is generally a shrubby marginal tangle of *Inga ingoides*.

*Maracá: The Biodiversity and Environment of an Amazonian Rainforest.*
Edited by William Milliken and James A. Ratter. © 1998 John Wiley & Sons Ltd.

In the east of the island there are a few areas of *campo* where the sparse tree/shrub layer is dominated by the two widespread species *Curatella americana* and *Byrsonima crassifolia*. These *campos* are outliers of the savanna vegetation found nearby on the mainland. There is some evidence that on Maracá these drier *campos* are being invaded by forest. Two categories of wet savanna were recognized. The first is found on the damper (seasonally inundated) parts of *campos* at the east of the island. The second, termed *vazante*, occurs along seasonal drainage lines where no banks have been cut, and consists of a rampant mass of soft, low shrubs, herbs and abundant herbaceous vines.

Some notes are given on vegetation of still and flowing waters and rock outcrops. There has been very little human disturbance on the island, but observations were made on colonizing species on a man-made causeway and in an area of *caapoeira* (secondary vegetation).

# RESUMO

São apresentados os resultados de um levantamento botânico da Ilha de Maracá, situada na zona de transição entre as florestas amazônicas (Hiléia) e as grandes áreas de campo seco (lavrado) que se encontram ao sul e ao leste da ilha (seguindo até as savanas do Rupununi na Guyana). Aproximadamente 90% da ilha se reveste com floresta, enquanto savana ocorre apenas em pequenos trechos na extremidade oriental da ilha. As florestas foram estudadas intensivamente por meio de interpretação de imagens de satélite, transectos e observação geral, e foram reconhecidas várias formas diferentes.

As florestas de *terra firme* da parte leste e no centro da ilha alcançam altura de c. 25–40 m, com árvores emergentes de até 50 m. Algumas áreas qualificam-se como 'florestas de tamanho excepcional', com área basal de árvores $\geq$10 cm dap (diâmetro na altura de peito) maior do que 40 $m^2$/ha. Árvores comuns do estrato mais alto incluem *Pradosia surinamensis*, *Licania kunthiana*, *Tetragastris panamensis*, etc., enquanto as palmeiras *Maximiliana maripa* e *Astrocaryum aculeatum* são abundantes no estrato médio. A família mais importante nesta área é Sapotaceae. Floristicamente, a floresta de *terra firme* nos morros da parte oeste da ilha é bastante diferente daquela do leste, incluindo espécies como *Hevea guianensis* e *Alexa canaracunensis*. A diversidade de espécies não é alta em termos amazônicos, em particular na parte leste da ilha, onde apenas 80 spp. $\geq$10 cm dap foram registradas em 1.5 ha de floresta. Estes dados, porém, são comparáveis aos obtidos em outras pesquisas na periferia da Hiléia em Mato Grosso e Tocantins.

Uma comunidade florística interessante, que denominamos floresta de *Peltogyne gracilipes*, é comum em toda a ilha, com exceção do extremo leste. É uma floresta alta dominada por árvores de *P. gracilipes* alcançando 40 m de altura, e geralmente mostra uma diversidade extremamente baixa. Uma proporção bastante alta das árvores do estrato superior nas florestas de *terra firme* são decíduas, às vezes mais do que 33%. Segundo os critérios de Beard (1944), estas áreas devem ser denominadas de 'floresta estacional semi-decídua', ou, onde a proporção é maior do que 66% (como ocorre em certas áreas de floresta de *Peltogyne gracilipes*), floresta estacional decídua. Nas áreas onde a proporção é menor de 33%, porém, segundo Beard (1944), a vegetação é denominada de 'floresta estacional sempre-verde'.

Foram reconhecidas várias categorias de floresta baixa, tais como mata de campina, carrasco etc. Estas geralmente ocorrem sobre solos arenosos, e possuem composições florísticas características. Vegetação dominada pela palmeira *Mauritia flexuosa* (buritizal) ocorre ao longo de riachos (igarapés) e em áreas de brejo, enquanto uma mata baixa dominada por *Clusia renggerioides* (manguezal) se encontra em outros lugares úmidos. As margens do Rio Uraricoera são geralmente de inclinação forte (abruptas) e a floresta de *terra firme* chega até a beira do rio, entretanto os bancos de inclinação suave caracterizam-se por uma comunidade de porte menor, com árvores de *Triplaris surinamensis* e *Cecropia* spp. Trechos com margens bem planas são revestidos por um matagal de *Inga ingoides*.

Na parte oriental da ilha há algumas manchas de campo onde o estrato ralo de árvores baixas e arbustos é dominado por *Curatella americana* e *Byrsonima crassifolia*. Estes campos são da mesma vegetação da savana que reveste a paisagem da área vizinha fora da ilha. Existem sinais que estes campos estão sendo invadidos por mata.

Foram reconhecidas duas categorias de savana úmida. Encontra-se a primeira nos lugares mais úmidos (sazonalmente inundados) dos campos no leste da ilha, enquanto a segunda, nomeada vazante, ocorre ao longo de linhas de drenagem que não têm margens bem definidas, e consta de uma vegetação densa e baixa, de arbustos pequenos e moles, ervas, e abundância de trepadeiras herbáceas.

São fornecidas algumas notas sobre a vegetação da água parada e água corrente e a dos afloramentos das rochas. A vegetação da ilha não tem sofrido muita influência humana recentemente, mas foram feitas observações sobre espécies colonizadoras ao longo de uma estrada nova e numa área de caapoeira (vegetação secundária).

# INTRODUCTION

The present paper is a summary of the observations of the vegetation survey team which consisted of the authors supported by the INPA technicians Luis and Dionísio Coelho, José F. Ramos, José Lima dos Santos and Cosme Damião A. da Mota. Dr Ray Harley, Gwilym Lewis, Brian Stannard and Peter Edwards of the Royal Botanic Gardens, Kew, and Robert Miller of INPA also participated in the project and made many important collections and observations.

PREVIOUS STUDIES

Only one major study of the vegetation of the island was undertaken prior to the work of the Maracá Rainforest Project. This was in June 1986 when a large team of INPA, New York Botanical Garden, and Museu Integral de Roraima staff led by Dr David Campbell (NYBG) visited Maracá as a part of Projeto Flora Amazônica. They set up 4 ha of permanent plots in the *terra firme* forest close to the Station in which all trees ≥10 cm dbh were inventoried. The results, which were kindly made available to us, are at present being prepared for publication, and it is planned that regular visits will be made in the future to monitor change. Much important information on Maracá comes from the research of Dr Debra Moskovits. Debra, a Brazilian biologist working for a doctorate at the University of Chicago, spent many months in 1980–82 tracking and studying two species of tortoise. The original limited trail network around the Ecological Station was cut to facilitate her work. Her doctoral thesis (1985) provided a rich source of basic data for the work of the Maracá Rainforest Project, and her research on the island is summarized in Chapter 13.

## PUBLICATIONS OF THE PROJECT

The initial reports of the Maracá Rainforest Project appeared early in 1989. Here we shall confine ourselves to those directly relevant to vegetation studies. A full report of the vegetation survey was provided by Milliken and Ratter (1989), including complete analyses of transect data, a checklist of all vascular plants recorded, and an appendix on soils and soil–plant relationships written by Dr Peter Furley. The work was also designed to serve as a guidebook to the vegetation of the island; the present chapter is essentially a summary of it. An illustrated guide to the Leguminosae of the island was also published during this period (Lewis and Owen, 1989).

Since these initial works, a series of publications has appeared. Many have been devoted to the savannas and the savanna–forest transition of the island (Furley and Ratter, 1990, 1994; Furley, 1992; Ross *et al.*, 1992; Thompson *et al.*, 1992b; Proctor, 1994) while others have dealt with the forest communities (Thompson *et al.*, 1992a, 1994; Nascimento, 1994; Nascimento *et al.*, 1997). A number have provided further information on soils and geomorphology (McGregor and Eden, 1991; Robison and Nortcliff, 1991, 1994) and have revealed the presence of pockets of surprisingly eutrophic forest soils in a predominantly extremely dystrophic landscape (Robison and Nortcliff, 1991). The interesting work of Furley *et al.* (1994) has refined vegetation mapping of the island by remote sensing. Lastly, two popular books have been published (Hemming *et al.*, 1988, and Hemming and Ratter, 1993). These were designed to communicate the work of the Maracá Rainforest Project to a wider public.

## METHODS

Quantitative ecological inventories were made in most of the woody vegetation types of the island. The sites studied are shown in Figures 5.1 and 5.2. In addition, general observation ('wide patrolling') and collecting were carried out over as much of the island as possible. In the main, the vegetation was sampled using point-centred quarter (PCQ) transects as described by Mueller-Dombois and Ellenberg (1974), but a 1.5 ha permanent transect made up of six contiguous 50 × 50 m quadrats was used in the *terra firme* forest at the east end of the island, and 50 × 50 m quadrats were sited in the *Curatella americana/Byrsonima crassifolia campo* at Santa Rosa and in the nearby forest.

The PCQ transects were run through as homogeneous vegetation as possible, and sampling points were generally sited at either 10 or 15 m intervals according to the density of the forest. On most transects parallel sets of data were recorded for trees of two size categories: 10 cm diameter at breast height (dbh) and $\geq$30 cm dbh. This provides a more representative picture of forest composition than if only one size class is used.

The data were analysed by an IBM computer using the FITOPAC program, written and developed by Dr George Shepherd of the University of Campinas (São

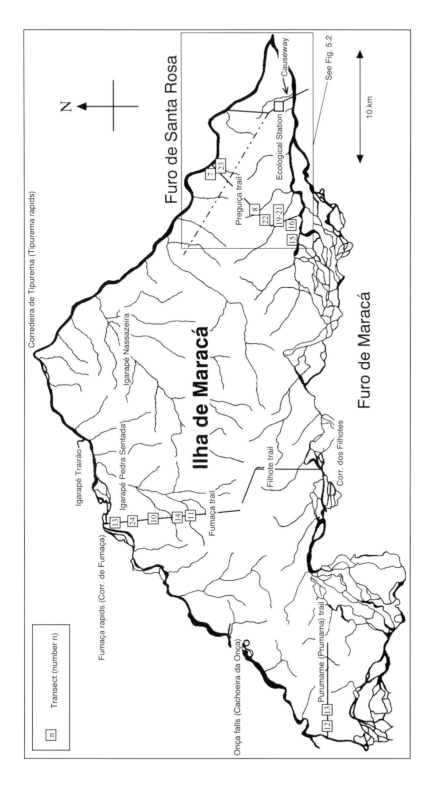

**Figure 5.1.** Locations of vegetation descriptions and transects on the Ilha de Maracá

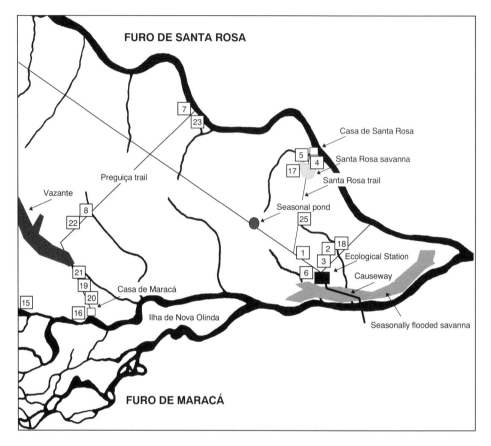

**Figure 5.2.** Locations of vegetation descriptions and transects at the eastern end of the Ilha de Maracá

Paulo State, Brazil). The following parameters were calculated: absolute density, absolute dominance, absolute frequency, relative density, relative dominance, relative frequency, importance value (IV) (= rel. dens. + rel. dom. + rel. freq.), and importance cover value (ICV) (= rel. dens. + rel. dom.). The last two indices are designed to give an estimate of the relative importance of species (or families) in the community. In addition, number of trees per hectare and total basal area (m$^2$) per hectare were calculated, as were diversity indices. Data from the transects and complete species lists for the vegetation types are presented in full in Tables 3–54 of Milliken and Ratter (1989), while Appendix 1 of the same work gives a checklist for all species of vascular plants recorded on Maracá.

Tree heights were measured by triangulation using a Suunto tree measurer. Remote sensing data from the LANDSAT TM, flown in October 1985, were used to estimate the extent and distribution of the vegetation types surveyed on the ground. Voucher specimens and other herbarium collections were made in replicate sets which are lodged in the herbaria of INPA, Royal Botanic Gardens, Kew (K),

Royal Botanic Garden Edinburgh (E), Museu Integral de Roraima (MIRR) and New York Botanical Garden (NY). Additional duplicates are being widely distributed in Brazil, Europe and the USA.

# RESULTS

## SPECIES DIVERSITY

A total of 1077 species of vascular plants were collected in the vegetation survey of the Maracá Rainforest Project and are listed together with their collection numbers in Milliken and Ratter (1989). Of these, 925 (86%) were identified to species level (or in some cases species affinity), 100 (9.3%) to the genus, 47 (4.4%) to family, while five (0.5%) remained completely undetermined. As in all extensive ecological surveys much of the material collected was sterile, and most of the specimens not identified to species fall into this category.

Many of the species collected are very widespread and comparatively few are new to science. Nevertheless, diversity of habitat means that the actual number of species occurring on the island is high. The level of diversity occurring in forest habitats is examined later in this chapter. Non-vascular cryptogams were also studied and 75 species of wood- and leaf-rotting fungi were recorded, as were 11 bryophytes. In addition, Drs Pedro Mera and André de Oliveira of INPA made extensive collections of aquatic (principally planktonic) algae, discovering a total of 232 species, comprising 176 species of Chlorophyta, 18 species of Cyanophyta, 16 species of Chrysophyta, 15 species of Euglenophyta, four species of Rhodophyta and three species of Pyrrophyta. No less than 140 species of the Chlorophyta were desmids. Mera and Oliveira (1989) give a brief résumé of this work which is still awaiting publication in detail.

## VEGETATION TYPES AND THEIR DISTRIBUTION

The island lies exactly at the junction of the Roraima–Rupununi savannas and the Amazonian forest (Hylaea). About 90% is forested and only a small area is occupied by the savanna formations which are dominant on the mainland to the south and east. Table 5.1 shows the classification of vegetation used in the survey, chosen after one year's field experience. The individual categories are described in the following pages.

Much work has been devoted to attempts to produce a vegetation map by analysis of satellite images taken from LANDSAT Thematic Mapper (TM) and their interpretation through fieldwork ('ground truthing'). A number of these maps have been published (Milliken and Ratter, 1989; Dargie and Furley, 1994; Furley et al., 1994) but although they can successfully differentiate savanna from forest, it is much more difficult to map the different forest types accurately. However, Furley et

**Table 5.1.** Classification of the vegetation of the Ilha de Maracá

| | |
|---|---|
| 1. | Forest |
| 1.1 | *Terra firme* forest of the eastern and central part of the island |
| 1.2 | *Terra firme* forest of the western part of the island |
| 1.3 | *Peltogyne gracilipes* forest |
| 1.4 | Low forest (*campina* forest, thicket (*carrasco*), etc.) |
| 1.5 | *Buritizal* and associated damp forest |
| 1.6 | Riverine forest |
| 2. | Savanna |
| 2.1 | *Curatella americana/Byrsonima crassifolia campo* |
| 2.2 | Seasonally flooded *campo* |
| 2.3 | *Vazante* |
| 3. | Aquatic vegetation |
| 3.1 | Still-water habitats |
| 3.2 | Flowing water habitats |
| 4. | Vegetation of rock outcrops |
| 5. | Colonizing vegetation and *caapoeira* |

**Table 5.2.** Estimated areas of vegetation zones on Maracá from satellite images taken from LANDSAT TM (from Furley *et al.*, 1994)

| Class | Area (ha) | % |
|---|---|---|
| Semi-deciduous closed canopy forest | 34 136.102 | 33.56 |
| Intermediate forest types | 14 709.961 | 14.46 |
| Evergreen closed canopy forest | 36 215.961 | 35.60 |
| Open canopy forest types | 2 979.090 | 2.93 |
| *Buritizal* | 1 855.260 | 1.82 |
| *Vazante* vegetation types | 1 318.050 | 1.30 |
| Unflooded savanna | 478.890 | 0.47 |
| Shallow-water/emergent vegetation | 4 108.590 | 4.04 |
| Trees overhanging water | 1 899.810 | 1.87 |
| Deep water | 2 380.410 | 2.34 |
| Cloud cover | 1 020.420 | 1.00 |
| Cloud shadow | 621.630 | 0.61 |
| Total | 101 723.852 | 100.00 |

*al.* (1994) succeeded in producing estimated areas of vegetation zones (Table 5.2). It is difficult to equate their forest categories with those recognized in the present survey (Table 5.1), but the semi-deciduous closed canopy forest probably corresponds in part to *Peltogyne gracilipes* forest, with the evergreen forest as the eastern, central and western *terra firme* forests, and the intermediate forests as the widespread mixed *P. gracilipes* type (see below).

To give an impression of the physiognomy of the vegetation, schematic profiles along parts of the Preguiça and Fumaça trails are given in Figures 5.3 and 5.4.

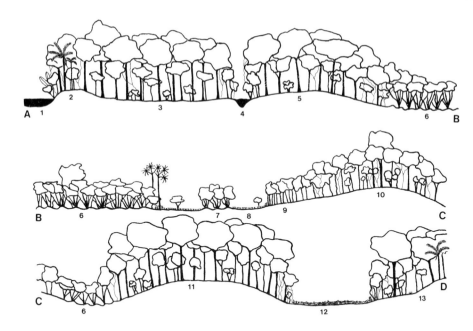

**Figure 5.3**. Schematic profile of the vegetation at the southern end of the Preguiça trail, Ilha de Maracá. A–D is a continuous SSE–NNW transect approximately 3 km in length. Key: 1 = river (Furo de Maracá); 2 = levee with mixed riverine forest; 3 = tall forest on damp ground, dominated by *Peltogyne gracilipes* and *Pradosia surinamensis*; 4 = stream; 5 = forest similar to 3, but taller and on drier ground; 6 = dense thicket on hummocked ground described in section 1.4; 7 = thicket similar to 6; 8 = *Curatella/Byrsonima campo*; 9 = marginal thicket (*carrasco*) vegetation described in section 1.4; 10 = low forest with *campina* influence; 11 = tall forest with extreme dominance of *Peltogyne gracilipes*; 12 = *vazante* vegetation; 13 = tall forest dominated by *Pradosia surinamensis*.

## 1. FOREST

### 1.1 *Terra firme* forest of the eastern and central parts of the island

The predominant vegetation encountered by anyone working from the Ecological Station is *terra firme* forest (Plate 3a). It covers the greater part of the flattish plateau of the extreme eastern end of the island. Further westwards, where the terrain is more undulating, this forest type tends to cover the upper slopes of the ridges. It was studied by a series of nine PCQ transects and by analysis of six 50 × 50 m quadrats, as well as by extensive general observation and collecting. Detailed descriptions, species lists, and tables of phytosociological parameters covering all transects are given in Milliken and Ratter (1989), so only a summary of salient features is given here. Table 5.3 gives a list of the commonest tree species, with information on stratification, etc., while Tables 5.4 and 5.5 give the importance values (IVs) of the first 20 species and families respectively on five representative transects.

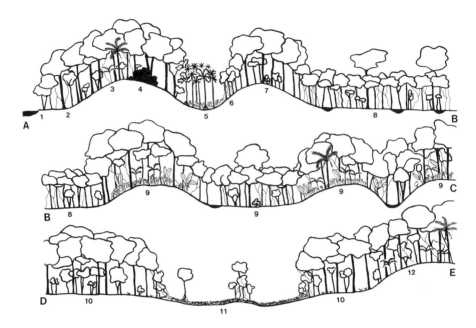

**Figure 5.4.** Schematic profile of the vegetation at the northern end of the Fumaça trail, Ilha de Maracá. Point A is the northernmost extremity of the trail. A–C is a continuous N–S transect of approximately 3–4 km. Point D lies approximately 1.5 km south of C. Key: 1 = river (Furo de Santa Rosa); 2 = tall forest dominated by *Peltogyne gracilipes*; 3 = tall mixed forest; 4 = granitic outcrops (see section 4); 5 = palm swamp dominated by *Mauritia flexuosa* and *Socratea exorrhiza*; 6 = low vine forest; 7 = tall forest with much *Centrolobium paraense* and *Peltogyne gracilipes*; 8 = low forest with emergent *P. gracilipes*; 9 = tall mixed forest with abundant herbaceous ground cover (see Transect 10, Tables 5.4–5.5); 10 = tall forest with extreme *P. gracilipes* dominance; 11 = *vazante* vegetation with clumps of trees of *Triplaris surinamensis*; 12 = tall mixed forest.

The height of the canopy is rather variable. Close to the Station it is generally 25–35 m with tall emergent trees reaching 40–45 m and occasionally more. The tallest emergents measured were 50 m specimens of *Astronium lecointei*[1] and *Enterolobium schomburgkii*, and a *Brosimum lactescens* fractionally lower than this. The thickest tree trunks are 1.2–1.5 m dbh, but the average for main canopy trees is only 30–50 cm. Many of the larger tree species have buttress roots which are sometimes very large (e.g. a specimen of *Couratari oblongifolia* was recorded with buttresses measuring 9 m across and extending 4 m up the trunk). The bombacaceous *Bombacopsis quinata* and *Ceiba pentandra* produce huge swollen-trunked trees with great buttresses and long, thick, superficial roots running along the soil surface for considerable distances; this is very striking in the former species which tends to grow in damper hollows in the forest. Buttressed species are marked with 'b' in Table 5.3.

The tall caesalpinoid *Peltogyne gracilipes* does not occur in the forest to the east of the Ecological Station but to the west of the Station it becomes an important

**Table 5.3.** Commonest tree (and shrub) species of the *terra firme* forest of the eastern and central parts of Maracá (b, buttress roots present; c, main canopy; e, emergent; s, strangler; s/t, shrub/treelet; u, understorey tree)

| | | | |
|---|---|---|---|
| *Abuta grandifolia* | s/t | *Licania kunthiana* | c |
| *Acosmium ?tomentellum* | c | *L. sprucei* | c/u |
| *Alseis longifolia* | c/u | *Lindackeria paludosa* | u |
| *Aniba taubertiana* | u | *Lueheopsis duckeana* | c/e, b |
| *Apeiba schomburgkii* | u | *Maximiliana maripa* | u |
| *Aspidosperma* cf. *eteanum* | c | *Mezilaurus lindaviana* | c |
| *Astrocaryum aculeatum* | u | *M.* sp. 1 (R5661) | c |
| *A. gynacanthum* | s/t | *Micropholis melinoniana* | c/e, b |
| *Astronium lecointei* | c/e | *Morinda tenuiflora* | s/t |
| *Bombacopsis quinata* | c/e, b | *Myrcia* cf. *splendens* | u |
| *Brosimum lactescens* | c/e | *Neea* sp. (R5508) | u |
| *Casearia sylvestris* | u | *Oenocarpus bacaba* | u |
| *Ceiba pentandra* | c/e, b | *Peltogyne gracilipes* | c/e, b |
| *Centrolobium paraense* | c | *P. paniculata* ssp. *pubescens* | c |
| *Clarisia racemosa* | c | *Phenakospermum guyannense* | s/t |
| *Couratari oblongifolia* | c/e, b | *Pouteria hispida* agg. | c, b |
| *Crepidospermum goudotianum* | u | *P. surumuensis* | c/u |
| *Duguetia lucida* | u | *P. venosa* ssp. *amazonica* | c/u |
| *Duroia eriopila* | u | *Pradosia surinamensis* | c, b |
| *Ecclinusa guianensis* | c | *Protium polybotryum* | c |
| *Enterolobium cyclocarpum* | c/e | *Quiina* cf. *rhytidopus* | s/t |
| *E. schomburgkii* | c/e | *Rinorea pubiflora* | s/t |
| *Ficus* spp. | s/b | *Rudgea crassiloba* | u |
| *Guatteria schomburgkiana* | u | *Ryania speciosa* var. *bicolor* | s/t |
| *Guettarda spruceana* | u | *Simaba paraensis* vel aff. | c/e |
| *Himatanthus articulatus* | u | *Sloanea garckeana* | c, b |
| *Hymenaea courbaril* | c/e | *Tabebuia uleana* | c |
| *Jacaranda copaia* ssp. *spectabilis* | c/e | *Tetragastris altissima* | c/e |
| *Laetia procera* | c | *T. panamensis* | c |
| *Lecythis corrugata* ssp. *rosea* | c/u | *Trattinnickia rhoifolia* | c/e |

emergent. It is found both in mixed forest and further towards the centre of the island in monodominant stands, termed *P. gracilipes* forest, which are treated as a separate forest type.

There is a well-developed understorey, usually about 12–20 m tall, in which the palms *Maximiliana maripa*, *Astrocaryum aculeatum* and, in some places, *Oenocarpus bacaba* are common. Below this there is a shrubby layer including characteristic species such as the widespread *Abuta grandifolia*, *Cheiloclinium cognatum*, *Rinorea pubiflora* and the Musaceous *Phenakospermum guyannense*. The last species varies greatly in quantity throughout the forest and often forms large banana-like clumps to *c.* 6 m tall, particularly where there is a greater penetration of light. It is extremely widespread throughout Amazonia and is regarded by Veloso (1966) as one of the two sure markers of the Amazonian phytogeographical province (the other being *Hevea*).

The ground vegetation consists of tree-saplings, young lianas and some species of small shrubs and herbs. Although it varies in density, it is seldom thick enough to

**Table 5.4.** Twenty most important species of trees ≥10 cm dbh in terms of IV (importance value) on five representative transects of eastern and central *terra firme* forest on the Ilha de Maracá

| | Eastern end of the island | | | | Preguiça trail | | Fumaça trail | |
|---|---|---|---|---|---|---|---|---|
| | Transect 1 | | Transect 2 | | Transect 3 | | Transect 8 | Transect 10 |
| Species | IV(%) | Species | IV(%) | Species | IV(%) | Species | IV(%) | Species | IV(%) |
| Pradosia surinamensis | 13.43 | Tetragastris panamensis | 19.22 | Brosimum lactescens | 11.42 | Oenocarpus bacaba | 14.93 | Peltogyne gracilipes | 8.67 |
| Myrcia cf. splendens | 7.54 | Brosimum lactescens | 10.20 | Pouteria hispida agg. | 8.73 | Tetragastris panamensis/altissima | 13.71 | Licania sprucei | 6.50 |
| Pouteria surumuensis | 6.74 | Pradosia surinamensis | 7.51 | Licania kunthiana | 7.33 | Licania kunthiana | 6.67 | Ecclinusa guianensis | 6.43 |
| Maximiliana maripa | 6.63 | Pouteria hispida agg. | 7.43 | Astrocaryum aculeatum | 6.60 | Couepia sp. 1 [R6240] 'Utirana' | 5.98 | Maximiliana maripa | 5.53 |
| Brosimum lactescens | 4.00 | Licania kunthiana | 6.18 | Tetragastris panamensis | 6.07 | Enterolobium cyclocarpum | 5.90 | Tetragastris panamensis/altissima | 4.92 |
| Jacaranda copaia ssp. spectabilis | 3.71 | Pouteria surumuensis | 5.15 | Duguetia lucida | 5.98 | Acosmium ?tomentellum | 4.97 | Maytenus sp. [M472] | 3.96 |
| Ocotea sandwithii | 3.39 | Lecythis corrugata ssp. rosea | 3.46 | Pouteria surumuensis | 3.47 | Leguminosae Indet. 1 [R6318] | 3.51 | Alseis longifolia | 3.90 |
| Guettarda spruceana | 2.92 | Astrocaryum aculeatum | 2.86 | Lecythis corrugata ssp. rosea | 3.20 | Protium polybotryum | 3.18 | Simaba paraensis vel aff. | 3.68 |
| Tabebuia uleanalcapitata | 2.86 | Mezilaurus sp. 1/indaviana | 2.31 | Crepidospermum goudotianum | 2.72 | Ecclinusa guianensis | 2.84 | Lecythis corrugata ssp. rosea | 3.07 |
| Pouteria hispida agg. | 2.85 | Ecclinusa guianensis | 1.99 | Himatanthus articulatus | 2.60 | Peltogyne paniculata ssp. pubescens | 2.45 | Hymenaea parvifolia/courbaril | 3.05 |
| Tetragastris panamensis | 2.73 | Myrcia cf. splendens | 1.93 | Pouteria venosa ssp. amazonica | 2.40 | Laetia procera | 2.42 | Himatanthus articulatus | 2.70 |
| Ecclinusa guianensis | 2.70 | Amaioua corymbosa | 1.93 | Lindackeria paludosa | 2.20 | Tapirira guianensis | 2.25 | Pouteria surumuensis | 2.61 |
| Himatanthus articulatus | 2.68 | Hymenaea parvifolia/courbaril | 1.84 | Duroia eriopila | 2.05 | Lecythis corrugata ssp. rosea | 2.23 | Micropholis melinoniana | 2.60 |
| Astrocaryum aculeatum | 2.39 | Lueheopsis duckeana | 1.70 | Sloanea garckeana vel aff. | 1.83 | Maximiliana maripa | 1.97 | Pradosia surinamensis | 2.39 |
| Rudgea crassiloba | 2.29 | Duguetia lucida | 1.67 | Pradosia surinamensis | 1.67 | Abarema jupunba | 1.96 | Quiina cf. rhytidopus | 2.39 |
| Lecythis corrugata ssp. rosea | 2.14 | Duroia eriopila | 1.66 | Lueheopsis duckeana | 1.44 | Peltogyne gracilipes | 1.81 | Inga obidensis | 2.37 |
| Guatteria schomburgkiana | 2.09 | Maximiliana maripa | 1.36 | Ficus sp(p.) incl. gomelleira | 1.41 | Pseudolmedia laevis | 1.77 | Trattinnickia rhoifolia | 2.24 |
| Cordia sellowiana | 1.95 | Crepidospermum goudotianum | 1.30 | Clarisia racemosa | 1.23 | Byrsonima spicata | 1.57 | Inga splendens | 2.10 |
| Alseis longifolia | 1.93 | Sloanea garckeana vel aff. | 1.29 | Simaba paraensis vel aff. | 1.10 | Trichilia sp. 'Gitó branco' | 1.56 | Matayba sp. 1 [M106] | 1.89 |
| Duguetia lucida | 1.92 | Ocotea glomerata | 1.23 | Rudgea crassiloba | 1.07 | Aniba taubertiana | 1.52 | Licania heteromorpha | 1.80 |

**Table 5.5.** Twenty most important families of trees ≥10 cm dbh in terms of IV (importance value) on five representative transects of eastern and central *terra firme* forest

| | Eastern end of the island | | | | | | Preguiça trail | | Fumaça trail | |
|---|---|---|---|---|---|---|---|---|---|---|
| Transect 1 | | Transect 2 | | Transect 3 | | | Transect 8 | | Transect 10 | |
| Family | IV(%) | Family | IV(%) | Family | IV(%) | | Family | IV(%) | Family | IV(%) |
| Sapotaceae | 25.46 | Sapotaceae | 22.56 | Sapotaceae | 17.45 | | Leguminosae | 20.78 | Leguminosae | 21.26 |
| Rubiaceae | 10.94 | Burseraceae | 20.75 | Moraceae | 14.33 | | Burseraceae | 17.25 | Sapotaceae | 15.37 |
| Palmae | 9.16 | Moraceae | 11.04 | Burseraceae | 10.46 | | Palmae | 16.43 | Chrysobalanaceae | 8.35 |
| Myrtaceae | 8.37 | Chrysobalanaceae | 6.71 | Annonaceae | 8.20 | | Chrysobalanaceae | 12.57 | Burseraceae | 7.21 |
| Bignoniaceae | 6.41 | Lecythidaceae | 5.48 | Chrysobalanaceae | 7.43 | | Sapotaceae | 5.97 | Palmae | 5.56 |
| Burseraceae | 5.77 | Palmae | 5.19 | Palmae | 6.87 | | Moraceae | 3.59 | Rubiaceae | 5.43 |
| Lauraceae | 4.48 | Rubiaceae | 4.39 | Rubiaceae | 5.96 | | Anacardiaceae | 3.51 | Celastraceae | 4.00 |
| Moraceae | 4.13 | Lauraceae | 3.86 | Apocynaceae | 3.72 | | Flacourtiaceae | 2.40 | Simaroubaceae | 3.70 |
| Apocynaceae | 4.02 | Annonaceae | 3.37 | Lecythidaceae | 3.57 | | Lauraceae | 2.35 | Apocynaceae | 3.62 |
| Annonaceae | 4.02 | Leguminosae | 2.83 | Lauraceae | 3.08 | | Lecythidaceae | 2.22 | Lecythidaceae | 3.09 |
| Lecythidaceae | 3.06 | Tiliaceae | 2.61 | Flacourtiaceae | 3.05 | | Euphorbiaceae | 1.78 | Meliaceae | 2.95 |
| Leguminosae | 2.88 | Myrtaceae | 2.42 | Tiliaceae | 2.44 | | Myristicaceae | 1.65 | Tiliaceae | 2.63 |
| Chrysobalanaceae | 2.80 | Anacardiaceae | 1.68 | Leguminosae | 2.36 | | Malpighiaceae | 1.57 | Quiinaceae | 2.41 |
| Boraginaceae | 2.01 | Elaeocarpaceae | 1.34 | Elaeocarpaceae | 1.87 | | Meliaceae | 1.56 | Sapindaceae | 1.90 |
| Flacourtiaceae | 2.00 | Bignoniaceae | 1.16 | Bignoniaceae | 1.45 | | Rubiaceae | 1.53 | Bignoniaceae | 1.69 |
| Tiliaceae | 1.38 | Apocynaceae | 0.93 | Simaroubaceae | 1.12 | | Bignoniaceae | 0.91 | Moraceae | 1.66 |
| Euphorbiaceae | 1.30 | Myristicaceae | 0.81 | Myrtaceae | 1.08 | | Annonaceae | 0.87 | Olacaceae | 1.61 |
| Simaroubaceae | 0.52 | Ochnaceae | 0.43 | Boraginaceae | 0.95 | | Violaceae | 0.77 | Lauraceae | 1.53 |
| Indet. | 0.47 | Meliaceae | 0.42 | Anacardiaceae | 0.76 | | Theaceae | 0.77 | Annonaceae | 1.51 |
| Elaeocarpaceae | 0.43 | Sapindaceae | 0.42 | Ulmaceae | 0.72 | | Indet. | 0.76 | Euphorbiaceae | 1.37 |

impede movement. Variegated-leaved species of *Calathea* and *Monotagma* (Marantaceae) are conspicuous and abundant, as are a number of species of bromeliads and a few forest grasses.

Woody lianas such as the undulate-stemmed *Bauhinia outimouta*, *Passiflora longiracemosa* (a striking cauliflorous passion-flower) and various Bignoniaceae are common, while the herbaceous climbing Swiss-cheese plant, *Monstera dubia*, is abundant. Epiphytes, on the other hand, are relatively infrequent, probably because the strongly seasonal climate with its severe, hot, dry season does not provide conditions which favour this synusia.

At the height of the dry season many of the trees are leafless, and this is especially the case with emergents. The proportion of deciduous trees of the uppermost storey is, however, probably less than one-third, so the forest qualifies as evergreen seasonal, rather than semi-evergreen seasonal, following the usage of Beard (1944). Species recorded as deciduous on 6 March 1987 are listed below; it is notable that most have compound leaves and many belong to the Tiliales (i.e. Tiliaceae, Bombacaceae, Elaeocarpaceae). In areas where trees of *Peltogyne gracilipes* are abundant their leafless, emergent crowns are very conspicuous when viewed from the air during this season.

Deciduous species

| | |
|---|---|
| *Apeiba schomburgkii* | *E. schomburgkii* |
| *Astronium lecointei* | *Lueheopsis duckeana* |
| *Bombacopsis quinata* | *Peltogyne gracilipes* |
| *Cassia moschata* | *Simaba paraensis* vel aff. |
| *Ceiba pentandra* | *Sloanea garckeana* vel aff. |
| *Centrolobium paraense* | *Spondias mombin* |
| *Cochlospermum orinocense* | *Tabebuia capitata* |
| *Cordia sellowiana* | *T. ulei* |
| *Enterolobium cyclocarpum* | |

*Angico slope forest*

A variant of this *terra firme* forest occurs on the slopes which descend from the Station plateau to the extensive wet *campo* at the southeast of the island. It consists of a low dense forest, only about 10 m tall, but with emergent trees of *Anadenanthera peregrina* ('Angico'), *Cochlospermum orinocense* and *Didymopanax morototoni* to about 15 m. Further up the slope there is a shift towards the more usual *terra firme* forest but with a preponderance of *Maximiliana maripa*, which had the highest IV on the small transect used to sample the area. This palm species was also found in other parts of the forest to be especially abundant on slopes.

*Anadenanthera peregrina* is a marker of this forest and has not been found in other areas. It is completely deciduous, and in the dry season its bare form is conspicuous on the slope between the Station plateau and the *campo* below. Forests with *Anadenanthera* in Central Brazil are characteristic of soils with higher levels of calcium and magnesium (Ratter *et al.*, 1973, 1978b) and Dr P. Furley's soil analyses demonstrated that the same occurs on the Ilha de Maracá. Fuller consideration of the Angico forest is given in Furley and Ratter (1990 and 1994).

*Variation in structure, diversity and floristics*

The eastern and central *terra firme* forests were intensively studied during the project and the following section compares the structure, diversity and floristics of the different study sites. In total the equivalent area sampled for the PCQ transects plus the actual area for quadrats totalled 3.9 ha for trees $\geq 10$ cm dbh, while the equivalent area for the PCQ transects for trees $\geq 30$ cm dbh was 7.1 ha. There were considerable differences in tree density and basal area between different sites. Density ranged from 309.4 to 504.6 trees/ha for those $\geq 10$ cm dbh and from 104.1 to 160.6 for those $\geq 30$ cm dbh. Basal areas for $\geq 10$ cm dbh ranged from 21.0 $m^2$/ha in Angico slope forest to 54.5 $m^2$/ha in tall eastern *terra firme* forest (in one exceptional case reaching 61.2 $m^2$/ha, but here the sample was only 0.115 ha and really too small to give a reliable extrapolation) and 16.4 $m^2$/ha to 38.6 $m^2$/ha for trees $\geq 30$ cm dbh.

The highest species diversity recorded for trees $\geq 10$ cm dbh was 80 species for a 1.46 ha quadrat and 54 species for a 50-point PCQ transect equivalent to 0.40 ha. As one would expect, there is a fair degree of variation in species composition and relative importance through the forest (see Table 5.4 showing the 20 most important species on a number of representative transects). Percentages of the most important species in common for these same transects are given in Table 5.6 and their Sørensen similarity indices in Table 5.7. As would be expected, the transects closest to each other (1, 2, 3) show the greatest floristic similarity, while in almost all cases there is a greater similarity of trees $\geq 30$ cm dbh than $\geq 10$ cm dbh, indicating that the number of species attaining greater stature is smaller (in all, 77 species of the former category were recorded on the transects and 125 of the latter). In general, the variation is related to (i) clumping of species in one area and their scarcity or absence in others (which may be related to local environmental factors or may be purely stochastic), (ii) perhaps more major distribution patterns over the island, and (iii) sampling factors such as the presence of a single, very large individual on a transect, which produces a high importance value (IV) for the species.

The species showing the highest IV for trees $\geq 10$ cm dbh on each of the 11 transects and quadrats were: *Pradosia surinamensis* (Sapotaceae), *Tetragastris panamensis* (Burseraceae), *Brosimum lactescens* (Moraceae), *Ecclinusa guianensis* (Sapotaceae), *Apeiba schomburgkii* (Tiliaceae), *Maximiliana maripa* (Palmae), *Alseis longifolia* (Rubiaceae), *Oenocarpus bacaba* (Palmae), *Peltogyne gracilipes* (×2, i.e. on two transects) (Leguminosae/Caesalpinoideae), and *Pouteria hispida* agg. (Sapotaceae). On the eight transects where the qualifying size of $\geq 30$ cm was used the top scoring species for IV were: *Pradosia surinamensis* (×2), *Tetragastris panamensis* (×3), *Alseis longifolia*, and *Peltogyne gracilipes* (×2). As already mentioned, *P. gracilipes* is a species which does not occur at the easternmost end of the island.

An estimate of family importance was calculated by taking the three most important families (in IV of individuals $\geq 10$ cm dbh) for each sampled area, giving them a score of three for each first place, two for second, and one for third, and

Table 5.6. Comparison of species composition of the most important tree species (in terms of IV) on five transects of eastern and central *terra firme* forest on the Ilha de Maracá

| Transects compared | Distance between transects (km) | Species in common (%) | | |
|---|---|---|---|---|
| | | 1st 10 spp. in IV order | 1st 20 spp. in IV order | 1st 30 spp. in IV order |
| 1 & 2 | 1.3 | 40 (70)* | 50 (70) | 60 (63) |
| 1 & 3 | 1.0 | 30 | 45 | 60 |
| 1 & 8 | 8.5 | 0 (40) | 20 (40) | 23 (34.5**) |
| 1 & 10 | 40.0 | 10 (60) | 40 (65) | 33 (50) |
| 2 & 3 | 0.5 | 70 | 65 | 60 |
| 2 & 8 | 9.7 | 40 (30) | 25 (35) | 27 (50**) |
| 2 & 10 | 41.0 | 30 (40) | 35 (55) | 30 (47) |
| 3 & 8 | 9.5 | 10 | 10 | 17 |
| 3 & 10 | 40.5 | 20 | 15 | 23 |
| 8 & 10 | 28.0 | 20 (40) | 25 (45) | 23 (41**) |

\* First figure is for trees 10 cm dbh, figure in parentheses is for trees ≥30 cm dbh.
\*\* Scored for only 29 species.

Table 5.7. Comparison of species composition for trees ≥10 cm dbh of five transects of eastern and central *terra firme* forest on the Ilha de Maracá using Sørensen similarity index* and, in parentheses, Dice association index**. The number of species recorded for each transect is given after the transect number. Distances between transects are given in Table 5.6

| | Station trail system | | | Preguiça trail | Fumaça trail |
|---|---|---|---|---|---|
| Transect no. | 1 (60 spp.) | 2 (54 spp.) | 3 (68 spp.) | 8 (50 spp.) | 10 (53 spp.) |
| 1 | | 0.67 (0.70) | 0.66 (0.70) | 0.33 (0.36) | 0.41 (0.43) |
| 2 | | | 0.70 (0.80) | 0.40 (0.42) | 0.45 (0.45) |
| 3 | | | | 0.29 (0.34) | 0.46 (0.36) |
| 8 | | | | | 0.35 (0.36) |

\* 2 × no. of spp. in common/no. of spp. at locality A + no. of spp. at loc. B (Sørensen, 1948).
\*\* No. of spp. in common/minimum no. of spp. at either locality (Dice, 1945).

adding all the scores for each family together. This gave the following score for the eastern end of the island: Sapotaceae 24, Burseraceae 8, Lecythidaceae 5, Palmae 5, Moraceae 4, Leguminosae 3, Rubiaceae 2, Tiliaceae 2, Bombacaceae 1; while the areas further west at the Preguiça and Fumaça trails showed: Leguminosae 20, Sapotaceae 19, Burseraceae 8, Moraceae 3, Rubiaceae 3, Sapindaceae 2, Tiliaceae 2, Chrysobalanaceae 1, Myrtaceae 1, Palmae 1. The importance of Leguminosae in the western part of the area (i.e. the more central part of the island) is almost entirely due to the abundance of *Peltogyne gracilipes*.

### 1.2 *Terra firme* forest of the western part of the island

The forest covering the areas of heavily dissected relief at the western end of the island was examined using PCQ transects. The region consists of a series of sharply peaked hills rising abruptly to *c*. 100 m from river level (reaching an altitude of

**Table 5.8.** Commonest tree (and shrub) species of the *terra firme* forest of the western part of Maracá (c, main canopy; e, emergent; s, strangler; s/t, shrub/treelet; u, understorey tree)

| | | | |
|---|---|---|---|
| *Albizia* sp. (*Pithecellobium* aff. *elegans*) | c | *Hevea guianensis* | c |
| *Alchorneopsis floribunda* | u | *Hymenolobium petraeum* | c/e |
| *Alexa canaracunensis* | u | *Jacaranda copaia* ssp. *spectabilis* | c |
| *Anacardium giganteum* | c/e | *Jessenia bataua* | u |
| *Anaxagorea acuminata* | s/t | *Laetia procera* | c/e |
| *Aspidosperma nitidum* | c | *Licania apetala* var. *apetala* | c |
| *Bocageopsis multiflora* | u | *Licaria chrysophylla* | u |
| *Brosimum rubescens* | c/e | *Maieta guianensis* | s/t |
| *B. utile* ssp. *ovatifolium* | c/e | *Maprounea guianensis* | c |
| *Caryocar villosum* | c/e | *Miconia lepidota* | u |
| *Cecropia* ?*palmata* | u | *M.* cf. *punctata* | u |
| *Chrysochlamys weberbauerii* | s/t | *M.* cf. *regelii* | u |
| *Clathrotropis macrocarpa* | c/u | *Minquartia guianensis* | u |
| *Clusia* sp. | s | *Ocotea bracteosa* | u |
| *Cordia nodosa* | s/t | *Oenocarpus bacaba* | u |
| *Dacryodes* spp. | c/u | *Osteophloem platyspermum* | c/e |
| *Dialium guianense* | c | *Parahancornia fasciculata* | c/e |
| *Duguetia cauliflora* | u | *Parkia pendula* | c/e |
| *D. marcgraviana* | u | *Phenakospermum guyannense* | s/t |
| *Enterolobium schomburgkii* | c/e | *Pourouma cucura* | c |
| *Eschweilera albiflora* | c | *Pouteria caimito* | c |
| *E. subglandulosa* | c | *Rheedia macrophylla* | u |
| *Euterpe precatoria* | u | *Rinorea* sp. | u |
| *Ficus guianensis* | s | *Sagotia racemosa* | u |
| *F.* cf. *pakkensis* | s | *Simaba paraensis* vel aff. | c |
| *Geonoma deversa* | s/t | *Siparuna guianensis* | u |
| *G. maxima* | s/t | *Socratea exorrhiza* | u |
| *Goupia glabra* | c | *Trattinnickia* cf. *glaviozii* | c |
| *Gustavia hexapetala* | u | *Trichilia* ?*septentrionalis* | u |
| *Helicostylis tomentosa* | u | Lauraceae (M673) | c |

300–350 m), separated by steep-sided valleys with small streams. Together with the hills in the centre of the island these make up the highest features on Maracá.

The main transect work was carried out on the upper slopes of the hills, avoiding the damp valley bottoms. Tree species composition was found to differ greatly from that of the forest studied further east. Of the species recorded in the $\geq 10$ cm dbh category only 26% had been encountered in the forest on the trail system around the Station, and a further 14% in other transects on the Preguiça and Fumaça trails further to the west. Corresponding figures for the $\geq 30$ cm category are 29% for the Station trail system and a further 12% from other transects. Of the overall list of 102 tree species for the area (including those that did not feature in the transects), 60 had not been recorded elsewhere on the island. Similar divergences were noted in the avifauna (see da Silva, Chapter 11). Table 5.8 gives a list of the commonest trees and shrubs; Tables 5.9 and 5.10 show the 20 most important species and families respectively in terms of IV for the two transects at the west of the island, while Table 5.11 gives similarity indices for forests in the eastern, central and western parts of the island.

Table 5.9. Twenty most important species of trees ≥10 cm dbh (in terms of IV%) on two transects (Nos 12 and 13) at the western end of the Ilha de Maracá

| | Transect 12 | IV(%) | Transect 13 | IV(%) |
|---|---|---|---|---|
| 1 | Clathrotropis macrocarpa | 7.57 | Eschweilera subglandulosa | 22.83 |
| 2 | Alexa canaracunensis | 6.62 | Clathrotropis macrocarpa | 14.16 |
| 3 | Simaba paraensis vel sp. aff. | 6.14 | Alexa canaracunensis | 8.31 |
| 4 | Trattinnickia cf. glaziovii | 4.54 | Sagotia racemosa | 6.59 |
| 5 | Licania apetala var. apetala | 4.36 | Lauraceae Indet. 1 (M649) | 4.49 |
| 6 | Hevea guianensis | 4.35 | Cecropia ?palmata | 4.04 |
| 7 | Brosimum utile ssp. ovatifolium | 3.75 | Licania apetala var. apetala | 3.80 |
| 8 | Oenocarpus bacaba | 3.56 | Dacryodes cf. roraimensis | 3.17 |
| 9 | Laetia procera | 2.99 | Alchorneopsis floribunda | 2.87 |
| 10 | Maprounea guianensis | 2.95 | Protium pedicellatum | 2.53 |
| 11 | Eschweilera albiflora | 2.84 | Gustavia hexapetala | 2.42 |
| 12 | Pouteria caimito | 2.70 | Myrciaria floribunda | 2.12 |
| 13 | Miconia cf. regelii | 2.50 | Ocotea sp. aff. amazonica | 2.08 |
| 14 | Hymenolobium petraeum | 2.30 | Goupia glabra | 1.96 |
| 15 | Eschweilera subglandulosa | 2.02 | Aniba? sp. 1 (M625) | 1.96 |
| 16 | Minquartia guianensis | 2.02 | Jacaranda copaia ssp. spectabilis | 1.93 |
| 17 | Jacaranda copaia ssp. spectabilis | 1.96 | Cupania cf. scrobiculata | 1.93 |
| 18 | Trichilia ?septentrionalis | 1.77 | Helicostylis tomentosa | 1.85 |
| 19 | Alchorneopsis floribunda | 1.59 | Oenocarpus bacaba | 1.85 |
| 20 | Lauraceae Indet. 2 (M673) | 1.52 | Trattinnickia cf. glaziovii | 1.82 |

Table 5.10. Twenty most important families of trees ≥10 cm dbh (in terms of IV%) on two transects (Nos 12 and 13) at the western end of the Ilha de Maracá

| | Transect 12 | IV(%) | Transect 13 | IV(%) |
|---|---|---|---|---|
| 1 | Leguminosae | 18.67 | Lecythidaceae | 24.90 |
| 2 | Euphorbiaceae | 10.57 | Leguminosae | 21.65 |
| 3 | Burseraceae | 7.39 | Lauraceae | 9.05 |
| 4 | Moraceae | 6.62 | Euphorbiaceae | 8.98 |
| 5 | Simaroubaceae | 6.28 | Burseraceae | 6.92 |
| 6 | Palmae | 5.24 | Moraceae | 6.28 |
| 7 | Lecythidaceae | 5.22 | Chrysobalanaceae | 3.93 |
| 8 | Chrysobalanaceae | 5.13 | Annonaceae | 3.91 |
| 9 | Lauraceae | 4.88 | Myrtaceae | 2.25 |
| 10 | Melastomataceae | 4.15 | Celastraceae | 2.10 |
| 11 | Sapotaceae | 4.02 | Sapindaceae | 2.06 |
| 12 | Flacourtiaceae | 3.09 | Bignoniaceae | 2.06 |
| 13 | Myristicaceae | 2.80 | Palmae | 1.98 |
| 14 | Annonaceae | 2.49 | Violaceae | 1.96 |
| 15 | Meliaceae | 2.26 | Monimiaceae | 1.96 |
| 16 | Olacaceae | 2.06 | | |
| 17 | Bignoniaceae | 2.00 | | |
| 18 | Anacardiaceae | 1.21 | | |
| 19 | Apocynaceae | 1.08 | | |
| 20 | Celastraceae | 1.07 | | |

# VEGETATION

**Table 5.11.** Comparison of species composition for trees ≥10 cm dbh of transects at the east, centre (Preguiça and Fumaça trails) and at the west of the island, using Sørensen similarity index and, in parentheses, Dice association index. The number of species recorded for each transect is given after the transect number

| Transect no. | East end<br>1 (60 spp.) | Preguiça trail<br>8 (50 spp.) | Fumaça trail<br>10 (53 spp.) | West end<br>12 (72 spp.) |
|---|---|---|---|---|
| 1 | | 0.33 (0.36) | 0.41 (0.43) | 0.15 (0.17) |
| 8 | | | 0.35 (0.36) | 0.21 (0.26) |
| 10 | | | | 0.18 (0.21) |

There are two other notable differences in composition between this forest and the *terra firme* forest described in the eastern and central parts of the island. The first is diversity: the highest number of species recorded in a 50-point PCQ transect in the east was 54 (Transect 1), whereas in an equivalent transect (no. 12) in the west 73 were present. Although the six contiguous 50 × 50 m quadrats on the Station trail system were found to contain 80 species ≥10 cm dbh, these represent an area almost four times as large as Transect 12, with three times as many trees. The other major difference is in the family composition of the forest. For instance, Sapotaceae, which are most important by a large margin in the majority of the eastern forests not dominated by *Peltogyne gracilipes*, are poorly represented in the west, and the reverse is the case for Euphorbiaceae (see Table 5.10). If the family lists (ranked in order of IV) of Transects 1 and 12 are compared, it can be seen that none of the first five families are in common, and only four of the first 10 (see Tables 5.5 and 5.10).

The structure of the forest is very similar to that at the eastern end of the island, but rather lower in stature, with the main canopy at about 20–25 m. The understorey at 10–20 m is dominated by *Alexa canaracunensis*, a papilionoid species not found in the eastern and central forests, followed in abundance by the palm *Oenocarpus bacaba*. Woody lianas are common, while *Philodendron melinonii* and *Heteropsis* sp. occur as epiphytes and *Ficus guianensis*, *F.* cf. *pakkensis* and a *Clusia* sp. as stranglers. The musaceous *Phenakospermum guyannense* is common, particularly on the tops of the hills, and the low vegetation contains various Marantaceae and small palms (such as *Geonoma* spp. and *Astrocaryum gynacanthum*). The leaf-litter on the upper slopes of the hills is relatively thick and bound by a layer of feeding roots not observed in the forests at the eastern end of the island.

Palms are very abundant in the vicinity of the streams on the lower ground, viz: *Euterpe precatoria*, *Oenocarpus bacaba*, the stilt-rooted *Socratea exorrhiza*, *Jessenia bataua*, *Geonoma deversa* and *G. maxima*. The steep stream banks support abundant ferns which are far more diverse than in the east of Maracá, and this is probably an indication of a damper climate.

## 1.3 *Peltogyne gracilipes* forest

This type of forest is easily recognized by the predominance of *Peltogyne gracilipes*, often accompanied by a remarkably small number of other species (Plate 3b). It

**Table 5.12.** Ten most important species of trees ≥30 cm dbh (in terms of IV%) on two transects (Nos 14 and 15) in *Peltogyne gracilipes* forest on the Fumaça (14) and Preguiça (15) trails on the Ilha de Maracá

| Transect 14 | IV(%) | Transect 15 | IV(%) |
|---|---|---|---|
| *Peltogyne gracilipes* | 70.38 | *Peltogyne gracilipes* | 31.18 |
| *Matayba* sp. 1 (M106) | 13.02 | *Bombacopsis quinata* | 29.54 |
| *Pradosia surinamensis* | 3.40 | *Matayba* sp. 1 (M106) | 7.71 |
| *Lonchocarpus margaritensis* vel aff. | 1.89 | *Lonchocarpus sericeus* | 4.28 |
| *Genipa americana* var. *caruto* | 1.83 | *Pouteria surumuensis* | 4.21 |
| *Hymenaea courbaril* | 1.82 | *Apeiba schomburgkii* | 4.18 |
| *Spondias mombin* | 1.73 | *Sapium* sp. | 4.08 |
| *Pouteria surumuensis* | 1.50 | *Pradosia surinamensis* | 4.07 |
| *Apeiba schomburgkii* | 1.49 | *Cochlospermum orinocense* | 2.38 |
| *Dialium guianense* | 1.49 | *Spondias mombin* | 2.15 |

**Table 5.13.** Most important families of trees ≥30 cm dbh (in terms of IV%) on two transects (Nos 14 and 15) in *Peltogyne gracilipes* forest on the Fumaça (14) and Preguiça (15) trails on the Ilha de Maracá

| Transect 14 | IV(%) | Transect 15 | IV(%) |
|---|---|---|---|
| Leguminosae | 75.11 | Leguminosae | 37.55 |
| Sapindaceae | 13.72 | Bombacaceae | 30.08 |
| Sapotaceae | 4.20 | Sapotaceae | 8.71 |
| Rubiaceae | 1.95 | Sapindaceae | 8.03 |
| Anacardiaceae | 1.84 | Tiliaceae | 4.39 |
| Tiliaceae | 1.60 | Euphorbiaceae | 4.29 |
| Myrtaceae | 1.59 | Cochlospermaceae | 2.49 |
| | | Anacardiaceae | 2.25 |
| | | Boraginaceae | 2.20 |

does not occur at the east end of the island, where *P. gracilipes* was never found, but is common on the Preguiça and Fumaça trails and abundant throughout the centre of the island where it is probably the predominant vegetation type. It can readily be identified from the air during the dry season because of the distinctive appearance of the deciduous crowns (see Plate 2a) which also give a characteristic pattern on remote sensing imagery.

Studies were made of *Peltogyne gracilipes* forest on the Fumaça trail and on the Preguiça trail. Table 5.12 gives the first 10 species ≥30 cm dbh in terms of IV for Transect 14 (Fumaça trail) and Transect 15 (Preguiça trail), while Table 5.13 shows the same information for families. The following description is from the Fumaça trail transect area, which represents this community in its extreme form.

The area is one of tall forest dominated by trees of *Peltogyne gracilipes* to *c*. 40 m tall. This species has a very characteristic appearance: it has well-developed buttress roots extending to 4 m up the trunk, which is very straight and to *c*. 1.1 m in diameter with smoothish pale grey bark. As in *Hymenaea*, the leaves consist of twin leaflets, and this together with its purple timber gives *Peltogyne* the vernacular name

'Jatobá roxa' (= purple *Hymenaea*) in much of Brazil; it is also known as 'Pau roxo' (= purple tree) in Roraima. The only other common tall tree in the transect area is a *Matayba* species (M106), which is of similar size to *P. gracilipes* but of very different appearance, with rough, knobbly, cratered bark and small buttresses. The understorey also contains very few species, the commonest of which is a shaggy-barked *Eugenia* species (R5868) reaching 15–18 m and *Apeiba schomburgkii* of similar height. Below this taller understorey there is a broken shrubby layer consisting principally of *Rinorea brevipes*, which sometimes reaches 7 m but is generally much lower.

The forest floor is covered by an enormous number of small saplings of *Peltogyne gracilipes* (Plate 3b) and *Eugenia* sp. (R5868) together with a few herbs such as *Calathea* sp., *Costus scaber*, *Adiantum pulverulentum* and occasional bromeliads. Analysis of a 2 × 2 m square of the forest floor showed 143 *P. gracilipes* saplings, 88 of *Eugenia* sp. (R5868), two of *Pradosia surinamensis*, two of *Rinorea brevipes*, one each of an *Inga* sp., a *Dalbergia*, and an unidentified Rubiaceae, three Malpighiaceae, three Calatheas and an *Adiantum pulverulentum*. The majority of the saplings were 30–80 cm tall with occasional specimens to 1 m and all were covered with epiphytic leafy liverworts. Bryophytes were also abundant on the bark of the *Eugenia* and *Apeiba* trees. The soil surface was covered by a considerable layer of dry leaves.

*Peltogyne gracilipes* forest in its extreme form, such as occurs in the area of this transect, is remarkable for its paucity of tree species. Only 15 species (10 represented by single individuals) were recorded out of the 80 trees in the ≥10 cm dbh category, and only 11 species (eight represented singly) out of the 80 trees of ≥30 cm dbh. In the latter category *P. gracilipes* constituted no less than 77.5% of the trees scored and represented 78% of the total basal area. We have never seen tropical *terra firme* forest elsewhere with so little species diversity. Amongst the notable absences there are no palms, while woody lianas are infrequent.

Tall *Peltogyne gracilipes* forest is notable for its impression of space and airiness, brought about by the height of the canopy, the tall, pale, pillar-like boles, the lack of encumbering lianas and the absence of dense undergrowth. As already mentioned, *P. gracilipes* is largely deciduous, and therefore where its presence exceeds one-third of the total top storey the vegetation cannot be classified as evergreen according to the criterion of Beard (1944). Instead it should be classified as semi-evergreen forest if one-third to two-thirds of the top-storey trees are deciduous, or as deciduous forest if the proportion is higher than two-thirds.

The foregoing description applies to the most extreme example of *Peltogyne gracilipes* forest encountered. Other forms occur showing a greater diversity of species and a floristic composition somewhat more similar to the eastern *terra firme* forests. These were studied by two transects (15 and 16) at the southern end of the Preguiça trail (Figure 5.2), in an area close to the river but separated from it by a levee. The soils are heavy and clayey, in contrast to the sandy soils found on the Fumaça trail transect, and for several months during the rainy season the water-table is at, or slightly above, the level of the soil surface. As a result there is very little leaf-litter on the forest floor, and the surface shows slight hummocking.

In both of these transects *Peltogyne gracilipes* is the dominant or co-dominant species, and a very conspicuous part of the community, with trees reaching 46 m in height and 2.4 m dbh. In Transect 16, however, it shares its supremacy with *Pradosia surinamensis*, the most common of the eastern *terra firme* forest species, but also found in large quantities associated with *P. gracilipes* in many other parts of the island. These two species account for over half of the trees ≥30 cm dbh on this transect. As in Transect 14 (Fumaça trail) there is a lower understorey layer dominated by *Rinorea brevipes*.

Transect 15, in a similar environment further west, shows less diversity of species. *Bombacopsis quinata*, another species often found associated with *Peltogyne gracilipes* on Maracá and one which appears to thrive on damp soils, is a conspicuous component of the forest. This deciduous, bat-pollinated species produces emergent trees of massive size, the stoutest of which (2.9 m dbh) was the largest tree we encountered on the island. As in the *P. gracilipes* forest of the Fumaça trail, *Matayba* sp. (M106) is also common amongst the larger trees and *Eugenia* sp. (R5868) is the dominant understorey species.

Transects 14 (Fumaça trail) and 15 (Preguiça trail) have Sørensen similarity indices of 0.58 for trees ≥30 cm dbh and 0.56 for those ≥10 cm dbh; the respective figures for the Dice association index are 0.64 and 0.60.

Westwards from the Casa de Maracá (see Figure 5.2) to the end of the island such forests make up the greater part of the vegetation bordering the Furo de Maracá and covering its numerous islands. Composition and diversity vary but *Peltogyne gracilipes*, *Pradosia surinamensis*, *Eugenia* sp. (R5868), *Rinorea brevipes* and *Spondias mombin* generally remain constant and conspicuous. The *Peltogyne gracilipes* forests of Maracá have created much interest (see Discussion), and since our initial survey Marcelo T. Nascimento has studied them in great detail (Nascimento, 1994; Nascimento *et al.*, 1997).

**1.4 Low forest**

A number of types of low forests were studied during the survey. They are not necessarily phytosociologically closely related but it is convenient to consider them in the same section.

Campina *forest*

The name of this forest type comes from common usage and does not imply any relationship, except in superficial appearance, with the *campinas* and *campinaranas* of the Manaus area and the Rio Negro. It occurs on sandy soils and is found particularly at the margins of *Curatella americana/Byrsonima crassifolia campos*. The area adjoining the Santa Rosa *campo* (Figure 5.2) was studied by means of a PCQ transect and by more general observations (wide patrolling). The forest is dense with trees generally about 8 (–12) m and frequently multi-trunked from the base (giving the appearance of coppice regeneration). The species listed in Table 5.14 are characteristic; most are typical of forest margin bordering *campo*. They are mixed with other species more widespread in the eastern and central *terra firme* forest (1.1) such as *Maximiliana maripa*, *Lecythis corrugata* ssp. *rosea*, *Pradosia surinamensis*, *Licania kunthiana*, *Ecclinusa guianensis* and abundant *Cochlospermum*

**Table 5.14.** Characteristic tree species of the *campina* forest at Santa Rosa on the Ilha de Maracá

| | |
|---|---|
| *Alchornea schomburgkii* | *Himatanthus articulatus* |
| *Andira surinamensis* | *Humiria balsamifera* |
| *Bauhinia ungulata* | *Isertia parviflora* |
| *Byrsonima schomburgkiana* | *Ormosia* sp. (R5626) |
| *Chomelia barbellata* | *Peltogyne paniculata* ssp. *pubescens* |
| *Clusia renggerioides* | *Swartzia grandifolia* |
| *Cupania rubiginosa* | *S. laurifolia* |
| *Didymopanax morototoni* | *Tapirira guianensis* |
| *Endlicheria dictifarinosa* | *Trichilia cipo* |
| *Erythroxylum rufum* | *Vismia cayennensis* |
| *Faramea crassifolia* | *Vitex schomburgkiana* |
| *Genipa americana* var. *caruto* | *Zanthoxylum* aff. *rigidum* |

**Table 5.15.** Characteristic tree species of the areas of thicket vegetation on the Ilha de Maracá

| | |
|---|---|
| *Byrsonima crassifolia* | *F. sessilifolia* |
| *B. schomburgkiana* | *Humiria balsamifera* |
| *Cupania rubiginosa* | *Swartzia grandifolia* |
| *Curatella americana* | *S. laurifolia* |
| *Eugenia* sp. (R6265) | *Tocoyena neglecta* |
| *Faramea crassifolia* | *Trichilia cipo* |

*orinocense*. Lianas are quite common and include *Abuta imene* and the festooning cyperaceous *Scleria secans*, while bromeliads are frequent in the ground vegetation.

A second area of *campina* forest was examined in a white sand hollow close to Transect 2. It is surrounded by tall forest and has no current association with *campo*. The trees are mainly 8–9 m and, as in the Santa Rosa *campina*, many are multi-trunked. The species correlate well with those of the Santa Rosa *campina* – about half are in common and they include a high percentage of the important marker species of the community listed in Table 5.14. If species occurring in the areas but not actually scored on the transects are taken into account, the number in common increases to at least 75%. It has been suggested that this area and others like it are remnants of small islands of *campo* (such as those found on the Preguiça trail, Figure 5.3) which have been encroached upon by the forest. Opinions on this are given by Proctor (1994) and the subject should be studied further.

*Thicket* (carrasco) *vegetation*

Two areas of thicket vegetation lying close to the two small strips of *Curatella americana*/*Byrsonima crassifolia campo* on the Preguiça trail (Figure 5.3) were examined. Short PCQ transects were sited in both, and because the trees were so small the qualifying diameter for scoring was reduced to 3 cm. Table 5.15 gives a list of the most characteristic tree species.

The first area consists of a very dense thicket of small trees and shrubs reaching 4–5 m tall, all of which are species very common in the wall-like edge of the forest where it meets the *campo*. The number of trees extrapolates to 4882 per hectare but even this does not give a true idea of the density of the thicket, since the spaces between the ≥3 cm dbh trees are packed with more slender individuals which were not scored – use of a machete is necessary to penetrate the vegetation. The trees are of low, shrubby, multi-trunked, deliquescent form. The vegetation was described as '*carrasco de campina*' by José Lima dos Santos (INPA/Botany), and forms a band of approximately 70 m between the *campo* edge and forest. The table provides a list of species of which *Eugenia* sp. (R6265) and *Faramea sessilifolia* are by far the most important. However, we also noted the presence of trees of *Humiria balsamifera* and *Curatella americana*. The latter species is quite common near the *campo* edge of the thicket and extends inwards up to 13 m from the boundary, probably indicating recent encroachment of thicket on to the *Curatella americana/Byrsonima crassifolia campo*. Some grasses are found amongst the tight-packed trees and the sedge *Scleria secans* climbs in the branches.

The second area studied is a band of thicket vegetation approximately 100 m wide lying between the two *campos*. Here the ground surface consists of a pattern of low mounds bearing trees and shrubs, with a network of little channels between. Unfortunately we took no measurements, but in general the tops of the mounds are probably no more than 30 cm above the bottom of the channels. In February, when we carried out our work, the area was dry but the channels are full of water for most of the rainy season. As in the first area, the vegetation produces a dense thicket, but the trees and shrubs are slightly lower; mainly about 3–4 m. The results of the transect show a not-surprising similarity to the previous one, which is only about 100 m away, although on better-drained soil. *Eugenia* sp. (R6265) is again by far the most important tree; of the others, *Byrsonima crassifolia* and *Tocoyena neglecta* are typical of the *Curatella americana/Byrsonima crassifolia campo*, and *B. schomburgkiana* and the swartzias of *campina*.

*Other low forest types*

Dwarf forms of eastern and central *terra firme* and *Peltogyne gracilipes* forest with main canopy at 7–10 m occur on the Preguiça and Fumaça trails. They can probably be related to topographic and drainage conditions. A number of transects were sited in these forests and they are reported in detail in Milliken and Ratter (1989). They are floristically similar to the tall forests to which they are related, although on the Preguiça trail elements of the nearby thicket vegetation are also present.

### 1.5 *Buritizal* and associated damp forest

Vegetation dominated by the tall palm *Mauritia flexuosa* (*buriti*) is frequently found along streams where there are no steep banks and the conditions are swampy. Such palmeries are known as *buritizais* (singular *buritizal*) and are found both within the forest and on *campo* areas, where the *buritis* tend to occur in lines following the drainage courses (Plate 1a).

The 'Corduroy bridge' *buritizal* on the track between the Ecological Station and Santa Rosa (Figure 5.2) will serve as an example. Here there is a wide, low depression ('*baixada*') through which a broad shallow stream flows during the rains, contracting to a narrow trickle of water or drying up completely during the dry season. The floor of the *baixada* is covered in small mounds on which the trees stand. The soil of these mounds is dry and covered in worm-casts during the dry season, whereas that of the channels between them is a plasticine-like muddy clay. The vegetation is dominated by the *buritis*, which are up to 23 m tall, and form a very broken top storey. A rather discontinuous 5–6 m understorey is made up of species such as *Virola surinamensis*, *Ficus* sp., *Tapirira guianensis*, *Didymopanax morototoni*, *Eschweilera pedicellata*, *Vismia cayennensis*, *Inga* sp., *Cecropia* sp., and, amongst the smaller tree/shrubs, *Isertia parviflora*, *Siparuna guianensis* and the small spiny palm, *Bactris maraja*. Tussocks of the erect, breathing roots of the *buritis* are found in the low muddy channels between the mounds, and they and the mounds themselves provide a substrate for herbaceous species.

There is a fairly dense ground layer which includes some tall herbs, e.g. the Scitamineae *Renealmia alpinia* (to 3 m), *Costus arabicus* (to 2 m), *C. scaber* and *Ischnosiphon arouma* (to 2 m), the grass *Olyra longifolia*, the sedge *Scleria stipularis* and the shrubby *Cordia nodosa*. *Monotagma plurispicatum*, *Psychotria poeppigiana* ssp. *barcellana*, *Justicia polystachya* and *Sauvagesia rubiginosa* are frequent lower herbs. Overall species lists for *buritizal* are given in Milliken and Ratter (1989).

Two other *buritizais* in the forest at the east of the island are very similar to that described above. That at the first *baixada* on the trail to Transect 2 (Figure 5.2) is slightly higher, with the larger *buritis* c. 28 m tall. The understorey is dense and reaches c. 10 m; its composition is almost the same as at the 'Corduroy bridge' except that *Phenakospermum guyannense*, *Cochlospermum orinocense*, *Himatanthus articulatus* and *Endlicheria dictifarinosa* are important.

The second valley on the same trail is much better-drained, with the stream running within steeper banks. Floristic composition includes almost all the species previously mentioned, but *Cochlospermum orinocense* is rather more abundant.

*Manguezais* (singular *manguezal*) sometimes occur on the seasonally wet soil of shallow inclines leading down to valley-bottom *buritizais* and streams. They consist of low woodland dominated by the *mangue* (*Clusia renggerioides*), a stilt-rooted species particularly adapted to damp soil conditions. A brief study, including the siting of a short PCQ transect, was made of the *manguezal* associated with the *buritizal* at the 'Corduroy bridge' and is reported in detail in Milliken and Ratter (1989).

The *manguezal* begins at a level approximately 50 cm above the bottom of the stream, which did not contain more than 3 cm of water at the height of the dry season when our study was made. It consists of trees mainly about 12–14 m tall, of which *Clusia renggerioides* is by far the most important, comprising more than 50% of individuals $\geq$10 cm dbh and of the total basal area, with some emergents of *Mauritia flexuosa* to c. 20 m making up a further 20% of basal area (obviously *manguezal* runs into *buritizal* and can be regarded as a marginal form of the latter).

The palm *Socratea exorrhiza*, with its conspicuous cone of spiny stilt roots to 2 m high, is a prominent member of the *buritizal* community in the western part of the island. It even replaces *Mauritia flexuosa* completely in some areas of the western hills. *Buritizais* also occur in habitats such as the margins of grassy ponds, edges of *vazantes*, etc., which will be considered in other sections. Obviously it is difficult to fit the vegetation of the wetter areas into exclusive categories when there is so much intergradation of habitat.

### 1.6 Riverine forest

The riverside margin of the Ilha de Maracá and the numerous islands in the Rio Uraricoera, particularly in the Furo de Maracá, must exceed 200 km, and it is only possible here to give a brief account of some salient points of the riverine forest. A more detailed account with species lists is given in Milliken and Ratter (1989).

*Eastern part of the island*

*Terra firme* forest comes right to the river edge where the banks are steep, but is usually flanked by a tangle of heliophilous vegetation, such as lianas and various shrubs. The result is usually a barely penetrable wall of vegetation broken only by the occasional gap of a recent treefall. Such banks surround the greater part of the eastern end of the island, and include those which form a levee along the riverside margin of the large expanse of seasonally flooded *campo* below the Station. A number of species of large trees are particularly conspicuous as one travels along the river by boat: the wide-crowned *Ceiba pentandra* and *Enterolobium cyclocarpum* (both deciduous during the dry season and difficult to distinguish at a distance) often extend over the water, while tall, elegant specimens of *Peltogyne gracilipes* are common westwards of the Station. Much *Cochlospermum orinocense* and *Maximiliana maripa* occur in the main canopy layer, where *Spondias mombin* is also common. Trees noted as characteristically overhanging the water include the three Leguminosae *Andira surinamensis*, *Macrolobium acaciifolium* and *Ormosia smithii* (all of which have a relatively low, wide-crowned form). *Bixa orellana*, *Bactris maraja*, *Phenakospermum guyannense* and, in some places, bamboos are common shrubby constituents of the riverside thicket, often showing distinct clumping. Lianas are abundant and belong to a host of species. Notable examples include *Entada polystachya*, which climbs high in the riverside trees and covers them with its large, straw-coloured, lomentaceous pods, *Uncaria guianensis* which trails down over the water from the riverside shrubs and can make passage perilous because of its hooked axillary thorns, and *Combretum rotundifolium*, whose handsome orange-red spikes of flowers often project close to the water's surface. In many places clumps of the bizarre, erect, unbranched aroid *Montrichardia linifera* occur at water level at the base of the bank. This species reaches about 2 m tall and has large, waxy, hastate leaves; it also occurs in other wet habitats.

The dioecious *Triplaris surinamensis* is a conspicuous riverside tree very common along many long stretches of the bank. It reaches about 20 m tall, and to the casual eye resembles a *Eucalyptus* in its pale bark and slender erect form. *Triplaris* is not a member of the normal *terra firme* community and is always associated with wet conditions, and the areas of bank it inhabits are usually somewhat lower and damper. In addition to *T. surinamensis*, such areas of bank often support a low forest of *Cecropia latiloba* and/or *C. palmata*, *Cochlospermum orinocense* and some *Gustavia augusta*, with a dense shrub layer in which *Tabernaemontana siphilitica* and, in places, *Trichanthera gigantea* (a bushy Acanthaceae reaching 5 m) are important species. Both the *Triplaris* and the *Cecropia* are myrmecophilous, a condition linked with their heliophilous, colonizing tendencies. In the areas where we encountered this *Triplaris/Cecropia* riverine vegetation the ground rose abruptly further inland to a second level 2.5–3 m higher and supporting normal *terra firme* forest. Sometimes the lower level was interrupted by a depression, flooded during the rainy season and dominated by *Cecropia* and *Bactris maraja*, with very little leaf-litter. This was the closest habitat to *várzea* forest found in the eastern part of the island.

In areas where the river bank shelves gradually there is a pure stand of *Inga ingoides*, producing a shrubby thicket to about 8 m tall. The bushy shrubs making up the tangle do not develop erect trunks but are based on semi-prostrate boughs lying more or less on the surface of the muddy bank. For much of the year this thicket is partly submerged in the river, and in the dry season clumps of dried-up fibrous roots hang from the boughs.

*Western and central parts of the island*

Westwards (upstream) from the Casa de Maracá (Figure 5.2), the vegetation bordering the river changes conspicuously. Perhaps the most obvious and important of these changes, as mentioned in section 1.3, is the sudden predominance of the leguminous tree *Peltogyne gracilipes*; a feature especially prominent during the dry season when the pale, almost bare crowns of this species can be seen from the river projecting above the forest canopy (Plate 2a).

The Furo de Maracá is particularly notable for its myriad islands, and the maze of narrow channels which separate them (Plate 2b). On entering these island complexes one has the impression of total isolation; vertical walls of dense vegetation rise to 20 m from the low banks of channels that may be no more than 5 m wide. Down these shady corridors the waters flow with remarkable force and rapidity, tumbling through fierce rapids as the river drops through the granitic formations which dominate the region.

One of the most conspicuous of the tree species common to this section of the river bank is the legume *Elizabetha coccinea* var. *oxyphylla*, whose scarlet and white flowers and large marzipan-coloured pods stand out from the surrounding green. Other, rather less flamboyant, species of note include *Virola surinamensis*, *Xylopia discreta*, *Calophyllum lucidum*, *Parinari excelsa*, *Homalium guianense*, *Macrolobium acaciifolium*, *Ormosia smithii*, *Acosmium tomentellum* and the omnipresent *Peltogyne gracilipes*. The palms *Maximiliana maripa* and *Euterpe precatoria* are also common. Closer to river level a dense mass of shrubs and smaller trees leans over the water, bound by numerous vines.

In other parts the river takes on a much more open appearance, bounded by broad sandbanks and expanses of exposed granite which give way to typical tall

*Peltogyne gracilipes*-dominated forest. Numerous small islands within the river, also largely of granite and sand, support a rather scrubby-looking community with trees of *Andira surinamensis, Astrocaryum jauari, Byrsonima schomburgkiana, Faramea crassifolia, Genipa spruceana* and *Macrolobium acaciifolium*. Buriti palms (*Mauritia flexuosa*) may be concentrated near the water's edge, where silty, semi-submerged banks support (and are probably bound by) dense concentrations of the tall aroid *Montrichardia linifera*.

Epiphytes are not abundant on Maracá, but are found in greater profusion in the type of riverside vegetation described above than anywhere else. A number of orchids were collected, of which the bright, purple-flowered *Cattleya violacea* is easily the most prolific, splashing the riverside with colour from the Casa de Maracá to the west of the island and far beyond throughout its long flowering period. There are also epiphytic bromeliads, the epiphytic cactus *Rhipsalis baccifera*, ferns, and various aroids, including the delicate *Anthurium gracile*. *Norantea guianensis*, a marcgraviaceous vine, produces its spectacular scarlet inflorescences during the dry season from amongst the riverside shrubs and trees, while the abundant, arboreal *Coussapoa villosa*, with its characteristic rhomboid leaves, supports itself on neighbouring trees by clasping adventitious roots.

Between the Casa de Maracá and the western point of the island the vegetation of the Furo de Maracá varies between that described above and broad open stretches of river bordered by tall *Peltogyne gracilipes* forest on steep banks to 3 m high (depending on season). The Furo de Santa Rosa west of the Tipurema rapids at the northernmost point of the island conforms closely to the description above, both in vegetation and topography. West of the Onça falls, however, the ground rises sharply from the river to the hills described above (see '*Terra firme* forest of the western part of the island'), and there is a corresponding change in the appearance of the vegetation. *Peltogyne gracilipes* becomes a very minor component along the banks, while the fronds of the palms *Socratea exorrhiza* and *Euterpe precatoria*, and the musaceous *Phenakospermum guyannense* project conspicuously from the riverside forest. Where streams run out from between the hills the level of the riverside vegetation drops to a low tangle of vines and shrubs, very similar to that described in section 2.3 as *vazante* vegetation.

## 2. SAVANNA

Savanna vegetation is found particularly at the eastern end of the island, but constitutes a rather low percentage (about 6% according to the estimates of Furley *et al.*, 1994) of the total vegetation cover. The following three categories were recognized during the survey.

### 2.1 *Curatella americana/Byrsonima crassifolia campo*

This is an enormously widespread vegetation type found from Mexico to Paraguay. It occurs on the mainland to the south and east of Maracá and extends right into Venezuela and the Rupununi Highlands of Guyana. On the island itself, however, it has been studied only on the Santa Rosa *campo* (Plate 1b), the drier marginal area

**Table 5.16.** Trees and shrubs ≥2 cm basal diameter in a 50 × 50 m square in the Santa Rosa *campo* on the Ilha de Maracá (total no. individuals = 134; individuals/ha = 536; basal area/ha = 52 487 cm$^2$)

|   |   | No. | AD | ADo | RD | RDo | ICV |
|---|---|---|---|---|---|---|---|
| 1 | *Curatella americana* | 79 | 316 | 45 025.2 | 58.9 | 85.8 | 144.7 |
| 2 | *Byrsonima crassifolia* | 20 | 80 | 6 251.2 | 14.9 | 11.9 | 26.8 |
| 3 | *Swartzia grandifolia* | 11 | 44 | 360.8 | 8.2 | 0.7 | 8.9 |
| 4 | *Vitex schomburgkiana* | 9 | 36 | 387.6 | 6.7 | 0.7 | 7.4 |
| 5 | *Vernonia brasiliana* | 6 | 24 | 371.2 | 4.5 | 0.7 | 5.2 |
| 6 | *Swartzia laurifolia* | 5 | 20 | 260.8 | 3.7 | 0.5 | 4.2 |
| 7 | *Cupania rubiginosa* | 2 | 8 | 106.8 | 1.5 | 0.2 | 1.7 |
| 8 = | *Casearia sylvestris* | 1 | 4 | 28.4 | 0.7 | 0.05 | 0.75 |
| 8 = | Melastomataceae Indet. | 1 | 4 | 28.4 | 0.7 | 0.05 | 0.75 |

AD, absolute density (individuals/ha); ADo, absolute dominance (basal area – cm$^2$/ha); RD, relative density; RDo, relative dominance; ICV, importance cover value (RD + RDo).

of the seasonally flooded *campo* in the southeast, and in a few tiny enclaves on the Preguiça trail to the north of the Casa de Maracá (Figure 5.2). A few other small islands of this vegetation occurring within the forest were detected by analysis of satellite imagery.

The Santa Rosa *campo* has been intensively studied and will serve as an example. The vegetation consists of 'grassland' with a patchy scattering of shrubs and small trees, rarely more than 5 m tall. Using the terminology of the *cerrado* vegetation of Central Brazil (of which this is an impoverished outlier) it is *campo sujo*. The great majority of the trees are of *Curatella americana* and *Byrsonima crassifolia*, both of which are typical, fire-resistant, savanna species with characteristic contorted form and leathery leaves. In fact, the leaves of *C. americana* are so silica-impregnated and hard that they can be used as sandpaper. The results of an inventory of trees and shrubs in a 50 × 50 m square are given in Table 5.16. They show the enormous predominance of *C. americana* and *B. crassifolia*: together the two species have a relative dominance of 97.7%.

The tallest tree recorded on the square was a *Curatella americana* 6.5 m tall, while the greatest basal diameter was 29 cm. A specimen of the same species 6 m × 45 cm dbh was seen in another part of the *campo*. In other areas scattered trees of *Humiria balsamifera* are found to 12 m tall. *Erythroxylum suberosum* and the small savanna form of *Casearia sylvestris* are other species occurring on this *campo* and characteristic of the *cerrado* vegetation of Central Brazil.

The ground layer is very variable in height and density, ranging from a low covering of Cyperaceae, Xyridaceae, *Cuphea antisyphilitica*, *Sauvagesia rubiginosa*, etc., on white sandy soils, to a metre-high sward of dense, rank grasses in areas which probably tend to be somewhat drier.

In the lower parts of the *campo*, drainage patterns are indicated by the dissection of the soil surface into a network of small channels, leaving mounds between them. For the greater part of the rainy season these channels are full of water, but are dry for the rest of the year. Generally the vegetation is restricted to the tops of the mounds which, obviously, are better drained.

Lines of *buriti* palms (*Mauritia flexuosa*) mark drainage lines across the *campo*, and near the point where the track to the Ecological Station enters the forest there is a *capão* (a small isolated piece of woodland). This is dominated by *buriti* palms to 14 m, with *Humiria balsamifera* to 12 m, and an understorey in which *Phenakospermum guyannense*, *Toulicia* sp. (R5728), *Endlicheria dictifarinosa*, *Clusia renggerioides*, *Alchornea schomburgkii* and *Siparuna guianensis* are important.

Two very small strips of *Curatella americana*/*Byrsonima crassifolia campo* occur on the Preguiça trail some 1.5 km northwest of the Casa de Maracá (Figure 5.2). They consist of *campo sujo* vegetation and, as at Santa Rosa, by far the most important tree/shrub is *C. americana* with *B. crassifolia* and *Erythroxylum suberosum* also represented.

A number of observations indicate that the forests on Maracá are expanding into areas formerly occupied by savanna: a process occurring in other parts of Brazil when it is not prevented by human influence (Ratter *et al.*, 1973, 1978a; Furley *et al.*, 1988). At various places around the Santa Rosa *campo*, trees of *Curatella americana* are found in the forest margin or a short distance into the forest itself, apparently demonstrating expansion of the forest. As already mentioned, the same occurs in the Preguiça trail *campo* area where trees of *C. americana* are found within the low *carrasco* thicket forest to about 13 m from the margin. This indicates considerable forest encroachment during the life of these trees which, judging from the evidence of those established on the causeway (see 'Colonizing species and *caapoeira*'), may be no more than eight to ten years old. In fact the small Preguiça trail *campos* probably represent the remaining vestiges of a formerly much more extensive area of this vegetation. Another locality where expansion of forest into *campo* has occurred is at the site of the old pig-sty to the east of the road at the base of the slope below the Station. Here an enclosure was set up on the edge of the *campo* in the early days of the Station, but is now almost filled with forest.

As is normal in savannas, the flora has a high degree of fire tolerance, but fires must have been of rare occurrence since the establishment of the Ilha de Maracá as a reserve. However, a fire burned several hectares of *campo* around December 1986, during track-making activity at Santa Rosa, and charring found on the bark of trees in other parts of the same *campo* testified to previous fires. One large specimen of *Curatella americana* showed charring of the bark to 2 m above ground level. Possibly much of the recent forest expansion into savanna (see Discussion) may be associated with the cessation of fires since the area has been protected as a reserve.

The transitional vegetation between *campo* and forest was studied intensively in many areas and is fully described in Milliken and Ratter (1989).

## 2.2 Seasonally flooded *campo*

A large expanse of seasonally flooded *campo*, traversed by the causeway, lies on the low ground between the Station and the Rio Uraricoera (Figure 5.2; Plate 4). This extends eastwards towards the tip of the island, and a similar patch is found on the Ilha de Nova Olinda further upstream. These campos are treeless apart from small islands of better-drained ground, supporting marginal scrub or forest, and lines of *buriti* palms. The dominant herbaceous family is the Gramineae, but interspersed with a variety of other species amongst which Malvaceae are prominent. A list of species is given in Milliken and Ratter (1989).

The majority of the *campo* is flooded during the rainy season, to a depth of up to one metre, but is dry for the rest of the year. There are patches, however, which remain damp throughout the year and support a vegetation more similar to that described in the following section (*vazante*). The tall marantaceous *Thalia geniculata* is common in these areas. There is a clear vegetational sequence running up the slope from these damp patches, through the grassy seasonally flooded *campo* to shrubby *Curatella americana/Byrsonima crassifolia campo*, and on into *terra firme* forest.

## 2.3 *Vazante*

A largely treeless vegetation called *vazante* runs through the forest as strips up to 100 m or more wide (Plate 4a). It occurs along drainage courses where the inclination is so slight that no stream bed has been cut, so that during the rainy season there is a slow-flowing marsh which dries up completely in the dry season.

The area of *vazante* studied most thoroughly lies on the Preguiça trail (Figure 5.2). The vegetation consists of tall herbs and low shrubs such as *Mimosa pellita*, *Senna alata*, *Melochia simplex*, *Canna glauca*, *Costus scaber*, *Thalia geniculata*, *Heliconia psittacorum* and *Scleria sprucei*, and grasses such as *Eriochloa punctata*, covered by a tangle of herbaceous vines. The whole produces a rolling mass 60–200 cm tall whose structure, or lack of it, resembles colonizing vegetation on abandoned areas of rich soil. There seem to be two components in the vegetation: semi-aquatic species which flourish during the rains and die back to their underground perennating organs during the dry season, and land plants which flourish during the drier periods. Notable amongst the former are *Thalia geniculata* and *Canna glauca*, while the dry phase is dominated to a large extent by tall grasses and herbaceous leguminous and convolvulaceous vines.

Trees of *Triplaris surinamensis* to *c*. 15 m occur here and there, presumably on better-drained spots and are especially common along the *vazante* margin. The vegetation at the margin of *vazante* and forest tends to be a particularly exuberant thicket. In addition to *Triplaris*, trees such as *Homalium guianense*, *Cecropia* sp., *Inga* sp., *Apeiba schomburgkii*, *Rheedia* sp., *Gustavia augusta*, *Vitex schomburgkiana*, and clumps of the small spiny palm *Bactris maraja* were noted, accompanied by an abundant growth of vines, *Desmoncus polyacanthos*, razor-grass (*Scleria sprucei*), *Olyra* sp., *Heliconia psittacorum* and *Costus scaber*. Isolated trees on *vazante* are often completely smothered with vines so that it is sometimes difficult to determine their species. This, however, does not seem to happen to trees of *Triplaris* which are perhaps kept free of vines by their resident ant colonies.

We were informed that the soils of the *vazante* were excellent for the cultivation of crops such as beans and melons. The method of cultivation is to cut and burn the standing vegetation at the beginning of the dry season and then to plant the crop, which allows plenty of time for harvesting before the next rains.

Similar areas of *vazante* were encountered on the Fumaça, Filhote and Anta trails, and many were seen from the air. *Vazante* is much more abundant and widely distributed on the island than *Curatella americana*/*Byrsonima crassifolia* campo.

## 3. AQUATIC VEGETATION

### 3.1 Still-water habitats

A number of still-water habitats were observed on the island, both in the forests and in the savannas (Plate 4b). The most obvious of these are the man-made borrow-pits, created 10 years prior to the survey, which border the causeway connecting the Ecological Station with the Furo de Maracá. Like all of the aquatic habitats on Maracá, these ponds are distinctly seasonal, some drying up entirely by the end of the dry season. We carried out no systematic survey of aquatic habitats but the following observations were made. Other studies have been reported by Volkmer-Ribeiro *et al.* (1989 and Chapter 21).

The borrow-pits below the Station lie within the seasonally flooded *campo* described previously, and are surrounded by the type of rank, herbaceous and shrubby vegetation typical of that habitat and the *vazante* margin. During the dry season the exposed mud around the remaining water supports various Cyperaceae and Gramineae, and the spiny herb *Hydrolea spinosa*. At this season the water is rarely deeper than 70 cm, and various herbaceous species grow in the shallows. These include *Thalia geniculata*, *Echinodorus scaber*, *Sagittaria* spp., *Eleocharis* spp., *Bacopa reflexa* and *Polygonum acuminatum* – the last found in dense, almost pure stands. Of the *Eleocharis* species, *E. acutangula* is by far the most common, dominating much of the surface of the water.

In the deeper regions of the borrow-pits there are conspicuous masses of *Eichhornia azurea*, the largest colonies of which measure *c.* 60 × 20 m. Around these there are tangled submerged masses of the aquatic weeds *Cabomba piauhiensis*, *Utricularia* spp. and *Mayaca fluviatilis*, of which the first is most abundant. Only their small and delicate flowers project above the water. The characteristic floating leaves of *Nymphaea* cf. *wittiana* are scattered over the surface.

The natural ponds within the savanna tend to dry up entirely during the dry season, leaving a large expanse of cracked mud. They support a flora essentially very similar to that of the borrow-pits, but lacking some of the species that presumably cannot tolerate seasonal drying. The larger lakes are often surrounded by dense stands of the giant aroid *Montrichardia linifera*, also common on the river banks.

Of the lakes occurring within the forest, the one most accessible and consequently the most studied lies at the western extremity of the Station trail system. It is surrounded by a band of *Mauritia flexuosa* forest (*buritizal*) approximately 30 m broad, very similar to that described above. The *buritizal* ends abruptly, giving way to a dense zone of *Montrichardia linifera*, *Thalia trichocalyx* and Cyperaceae. Beyond this lies the open expanse of the lake, roughly 200 m across.

During the wet season the *buritizal* surrounding the lake is partially flooded. A large patch of the fern *Blechnum serratulum* was found growing on the pneumatophore hummocks of the *Mauritia* palms, which protrude above the water's surface. The lake itself is flooded to a depth of up to 1.5 m. Various herbaceous species grow in abundance in the shallower parts, amongst which *Ludwigia torulosa* and Gramineae (including *Luziola subintegra*) are common. Further towards the centre there are patches of open water with *Nymphaea* sp(p). and dense submerged tangles of *Utricularia foliosa*, and floating mats of vegetation. These mats are 30–40 cm thick and composed of a mixture of tightly bound herbaceous material which will support the weight of a man. *Xyris laxifolia* is a conspicuous constituent. On similar mats on a lake further to the west there are large quantities of the fern *Cyclosorus interruptus*.

Apart from scattered individuals of the onagraceous *Ludwigia nervosa*, few shrubs are present within the lake. During the dry season there is little or no water apart from a small central area, and the *Ludwigia*, with its large primrose-yellow flowers, emerges conspicuously from amongst the surrounding grasses and herbs.

## 3.2 Flowing water habitats

We made no systematic observations of the vegetation of flowing water. However, this habitat was studied by Drs Pedro Mera and André de Oliveira (INPA) whose data are mentioned in Milliken and Ratter (1989).

During our travels on the Rio Uraricoera we noted the abundance of two Podostemaceae, *Mourera fluviatilis* and *Apinagia tenuifolia*, growing on the rocks of rapids. The former is particularly spectacular, with long, trembling spikes of pink flowers emerging from the turbulent waters. Dislodged clusters of *Eichhornia crassipes*, which grows in places along the river banks, were frequently seen floating down the river during the rainy season.

## 4. VEGETATION OF ROCK OUTCROPS

Outcrops of rock within the forest were found in a number of locations, but those on the hill at the northern end of the Fumaça trail are far more extensive than any others. They are granitic, and appear to have been river-worn in the past, and the largest of the outcrops found was approximately 15 m tall. Although some of the species growing on these rocks were also found in the surrounding forest, the majority were not. Species collected in this habitat are listed below. Those which have not been collected elsewhere are marked with an asterisk.

ARACEAE
 *Anthurium clavigerum*
 *Monstera dubia*
 *Philodendron melinonii*
CYPERACEAE
 *Cyperus simplex*

GUTTIFERAE
 *Clusia minor**
HAEMODORACEAE
 *Xiphidium caeruleum**
MARANTACEAE
 *Maranta protracta**

MELIACEAE
*Trichilia surumuensis*
MORACEAE
*Ficus* vs. *eximia**
MUSACEAE
*Heliconia bihai**
ORCHIDACEAE
*Oceoclades maculata**
*Vanilla* sp.
PIPERACEAE
*Peperomia alata**

URTICACEAE
*Urera baccifera**
*Urera caracasana**
PTERIDOPHYTA
*Anetium citrifolium**
*Asplenium serratum**
*Campyloneurum latum**
*Gymnopteris rufa*
*Hemionitis palmata**

Other similar areas of bare granitic rock occur beside the Rio Uraricoera, weathering in the 'onion-skin' manner typical of the tropics. The largest of these outcrops, covering at least a hectare, was observed close to Fazenda União. Numerous species were noted in this area including a large *Agave*, the 2-m-tall yellow-flowered orchid *Cyrtopodium poecilum*, the tall branching cactus *Cereus hexagonus*, and large numbers of *Melocactus smithii*, the last not encountered anywhere else on the island. Herbs and small shrubs grow in the cracks and pockets of sand in the rock and include *Cnidoscolus* sp. aff. *urens*, *Chamaecrista rotundifolia*, *Sinningia incarnata* and numerous other species.

## 5. COLONIZING VEGETATION AND *CAAPOEIRA*

The causeway from the landing on the Rio Uraricoera to the Station provides an excellent opportunity to study colonizing species. This 2 km strip of *terra firme* was created about 1½ years before our observations to provide an all-weather road to the Station. The material for the causeway was extracted from the wet *campo*, thus producing borrow-pits lying on either side, and these in turn have produced interesting aquatic habitats (see above).

Species colonizing the well-drained soil of the causeway are found in a number of natural habitats on the island, such as *terra firme* forest (e.g. *Jacaranda copaia*, *Rollinia exsucca*), *Curatella americana*/*Byrsonima crassifolia campo* (e.g. *Curatella americana*) and, particularly, the scrubby marginal vegetation which lies between these two habitats (e.g. *Bauhinia ungulata*). In addition there are a few introduced weeds (e.g. *Cenchrus echinatus*, *Emilia coccinea*, *E. fosbergii* and *Mimosa pudica*). Most of the species present are well known as colonizers.

The tallest tree on the causeway (in March 1988) was a specimen of *Jacaranda copaia* ssp. *spectabilis* 13 m tall, but specimens of *Cochlospermum orinocense* approached this height, while *Simarouba amara* and the *Cecropia* species reached 8 m. There were many small trees of *Curatella americana*, especially at the Station end of the causeway, some of which reached 4.5 m in height. Vines were abundant and included *Securidaca diversifolia*, *Uncaria guianensis* and a number of species of Bignoniaceae. Loranthaceae were also abundant on the trees of the causeway – presumably because the habitat, lying in the centre of open *campo*, provides a good place for birds to perch.

The margins of the Station clearing also demonstrate colonizing species, of which *Cecropia palmata, C. latiloba, Guazuma ulmifolia, Casearia sylvestris* and *Coursetia ferruginea* are conspicuous.

Close to the river at Santa Rosa there is an old clearing formerly used for growing bananas, coffee, citrus, etc., but abandoned 10 years prior to our study (see the discussion of human occupation on Maracá by Proctor and Miller, Chapter 23). This was occupied by low *caapoeira* vegetation approximately 6–10 m tall in 1988. No systematic studies were made here, nor at the similar abandoned area slightly to the south of the Santa Rosa *campo*, but in general they appear to contain species of the adjoining forest, probably established mainly by coppice regeneration, as well as a number of typical colonizers. *Cochlospermum orinocense, Swartzia laurifolia, Crepidospermum goudotianum, Genipa americana, Apeiba schomburgkii* and *Xylopia aromatica* were noted as common tree species. This observation corresponds to those made by Miller and Proctor (Chapter 6), who conducted a more detailed (quantitative) study of an area of *caapoeira* approximately 13 years old, close to the forest area sampled by Transect 3 in this study.

The area of the river bank immediately upriver of the landing was also presumably cultivated in the past, but now supports a low forest mainly 8–10 m tall at the time of study. Floristically it is fairly typical of the vegetation of the well-drained areas of river bank at the east of the island. *Cochlospermum orinocense*, reaching *c*. 14 m, is an important species.

## GENERAL DISCUSSION

A wealth of information was derived from the vegetation studies of the Maracá Rainforest Project and much of it has already been reported in detail in a series of studies. The present general discussion is confined to a few salient points; more detailed considerations, inappropriate for the present publication, can be obtained in the literature already published.

### FOREST BIOMASS

It is useful to compare the biomass of the forests we studied on Maracá with those occurring in other areas of the Amazon. Pires and Prance (1985), in their description of vegetation types of the Brazilian Amazon, give figures for biomass expressed as basal area of trees per hectare, using individuals $\geq 30$ cm in circumference (corresponding closely to the $\geq 10$ cm dbh category used in our studies). According to these authors 'exceptionally large forests' can exceed 40 $m^2$ of basal area per hectare. Extrapolating our figures to give basal area/ha, two of the eastern *terra firme* forest areas sampled by transects fall into this category with 41.1 $m^2$/ha (Transect 2A) and 54.5 $m^2$/ha (Transect 7A), as did *Peltogyne gracilipes* forest on the Fumaça trail in the centre of the island (44.7 $m^2$/ha, Transect 14A) and towards the eastern end (40.4 $m^2$/ha, Transect 16A). In continuing studies on mixed *P. gracilipes* forest on the Preguiça trail, basal areas per hectare up to 38.3 $m^2$ have been recorded by Nascimento *et al.* (1997). *Terra firme* forest at the western end of the island practically reaches the 'exceptionally large forest' category with 38.3 m$^2$/ha (Transect 13A). Most of the other *terra firme* areas sampled fall into Pires and Prance's 'dense forest' biomass category ($\geq$24 m$^2$/ha), but the angico slope forest (21 m$^2$/ha, Transect 6), an area on the Fumaça trail in the centre of the island (21.5 m$^2$/ha, Transect 10A) and the eastern semi-evergreen forest reported by Thompson *et al.* (1994) (21.6 m$^2$/ha) have a lower biomass more characteristic of their 'open or vine forests' (18–24 m$^2$). In general, the 'low forests' of Maracá fall well below the biomass usually found in open forests.

## SPECIES DIVERSITY

The highest number of tree species recorded in this study (for individuals $\geq$10 cm dbh) was 80, belonging to 31 families on a 1.5 ha quadrat in eastern *terra firme* forest (Transect 3). However, a western-end PCQ transect (No. 12A) approached this, with 73 species of 27 families on an area equivalent to 0.378 ha. No doubt the diversity of the western-end forest would have been the higher of the two had the sample areas been equal (two PCQ transects close to Transect 3 show only 54 species on an area equivalent to 0.396 ha and 51 species on 0.519 ha respectively). The majority of transects, particularly those of low forests, show much lower species diversity than this.

It is interesting to compare these figures with others from *terra firme* forest in Amazonia. Campbell *et al.* (1986) give an extremely useful compilation of 20 phytosociological inventories made in diverse parts of Amazonia which provides a good basis for comparison, while some more recent surveys are listed in Silva *et al.* (1992). Gentry (1988) records the most species-rich forest yet discovered anywhere in the world with *c.* 300 spp. $\geq$10 cm dbh in a single hectare plot near Iquitos, Peru, while Silva *et al.* (1992) working in Amazonas, Brazil, record 271 species of the same size category on hectare plots at Munguba, and 260 species at Jaraqui. Another exceptionally diverse forest is recorded by Campbell *et al.* (1986) who recorded 265 spp. $\geq$10 cm dbh on 3 ha at the Rio Xingu, Brazil. Obviously our figures are very low compared to these extremes. However, judging from the figures compiled by Campbell *et al.*, the average for Amazonia is probably closer to 120–150 spp./ha, while there are a number of records close to ours. For example, Black *et al.* (1950) found 87 species of 31 families in 1 ha of *terra firme* forest at Belém, Pará. Figures for fringe Amazonian forest (dry forest) in Mato Grosso and Tocantins (not cited in Campbell *et al.*) show a close correspondence to those from Maracá. Approximately 80 spp. were recorded on 0.35 ha at Suiá-Missu, Mato Grosso (Ratter *et al.*, 1978a), and 64 and 39 spp. on two 50-point PCQ transects (i.e. 200 trees each) at the Parque Nacional do Araguaia, Ilha do Bananal, Tocantins (Ratter, 1987).

Not only does the species diversity of the fringe Amazonian forest of the Ilha de Maracá appear to be similar to that of the dry forests of Mato Grosso and Tocantins, its equivalent in the south of Amazonia, but many of the species are in common. Examples are *Enterolobium schomburgkii*, *Buchenavia tetraphylla*, *Spondias mombin*, *Cochlospermum orinocense*, *Aspidosperma nitidum*, *Didymopanax morototoni*, *Licania kunthiana* and many others. These are very widespread Amazonian species, many of them pioneers, which also extend into Central Brazil. Pires and Prance (1985) mention the physiognomic similarity between the 'dry forests' of southeastern Amazonia and Roraima, but stress the differences in some of their most important species.

The presence of large numbers of tall trees of pioneer species such as *Jacaranda copaia*, *Didymopanax morototoni* and *Maximiliana maripa* in the eastern *terra firme* forests of Maracá perhaps suggests that these forests are relatively recently established. This would agree with the observations already mentioned of forest on the island expanding into savanna vegetation. It also conforms with observations from Mato Grosso and the Distrito Federal indicating a general current trend for forest to expand into savanna (Ratter *et al.*, 1973, 1978a; Ratter, 1985, 1986, 1992). In addition there is evidence that there was considerable human disturbance of the east end of the island in the past and particularly before expulsion of the Indian population in 1880 (see Proctor and Miller, Chapter 23), and this clearly would provide conditions for establishment of forest rich in pioneer species. However, the only places where recent obviously secondary forest (*caapoeira*) was observed were at Santa Rosa, where cultivation ended only about 1977, and in the area of river bank close to the Ecological Station landing stage, both at the eastern end of the island.

The differences in composition of the eastern and western *terra firme* forests of the island have already been discussed. INPA collectors commented that the western forests resembled those around Manaus more closely than did those in the east, and it seems probable that they represent a more mature and typical Amazonian forest since they lie further from the forest–savanna transition.

As already discussed, areas of the *Peltogyne gracilipes* forest fall into the category of exceptionally large Amazonian forest, but in its tallest and purest form, as found on the Fumaça trail in the centre of the island, it has a remarkably low species diversity. In this extreme form only 11 species of trees ≥30 cm dbh were recorded out of 80 individuals and eight of these occurred only once, while *Peltogyne gracilipes* made up 77% of the individuals (relative density) and 78% of the basal area (relative dominance). Such monodominant moist forests have now been recorded in various parts of the tropics. The *Cynometra alexandri* forest in Uganda provides a comparable example (D.L. Hafashimana, pers. comm.), as do the Mora and Wallaba forests of Guyana, dominated by the leguminous genera *Mora* and *Eperua* respectively (see Davis and Richards, 1934; Steege, 1990). The anomalous characteristics of these forests are clearly of great interest. Since the initial surveys on Maracá in 1987–88, Dr Nascimento has carried out a study of these *Peltogyne gracilipes* forests which has yielded some interesting results, including the discovery of substantially higher levels of magnesium in the soils underlying the '*Peltogyne*-rich forests' than in neighbouring forests in which the species is sparse or absent (Nascimento, 1994; Nascimento *et al.*, 1997).

## *CAMPINA* AND *CAMPO*

The low campina forests of the Ilha de Maracá occur on areas of sandy soils but, apart from this and their stature, seem to have little affinity with those of the region of the Rio Negro (also known as '*caatingas* of the Rio Negro') described by Takeuchi (1960b), Rodrigues (1961), and by other workers. Judging from the species lists of these authors there is hardly a species in common with the Maracá *campinas*.

The *Curatella americana/Byrsonima crassifolia campo* on Maracá represents the fringe of an area of 54 000 km$^2$ of this vegetation occupying the northeast of Roraima and running into the Rupununi Highlands of Guyana (McGill University, 1966; Pires, 1974). It is a characteristic hydrologic savanna (Denevan, 1968) of the type widespread in South and Central America. Such vegetation is known by a variety of terms which include edaphic and topographical savannas and corresponds to Sarmiento's (1992) hyperseasonal and semiseasonal savannas. Occurrence of such savannas is determined by a fluctuating water-table that is very high during the rainy season (often producing flooding) and low during the dry season. This contrasts with true Brazilian *cerrado* vegetation where soils are always well drained (Eiten, 1972). The area of *campo* at Santa Rosa and the region of savanna–forest transition at its margins have been intensively studied, particularly with regard to water-table fluctuation (Milliken and Ratter, 1989; Furley and Ratter, 1990, 1994; Furley, 1992; Thompson *et al.*, 1992b; Proctor, 1994).

The *campo* of the island is floristically more depauperate than that seen nearby on the mainland. In the *campos* close to the south bank of the Rio Uraricoera only a few kilometres from the island there are trees of *Bowdichia virgilioides* and *Xylopia aromatica*. The former is a widespread savanna (*cerrado*) species not found on the island, and the latter is also widespread but, although found in *caapoeira* (secondary scrub) and other vegetation on the island, did not occur in the *Curatella americana/Byrsonima crassifolia campo*. In addition to these species, we also noted trees of *Byrsonima coccolobifolia*, *Anacardium occidentale*, *Himatanthus articulatus*, *Roupala montana* and the very characteristic *cerrado* subshrub *Byrsonima verbascifolia* in the savannas close to Boa Vista. Takeuchi (1960a) studied an area of savanna a little to the south of the Boa Vista–Maracá road and published observations similar to ours. He records *C. americana* and *B. crassifolia* as the dominant trees with *Swartzia diphylla* in third place. Our results from Maracá have *Swartzia grandifolia* and *S. laurifolia* as amongst the most important species after *C. americana* and *B. crassifolia*. It seems probable, however, that we are using a different name for the same *Swartzia* species observed by Takeuchi (1960a).

## CONCLUSION

The Ilha de Maracá affords a unique area of some 100 000 ha of natural and, apart from some areas in the east of the island, almost completely undisturbed vegetation. The present study provides basic descriptions of the vegetation and suggests the action of dynamic processes: the invasion of *campo* by forest, and perhaps the maturation in composition of a forest which still contains many colonizing species. These processes should be closely monitored and allowed to continue as a vital part of the Reserve's role as a living laboratory. Management of the Reserve should not include manipulation of habitats (with the possible exception of carefully controlled burning of areas of *Curatella americana/Byrsonima crassifolia campo*).

A particular value of the Reserve is that it contains a large area of marginal Amazonian *terra firme* forest (*mata seca*). This is probably the most endangered of all types of Amazonian tall forest and is of great interest since it may resemble a form of the Hylaea prevalent during periods of the Pleistocene when conditions were less favourable for rainforest growth.

## ACKNOWLEDGEMENTS

The participation of the INPA technicians Luis Coelho, Dionísio Coelho, José Lima dos Santos, Cosme Damião A. da Mota and José F. Ramos was essential for the survey and we thank them not only for sharing their knowledge but also for their friendship. We worked in collaboration during much of the project with Drs Peter Furley, John Proctor, Duncan Scott and Jill Thompson and the Kew botanists Dr Ray Harley, Gwilym Lewis, Peter Edwards and Brian Stannard and their help is much appreciated. Many of the plant collections were identified by specialists, who are listed in Appendix 1 (p. 443).

WM's visit to Maracá was funded by the following sponsors (in addition to RGS): Peter Nathan Charitable Trust, Godinton Charitable Trust, Percy Sladen Memorial Fund, D.M. Charitable Trust, Welconstruct Trust and Ever Ready (UK). On return to UK he was employed by the Royal Botanic Garden Edinburgh, Royal Botanic Gardens Kew and by the Overseas Development Administration to work up the results of the fieldwork. We wish to acknowledge Grenville Lucas (Keeper of the Herbarium, Kew), Professor John McNeill (then Regius Keeper, RBG Edinburgh) and the administrators of ODA for their support.

## NOTE

1. A list of all species mentioned with authorities is given in Appendix 1, p. 443.

## REFERENCES

Beard, J.S. (1944). Climax vegetation in tropical America. *Ecology*, 25, 127–158.

Black, G.A., Dobzhansky, T. and Pavan, C. (1950). Some attempts to estimate species diversity and population density of trees in Amazonian forests. *Botanical Gazette*, 111, 413–425.

Campbell, D.G., Daly, D.C., Prance, G.T. and Maciel, U.N. (1986). Quantitative ecological inventory of *terra firme* and varzea tropical forest on the Rio Xingu, Brazilian Amazon. *Brittonia*, 38, 369–393.

Dargie, T. and Furley, P.A. (1994). Monitoring change in land use and the environment. In: *The forest frontier: Settlement and change in Brazilian Roraima*, ed. P.A. Furley, pp. 68–85. Routledge, London.

Davis, T.A.W. and Richards, P.W. (1934). The vegetation of Moraballi Creek, British Guiana: an ecological study of a limited area of tropical rain forest. Part II. *Journal of Ecology*, 22, 106–155.

Denevan, W.M. (1968). *The ecology of the forest/savanna boundary – Proceedings of the IGU Humid Tropics Commission Symposium, Venezuela, 1964*, eds T.L. Hills and R.E. Randall, pp. 45–49. McGill University, Montreal.

Dice, L.R. (1945). Measures of the amount of ecological association between species. *Ecology*, 26, 297–302.
Eiten, G. (1972). The cerrado vegetation of Brazil. *Botanical Review*, 38, 201–341.
Furley, P.A. (1992). Edaphic changes at the forest–savanna boundary with particular reference to the neotropics. In: *Nature and dynamics of forest–savanna boundaries*, eds P.A. Furley, J. Proctor and J.A. Ratter, pp. 91–117. Chapman & Hall, London.
Furley, P.A. and Ratter, J.A. (1988). Soil resources and plant communities of the central Brazilian cerrado and their development. *Journal of Biogeography*, 15, 97–108.
Furley, P.A. and Ratter, J.A. (1990). Pedological and botanical variations across the forest–savanna transition on Maracá Island. *Geographical Journal*, 156, 251–266.
Furley, P.A. and Ratter, J.A. (1994). Soil and plant changes at the forest–savanna boundary on Maracá Island. In: *The rainforest edge – Plant and soil ecology of Maracá Island, Brazil*, ed. J.H. Hemming, pp. 92–114. Manchester University Press, Manchester.
Furley, P.A., Ratter, J.A. and Gifford, D.R. (1988). Observations on the vegetation of eastern Mato Grosso, III. Woody vegetation and soils of the Morro de Fumaça, Torixoreu, Brazil. *Proceedings of the Royal Society of London* B, 235, 259–280.
Furley, P.A., Dargie, T.C.D. and Place, C.J. (1994). Remote sensing and the establishment of a geographic information system for resource management on and around Maracá Island. In: *The rainforest edge – Plant and soil ecology of Maracá Island, Brazil*, ed. J.H. Hemming, pp. 115–133. Manchester University Press, Manchester.
Gentry, A.H. (1988). Tree species richness of upper Amazonian forests. *Proceedings of the National Academy of Sciences USA*, 85, 156–159.
Hemming, J.H. and Ratter, J.A. (1993). *Maracá: rainforest island*. Macmillan, London.
Hemming, J.H., Ratter, J.A. and Santos, A. dos (1988). *Maracá*. Empresa das Artes, São Paulo.
Lewis, G.P. and Owen, P.E. (1989). *Legumes of the Ilha de Maracá*. Royal Botanic Gardens, Kew.
McGill University. Savanna Research Project (1966). *South American savannas; comparative studies, Llanos and Guyana*. Technical Report, 5.
McGregor, D.F.M. and Eden, M.J. (1991). Geomorphology and land development in the Maracá area of Northern Roraima, Brazil. *Acta Amazonica*, 21, 391–407.
Mera, P.A.S. and Oliveira, A.N.N. de (1989). Estudo dos fatores ecológicos, variações estacionais e sazonais das populações fitoplanctônicas e macrófitas aquáticas na Ilha de Maracá (RR.). In: *The vegetation of the Ilha de Maracá*, eds W. Milliken and J.A. Ratter, Appendix 6, p. 277. Royal Botanic Garden Edinburgh.
Milliken, W. and Ratter, J.A. (1989). *The vegetation of the Ilha de Maracá*. Royal Botanic Garden Edinburgh.
Moskovits, D.K. (1985). The behavior and ecology of two Amazonian tortoises, *Geochelone carbonaria* and *G. denticulata* in northwestern Brasil. Doctoral thesis, Div. of Biology, University of Chicago, Illinois, USA.
Mueller-Dombois, D. and Ellenberg, H. (1974). *Aims and methods of vegetation ecology*. John Wiley & Sons, New York.
Nascimento, M.T. (1994). A monodominant rainforest on Maracá Island, Roraima, Brazil. PhD thesis, University of Stirling.
Nascimento, M.T., Proctor, J. and Villela, D.M. (1997). Forest structure, floristic composition and soils of an Amazonian monodominant forest on Maracá Island, Roraima, Brazil. *Edinburgh Journal of Botany*, 54 (1), 1–38.
Pires, J.M. (1974). Tipos de vegetação da Amazônia. *Brasil Florestal*, 5 (17), 48–58.
Pires, J.M. and Prance, G.T. (1985). The vegetation types of the Brazilian Amazon. In: *Key environments: Amazonia*, eds G.T. Prance and T.E. Lovejoy, pp. 109–145. Pergamon Press, Oxford.
Proctor, J. (1994). The savannas of Maracá. In: *The rainforest edge – plant and soil ecology of Maracá Island, Brazil*, ed. J.H. Hemming, pp. 8–18. Manchester University Press, Manchester.

Ratter, J.A. (1985). *Notas sobre a vegetação da Fazenda Água Limpa (Brasília, DF, Brasil)*. Royal Botanic Garden, Edinburgh.
Ratter, J.A. (1986). Notas sobre a vegetação da Fazenda Água Limpa (Brasília, DF, Brasil). Univ. de Brasília, *Texto Universitário No. 3*. Brasília, Brazil.
Ratter, J.A. (1987). Notes on the vegetation of the Parque Nacional do Araguaia (Brazil). *Notes from the Royal Botanic Garden Edinburgh*, 44, 311–342.
Ratter, J.A. (1992). Transitions between cerrado and forest vegetation in Brazil. In: *Nature and dynamics of forest–savanna boundaries*, eds P.A. Furley, J. Proctor and J.A. Ratter, pp. 417–429. Chapman & Hall, London.
Ratter, J.A., Richards, P.W., Argent, G. and Gifford, D.R. (1973). Observations on the vegetation of northeastern Mato Grosso. 1. The woody vegetation types of the Xavantina–Cachimbo Expedition area. *Philosophical Transactions of the Royal Society, London* B, 266, 449–492.
Ratter, J.A., Askew, G.P., Montgomery, R.F. and Gifford, D.R. (1978a). Observations on the vegetation of northeastern Mato Grosso. 11. Forests and soils of the Rio Suiá–Missu area. *Proceedings of the Royal Society, London* B, 203, 191–208.
Ratter, J.A., Askew, G.P., Montgomery, R.F. and Gifford, D.R. (1978b). Observations on forests of some mesotrophic soils in central Brazil. *Revista brasileira de Botânica*, 1, 47–58.
Robison, D.M. and Nortcliff, S. (1991). Os solos da Reserva Ecológica de Maracá, Roraima: segunda aproximação. *Acta Amazonica*, 21, 409–424.
Robison, D.M. and Nortcliff, S. (1994). A tentative interpretation of the Quaternary geomorphology of Maracá Island, based on an analysis of soils developed on residua and drift deposits. In: *The rainforest edge – Plant and soil ecology of Maracá Island, Brazil*, ed. J.H. Hemming, pp. 158–172. Manchester University Press, Manchester.
Rodrigues, W.A. (1961). Aspectos fitossociológicos das catingas do Rio Negro. *Boletim Museu Paraense Emílio Goeldi, nova Série Botânica*. No. 15.
Ross, S.M., Luizão, F.J. and Luizão, R.C.C. (1992). Soil conditions and soil biology in different habitats across a forest–savanna boundary on Maracá Island, Roraima, Brazil. In: *Nature and dynamics of forest–savanna boundaries*, eds P.A. Furley, J. Proctor and J.A. Ratter, pp. 145–170. Chapman & Hall, London.
Sarmiento, G. (1992). A conceptual model relating environmental factors and vegetation formations in the lowlands of tropical South America. In: *Nature and dynamics of forest–savanna boundaries*, eds P.A. Furley, J. Proctor and J.A. Ratter, pp. 583–601. Chapman & Hall, London.
Silva, A.S.L. da, Lisboa, P.L.B. and Maciel, U.N. (1992). Diversidade florística e estrutura em floresta densa da bacia do Rio Juruá-AM. *Boletim do Museu Paraense Emílio Goeldi, série Botânica*, 8, 203–258.
Sørensen, T. (1948). A method of establishing groups of equal amplitude in plant sociology based on similarity of species content, and its application to analyses of the vegetation on Danish commons. *Det Kongelige Danske Videnskabernes Selskab Biologiske Skrifter*, 5, 1–34.
Steege, H. ter (1990). *A monograph of wallaba, mora and greenheart*. Tropenbos Technical Series 5. Tropenbos Foundation, Wageningen.
Takeuchi, M. (1960a). A estrutura da vegetação na Amazônia II – as savanas do norte da Amazonia. *Boletim do Museu Paraense Emílio Goeldi, nova série Botânica*. No. 7.
Takeuchi, M. (1960b). A estrutura da vegetação na Amazônia III – a mata de campina na região do Rio Negro. *Boletim do Museu Paraense Emílio Goeldi, nova série Botânica*. No. 8.
Thompson, J., Proctor, J., Viana, V., Milliken, W., Ratter, J.A. and Scott, D.A. (1992a). Ecological studies on a lowland evergreen rainforest on Maracá Island, Roraima, Brazil. 1. Physical environment, forest structure and leaf chemistry. *Journal of Ecology*, 80, 689–703.
Thompson, J., Proctor, J., Viana, V., Ratter, J.A. and Scott, D.A. (1992b). The forest–savanna boundary on Maracá Island, Roraima, Brazil: An investigation of two contrasting transects. In: *Nature and dynamics of forest–savanna boundaries*, eds P.A. Furley, J. Proctor and J.A. Ratter, pp. 367–392. Chapman & Hall, London.

Thompson, J., Proctor, J. and Scott, D.A. (1994). A semi-evergreen forest on Maracá Island 1: Physical environment, forest structure and floristics. In: *The rainforest edge – Plant and soil ecology of Maracá Island, Brazil*, ed. J.H. Hemming, pp. 19–29. Manchester University Press, Manchester.

Veloso, H.P. (1966). *Atlas florestal do Brasil*. Ministério da Agricultura, Rio de Janeiro.

Volkmer-Ribeiro, C., Mansur, M.C.D., Ross, S.M. and Mera, P.A.S. (1989). Biological indicators of the quality of water on the Island of Maracá, Roraima, Brazil. In: *Maracá Rainforest Project, invertebrates and limnology, preliminary report*, eds J.A. Ratter and W. Milliken, pp. 65–74. Royal Botanic Garden Edinburgh.

# 6 Further Contributions to the Botany of the Ilha de Maracá

## The pteridophytes of the Ilha de Maracá

PETER J. EDWARDS

*Royal Botanic Gardens, Kew, Richmond, UK*

### SUMMARY

Fifty species of pteridophytes were recorded on the Ilha de Maracá during the project. Thirty-two were apparently new species records for Roraima State of which nine were also new for Brazilian Amazonia. This represents an increase in the known pteridophyte flora of Roraima State and Brazilian Amazonia of 36% and 3.2% respectively. The records are given in the form of an annotated list, and are placed in the wider context of Brazilian Amazonia. Some ecological and phytogeographic aspects are considered. Knowledge of the pteridophyte flora of the state is still at a rudimentary stage. It seems that the pteridophyte flora of Roraima State will prove to be much larger than is currently known, even for the lowlands.

### RESUMO

Coleções e observações de pteridófitas feitas na Ilha de Maracá durante o projeto resultaram na documentação de 50 espécies. Aparentemente 32 foram registradas pela primeira vez no Estado de Roraima, e nove representaram ocorrências novas para a Amazônia Brasileira. Isto representou um aumento na flora pteridofítica de Roraima e da Amazônia Brasileira de 36% e 3.2%, respectivamente. Os registros são apresentados na forma de uma lista anotada, colocando-os no contexto mais amplo da Amazônia Brasileira. Aspectos ecológicos e fitogeográficos também são considerados. Foi concluído que o conhecimento da flora pteridofítica do estado estava num estágio de conhecimento muito rudimentar, e que esta flora vai se mostrar muito maior do que era conhecida até agora, mesmo nas planícies.

### INTRODUCTION

Lowland Brazilian Amazonia as defined here (the land below *c.* 500 m within the states of Amapá, Pará, Amazonas, Roraima, Rondônia and Acre) is a huge region covering approximately 3 108 000 km², with a very low pteridophyte diversity and

---

*Maracá: The Biodiversity and Environment of an Amazonian Rainforest.*
Edited by William Milliken and James A. Ratter. © 1998 John Wiley & Sons Ltd.

exceptionally low endemism. Tryon and Conant (1975) listed just 279 species (283 taxa if sub-species are included) and four endemics with no endemic Filices. This is in marked contrast to the high endemism in many flowering plant genera. Tryon (1986) attributed these very low figures to the poorly developed ecological diversity of most of the Amazonian region, 'with its few broad and often continuous habitats providing suitable conditions for relatively few species and little opportunity for speciation'. The species known on Maracá are, indeed, mostly widely distributed in Tropical America.

However, Tryon and Conant's paper was written at a time when exploration of much of Brazilian Amazonia was in its infancy, not least because of the difficulty of access. For Acre, Rondônia and Roraima this was particularly true, perhaps more especially so for pteridophytes, which often need specialist collectors if their representation in the flora of an area is to be accurately represented.

The pteridophyte collections made as part of the biological inventory of the Ilha de Maracá between 1986 and 1988, as part of Projeto Maracá, consisted of just 80 numbers. Despite this small total, these specimens accounted for 48 species (a further two distinctive species were identified by sight although not collected), of which 32 represented additional species for Roraima – a 36% increase on the 91 species already listed for the state in Tryon and Conant (1975). Nine of the species were also apparently new for lowland Brazilian Amazonia, representing a 3.2% increase on the 1975 list.

These are considerable increases when one considers that collections were only made in *c.* 10 000 ha of the island's 110 000 ha. On the other hand, 45 species were collected on the *c.* 1000 ha of lowland rainforest of the Reserva Ducke near Manaus, but this reserve has been much more intensively studied than Maracá. Experience in wet tropical forests elsewhere generally shows that the number of pteridophyte species constitutes about 5–10% of the total vascular plant flora, and the figure for Maracá is about 6%. One advantage of the general inventory approach is that sterile material is collected, as 'vouchers', whereas it would usually be avoided by collectors. Several of the species recorded are represented only by such sterile voucher specimens, e.g. *Polybotrya osmundacea*.

Does the ecological diversity within Maracá (see Table 6.1) help to explain its apparent pteridophyte diversity? The range of physiographic features present includes, for example, granitic outcrops in the forests to 15 m high (see Martini, Chapter 2). These were used by seven species of pteridophytes (in four families), six of which were not found in any other habitat. Five species, including *Anetium citrifolium*, *Hemionitis palmata*, *Gymnopteris rufa* and *Campyloneurum latum*, were new for Roraima State. All of these could reasonably have been expected to occur in wetter parts of the island, and this proved to be the case. Also on granitic rock outcrops (in rapids at the eastern end of the island) was *Doryopteris collina*, a species which Tryon (1942, 1972) indicates as one of the 16% of homosporous ferns that show the affinity between the fern floras of the 'Guayanan secondary centre' and the 'Brazilian primary centre' of Tryon (1972), which are separated by a distance of around 2900 km.

Maracá is in fact almost in the centre of Maguire's geophysical province of 'Guayana', and just south of the superimposed Roraima sediments (at the *meseta* of Tepequém). The granitic boulder habitat produced a new species for Roraima State which was also new for Brazilian Amazonia. It was a *Pecluma* (Figure 6.1) which provides a further possible link with the Brazilian primary centre of fern diversity. It most closely matches *P. recurvata* (Kaulf.) M.G. Price. These two finds also help to support the phytogeographic relationship between the northern portion of the Amazon basin, where sandstone and sandy soils occur, and the limited disjunct ones within southeastern Brazil (Tryon and Tryon, 1982). Such rock outcrops are rare in lowland Brazilian Amazonia. Through their special environmental conditions these outcrops provide a valuable fern habitat which is characteristic of the northern rocky parts of Roraima State and largely lacking from most of lowland Brazilian Amazonia. Maracá (and Roraima State generally) has little of the annually inundated lands that are probably partially responsible for the paucity of terrestrial pteridophytes in lowland Brazilian Amazonia (Tryon, 1985).

Even when we look at the commonest habitat on Maracá (and in Brazilian Amazonia), i.e. that of *terra firme* forest, out of the total of 35 species that are not lithophytes, 15 were new for Roraima State, of which five were also new for lowland Brazilian Amazonia. *Tectaria plantaginea* and *Adiantopsis radiata* are both new for Brazilian Amazonia, and both are conspicuous and would draw the attention even of the general collector elsewhere in this huge region. *Selaginella umbrosa*, a large species with red main axes, is also new for the region. One can predict that thelypteroid ferns will be overlooked, except by specialists, so the record of *Thelypteris tetragona* as new for Brazilian Amazonia is no surprise. Nevertheless it increases the total of this large and natural family for the region to (only) 13, which must represent considerable under-recording.

Just west of Maracá six more pteridophyte collections were made, of which three were new for Roraima State. All three were common neotropical species.

All the above finds lead me to the obvious conclusion that the pteridophytes of lowland Brazilian Amazonia, and more especially Roraima State, are very poorly known. This is yet further attested to by an August 1987 collection by G.P. Horton (of only 14 numbers) made in the far south of Roraima State, close to the Rio Negro and in a readily accessible area of forest. Of these 14 collections, four proved to be first records for the State. The species were *Trichomanes tanaicum* Sturm, *T. ankersii* Parker ex Hook. and Grev., *Hecistopteris pumila* (Spreng.) J. Sm. and *Triplophyllum funestum* (Kunze) Holttum var. *funestum*. The last named, despite being an attractive, conspicuous, and eminently collectable fern, proved also to be the first published record for Brazilian Amazonia. This collection would have had six 'firsts' for the State if it had not been anticipated by the Maracá records (made in March of the same year) of *Pityrogramma calomelanos* and *Nephrolepis biserrata* – two very common and obvious ferns with huge ranges – which one would expect to be gathered even by the general collector.

The statements above regarding species being 'new' for Roraima State and lowland Brazilian Amazonia should be taken with some qualification. They are 'new' in the context of the species and totals in the Tryon and Conant (1975) paper, consolidated by the addition of what few other published records I have discovered, these being Windisch (1978), Alston *et al.* (1981), Nauman (1985) and

Moran (1987). For the purposes of updated totals I have also included the new records in the collections of Milliken *et al.* from just west of Maracá, and those from the south of the state by Horton. Together these only raise the total for the state by seven and for the rest of lowland Brazilian Amazonia by one. The published total number of pteridophyte species for Roraima State is now raised to 129, of which 116 were homosporous ferns, an increase of 30% and 35% on the previous totals of 101 and 86. For lowland Brazilian Amazonia the published pteridophyte total now stands at 304, of which 255 were homosporous ferns. (Compare with the consolidated totals 295 (280 in Tryon and Conant) and 246 (236 in Tryon and Conant).)

Herbarium searches would no doubt add another 20 or more species, as indicated by a brief survey at Kew. A computer-printed list by Hopkins (1986) includes three further records, but these have not been included as the plants were from 1200 m on Serra Tepequém, which does not fall within the sub-500 m definition of lowland Brazilian Amazonia used here. If records from the Guayana highland parts of Roraima State were available, the species totals for the State would rise dramatically. As far as I can ascertain no expeditions have visited this remote area. Even under the ambitious Projeto Flora Amazonica only one expedition had by 1984 penetrated near there, and very few pteridophytes were collected (see Prance and Nelson, 1984).

The flora of those Roraima sandstone sediments is much richer than that of lowland Brazilian Amazonia. They have much of 'the greater ecological diversity developed in a mozaic found in wet tropical mountain regions' (Tryon, 1986), which is so important for pteridophyte persistence and speciation. Collections made within Guyana on the Kew Expedition to Mt Roraima in 1978 (Holttum and Edwards, 1983) yielded over 100 species of pteridophytes (nearly all homosporous ferns) during just seven days in an area of no more than 0.5 km$^2$ at *c.* 1300 m (generally recognized as the optimum altitude for pteridophyte diversity) which lies no more than 1.5 km from the Brazil–Venezuela frontier. Only *c.* 10% of those species figure on the current Roraima State list. This gives some indication of the number of pteridophytes likely to occur in Roraima State, which I tentatively estimate will prove to be between 300 and 350, close to that of Surinam (Kramer, 1978). In fact many species of lowland Venezuela and the Guianas are likely to be on Maracá. So far this riverine island has given a tantalizing glimpse of the pteridophyte flora of the region, with its few Andean and Guayanan–Brazilian species (*Doryopteris collina*, *Hemidictyum marginatum*, *Ctenitis refulgens*, *Pteris pungens*, *Pecluma* sp. aff. *recurvata*) amongst its other mostly widely ranging tropical American species.

# SOME ECOLOGICAL CONJECTURES

The western half of the island, and more especially the western end, has a much more diverse pteridophyte flora than in the east. This is reflected in the flowering plants also. The western end almost certainly receives a higher and more even

Table 6.1. Diversity of pteridophytes in habitats on the Ilha de Maracá

| Habitat | No. species |
|---|---|
| Aquatic riverine (facultative rheophytes) | 3 |
| *Buritizal* (*Mauritia flexuosa* swamp) | 7 |
| Seasonally inundated grassy savanna | 1 |
| Savanna with *Curatella americana* trees | 2 |
| Forest | 34 |
| Granitic outcrops in forest | 8 |
| Marginal vegetation (excluding riverside) | 1 |
| Riverine/riverside vegetation | 12 |
| Secondary vegetation | 2 |
| *Vazante* vegetation (seasonally inundated herbaceous) | 2 |
| Epiphytic (total and partial epiphytes) | 11 |
| Epilithic | 9 |

rainfall and is set amongst tropical evergreen forest in all directions. The generally damper climate, coupled with the microclimates and substrates provided on and amongst dark damp stream gullies and rock outcrops, plus the sharply peaked hills (to 350 m) separated by steep-sided valleys with small streams, provide an almost ideal range of environmental conditions, at least within the context of lowland tropical forest – a more favourable habitat for pteridophytes than most of lowland Brazilian Amazonia. Unable to visit these fascinating areas myself, I can only speculate upon what would be present but still undetected: mixed populations of *Selaginella*, difficult-to-reach trunk epiphytes including *Hymenophyllum* and hemiepiphytic *Trichomanes* species. One or two *Elaphoglossum* species (as branch epiphytes) also seem likely candidates, while *Lindsaea* and *Asplenium* are both likely as singletons or in discrete colonies. Certainly more *Cyathea* species will be found, as well as interesting *Doryopteris* Guayanan–Brazilian disjuncts, and so on.

Research carried out (and still progressing) on Maracá indicates that forest is expanding at the expense of *campo*/savanna (see Milliken and Ratter, Chapter 5). These environmental changes may have a long history in the area of Maracá and make this beautiful biological reserve a particularly fruitful site for investigating and assessing the role of such changes in determining the range of species and their genetic variability, and hence capacity and opportunity to speciate. Some of the 'Guayanan' and Amazon basin species may be on the edge of their geographic range in the vicinity of Maracá. The occurrence at the eastern end of the island of partly desiccated sporelings of species not encountered there perhaps provides evidence for this. For instance, some of these sporelings were definitely of *Asplenium* species, a genus known only as adult sporophytes at the west of the island.

## ANNOTATED CHECKLIST

The previously known distribution within Brazilian Amazonia (*sensu* Tryon and Conant, 1975) is given at the end of, or within, each species account. The total distribution, obtained from a wide variety of sources, is given (inset) below this. Locations are shown in Figure Pr. 1. Collector, habit and habitat codes are as follows:

*Collectors*: BLS, Stannard; H, Harley; HO, Hopkins; L, Lewis; M, Milliken; PJE, Edwards; R, Ratter.

*Habits*: E, epiphyte; H, herb; L, epilithic; Ts, small tree; V, vine.

*Habitats*: AQR, aquatic riverine; B, *buritizal*; CA, seasonally inundated grassy savanna; CAC savanna with *Curatella americana*; F, forest; G, granitic outcrops in forest; M, marginal vegetation; R, riverine vegetation; Se, secondary vegetation; Va, *vazante* (see Milliken and Ratter, Chapter 5).

Commonly used synonyms are given in parentheses. Plants are terrestrial unless otherwise stated. The families and genera and sequence thereof are as currently used in the Kew herbarium.

§ = first (published) record for Brazilian Amazonia.
\* = first (published) record for Roraima State.

### LYCOPSIDA

**Lycopodiaceae**

***Huperzia dichotoma*** (Jacq.) Trevis. (*Lycopodium dichotomum* Jacq.)     E;F*

Epiphyte high in trees, in seasonal evergreen forest on a riverine island, east end of Maracá. This species is little collected in Brazil but appears to be widespread. (M84).

  Amapá, Pará, Rondônia.
  Florida, Mexico, Panama, Brazil, West Indies.

**Selaginellaceae**

***Selaginella umbrosa*** Lem. ex Hieron. (*S. erythropus* (Mart.) Spring var. *major* Spring.)     H;R§

Common on sandy alluvium near or on river and stream banks. Perhaps the commonest pteridophyte in this habitat. PJE2629A was on the top and sides of a vertical stream bank with *Tectaria incisa* and *Trichomanes pinnatum*, forming a dense pteridophyte-dominated colony which was apparently sustained by recruitment of sporelings of the last two species within the supportive matrix of the *Selaginella*, which seems capable of retaining a firm hold on the bank whatever the conditions. (M195, BLS766, PJE2629A).

  This large and beautiful plant with ruby red to pale pink main axes was hitherto known from Mexico to Colombia and Venezuela. Being of such striking appearance it is unlikely to be overlooked by collectors. The closely related *S. erythropus* is recorded by Tryon and Conant (1975) only from Roraima State in Brazilian Amazonia. Sandy alluvium was seen to wash straight through colonies of this species, with apparently little accumulating within the colony. This is presumably a flooded forest/streamside adaptation.

PTERIDOPHYTES

*Selaginella seemanii* Baker　　　　　　　　　　　　　　　　　　　　H;B§

Occasional in the seasonally flooded and rather open-canopied vegetation dominated by the *buriti* palm *Mauritia flexuosa*. Horizontal in muddy channels and amongst *Olyra longifolia*. Easily overlooked and probably widespread on the island on mud. (PJE2541).

  Costa Rica south to Peru, Surinam, French Guiana.

*Selaginella brevifolia* Baker (*Selaginella cryptogaea* Baker)　　　　　　H;CAC*

Apparently rare, and even more likely to be overlooked than the previous species! Horizontal on very dry soil near *Curatella americana* shrubs in a small open *campo*. In Alston *et al.* (1981) it was recorded in Brazil only for western Amazonas, but recent collections (at BM) from Pará and even as far afield as Bahia show that this species has an extensive distribution. (PJE2489).

  Colombia, Venezuela, Peru, Brazil.

## FILICOPSIDA

### Ophioglossaceae

*Ophioglossum ellipticum* Hook. & Grev.　　　　　　　　　　　　　　H;R*

On a rocky island in the Furo de Maracá, growing in a damp sandy/mossy depression in the rocks. The smallest fern recorded on Maracá, with sterile lamina 4–5 mm × 1.5–3.5 mm, fertile spike 40–65 mm. (M433).

  Pará, Amazonas.
  Southern Mexico south to Bolivia, the Guianas.

### Schizaeaceae

*Anemia oblongifolia* (Cav.) Sw.　　　　　　　　　　　　　　　　　　H;F§

On a vertical sandy face of a small rocky island in the Furo de Maracá. (M412).

  Mexico south to Bolivia, The Guianas, Goiás, Minas Gerais.

*Lygodium venustum* Sw. (*L. polymorphum* (Cav.) Kunth)　　　　　V;CAC,F,Se

Common climber in *campo*, forest and in the secondary vegetation of the road to the Ecological Station. Sporelings common in the latter habitat. (PJE2480).

  Amapá, Pará, Amazonas, Rondônia.
    Tropical America.

*Lygodium volubile* Sw.　　　　　　　　　　　　　　　　　　　　　V;Va,B

Apparently restricted to *buritizal*, where it is often common and climbs to 4 m and more. Often sterile in lower and accessible parts, but nevertheless readily identified. (PJE2543).

  Amapá, Pará, Amazonas, Rondônia.
    Southern Mexico south to Argentina and Brazil, Greater Antilles.

### Pteridaceae

***Pteris pungens*** Willd.   H;F*

One collection, from *terra firme* forest, adjacent to a stream. Previously recorded in Brazilian Amazonia only from Acre. Frequent in the Guayanan regions of Venezuela and the Guianas and considered by Tryon and Conant (1975) as one of the 15 'Andean elements' of the Brazilian Amazonian pteridophyte flora. (M541).

Mexico south to Peru, West Indies.

NB The very different *Pteris pearcei* Baker could be expected on the island.

### Adiantaceae

***Pityrogramma calomelanos*** (L.) Link   H;R*

Seen on several almost vertical banks of the Furo de Santa Rosa by PJE but collection not possible. This fills the 'Roraima gap' for this species, which is recorded for most of tropical and subtropical America. Widely naturalized in the Old World.

***Hemionitis palmata*** L.   H(L);F,G*

A widespread tropical American species only recorded on rocks in forest on Maracá. The epilithic situation is the normal (and perhaps obligatory) habitat throughout its range. (M504).

Pará.
Mexico south to Bolivia, southern Brazil, West Indies.

***Gymnopteris rufa*** (L.) Bernh. ex Underw. (*Hemionitis rufa* (L.) Swartz)   H(L);F,R,G*

On rocks and elevated ground close to the river. Another widespread and conspicuous tropical American species that is greatly under-recorded in Brazilian Amazonia. (M124).

Amazonas.
Mexico south to Peru, Surinam, Brazil, West Indies.

***Doryopteris collina*** (Raddi) J. Sm.   H(L);R,G

In dark crevices in a granitic rock outcrop situated on an island in the rapids of the Furo de Maracá (east of the island). Only recorded from Roraima State in Brazilian Amazonia, which may reflect its true distribution, for this is one of the 16% of species that the Guayanan and Brazilian regional centres of speciation have in common (Tryon, 1972). (M396).

Guyana, Surinam, southern Brazil, Paraguay.

***Adiantopsis radiata*** (L.) Fée   H;F§

Terrestrial in forest near Ecological Station. A widespread and often common tropical American species, but nevertheless this is the first published record for Brazilian Amazonia. (H0661).

Tropical America to northern Argentina.

*Adiantum latifolium* Lam. *s.l.*   H;F

Common in *terra firme* forest at the east of the island. An archetypal Amazon basin fern which is almost certainly a complex of species. Considerable variation is present on Maracá, apparently even within the same colony. A glaucous undersurface to the pinnules was barely discernible on many fronds. (PJE2466).

Pará, Amazonas, Rondônia, Acre.
Tropical America.

*Adiantum lucidum* (Cav.) Sw.   H;F*

Uncommon in *terra firme* forest near the Station. (PJE2495).

Amapá (Nauman, 1985).
Central and South America to Peru, Trinidad and Tobago.

*Adiantum petiolatum* Desv.   H;F*

In *terra firme* forest at east end of the island. These collections fill the 'Roraima gap' as the species is recorded for all the other Brazilian Amazonian States. (PJE2467).

Tropical America.

*Adiantum pulverulentum* L.   H;F

In *terra firme* forest including the open-canopied *Peltogyne gracilipes* forest. (PJE2462).

Pará, Amazonas, Rondônia, Acre.
Tropical America.

*Adiantum serratodentatum* Willd.   H;F,M,Se*

This is the *Adiantum* with the widest ecological range on Maracá, occurring on the road bank (causeway), at the junction of low forest and grassy *campo* and in dense *terra firme* forest. (PJE2487).

Amapá, Pará, Amazonas.
Panama south to Bolivia, the Guianas, Trinidad.

*Adiantum terminatum* Kunze ex Miq.   H;CA

The common fern of the seasonally flooded *campo* southwest of the Ecological Station, but mostly occurring on slightly drier margins with *Curatella americana*. (PJE2552).

Amapá, Pará.
Central America.

**Vittariaceae**

*Ananthocorus angustifolius* (Sw.) Underw. & Maxon (*Vittaria costata* Kunze)   H(L);R

On rocks beside Cachoeira de Purumame at the far western end of the island. Usually recorded as an epiphyte throughout its range, but epiphytes of riverine forests can and often do have the facility to grow on adjacent rocks, and for a short time at least live as facultative rheophytes. Surprisingly, no Amazonas State records exist. (M561).

Amapá, Pará, Rondônia, Acre.
Central and tropical South America.

*Vittaria lineata* (L.) Sm.  E;F*

In moss on the trunk of a *Maximiliana maripa* palm. (R6253). This record for Roraima State completes the representation for all the Brazilian Amazonian states.

Georgia to Uruguay and northeast Argentina.

*Anetium citrifolium* (L.) Splitg.  H(L);R,G*

Recorded only from a boulder in riverine forest, but doubtless it also occurs in its typical niche as a trunk epiphyte. A monotypic genus. The species often persists as juvenile plants in low density. (M500).

Amapá, Amazonas, Acre.
Central and South America to Bolivia, West Indies.

### Hymenophyllaceae

*Trichomanes diversifrons* (Bory.) Mett. ex Sadeb.  H;AQR,F*

On banks of a stream, a facultative rheophyte. Doubtfully distinct from *Trichomanes trollii* Bergdolt. (M647).

Amapá, Amazonas.
Central America to Venezuela, Amazonian Brazil, the Guianas.

*Trichomanes elegans* L.C. Rich. (*Trichomanes prieurii* Kunze)  H;F

On stream banks in forest. One of the largest species of *Trichomanes*, with fronds 30 × 30 cm or more. (M559).

Amapá, Amazonas, Roraima, Acre.
Costa Rica through Tropical South America to southern Brazil, Lesser Antilles.

*Trichomanes pinnatum* Hedwig  H;F,AQR*

Common on stream banks in forest. At least on Maracá, this species behaves as a facultative rheophyte. See comments under *Selaginella umbrosa* (above). Now recorded for all the Brazilian Amazonian states. (PJE2630).

Mexico, tropical South America, Lesser Antilles.

### Cyatheaceae

*Cyathea oblonga* (Klotzsch) Domin (*Trichipteris procera* (Willd.) Tryon *s.l.* Barrington, 1978)  Ts;F§

In a damp low area (*baixada*) with *Socratea exorrhiza*, close to a stream. This collection, with its pale brown to chestnut bullate scales on the lower surface of the costules, and with some veins forking at the sori, closely matches material from 1400 m on the north side of Mt Roraima, Guyana (see Holttum and Edwards, 1983). (M532).

Distribution of *C. oblonga* is essentially the Guayana Highland of Venezuela and the Guianas, effectively a 'Guayanan endemic'. Distribution of *Trichipteris procera s.l.* (fide Barrington 1978): from the West Indies to Venezuela, the Guianas, Central Mato Grosso, Amazonas, Goiás, Brazilian Highlands, Andes south to Bolivia.

## Metaxyaceae

***Metaxya rostrata*** (Willd.) Presl. H(incl. L);F,R,G

M543 was in *terra firme* forest on a hillside slope near a stream, a typical habitat for this monotypic genus. The one-pinnate pinnatisect sporelings were seen on river rocks at several sites along the Furo de Santa Rosa. It is conceivable that these small plants are occasionally washed onto suitable permanent sites. M249, from just west of Maracá at the Cajumá rapids, was common on rocks in forest close to the river, giving some credibility to the above theory. (M534).

Central and South America to Bolivia and French Guiana, Lesser Antilles.

## Dennstaedtiaceae

***Saccoloma inaequale*** (Kunze) Mett. H;F

On banks of streams in *terra firme* forest at the far west end of the island. (M560).

Pará, Amazonas, Roraima, Acre.
Central and South America, West Indies.

***Lindsaea portoricensis*** Desv. H;B*

Common in *buritizal* habitats at the east end of the island, mostly growing at the bases of *Mauritia flexuosa* palms. (M74).

Other *Lindsaea* species are doubtless present on Maracá (there are 22 species known for Brazilian Amazonia, making it the fourth largest genus), but plants of many species often occur singly and widely separated, and are therefore often overlooked.

Amazonas, Rondônia.
Mexico south to Bolivia, the Guianas, Trinidad, Greater Antilles.

## Woodsiaceae

***Hemidictyum marginatum*** (L.) C. Presl. (*Diplazium marginatum* (L.) Diels) H;R*

A remarkable large handsome fern with a disposition so characteristic that I had no trouble identifying it from 10 to 50 m in a moving boat. Some plants along the margins of the Furo de Santa Rosa were at least 2 m tall (PJE sight records only).

This was considered as one of the 15 'Andean elements' in the Brazilian Amazonian pteridophyte flora by Tryon and Conant (1975), and together with *Pteris pungens* (also on Maracá) was at that time only known from Acre State.

Acre.
Central and South America, West Indies.

## Dryopteridaceae

***Ctenitis refulgens*** (Kl. ex Mett.) C. Chr. ex Vareschi H;F*

In damp low areas (*baixadas*) with *Socratea exorrhiza* palms close to streams and in drier soils in forest, including near the Station. (M122).

Amapá (Nauman, 1985), Amazonas.
Panama to Colombia and Venezuela, Surinam and French Guiana.

*Polybotrya osmundacea* Willd. H–E;F*

A hemi-epiphyte in *terra firme* forest near the Station. The sterile specimen from *terra firme* forest near the Station constitutes the first record for Roraima State, the nearest previous record being from the Venezuela–Brazilian border in Venezuelan Amazonas 400 km to the west (Moran, 1987). (PJE sn 20/3/1987 sterile voucher).

Pará, Amazonas.
Honduras south to Bolivia, Venezuela, the Guianas.

*Tectaria incisa* Cav. (*Tectaria martinicensis* (Spreng.) Copel.) H;F

Perhaps the commonest, or at least the most conspicuous fern in *terra firme* forest in the north and west of the island. Recorded from all Brazilian Amazonian states. (M126).

Mexico south to Bolivia and Brazil, West Indies.

*Tectaria plantaginea* (Jacq.) Maxon H;F,AQR§

On damp rocks by streams in the far west of the island. A *Tectaria* which is certainly a facultative rheophyte, its linear–lanceolate simple fronds and tenaciously clinging rhizome fitting it well to that role. (M542).

Central America to Peru and Brazil, West Indies.

**Oleandraceae**

*Nephrolepis biserrata* (Sw.) Schott E;F*

Epiphytic in crowns of *Maximiliana maripa* palms. *Terra firme* forest near Station. (M98).

Amapá, Pará, Amazonas.
Pantropical.

**Adiantaceae**

*Lomariopsis japurensis* (Mart.) J. Sm. V;F*

Scandent climber to 2 m. *Terra firme* forest in west of the island. (M518).

Amapá, Amazonas.
Mexico south to Bolivia, Guyana, Trinidad, Brazil.

*Asplenium serratum* L. H(L);F,G

On boulders on low hill in *terra firme* forest in the west of the island. (M509).

Common throughout Brazilian Amazonia.
Mexico south to northern Argentina, Paráguay, the Guianas, Brazil, West Indies, Florida.

**Blechnaceae**

*Blechnum serrulatum* Rich. H (incl. E);B

On damp soil and on pneumatophores of *Mauritia flexuosa* in *buritizal* habitat in the east of the island. In dense large stands. (M56).

Mexico south to northern Argentina, the Guianas, Brazil, West Indies, Florida. A common swamp fern of low altitudes.

PTERIDOPHYTES 125

**Thelypteridaceae**

*Cyclosorus interruptus* (Willd.) H. Itô (*Thelypteris interrupta* (Willd.) K. Iwats., *Thelypteris totta* (Thunb.) Schelpe, *Cyclosorus gongylodes* Schkuhr)  H;Va*

Forming part of a floating mat of vegetation in open swampland. (R6202).

Amapá, Pará, Amazonas.
A highly variable species found in marshes throughout the tropics of the New and Old World.

*Goniopteris abrupta* (Desv.)-(*Thelypteris abrupta* (Desv.) Proctor)  H;R*

Common in riverine forest, probably throughout the island. BLS829B is slightly atypical in the narrowness of the pinnae (BLS829A, BLS829B).

Pará.
Venezuela, the Guianas, Trinidad, East and Central Brazil, West Indies.

*Goniopteris tetragona* (Sw.) C. Presl (*Thelypteris tetragona* (Sw.) Small)  H;F§

In *terra firme* forest close to the river in the southeast of the island. (M332).

Apparently the first published record for Amazonian Brazil.
Mexico south to Ecuador, the Guianas, Trinidad, West Indies, Florida.

There appear to be no previous published records of Thelypteridaceae in Roraima State. Only 12 species were recorded for the whole of Brazilian Amazonia in Tryon and Conant (1975). This collection brings the total to 13, and doubtless many more species remain to be discovered. This is a large family, some members of which will be colonizing much of the recently created roadside and deforested land in Amazonia.

**Polypodiaceae**

*Campyloneurum phyllitidis* (L.) C. Presl (*Polypodium phyllitidis* L.)  E;F

In *terra firme* forest, common as a trunk epiphyte on moss-covered small trees. Roots forming a dense mass to 40 cm across. (PJE2633).

Amapá, Pará, Amazonas, Roraima, Acre.
Mexico south to Bolivia, Venzuela, the Guianas, Central Brazil, West Indies, Florida.

*Campyloneurum latum* T. Moore (*Polypodium latum* (T. Moore) T. Moore ex Sodiro)  H(L);F,G*

On boulders in *terra firme* forest at the far west end of the island. (M506).

Rondônia (Lellinger, unpublished database, December 1987).
Mexico south to Bolivia, Venezuela, Rondônia, Trinidad, West Indies.

See Lellinger (1988) regarding the relationships between *C. latum* and *C. phyllitidis*.

*Campyloneurum repens* (Aubl.) C. Presl  E;F

Trunk epiphyte in *terra firme* forest at far west end of the island. (M671).

Amapá, Roraima, Acre.
Mexico south to Bolivia, Venezuela, the Guianas, Central Brazil, West Indies.

*Microgramma persicariifolia* (Schrad.) C. Presl (*Polypodium persicariifolium* Schrad.)  E;R,B,F

Common branch and trunk epiphyte on the island, behaving (on streamsides at least) as a hemi-epiphyte also. (PJE2610).

Recorded for all the Brazilian Amazonian states.
Northern South America, Trinidad.

*Pecluma plumula* (Humb. & Bonpl. ex Willd.) M.G. Price  E;R

Trunk epiphyte in riverside vegetation on small riverine islands in the Furo de Maracá at the east end of the island. (M437).

Amazonas, Roraima, Acre.
Mexico to Peru, Bolivia, West Indies, Florida.

*Pecluma* sp. aff. *recurvata* (Kaulf.) M.G. Price  H(L);F§

On large boulders in low seasonally flooded forest at the east end of island. Most closely matching *recurvata*, which is known only from southern Brazil. The most obvious differences from the southern Brazilian material is the reduction of the lowest segments. The indumentum is similar. *Pecluma* is in need of a thorough revision, many 'species' being poorly characterised. (M156 – See Figure 6.1).

*Pleopeltis macrocarpa* (Bory ex Willd.) Kaulf. *s.l.* (*Pleopeltis lanceolata* (L.) Kaulf., *Polypodium lanceolatum* L.)  E; F,B§

Trunk and branch epiphyte on trees in the experimental plots near the Station; recorded on *Pouteria hispida* Eyma agg., *Duguetia* sp. and *Mauritia flexuosa* palms. Commonly occurring in the vicinity but with only a few plants on each tree. Associated (on large branches of rougher-barked trees) with *Polypodium polypodioides*. (PJE2612).

A cytologically complex 'species' known from Tropical America, South Africa, Madagascar, Ceylon and India. The complex needs revision. The Maracá plants closely match a collection from Guyana (A.C. Smith 2250, Rupununi River (K)).

*Polypodium polypodioides* (L.) Watt.  E; B,F

A common trunk and branch epiphyte at the east end of the island, recorded from *Duguetia* sp., *Pouteria hispida* agg. and *Mauritia flexuosa*. Whole colonies were seen to fall off the trees in dry weather, and when exposed to full sunlight by felling of a few adjacent trees. Colonies on trees not so exposed remained attached. (PJE 2544).

Recorded from all Brazilian Amazonian states.
USA to northern Argentina, West Indies.

---

**Figure 6.1**. (opposite) *Pecluma* sp. nr *recurvata* (M156, Ilha de Maracá, Brazil)

(a) Portion of plant ($\times$ 0.4); (b) detail of base of a medial segment from a mid frond ($\times$ 11); (c) median segment ($\times$ 2); (d) sporangium and typical hair from receptacle ($\times$ 60); (e) a range of hairs and scales from the stipe and rachis of a young frond ($\times$ 60); (f) scale from the side of a rhizome, dorsal view ($\times$ 40); (g) scale from the side of a rhizome, ventral view; the hatched area is the point of attachment to the rhizome ($\times$ 40)

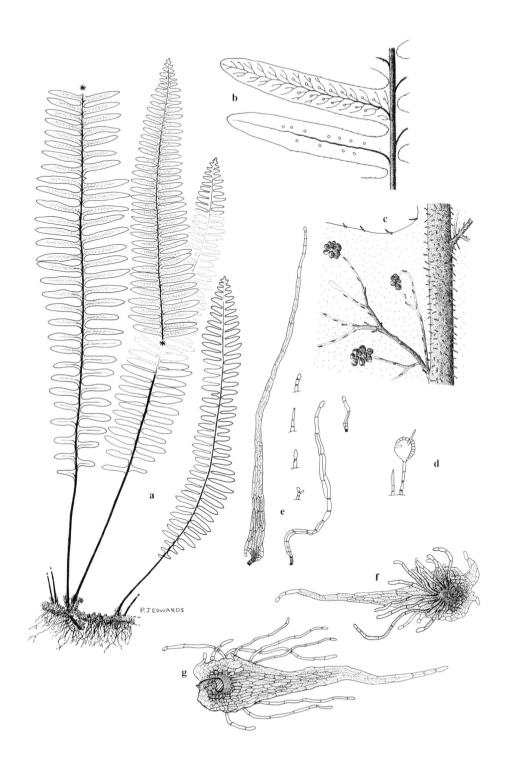

## SPECIES OCCURRING JUST WEST OF THE ISLAND (AND LIKELY TO BE FOUND ON THE ISLAND ITSELF)

These species are not included in Table 6.1, but the three new state records here were added to the totals against which Maracá records are compared.

### Cyatheaceae

*Cyathea microdonta* (Desv.) Domin. (*Trichipteris microdonta* (Desv.) R.M. Tryon     Ts;R*

A 2 m high spiny-trunked tree fern collected 120 km west of the western tip of the Ilha de Maracá. In riverine vegetation. (M278).

Amapá, Pará, Amazonas, Acre.
Central and tropical South America, Greater Antilles.

### Aspleniaceae

*Asplenium auritum* Sw. s.s.     E,L;F

On rocks and liana stems bordering streams in *terra firme* forest 90 km west of the Ilha de Maracá. (M282). The brown reniform spores and pinna characters place this most closely to *A. auritum*. As the segregate species of *auritum* are not adequately known, it is not possible to give a distribution list, but *auritum* s.s. is probably the common species of low altitudes in Central America to Venezuela, Guyana, Jamaica.

### Gleicheniaceae

*Dicranopteris pectinata* (Willd.) Underw.     H;R*

In riverine vegetation 120 km west of the Ilha de Maracá, almost entirely covering a steep section of bank and with thicket-forming trailing fronds up to 6 m long. This is a typical habitat for the species. It is likely to become one of the predominant colonizers of many severely disturbed areas. (M277).

Amapá, Pará, Amazonas.
Mexico south to Bolivia, the Guianas, Brazil, Trinidad, West Indies.

### Polypodiaceae

*Microgramma percussa* (Cav.) de la Sota (*Polypodium percussum* Cav.)     E;F

Epiphytic on large trees close to the river 180 km west of the Ilha de Maracá, at Cajumá rapids. Not a typical *Pleopeltis*, perhaps best placed in *Microgramma*. (M245).

Recorded for all Brazilian Amazonian states.
Southern Mexico, south to Bolivia, Brazil, the Guianas, West Indies.

*Niphidium crassifolium* (L.) Lellinger. (*Polypodium crassifolium* L.)     E;F

Epiphytic in forest close to the river 180 km west of the Ilha de Maracá, at Cajumá rapids. (M240).

Pará, Amazonas, Roraima, Rondônia.
Mexico south to Bolivia, Brazil, the Guianas, Trinidad, West Indies.

*Polypodium triseriale* Sw.  E;R*
Epiphyte at Igarapé Aracaça, 170 km west of the Ilha de Maracá. (M256).

Pará, Amazonas.
Mexico, south to Bolivia, the Guianas, Brazil, West Indies.

## ACKNOWLEDGEMENTS

Specialist determinations/confirmations were made by J.C. Camus (*Selaginella*), D.B. Lellinger (*Campyloneurum*), J.T. Mickel (*Anemia* and *Microgramma*), R. Moran (*Polybotrya*), B. Øllgaard (*Huperzia*), J. Prado (*Pteris*, *Doryopteris*), M.G. Price (*Pecluma*), A.R. Smith (Thelypteridaceae) and B. Zimmer (*Adiantum*), all of whom are gratefully acknowledged.

## REFERENCES

Alston, A.H.G., Jermy, A.C. and Rankin, J.M. (1981). The genus *Selaginella* in Tropical South America. *Bull. British Museum (Natural History), Botany Series*, 9 (4), 233–330.
Barrington, D.S. (1978). A revision of the genus *Trichipteris*. *Contrib. Gray Herb.*, 208, 3–93.
Holttum, R.E. and Edwards, P.J. (1983). The tree-ferns of Mount Roraima and neighbouring areas of the Guayana highlands with some comments on the family Cyatheaceae. *Kew Bull.*, 38(2), 155–188.
Hopkins, M. (1986). Unpublished list of collections from the Ilha de Maracá and nearby areas. New York Botanical Gardens.
Kramer, K.U. (1978). The pteridophytes of Suriname. *Natuurwetenschappelijke studiekring voor Suriname en de Nederlandse Antillen, Utrecht*, no. 93.
Lellinger, D.B. (1988). Some new species of *Campyloneurum* and a provisional key to the genus. *Amer. Fern. J.*, 78 (1), 14–35.
Moran, R.C. (1987). Monograph of the neotropical fern genus *Polybotrya* (Dryopteridaceae). *Illinois Natural History Survey Bulletin*, 34, Article 1.
Nauman, C.E. (1985). New pteridophyte records for the Territory of Amapá Brazil. *Acta Amazonica*, 15 (3–4), 303–305.
Prance, G.T. and Nelson, B.W. (1984). Projeto Flora Amazonica: eight years of binational botanical expeditions. *Acta Amazonica*, 14 Suppl., 5–29
Tryon, R.M. (1942). A revision of the genus *Doryopteris*. *Contrib. Gray Herb.*, 143, 1–80.
Tryon, R.M. (1972). Endemic areas and geographic speciation in tropical American ferns. *Biotropica*, 4 (3), 121–131.
Tryon, R.M. (1985). Fern speciation and biogeography. *Proc. Royal Soc. Edinburgh*, 86B, 353–360.
Tryon, R.M. (1986). The biogeography of species, with special reference to ferns. *Botanical Review*, 52 (2), 117–156.
Tryon, R. and Conant, D. (1975). The ferns of Brazilian Amazonia. *Acta Amazonica*, 5 (1), 23–34.
Tryon, R.M. and Tryon, A.F. (1982). *Ferns and allied plants, with special reference to Tropical America*. Springer, New York.
Windisch, P.G. (1978). Adições ao inventário das pteridófitas do Acre. *Bradea*, 5, 29–30.

# A small area of young secondary forest on the Ilha de Maracá: structure and floristics

## ROBERT P. MILLER[1] AND JOHN PROCTOR[2]

[1]University of Florida, Gainesville, USA; [2]University of Stirling, Stirling, UK

## SUMMARY

A survey of trees $\geq 10$ cm dbh was conducted in a small patch of secondary forest (0.25 ha) on the Ilha de Maracá. The forest was approximately 13 years old, and had regenerated after previous felling and burning. It contained 22 species, with an equivalent of 600 trees per hectare and a basal area of 16.4 $m^2$ $ha^{-1}$. When compared with the older forest in the near vicinity, this regrowth demonstrated significantly lower basal area and species diversity, and significantly higher tree density. Multi-trunked trees, resulting from coppice regeneration, were much more abundant in the regrowth than in the old forest. Considerable floristic differences were found between the two forest types, with an increased presence of pioneer species in the regrowth. The most abundant species in the plot, *Ocotea glomerata* (Lauraceae), *Cecropia* sp. (Moraceae) and *Apeiba schomburgkii* (Tiliaceae), are present but not common in the surrounding forest.

## RESUMO

Foi realizado um levantamento das árvores $\geq 10$ cm dap numa amostra de 0.25 ha de caapoeira (mata secundária) na Ilha de Maracá. A mata tinha aproximadamente 13 anos de idade, e a área havia sido queimada depois da derrubada original. Conteve 22 espécies, com um equivalente de 600 árvores por hectare e uma área basal de 16.4 $m^2$ $ha^{-1}$. Comparada com a floresta mais velha nas proximidades, esta caapoeira mostrou uma diversidade e área basal consideravelmente menores, e uma densidade maior. Arvores com troncos múltiplos, resultado de brotação de tocos ou de raízes, foram bem mais abundantes na caapoeira do que na floresta velha. Diferenças florísticas consideráveis foram encontradas entre os dois tipos de mata, com uma presença maior de espécies pioneiras na caapoeira. As espécies mais abundantes nesta mata, *Ocotea glomerata* (Lauraceae), *Cecropia* sp. (Moraceae) e *Apeiba schomburgkii* (Tiliaceae), foram presentes mas não abundantes na floresta circundante.

## INTRODUCTION

The forests of the Ilha de Maracá have been described in detail by Milliken and Ratter (1989 and Chapter 5) from a series of transects across the island, and also on a more localized basis by Thompson *et al.* (1992) from six 0.25 ha plots near its eastern end. All of these plots and transects contain primary or old secondary (> 100 years) growth (see Proctor and Miller, Chapter 23). In 1988, one of us (RPM) identified a small patch (*c.* 1 ha) of much more recent regrowth (*caapoeira* or *capoeira*). The patch was within 100 m of the nearest plot of Thompson *et al.*

(1992), and local informants claimed that it had been felled and burned by a farmer called Antonio Dantas in 1976. It was apparently abandoned after a year, after crops including maize had been grown there. There are traces of a path leading to ruined farm buildings on the edge of a savanna about 1 km from the plot.

The patch of regrowth is important for two reasons. Firstly, it demonstrates a much earlier stage of secondary succession after shifting cultivation than has previously been described on Maracá. Secondly, it provides a fortuitous additional treatment for experiments set up in 1987 in adjacent forest and clear-felled (but unburned) plots (Luizão et al., 1997; Thompson et al., 1997).

The aim of the present paper is to describe a plot in the recent regrowth and to contrast it with the much older surrounding forests. However, the description of this burned plot will also provide an interesting comparison with the 1987 unburned fellings when they reach the same age. The regrowth in these fellings, which were created for a long-term forest regeneration study, has been monitored at intervals since their establishment and the results are to be published shortly (Thompson et al., 1997).

## MATERIALS AND METHODS

A plot of 50 × 50 m was located in the young secondary forest area which occurs adjacent to the plots described by Thompson et al. (1992) on a virtually flat plateau about 130 m above sea level (see Figure 23.2, p. 436). The plot was carefully located to avoid parts which had not been felled and burned. This was difficult since the farmed area was of irregular shape. No replication of the plot was possible. The plot was marked out and subdivided into twenty-five 10 × 10 m sub-plots which were used as a sample grid.

The girths of all trees (≥10 cm dbh) were measured in May 1991. The measurements were made at breast height (1.3 m), and where the trees had multiple stems each stem was measured separately. The trees were preliminarily identified in the field and confirmed from collected specimens. A 20 × 7.5 m transect of a typical part of the plot was selected for a profile diagram of trees over 6 m high.

## RESULTS

An overall impression of the forest is given by the profile diagram (Figure 6.2). Data on some forest structural features, including the numbers and basal area contributions of each tree species, are given in Tables 6.2 and 6.3.

## DISCUSSION

The young secondary forest shows many contrasts with the adjacent older forest described by Thompson et al. (1992), which is itself believed to be old secondary

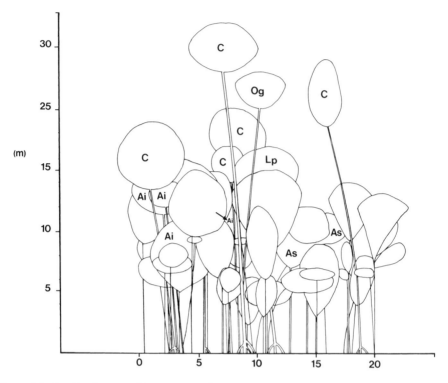

**Figure 6.2**. Profile diagram (20 × 7.5 m) of a recent secondary forest on the Ilha de Maracá. Trees less than 6 m high are excluded. Symbols for trees over 10 cm dbh: Ai, *Aegiphila integrifolia*; As, *Apeiba schomburgkii*; C, *Cecropia* sp.; Lp, *Lindackeria paludosa*, Og, *Ocotea glomerata*

growth (Proctor and Miller, Chapter 23). The density was higher in the young secondary forest with an equivalent of 600 trees per hectare. The corresponding mean value for the six plots in the older forest was 419 (range 340–476). The basal area of the young secondary forest was equivalent to 16.4 m$^2$ ha$^{-1}$, well below the corresponding mean of 23.8 (range 21.7–26.7) for the older forest. However, a substantial proportion (23.5%) of the basal area of the younger forest was contributed by five large pre-cultivation survivors: two individuals of *Trattinickia rhoifolia* and one each of *Ecclinusa guianensis*, *Himatanthus articulatus*, and a *Pouteria* sp. An equivalent of 64 individuals per hectare possessed multiple stems in the younger forest compared with four per hectare in the older forest. This presumably reflects sprouting after the slash and burn activity. Only 22 tree species were recorded from the young plot (Table 6.2) compared with a mean of 40 (range 33–47) from six plots of similar size in the adjacent forest.

The two types of plots have little in common floristically, with only the palm, *Astrocaryum aculeatum*, being a leading species in both (Table 6.4). Several of the species in the younger plot are likely to be pioneers, e.g. *Cecropia* sp. and

**Table 6.2.** Tree (≥10 cm dbh) features in a 50 × 50 m plot of c. 13-year-old secondary forest on the Ilha de Maracá

| Density (0.25 ha$^{-1}$) | Basal area (m$^2$ 0.25 ha$^{-1}$) | Individuals with multiple stems (0.25 ha$^{-1}$) | No. of species (0.25 ha$^{-1}$) |
|---|---|---|---|
| 150 | 4.13 | 16 | 22 |

**Table 6.3.** The tree species (≥10 cm dbh) recorded from a c. 13-year-old secondary forest on the Ilha de Maracá, their numbers in a range of dbh size classes, and proportional contribution to plot basal area

| Species | No. in dbh size class | | | | Basal area (%) |
|---|---|---|---|---|---|
| | 10–19.9 | 20–29.9 | 30–39.9 | ≥40 | |
| ANNONACEAE | | | | | |
|   *Rollinia exsucca* (Dun.) A.DC. | 5 | | | | 2.59 |
|   *Xylopia frutescens* Aubl. | 2 | | | | 0.65 |
| APOCYNACEAE | | | | | |
|   *Himatanthus articulatus* (Vahl) Woods. | 4 | | | 1 | 4.83 |
| BIGNONIACEAE | | | | | |
|   *Jacaranda copaia* (Aubl.) D. Don ssp. *spectabilis* (Mart. ex DC.) A. Gentry | 2 | 1 | | | 3.48 |
|   *Tabebuia* sp. | 1 | | | | 0.20 |
| BORAGINACEAE | | | | | |
|   *Cordia* cf. *sellowiana* Cham. | 3 | | | | 0.74 |
| BURSERACEAE | | | | | |
|   *Trattinickia rhoifolia* Willd. | 2 | | | 2 | 11.85 |
| FLACOURTIACEAE | | | | | |
|   *Banara guianensis* Aubl. | 4 | | | | 2.85 |
|   *Lindackeria paludosa* (Benth.) Gilg. | 1 | | | | 0.25 |
| LAURACEAE | | | | | |
|   *Ocotea glomerata* (Nees) Mez | 37 | 4 | | | 20.93 |
| LEGUMINOSAE | | | | | |
|   *Fissicalyx fendleri* Benth. | 6 | | | | 2.63 |
|   *Inga alba* Willd. | 7 | 2 | | | 7.17 |
| MORACEAE | | | | | |
|   *Cecropia* sp. | 17 | 5 | | | 13.38 |
| MYRISTICACEAE | | | | | |
|   *Virola sebifera* Aubl. | 4 | | | | 1.43 |
| PALMAE | | | | | |
|   *Astrocaryum aculeatum* G.F.W. Mey. | 1 | 4 | | | 3.87 |
| SAPOTACEAE | | | | | |
|   *Ecclinusa guianensis* Eyma | | | 1 | | 5.60 |
|   *Pouteria* sp. | | | | 1 | 2.84 |
| STERCULIACEAE | | | | | |
|   *Guazuma ulmifolia* Lam. | 3 | 1 | | | 4.90 |
| TILIACEAE | | | | | |
|   *Apeiba schomburgkii* Szysz. | 16 | | | | 6.39 |
| VERBENACEAE | | | | | |
|   *Aegiphila integrifolia* (Jacq.) Jacks. | 11 | | | | 2.82 |
| Unidentified | 2 | | | | 0.48 |

**Table 6.4.** The most numerous tree species in (a) 50 × 50 m of c. 13-year-old secondary forest and (b) six plots of similar size in older secondary forest on the Ilha de Maracá. The relative contribution to total tree numbers in the plots (150 in a, 631 in b) are given for both forest types

| Leading species in (a) | % in (a) | % in (b) | Leading species in (b) | % in (a) | % in (b) |
|---|---|---|---|---|---|
| *Ocotea glomerata* | 27.3 | 1.0 | *Brosimum lactescens* | 0.0 | 9.2 |
| *Cecropia* sp. | 14.7 | 0.2 | *Duguetia lucida* | 0.0 | 7.6 |
| *Apeiba schomburgkii* | 10.7 | 1.0 | *Astrocaryum aculeatum* | 3.3 | 7.3 |
| *Aegiphila integrifolia* | 7.3 | 0.0 | *Pouteria hispida* | 0.0 | 8.2 |
| *Inga alba* | 6.0 | 0.0 | *Licania kunthiana* | 0.0 | 4.9 |
| *Fissicalyx fendleri* | 4.0 | 0.2 | *Tetragastris panamensis* | 0.0 | 4.1 |
| *Astrocaryum aculeatum* | 3.3 | 7.3 | *Pouteria surumuensis* | 0.0 | 3.8 |
| *Himatanthus articulatus* | 3.3 | 2.9 | *Crepidospermum goudotianum* | 0.0 | 3.5 |
| *Rollinia exsucca* | 3.3 | 1.1 | *Lecythis corrugata* | 0.0 | 3.2 |

*Jacaranda copaia* which are rare in the older forest. The more shade-tolerant species of the older forest have not managed to reinvade the younger plot.

The younger plot is now to be monitored along with the experimentally felled and control plots and future changes in it will be reported over the next few years.

## ACKNOWLEDGEMENTS

Cristiane V. Teixeira is thanked for her assistance in the collection of data, and José L. dos Santos for his help with specimen identification.

## REFERENCES

Luizão, F.J., Proctor, J., Thompson, J., Luizão, R.C.C., Marrs, R.H., Scott, D.A. and Viana, V. (1997). Rain forest on Maracá Island, Roraima, Brazil: soil and litter process response to artificial gaps. *Forest Ecology and Management* (in press).

Milliken, W. and Ratter, J.A. (1989). *The vegetation of the Ilha de Maracá: First Report of the Vegetation Survey of the Maracá Rainforest Project*. Royal Botanic Garden Edinburgh.

Thompson, J., Proctor, J., Viana, V., Milliken, W., Ratter, J.A. and Scott, D.A. (1992). Ecological studies on a lowland evergreen rain forest on Maracá Island, Roraima, Brazil. I. Physical environment, forest structure and leaf chemistry. *Journal of Ecology*, 80, 689–703.

Thompson, J., Proctor, J., Scott, D A., Fraser, P.J., Marrs, R.H., Miller, R.P. and Viana, V. (1997). Rain forest in Maracá Island, Roraima, Brazil: artificial gaps and plant response to them. *Forest Ecology and Management* (in press).

# Phenological observations on a savanna–forest boundary on the Ilha de Maracá

ROBERT P. MILLER[1], JOÃO FERRAZ[2] AND JOSÉ L. DOS SANTOS[2]

[1]University of Florida, Gainesville, USA; [2]Instituto Nacional de Pesquisas da Amazônia, Manaus, Brazil

## SUMMARY

Phenological data are presented for a forest–savanna–forest transect. The 130 species of plants recorded along the transect were divided into five groups: trees, shrubs, graminoids, perennial herbs and seasonal herbs. With the exception of the savanna trees, which probably have access to the water-table, all groups showed reduced reproductive activity during the dry season. The reproductive activity observed within the other groups during the dry season, albeit reduced, suggests the existence of individual adaptations for water-loss control and/or the ability to make optimal use of the few rains which do fall during that period.

## RESUMO

Dados fenológicos são apresentados para um transecto floresta–savana–floresta. As 130 espécies de plantas encontradas ao longo do transecto foram divididas em cinco grupos: árvores, arbustos, graminoides, ervas perenes e ervas sazonais. Com a exceção das árvores da savana, que provavelmente tem acesso ao lençol freático, todos os grupos mostraram reduzida atividade reprodutiva durante a estação seca. A atividade reprodutiva observada nos outros grupos na estação seca, apesar de ser reduzida, sugere a existência de adaptações individuais para controlar a perda de água e/ou a habilidade de tirar proveito das poucas chuvas que caem nessa época.

## INTRODUCTION

### STUDY AREA

The area of savanna discussed in this study, the Santa Rosa savanna (see Plate 1b), is situated close to the eastern end of the Ilha de Maracá approximately 5 km north of the Ecological Station (Figure Pr.1, p. xvii). It is separated from the Furo de Santa Rosa by a strip of forest approximately 800 m wide. Observations during the study period suggest that this patch of savanna fits Sarmiento and Monasterio's classification of 'hyperseasonal savanna', i.e. savanna subject to both seasonal drought and seasonal waterlogging (Sarmiento and Monasterio, 1975). A detailed description of the physiognomy and composition of this vegetation unit is given by Milliken and Ratter (1989 and Chapter 5).

## HUMAN INFLUENCE IN THE STUDY AREA

Human occupation of the Ilha de Maracá, prior to its conversion to a reserve, is discussed at length by Proctor and Miller (Chapter 23). The wooden house at Santa Rosa was constructed by SEMA alongside one of the sites of relatively recent habitation (to 1978) at the eastern end of the island, at which had been planted banana, mango, lime and orange trees. Between this house and the study area were found several tracts of secondary vegetation, and the remains of a *casa de farinha* for the production of manioc flour. In one of the adjacent islands of savanna vegetation there was evidence of vegetational disturbance and previous habitation.

It is highly likely that the Santa Rosa savanna was periodically burned in the past by the inhabitants of the area. This practice, which is common in the region, is intended to clear the area of animal pests, remove the vegetation in the environs of the houses, and renovate pasture for the rearing of cattle. It is not known whether or not there was further burning following the island's transformation to a reserve, but a few months prior to commencement of the Maracá Rainforest Project it was partially burned by employees of SEMA (the agency then responsible for management of Maracá), during the clearing of the track from the Ecological Station to the Santa Rosa house.

## METHODOLOGY

A transect 305 m long was set up across the Santa Rosa savanna, crossing the transition zone and entering the forest at each of its extremities. The vegetation along the transect was divided into nine physiognomically distinct zones (Figure 6.3 and Table 6.5). The first vegetation survey was carried out in June 1987, when collections were made (fertile where possible) of each of the species present, and records were made of the vegetational zones in which they occurred.

The phenological studies of the transect were begun in November 1987, and continued until July 1988 with an additional recording in November 1988. During this period the transect was checked monthly and phenological data were recorded for each of the listed species. Plants which had not been recorded previously were collected and added to the list, and species seen in flower or fruit for the first time were also collected to facilitate identification. The specimens were identified and deposited in the INPA herbarium, Manaus.

Plant species observed along the transect were allocated to the following life-form categories:

TR  Trees – woody plants with a main stem $\geq 5$ cm dbh (diameter at breast height).
SR  Shrubs – woody plants with multiple stems or a single stem $<5$ cm dbh when mature.
GR  Graminoids – Graminae and other grass-like plants (e.g. Cyperaceae and Eriocaulaceae).
PH  Perennial herbs – non-woody plants with living aerial shoots throughout the year.
SH  Seasonal herbs – non-woody plants with aerial shoots dying off during the dry season.

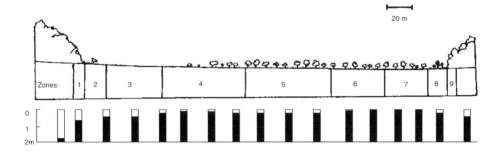

**Figure 6.3.** Zonation of vegetation across the Santa Rosa savanna, Ilha de Maracá (numbers refer to the vegetation types described in the text), and depth of water-table in the dry season of 1988

**Table 6.5.** Physiognomy of the vegetation along a forest–savanna–forest transect, Ilha de Maracá

| Zone |
| --- |
| 1. Transition from low forest through dense tangle of shrubs and climbers to open savanna. |
| 2. Grasses, herbs, scattered *Curatella americana*. |
| 3. Low grasses, *Bulbostylis lanata*, and bare soil with crust of blue–green algae. |
| 4. Grasses, herbs, scattered *Curatella americana* and *Byrsonima crassifolia*. |
| 5. Grasses, herbs, shrubs and scattered trees. Greatest density of *Curatella americana*, with individuals of all sizes. |
| 6. Grasses, herbs, shrubs and scattered trees. Tree density somewhat less than in zones 5 and 7. Some surface drainage channels. |
| 7. Grasses, herbs, shrubs, climbers and scattered trees. Microrelief dominated by drainage channels and small mounds (20 cm height). |
| 8. Dense, high grasses, shrubs, woody climbers and scattered trees. Micro-relief with drainage channels and mounds. |
| 9. Abrupt transition to forest. |

Due to their scarcity on the transect, woody climbers were classified with the perennial herbs.

## RESULTS AND DISCUSSION

Phenological data are presented as phenograms (Figures 6.4 and 6.5) for each of the five life-form groups, along with mean monthly precipitation figures. The arboreal species were further subdivided into savanna species (*Curatella americana* L., *Byrsonima crassifolia* (L.) Kunth and *Isertia parviflora* Vahl), and species of the forest transition zone.

The reproductive activity of the arboreal species of the savanna was found to be concentrated within the first half of the year, whereas that of the forest transition

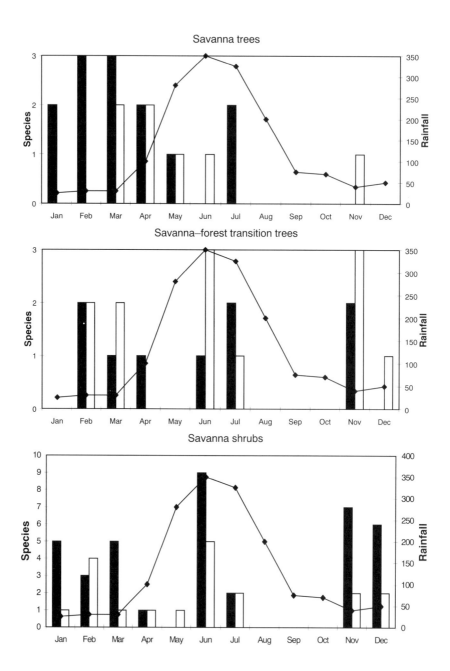

**Figure 6.4.** Phenology of trees and shrubs on the Santa Rosa savanna–forest transition in 1988, with precipitation data for the same period. Black column = flowering, white = fruiting

# PHENOLOGICAL OBSERVATIONS ON A SAVANNA–FOREST BOUNDARY 139

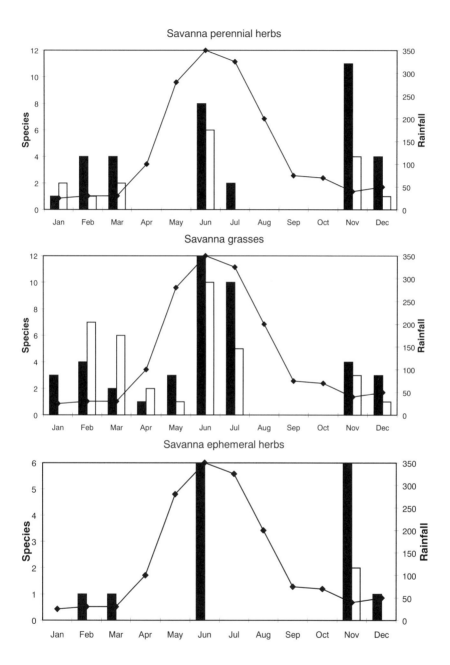

**Figure 6.5.** Phenology of herbaceous plants on the Santa Rosa savanna–forest transition in 1988, with precipitation data for the same period. Black column = flowering, white = fruiting

species was distributed more or less evenly throughout. The majority of the other life-forms exhibited greatest reproductive activity during the rainy season and, with the exception of the graminoids, in the first half of the dry season. The period of lowest overall reproductive activity was between April and May.

The different patterns of phenology shown by the various groups can to a certain extent be explained by the soil water conditions, given that all reproductive processes and tissue growth require consumption of water (be it stored or directly taken up from the soil). In the rainy season conditions became extremely damp in the savanna, and blue–green algae (Cyanophyta) were seen growing on the sandy surfaces of the exposed soil. In the lower reaches of the transect (zones 7 and 8), aquatic species (e.g. *Nymphaea* sp.) were found growing in the runoff channels between the vegetated tussocks. A fish was found in soil profile pit 14 (zone 6). Water levels in this season in soil profile pits (dug in the dry season) are shown in Figure 6.3.

During the dry season it appears that the savanna tree species, in spite of the drying of the soil to a depth of 1.8 m and the low relative humidity of the air, undergo no significant water stress. Similarly, on the savannas between the Ilha de Maracá and Boa Vista, rapid leaf regrowth was observed in these species after the annual burning at the end of the dry season. It seems that these species, thanks to the depth of their root systems, are able to reach the water-table during the dry season, as do the *cerrado* species of Central Brazil (Goodland and Ferri, 1979). Another possibility is that they are able to make use of the occasional showers which fall during the dry months. In attempting to understand why savanna trees do exhibit reproductive activity in the dry season, various possible explanations can be advanced, ranging from greater availability of dispersal agents (both *Curatella americana* and *Byrsonima crassifolia* have a fleshy aril or pulp surrounding bird-dispersed seeds (Roosmalen, 1985)), to the need for seed germination to occur in the beginning of the rainy season in order to ensure seedling survival. Differences between the reproductive strategies of the forest transition species and the savanna species must best be examined both in view of differences in microclimate (chiefly light, humidity and temperature) and their importance to seedling survival, as well as the fact that these vegetation types reflect differences in soil and water-table conditions (see Ross *et al.*, 1992; Thompson *et al.*, 1992).

In the non-arboreal species, those which were observed to reproduce during the dry season were probably able to do so as a result of the occasional rains, or thanks to individual adaptations for the control of water loss and tolerance of dry soil conditions. Within the life-form groups, such individual adaptations are probably responsible for the variety of reproductive behaviour observed. For example, within one family of shrubs, the Melastomataceae, there was a wide range of phenological behaviour: *Tibouchina aspera* Aubl. var. *asperrima* Cogn. did not flower during the study period, whilst *Comolia* sp. flowered during seven of the nine months of the study period. Other melastomes exhibited intermediate behaviour. Meanwhile the graminoids tend to enter a phase of progressive semi-dormancy, during which the greater part of their green tissues dry up (Sarmiento, 1984).

All of the above groups, with the exception of the savanna trees, showed minimal reproductive activity between the months of April and May. This coincided with a particularly dry part of the year; in 1988 there was no rain whatsoever during April and the rains only arrived at the beginning of May. According to G.M. de Oliveira, administrator of the Ilha de Maracá, this was the most prolonged dry season which he had seen during 10 years on the island. This phenomenon makes it difficult to distinguish between what Sarmiento and Monasterio (1983) term 'dry season annual flora', which grow and reproduce in waterlogged areas, and those annuals which flower at the beginning of the dry season before the soil dries out completely. The protracted dry season simply meant that, as observed, all activity of seasonal herbs on the transect ceased for a two month period.

The above observations show that besides the regular, yearly fluctuations in the environment, which govern the phenological strategies of different plant groups, there are also larger fluctuations from year to year. The following dry season (end of 1988–beginning of 1989), for example, could scarcely be called such, as it rained periodically throughout (G.M. de Oliveira, pers. comm.). Consideration of the relative advantages of different phenological strategies in plants which do not have access to the water-table must take into account the effect on savanna phenology of such irregularly alternating periods of drought and waterlogging.

Another environmental factor, of importance equal to that of drought or waterlogging in savannas, is fire. This study did not examine the effect of fire on savanna phenology, given its absence during the study period. Use of a controlled burn was discarded, in view of possible damage to the scientific value of the Santa Rosa savanna as well as to other research efforts being carried out there. The question of fire, however, deserves further attention. At one extreme, the excess use of fire in savanna ecosystems leads to the deterioration of the vegetation, while less frequent fires may have beneficial effects on certain groups of plants, due to the greater availability of nutrients following burning, as well as the presence of bare soil important for the germination and establishment of plants with wind-dispersed seeds (Coradin, n.d.; Coutinho, 1990). Sarmiento and Monasterio (1983) argue that a late burning (at the end of the dry season) does not alter drastically the normal course of phenological events in savannas, other than causing a greater synchronization of phenological activities, which may in fact be highly beneficial to plants which require cross-pollination (Coutinho, 1990). Given the existence of these beneficial effects of fire, an important question to ask in the case of the Santa Rosa savanna concerns the possible effects of fire exclusion on plant diversity – will diversity decline over time? While at the moment there are no plans for research on the effects of fire, continued monitoring of the study area could yield interesting information on the effects of fire exclusion, especially in the case of the seasonal herb strata, in which what we see as phenological events are in fact entire life cycles.

## ACKNOWLEDGEMENTS

The authors would like to acknowledge Dra M.F. da Silva, W. Milliken and J.A. Ratter who assisted with the identification of the botanical material.

# REFERENCES

Coradin, L. (n.d.) *The grasses of the natural savannas of the Federal Territory of Roraima, Brazil*. Unpublished Masters thesis, Lehman College, City University, New York.

Coutinho, L.M. (1990). O cerrado e a ecologia do fogo. *Ciência Hoje*, 12 (68), 20–30.

Goodland, R.J.A. and Ferri, M.G. (1979). *Ecologia do Cerrado*. Itatiaia/EDUSP, Belo Horizonte.

Milliken, W. and Ratter, J.A. (1989). *The vegetation of the Ilha de Maracá: First Report of the Vegetation Survey of the Maracá Rainforest Project*. Royal Botanic Garden Edinburgh.

Roosmalen, M.G. van. (1985). *Fruits of the Guianan flora*. Inst. of Systematic Botany, Utrecht.

Ross, S.M., Luizão, F.J. and Luizão, R.C.C. (1992). Soil conditions and soil biology in different habitats across a forest–savanna boundary on Maracá Island, Roraima, Brazil. In: *Nature and dynamics of forest–savanna boundaries*, eds P.A. Furley, J. Proctor and J.A. Ratter, pp. 145–170. Chapman & Hall, London.

Sarmiento, G. (1984). *The ecology of Neotropical savannas*. Harvard University Press, Cambridge, Massachusetts.

Sarmiento, G. and Monasterio, M. (1975). A critical consideration of the environmental conditions associated with the occurrence of savanna ecosystems in tropical America. In: *Tropical ecological systems*, eds F.B. Golley and E. Medina, pp. 223–250. Springer-Verlag, New York.

Sarmiento, G. and Monasterio, M. (1983). Life forms and phenology. In: *Ecosystems of the world, Vol. 13, tropical savannas*, ed. F. Bourlière, pp. 79–108. Elsevier, Amsterdam.

Thompson, J., Proctor, J., Viana, V., Ratter, J.A. and Scott, D.A. (1992). The forest–savanna boundary on Maracá Island, Roraima, Brazil: an investigation of two contrasting transects. In: *Nature and dynamics of forest–savanna boundaries*, eds P.A Furley, J. Proctor and J.A. Ratter, pp. 367–392. Chapman & Hall, London.

# 7 Primates of the Ilha de Maracá[1]

ANDREA P. NUNES, JOSÉ MARCIO AYRES, EDUARDO S. MARTINS AND JOSÉ DE SOUSA E SILVA

*Museu Paraense Emílio Goeldi, Belém, Brazil*

## SUMMARY

Preliminary surveys of the primates of northeastern Roraima, Brazil, were carried out during the second half of 1987. The present census was made on the Ilha de Maracá. Surveys revealed the occurrence of five species of primates in the study area, the most abundant being the spider monkey *Ateles belzebuth belzebuth*. The geographical distributions and conservation status of the primate fauna of Roraima are discussed.

## RESUMO

Durante o segundo semestre de 1987 foi feito um levantamento preliminar dos primatas do nordeste do Estado de Roraima, dando-se ênfase à Ilha de Maracá. Nos levantamentos na ilha, foram detectadas a presença de pelo menos cinco especies de primatas, das quais os coatás (*Ateles belzebuth belzebuth*) foram os mais abundantes. As distribuições geográficas dos primatas de Maracá, bem como a situação atual das espécies em relação à ocupação humana na região, são discutidas.

## INTRODUCTION

Until very recently the fauna of Roraima was one of the least known within the Amazon basin. The State makes up approximately 6.7% of the area of the Brazilian Amazon, and the faunal distribution within the region is of considerable interest due to the great variety of vegetation types found. These range from dense Amazonian lowland forests to flooded habitats, montane forests, open savannas of the Rio Branco–Essequibo depression and the grassland plateaus near the Venezuelan and Guyanan borders.

According to recent surveys of the distribution of Amazonian primates (e.g. Hershkovitz, 1977, 1983, 1987a, b), at least 11 species of primates occur in the State of Roraima including *Saguinus midas midas*, *Saimiri sciureus sciureus*, *Callicebus torquatus lugens*, *Aotus trivirgatus*, *Cebus apella*, *Cebus nigrivittatus*, *Pithecia pithecia chrysocephala*, *Chiropotes satanas chiropotes*, *Alouatta seniculus*, *Ateles belzebuth belzebuth* and *Ateles paniscus paniscus*. These represent nearly 30% of the Amazonian primate fauna.

---

*Maracá: The Biodiversity and Environment of an Amazonian Rainforest.*
Edited by William Milliken and James A. Ratter. © 1998 John Wiley & Sons Ltd.

The present chapter deals with a preliminary survey of the resident primate fauna of the Ilha de Maracá, with comparative reference to studies carried out contemporaneously in two other sites in Roraima: Colônia Apiaú to the southwest of the island and Serra de Pacaraima to the northeast. The primary object of the survey was to produce a preliminary assessment of the densities, distributions and conservation status of the primate populations within the Maracá reserve.

## STUDY AREA AND METHODS

A systematic census of approximately 297.5 km was carried out on Maracá between August and December 1987 in an area of *terra firme* forest at the eastern end of the island (see Table 7.1). An already established trail system for a previous tortoise study (Moskovits, Chapter 13) was used for the primate surveys (see Figure 23.2, p. 436). In general these were done in the early part of the morning (0630–1100 hrs) or later in the afternoon (1500–1830 hrs) – the periods when primates are usually more active. Standard transect census techniques were employed such as King's method (National Research Council, 1981), Fourier series (Burnham et al., 1980) and/or the simple counting of groups per distance travelled (Emmons, 1984). Trails within the forest were walked a number of times, but where possible no section of trail was walked more than once within a single day. Distances travelled were measured with a previously calibrated pedometer, and for each group of monkeys encountered, the following were recorded: species, vegetation type, troop size and troop composition (where possible).

## PRIMATE SPECIES OF MARACÁ

During our surveys we were able to confirm the presence of five species of primates on the Ilha de Maracá, a further two species having been recorded on the island by previous workers. In the next section we analyse the information gathered during surveys and that available in the literature.

### SPECIES RECORDED ON THE ILHA DE MARACÁ DURING THE PRESENT SURVEY

*Saimiri sciureus sciureus* **(Linnaeus 1758)**

There are four collection localities for this squirrel monkey in Roraima, all near Boa Vista. It probably occurs over most of Roraima, especially in or near riverine habitats, flooded forests and swampy areas, but is not found in the mountain complexes to the north and west of the State. On the Ilha de Maracá it was the fifth most abundant species (1.1 groups/km$^2$). These densities are lower than for most other areas surveyed for this species (Robinson and Janson, 1987), probably because our surveyed areas included only *terra firme* forest. Groups varied between 20 and 30 individuals in the census areas in Roraima.

Table 7.1. Primate densities on the Ilha de Maracá (81 troops observed)

| Species | Groups per 10 km | King's method (groups/km$^2$) | Fourier method (groups/km$^2$) | Individuals per km$^2$ (Fourier) | Group size (range) |
|---|---|---|---|---|---|
| *Saimiri sciureus* | 0.1 | 1.1 | 1.1 | 26.3 | (20–30) |
| *Cebus apella* | 0.5 | 4.3 | 3.8 | 23.4 | 6.1 (1–14) |
| *Cebus nigrivittatus* | 0.3 | 2.3 | 1.7 | 4.0 | 2.3 (2–3) |
| *Alouatta seniculus* | 0.6 | 5.2 | 4.5 | 14.9 | 3.3 (1–5) |
| *Ateles belzebuth** | 1.2 | 8.6 | 8.7 | 34.0 | 3.9 (1–9) |
| Not identified | 0.1 | – | – | – | – |
| Total distance surveyed (km) | 297.5 | 149.5 | 149.5 | 149.5 | |
| Number of groups observed | 81 | 57 | 57 | 57 | |

* Subgroups.

## *Cebus apella* (Linnaeus 1758)

The black-capped capuchin probably occurs in most areas of Roraima, and was observed on the Ilha de Maracá. Mixed bands of this species with *Cebus nigrivittatus* were not uncommon. Troops varied in size up to 14 individuals and their densities were higher than for its congener on the island. There are no specimens of this species at the Museu Goeldi from the study area.

## *Cebus nigrivittatus* (Wagner 1848)

This species probably occurs sympatrically with *Cebus apella* in most areas of Roraima. It has been observed in relatively small troops on the Ilha de Maracá. Troops usually vary from a few to 10 individuals, but on the Ilha de Maracá the species was observed in relatively small groups of two or three only. This was the fourth most abundant primate on Maracá, where it is usually found in *terra firme* forest. Densities on the island were much lower than those found for the same species in the Venezuelan *llanos* by Robinson (1986).

## *Alouatta seniculus* (Linnaeus 1766)

Red howlers occur throughout Roraima, including into the montane forests up to 1200 m. They were the second most abundant primate on the Ilha de Maracá. However, densities found there were lower than for most other areas surveyed for this species (Crockett and Eisenberg, 1987). Troop sizes ranged from one to five individuals. Together with *Ateles* spp. they are the largest primates in Roraima.

## *Ateles belzebuth belzebuth* (E. Geoffroy 1806)

This species of spider monkey is found in most forested areas of the State, west of the Rio Branco. On the Ilha de Maracá it was the most numerous primate, found in subgroups varying from one to nine individuals. They are mainly frugivorous, but preliminary observations on Maracá (APN, pers. obs.), and two stomach contents examined, indicated higher proportions of leaves in the diet than in other Amazonian areas (Robinson and Janson, 1987). The densities found on the island are comparable to those found in Cosha Cashu, Manu, Peru (Emmons, 1984).

## SPECIES REPORTED TO OCCUR ON THE ISLAND BUT NOT ENCOUNTERED DURING THE PRESENT SURVEY

### *Aotus trivirgatus trivirgatus* (Humboldt 1812)

Night monkeys have not been observed in the few night surveys conducted during this study. Locals, however, report their presence on the Ilha de Maracá. There are only two localities recorded for this subspecies in the literature (Hershkovitz, 1983). The only specimen in the Museu Goeldi was collected by B. Albert on the Rio Toototobi, a tributary of the Rio Demini in the State of Amazonas (near the Roraima border).

### *Pithecia pithecia chrysocephala* (I. Geoffroy 1850)

Hershkovitz (1987a) indicates two localities for this subspecies of saki in Roraima: Maracá (same coordinates as the island) and Forte do Rio Branco (Boa Vista). No individual of this species was observed during our surveys and there is no indication that it occurs west of the Rio Branco in the central and northern areas of Roraima. The 'Maracá' specimen could be mislabelled.

## OTHER SPECIES

The following species have been reported to occur in Roraima, but have not been recorded on the Ilha de Maracá.

### *Chiropotes satanas chiropotes* (Humboldt 1812)

There are six skins from five localities reported in the literature, indicating the presence of bearded sakis in Roraima (Ayres, 1981; Hershkovitz, 1985). Specimens were collected on the lower Rio Mucajaí, upper Rio Catrimani, Agua Boa stream and Rio Toototobi (B. Albert, pers. obs.). They probably do not occur in the eastern part of the Ilha de Maracá, but there are reports that this species is found in the forests surrounding the island (D. Moskovits, pers. comm.). Troops range in size between 20 and 30 individuals, but no precise counts were obtained.

### *Saguinus midas midas* (Linnaeus 1758)

The Midas tamarin is perhaps the only callithrichid in Roraima. No individuals of this species were observed during our surveys. Hershkovitz (1977), however, indicated the presence of this species throughout most of the State, with a distribution extending west of the left bank of the Rio Negro and north to the Parima and Pacaraima mountain complexes. Since this callithrichid is quite an abundant primate in other areas of Brazilian Amazonia (e.g. Martins and Ayres, 1987), we suspect that its absence in the areas surveyed strongly suggests its non-occurrence.

### *Callicebus torquatus lugens* (Humboldt 1812)

Titis are not among the most common primates in Roraima, at least in the areas so far surveyed. This species does not occur either on the Ilha de Maracá or in the elevations to the north, but two groups, both with few individuals, have been seen in the Rio Apiaú area. One of the troops was seen in a sandy soil area of tall primary forest, while the other was in vegetation characterized by the abundance of the *ubim* palm, *Geonoma baculifera*, over a silty swampy soil. The only skins at the Museu Goeldi were collected near the mouth of the Rio Mucajaí by E. Dente in the 1950s.

*Ateles paniscus paniscus* (Linnaeus 1758)

This species probably occurs east of the Rio Branco, according to the sources of animals kept in captivity in Boa Vista, Colônia Confiança and Marco BV-8. This and *Ateles belzebuth* are the largest primates of Roraima. Reliable informants also indicate the Rio Branco as the limit between the two species, as suggested by Konstant *et al.* (1985). Kellogg and Goldman (1944) list no specimens from Roraima, but there is a Guyanan collecting locality (Maripa) close to the Brazilian border which suggests the presence of the species within the State.

# DISCUSSION

## PRIMATE DENSITIES ON THE ILHA DE MARACÁ

Primate densities recorded on the Ilha de Maracá (Table 7.1) were very similar to those found in the unprotected forest area surveyed in the Apiaú basin to the south, although the primate composition was slightly different due to the presence of *Chiropotes satanas chiropotes* and *Callicebus torquatus lugens* in the latter area. These sites, however, were different in recorded abundance from those in the northern part of Roraima around the mountains of the Serra de Pacaraima, where primate densities were found to be substantially lower. This is probably not a consequence of hunting pressure in the mountainous areas, but of other environmental or ecological factors.

The overall primate densities found on Maracá are quite low when compared with sites in the western part of Amazonia, but similar to other sites of the crystalline basement (Table 7.2). One of the most interesting findings from our study was the low densities of *Saimiri*, *Cebus* and *Alouatta* spp. when compared with other surveys (Crockett and Eisenberg, 1987; Robinson and Janson, 1987), whereas *Ateles belzebuth* presented unusually high densities for a frugivore of its size. Densities found for this genus were among the highest recorded within Amazonia (Robinson and Janson, 1987) and are similar to those found in Cocha Cashu, Peru (Emmons, 1984).

## CONSERVATION STATUS OF PRIMATES IN THE MARACÁ AREA

Until the 1960s most of the population of Roraima was made up of Amerindian groups (e.g. Yanomami, Macuxi, Taurepang, Wapixana, Wai-Wai) (Oliveira, 1983), and the human density in the State was the lowest within the Brazilian Amazon. However, as a result of increased accessibility (the Manaus–Caracaraí highway), deliberate colonization and development policies, and the gold and diamond rushes of the late 1980s and early 1990s, this has been followed by a period of rapid demographic growth.

Despite the increasing habitat disturbance in Roraima, until 1978 the rate of deforestation was the lowest within Amazonia (Fearnside, 1984). This was largely due to the fact that over 60.5% of the total human population remained concentrated in urban areas, especially in Boa Vista and Caracaraí (Figure 1.1), partly as a

Table 7.2. Comparative densities of primates at several Amazonian localities (after Nunes et al., 1988)

| Locality | Groups per 10 km | No. of spp. | Distance surveyed (km) |
|---|---|---|---|
| Ilha de Maracá* | 2.8 | 6 | 297.5 |
| R. Apiaú* | 2.6 | 6 | 77.0 |
| R. Xingu–Iriri | 3.8 | 8 | 96.0 |
| R. Jarí | 1.8 | 7 | 120.0 |
| R. Jamarí | 4.0 | 7 | 97.0 |
| R. Japurá (terra firme) | 5.7 | 9 | 655.0 |
| R. Japurá (várzea) | 10.2 | 4 | 249.0 |
| R. Aripuanã | 2.2 | 10 | 82.0 |
| R. Tapajós | 3.1 | 9 | 34.0 |
| Manaus (north) | 1.1 | 7 | 80.0 |
| R. Manu | 8.2 | 9 | 67.0 |
| Tambopata | 2.5 | 8 | 52.0 |
| CMSE** (N of Manaus) | 3.0 | 6 | 49.0 |

\* Data from present study.
\*\* CMSE – Critical Minimum Size Ecosystem (WWF/Smithsonian project).

consequence of colonization projects failing on account of the poor quality of the soils. Nevertheless, population increase and habitat destruction do offer some threat to the primate populations of Roraima.

The species most at risk from hunters are the red howler (*Alouatta seniculus*) and the spider monkeys (*Ateles* spp.). In addition, medium-sized primates such as *Cebus* and *Chiropotes* are hunted to a lesser degree. The conservation status of the Ilha de Maracá should theoretically ensure that the primate populations of the island itself will continue to remain relatively high (see Table 7.2). However, the presence of wildcat gold prospectors (*garimpeiros*) around the periphery of the island since this survey was carried out, and the approach of the Trairão colonization project and other developments around Maracá (see Preface), may already have had some detrimental effect on primate populations through the consequent increase in hunting pressure, and will probably continue to do so.

## ACKNOWLEDGEMENTS

Celso Morato de Carvalho and José Antonio Gomes from INPA/Roraima provided invaluable help with logistics and their knowledge of the region. Natalia Inganaki, Manuel Santa Brigida, José Nilton Santa Brigida and Dionísio Corrêa Pimentel from Museu Goeldi were extremely helpful in the field. The collaboration of José Ribeiro from INPA was much appreciated.

## NOTE

1. This chapter is an adapted version of a paper published in the *Boletim do Museu Paraense Emílio Goeldi, série Zoologia*, 4 (1).

# REFERENCES

Ayres, J.M. (1981). *Observações sobre a ecologia e o comportamento dos Cuxius* (Chiropotes albinasus *e* Chiropotes satanas, *Cebidae*). FADESP, Belém.

Burnham, K.P., Anderson, D.R. and Laake, J.L. (1980). Estimation of density from line transect sampling of biological populations. *Wildlife Monographs*, 72, 1–202.

Crockett, C.M. and Eisenberg, J.F. (1987). Howlers: variation in group size and demography. In: *Primate societies*, pp. 54–68. University of Chicago.

Emmons, L. (1984). Geographic variation in densities and diversities of non-flying mammals in Amazonia. *Biotropica*, 16, 210–222.

Fearnside, P.M. (1984). A floresta vai acabar? *Ciência Hoje*, 2, 42–52.

Hershkovitz, P. (1977). *Living new world monkeys (Platyrrhini) with an introduction to primates*. Vol. 1. University of Chicago.

Hershkovitz, P. (1983). Two new species of night monkeys, genus *Aotus* (Cebidae, Platyrrhini). A preliminary report on *Aotus* taxonomy. *American Journal of Primatology*, 4, 209–243.

Hershkovitz, P. (1985). A preliminary taxonomic review of South American bearded saki monkeys, genus *Chiropotes* (Cebidae, Platyrrhini) with the description of a new subspecies. *Fieldiana Zool.*, 27, 1–46.

Hershkovitz, P. (1987a). The taxonomy of South American sakis, genus *Pithecia* (Cebidae, Platyrrhini): a preliminary report and critical review with the description of a new species and a new subspecies. *American Journal of Primatology*, 12, 387–468.

Hershkovitz, P. (1987b). Uacaries, new world monkeys of the genus *Cacajao* (Cebidae, Platyrrhini): a preliminary taxonomic review with the description of a new subspecies. *American Journal of Primatology*, 12, 1–53.

Kellogg, R. and Goldman, E.A. (1944). Review of the spider monkeys. *Proc. United States Nat. Mus.*, 96, 1–45.

Konstant, W., Mittermeier, R.A. and Nash, S.D. (1985). Spider monkeys in captivity and in the wild. *Primate Conservation*, 5, 82–109.

Martins, E.S. and Ayres, J.M. (1987). *Levantamento faunístico do Rio Jari: área da influência da UHE da Cachoeira de Santo Antônio*. 37 pp.

National Research Council (1981). *Techniques for the study of primate population ecology (National Academy of Sciences)*. Washington, National Academic Press.

Nunes, A.P., Ayres, J.M., Martins, E.S. and Sousa e Silva, J. de (1988). Primates of Roraima (Brazil). I. Northeastern part of the territory. *Bol. Mus. Para. Emílio Goeldi, sér. Zool.*, 4 (1), 87–100.

Oliveira, A.E. (1983). Ocupação humana. In: *Amazônia: desenvolvimento, integração, ecologia*. CNPq, Brasília.

Robinson, J. (1986). Seasonal variations in the use of time and space by the wedgecapped capuchin monkey, *Cebus olivaceus*: implications for foraging theory. *Smithsonian Contributions to Zoology*, 431, 1–60.

Robinson, J. and Janson, C.M. (1987). Capuchins, squirrel monkeys, and atelines: Socioecological convergence with old world primates. In: *Primate societies*, pp. 69–82. University of Chicago.

# 8 White-lipped Peccaries and Palms on the Ilha de Maracá

JOSÉ M.V. FRAGOSO
*University of Florida, Gainesville, USA*

## SUMMARY

Studies were made of the population, behaviour and feeding ecology of white-lipped peccaries (*Tayassu pecari*) at the eastern end of the Ilha de Maracá, and of their interactions with palms. Groups of up to 48 individuals were observed, and densities of up to 542 individuals per km$^2$ were estimated in the rainy season. Populations in the area fell substantially during the dry season. Greatest densities coincided with the masting of the *inajá* palm (*Maximiliana maripa*) and the period of greatest fruit abundance in the forest. During the dry season the *buriti* palm (*Mauritia flexuosa*), which fruits throughout the year, appears to act as the key resource influencing peccary movements. The possibility that the range of these animals increases during the dry season, that the distribution of certain palm species is influenced by peccary activity, and that in turn the large-scale migrations of peccaries are influenced or determined by palm phenology, is discussed.

## RESUMO

Foram realizados estudos de população, comportamento e ecologia alimentar de queixadas (*Tayassu pecari*) na parte oriental da Ilha de Maracá, e de suas interações com palmeiras. Grupos de até 48 indivíduos foram observados, tendo-se estimado densidades alcançando 542 indivíduos por km$^2$ durante a estação chuvosa. As populações na área diminuiram nitidamente na estação seca. Densidades maiores corresponderam com a frutificação da palmeira 'inajá' (*Maximiliana maripa*), e com o período de maior abundância de frutos na floresta. Durante a estação seca, a palmeira buriti (*Mauritia flexuosa*), que frutifica durante o ano todo, parece funcionar como um dos principais recursos influenciando distribuição/movimentação de queixadas. São discutidas as possibilidades de que a área de forragear destes animais aumenta durante a estação seca, que a distribuição de certas espécies de palmeiras é influenciada pela atividade de queixadas, e que as migrações de grande escala deles são, por sua vez, influenciadas ou determinadas pela fenologia das palmeiras.

## INTRODUCTION

White-lipped peccaries (*Tayassu pecari*) have an unusual socio-ecology for a tropical forest mammal: individuals form herds with hundreds and possibly thousands of members, and a herd may range over hundreds and perhaps thousands of square kilometres (Mayer and Brandt, 1982; Kiltie and Terborgh, 1983; Sowls, 1984; Mayer and Wetzel, 1987). This is rare behaviour for a mammal that inhabits tropical forests (Jarman, 1974; Leuthold, 1977), and probably results in peccaries having unique ecological interactions with plants (Janzen, 1974; Smythe, 1989).

Here, I report some preliminary results from a two-year study of white-lipped peccaries and palms on the Ilha de Maracá. I examine peccary population densities, seasonal movements, feeding and behaviour, consider their ecological relationships with *inajá* (*Maximiliana maripa* (Correa) Drude) and *buriti* (*Mauritia flexuosa* L.f.) palms, and speculate on their interactions with tropical plant communities.

White-lipped peccaries are one of the three species of peccaries (Tayassuidae). They range throughout the wet and dry tropical forests from southern Mexico to northern Argentina (Sowls, 1984; Mayer and Wetzel, 1987; Eisenberg, 1989). Although of economic and ecological importance, we know little of their ecology (see Kiltie, 1980, 1981a, 1981b, 1981c, 1982; Kiltie and Terborgh, 1983; Sowls, 1984). Most of what is known is speculative, inferential or anecdotal in nature: they probably require large areas of undisturbed habitats; they may be migratory; their food resources are probably patchy and widely distributed; their migrations may be related to a fruiting cycle across huge landscapes; and they may require a certain minimal herd size to reproduce successfully (Sowls, 1984; Smythe, 1986, 1987; Mayer and Wetzel, 1987; Bodmer, 1989). The only other large mammal with a similar socio-ecology is the bearded pig (*Sus barbatus*) of Borneo (Caldecott, 1988).

The white-lipped peccary is a key consumer of the fruits and seeds of tropical forest trees (Kiltie and Terborgh, 1983; Bodmer, 1989). They eat both freshly fallen fruits and seeds, and seeds whose exocarps have been consumed by other frugivores (Kiltie and Terborgh, 1983). They co-occur with collared peccaries (*Tayassu tajacu*) over most of their range. However, they are generally bigger (25–40 kg) than the collared peccary (15–25 kg) (Eisenberg, 1981; Robinson and Redford, 1986). There is much overlap in the diet of the two species, but only the white-lipped peccary can break open the seed wall of very hard palm nuts (Kiltie, 1981c, 1982; Bodmer, 1989). Rodents are also capable of gnawing open some of these hard palm seeds (Hallwachs, 1986; Janzen, 1986; Smythe, 1989).

At least 16 species of palms occur on the Ilha de Maracá (Milliken and Ratter, 1989), covering approximately 17% of the area around the Ecological Station (Moskovits, 1985). Palm forests (mostly *Maximiliana maripa*) are common on the high ridges of Maracá, and stands of *buriti* (*Mauritia flexuosa*) are found in poorly drained areas. Although palm seeds and fruit pulp are an important food for many animals and for man, the ecology of most Amazonian palm species is poorly known (Moore, 1977; Kahn and Castro, 1985).

## METHODS

### STUDY SITE

The study site consisted of a trail system (Figure 23.2, p. 436) located in the forest at the eastern end of Maracá, originally established by Moskovits (Chapter 13) for her studies of *Geochelone* tortoises. The dominant vegetation of the region is tropical evergreen lowland rainforest (Moskovits, 1985; Milliken and Ratter, 1989). The forest is contiguous with the main body of Amazonian rainforest, and supports populations of animals typical of areas undisturbed by Europeans or their descendants (Moskovits, 1985). Hunting in the reserve is illegal. However, unlawful hunting does occur in outlying regions, although hunters avoid the area around the Ecological Station.

## STUDY PERIOD

Data from a wet season (12 June to 3 July 1988), and a dry season (23 December 1990 to 30 April 1991), are compared. The former period represents a preliminary survey and the latter the beginning of a two-year study of white-lipped peccary ecology and the animals' interactions with the palm community on Maracá.

## PECCARY POPULATION DENSITIES

Peccary densities were estimated using a line transect census method (Davis and Winstead, 1980). Transect lines followed pre-existing trails extending for approximately 50 km in a grid pattern. One line transect usually included a few adjoining trails. However, none were walked more than once during a transect. Upon encountering peccaries I recorded the following: estimated distance to herd, azimuth from the transect line, number, sex and behaviour of adults and juveniles, time and location, total time and total distance (km) walked. In some instances azimuths were unrecorded. Thus I present densities derived using two census methods: King's method which considers only the distance of the animal from the observer; and the Webb method which also requires the azimuth of the animal's location relative to the transect line (Davis and Winstead, 1980). The Webb method is more appropriate if animals are detected before they are flushed. Some researchers (e.g. Ayres, 1986) have shown that the Webb method estimates actual animal populations more accurately than King's method. Generally two people walked the transect lines, but occasionally only one. The assistance of a local guide, and our familiarity with the trail system, helped us to avoid counting groups more than once. Transects were walked between 0700 hrs and sunset (approximately 1800 hrs). Some of the statistical analyses included data collected outside the line transects and information provided by IBAMA employees on the island.

## HUNTING SURVEYS

To determine the importance of peccaries to the diet of local people, I interviewed the heads of three of 13 households in a Macuxi Indian village near Maracá in the wet season, and chiefs of two other villages and managers of eight ranches in the dry season.

PALM PHENOLOGY AND ECOLOGY

To ascertain the phenology of *inajá* and *buriti*, we surveyed 50 trees of each species during the 1991 study period: 10 trees were randomly selected from each of five 250 × 10 m transects. The transects were randomly chosen from 34 *inajá* and 18 *buriti* patches (*buritizais*), which were identified within the study region by searching trails and a satellite image of the vegetation of Maracá. I noted fruit availability on the ground in the transects, and incidentally while conducting peccary surveys. I obtained additional information on palm phenology and fruit availability from employees of the Ecological Station, a biological technician who had studied fruit-eating by spider monkeys (*Ateles belzebuth*) on Maracá for two years, and data presented by Moskovits (1985) and Milliken and Ratter (1989). To learn what fruits and fruit parts were eaten by peccaries I watched them feed, noted foraging signs, and recorded fruit and seed abundance on the ground before and after peccaries passed through palm patches.

## RESULTS

I searched for peccaries for 108 hours (17 days) in the wet season, and 308 hours (48 days) in the dry season. Thirty-six groups (603 individuals) were encountered during the wet season, and peccaries were seen on 16 of 17 search days. Seventeen of the 34 wet season groups for which there are age data contained juveniles. They formed 7% of the total peccary population (Table 8.1). Only two peccary groups (42 individuals) were seen in the dry season, and neither contained juveniles. It seems that peccaries may have migrated away from the eastern part of the island in the dry season. Local people and employees of the Ecological Station reported that peccaries leave in the dry season and return at the beginning of the wet season (May to June). I analysed only the wet season density data in detail, since few herds were seen in the dry season of 1991.

In the wet season 21 peccary groups (378 individuals) were encountered on the 121.71 km of transects for which there are azimuth data. I also walked 13.95 km for which there are no azimuth data, from which four peccary groups were observed. The density of peccaries in southeastern Maracá island derived by the Webb method is greater than that obtained using King's method (Table 8.2).

BEHAVIOUR

Twenty-four of the 34 peccary groups observed in the wet season were foraging, six were travelling, two were rolling in clay or sand, and two consisted of females suckling young. Of the two groups seen in the dry season, one was foraging in a *buritizal* (*Mauritia* palm stand) and the other travelling through riverine forest. The spatial arrangement of foraging peccaries differed greatly from that of travelling individuals. Those in the former groups were spatially distant from and frequently beyond each other's view, but they generally moved in the same direction. In

Table 8.1. Population data for 36 white-lipped peccary groups encountered in the wet season (1988) in southeastern Maracá, Brazil

| Peccaries | Total no. | Mean no. per group | s.d. |
|---|---|---|---|
| Adults | 558 | 17.0 | 14.8 |
| Juveniles | 45 | 3.5 | 2.1 |
| Total | 603 | 18.5 | 15.6 |

Table 8.2. The density of white-lipped peccaries on southeastern Maracá, estimated using two methods

| Method | Groups per $km^2$ | Peccaries per $km^2$ | Mean no. per group |
|---|---|---|---|
| King's method | 7.0 | 138.8 | 19.9 |
| Webb method | 30.1 | 542.3 | 18.0 |

contrast, members of travelling groups moved in a tightly cohesive unit, lined up in single file. The latter formation created distinctive trails which extended for long distances.

## COMMUNICATION

White-lipped peccaries emit a variety of sounds. I briefly describe some of the more frequent noises recorded, their context, and possible function.

### Teeth clacking

A very loud, sharp clacking sound made by snapping the teeth together. Peccaries often clacked when they became aware of my presence. The frequency of clacking was highest when the animals were startled, or when they approached us, and lowest when they foraged unaware of our presence. In the latter context clacking was directed by one peccary towards another. The target animal responded by squealing (similar to that made by a frightened domestic pig) and quickly moving away from the clacker. The context of clacking suggest that it functions as an aggressive warning.

### Squealing

Described above. Young animals often squealed when startled, or chased by an adult. Adults squealed during confrontations in response to aggressive behaviour from other adults. The context suggested an appeasement or submissive function.

### Barking

Similar to the low bark of a dog, this vocalization occurred most often during intra-group feeding interaction. A bark was sometimes directed by an adult towards a juvenile. The juvenile usually responded with a squeal. Adults also barked at one another. The context of the call suggested a mildly aggressive warning, since the receiving animal often ignored the bark.

**Whoofing**

This sound was the first emitted by a startled animal, before it began clacking. Whoofing sounds were also made during foraging periods. Group members responded to whoofing with a whoof. The whoof seems to serve a 'take notice' function and/or to locate other group members. Whoofing may help maintain group cohesion when the animals forage through thick understorey vegetation.

## BIRTHING PERIOD

Two size-classes of juvenile peccaries were evident in the wet season. One consisted of small individuals approximately 30 cm in length. Individuals in the second size class were slightly larger (approximately 45 cm in length). In contrast, adults were approximately 90 cm in length. Local people reported that white-lipped peccaries give birth throughout the year. The two different juvenile size-classes encountered suggest that there are at least two breeding and birthing periods per year. My guide indicated that a female bears young only once a year.

## GROUP SIZE

Peccary group counts ranged from 1 to 48 individuals, but there were always animals hidden by vegetation. The most frequently encountered group sizes were 4, 30 and 48 individuals. The data on group size are preliminary and rudimentary in nature. Reliable local people reported seeing groups of over 200 individuals. This estimate is similar to that recorded in the literature (Kiltie and Terborgh, 1983; Sowls, 1984). In this study, counts averaged 18.3 members.

## PECCARY FORAGING AND SIGNS

In the wet season, foraging ($n = 24$) and non-foraging ($n = 12$) peccaries were seen only in *terra firme* forest. Peccaries furrowed (rooted) huge areas of soil when searching for food. These were fairly distinctive in appearance since the peccaries furrowed and turned over soil and leaf-litter. Rooted areas were most common below trees dropping fruit, around the base of *inajá* palms (fruiting and non-fruiting), near rotting logs, and in *buritizais*. This activity exposed the soil beneath the leaf-litter and eliminated established seedlings by accidentally uprooting them.

Of the two herds encountered in the dry season, one foraged in a *buritizal* and the other was travelling through riverine forest. During this season, peccary rooting was seen only in *buritizais*, and every *buritizal* ($n = 18$) had been rooted by January. From January to March 1991, I found fresh rooting three times in one *buritizal*, and twice in another. These observations suggest that peccaries selectively searched for *buritizais*. However, there are reports of peccaries feeding in other low, wet areas in the dry season (J. da Silva, pers. comm.). I visited one *buritizal* almost daily during the peak *buriti* mast, when many trees had 30–90 freshly fallen fruits below their crowns. Indirect evidence (tracks, teeth-marks on fruit, and faeces) and actual sightings indicated that tapirs (*Tapirus terrestris*), tortoises (*Geochelone carbonaria* and *G. denticulata*), agouti (*Dasyprocta agouti*), paca (*Agouti paca*), and brocket deer (*Mazama americana*) fed daily on the fruit pulp, but perhaps because of their low densities they consumed little of the amount available. However, when a peccary herd foraged in the *buritizal* they stayed for approximately four hours, and consumed the pulp from almost every fruit in their path. They did not appear to eat the seeds at this time.

PALMS

The phenologies and habitat distribution of *inajá* and *buriti* differed. *Inajá* occurred in dense patches ($n = 5$, mean density = $20.4/2500$ m$^2$, s.d. = 7.9), and as solitaries, in *terra firme* areas and river levees. Large patches were common on gently sloping hillsides adjacent to *buritizais*. In January 1991, most trees were in the early stages of flowering. By March and April, 74% (37) of the 50 *inajá* trees surveyed were in flower, 10% (5) had green fruit, and 16% (8) were non-reproductive. *Inajá* trees and patches mast synchronously. Generally, fruit-drop peaks from June through August, and fruit-fall ends in September (G. de Oliveira, pers. comm.). However, in 1987, Latham (in Milliken and Ratter, 1989) noted ripe *inajá* fruit falling from August to October in eastern Maracá. Thus the timing of the *inajá* mast appears to vary slightly between years, but in general it occurs soon after the onset of the wet season.

I observed white-lipped peccaries consuming *inajá* seeds. They and perhaps collared peccaries are probably the only terrestrial vertebrates on Maracá capable of cracking *inajá*. The other ungulates cannot crack open hard seeds (Bodmer, 1989), the rodents gnaw them open (Smythe, 1986), while the local monkey species are probably incapable of cracking mature seeds (G. da Silva, pers. comm.).

*Buriti* occurred in dense patches ($n = 5$, mean density = $62.4/2500$ m$^2$, s.d. = 40.2). These were restricted to seasonally inundated areas, but patches were common throughout the forest. The *buriti* mast peaked in January and February 1991. By late March and April 90% (45) of the 50 trees surveyed were non-reproductive, 4% (2) had ripe fruit, 2% (1) had green fruit, and 4% (2) were in flower. Moskovits (1985) reported that *buriti* drops ripe fruits throughout the year. My observations indicate a peak in fruit fall at the end of the dry season. According to Moskovits (1985), *buriti* flowers and fruits synchronously within a patch (*buritizal*), but asynchronously between patches. I observed synchronous flowering and fruiting between patches of *buriti*.

Peccaries ate the pulp off the *buriti* fruit but dropped the seeds undamaged. According to local hunters they consume only seeds that have germinated. As the herd moved through the *buritizal* the peccaries trod *buriti* seeds into the moist soil incidentally.

**Table 8.3.** Numbers of family members of hunters of peccaries around the Ilha de Maracá, and the number of white-lipped peccaries killed per hunt and per month

| Hunter number | No. of family members | Peccaries killed/hunt | Peccaries killed/month |
|---|---|---|---|
| 1 | 7 | 4 | 8 |
| 2 | 5 | 2 | 4 |
| 3 | 6 | 3 | 3 |
| Total | 18 | 9 | 15 |

## HUMAN HUNTING OF PECCARIES

No peccaries were killed in the dry season of 1991. The three men interviewed in 1988 regularly hunted peccaries in the wet season, and two obtained most of their meat from this species. The hunters killed an average of three peccaries per hunt, and five per month (Table 8.3). They reported no sex bias in the killing of peccaries. Hunting frequency varied in relation to success. For example, one man hunted for four full days per week during unsuccessful periods, and only once per two weeks when successful. The other hunted for half a day approximately once per week. The men believed white-lipped peccaries were more common five years ago. Gold prospectors also hunted peccaries extensively, with a former prospector reporting kills of over 50 peccaries during single hunts.

## DISCUSSION

### WHITE-LIPPED PECCARIES AND PALMS

In 1988, eastern Maracá had the highest densities of peccaries ever recorded (see Kiltie and Terborgh, 1983; Sowls, 1984). These densities, observed during the wet season, correlated with the masting of *inajá* and the time of greatest fruit abundance on eastern Maracá (Moskovits, 1985; A. Nunes, pers. comm.). I caution that the data on peccary densities were derived from a local area over a short time period and may not represent overall densities. Since I was limited to a few pre-existing trails, it is likely that I counted some herds more than once during the course of the study. Davis and Winstead (1980) discuss this and other problems associated with the line transect method of estimating animal densities.

The predictability of the *inajá* mast may influence the direction and timing of peccary movements, since it provides a relatively reliable cue to the location of superabundant food. In the dry season *buriti* appeared to replace *inajá* as the key resource influencing peccary movements. Many palm species mast abundantly on a yearly basis, but non-palm tree species (dicotyledons) often fluctuate yearly in their production of fruit, sometimes skipping a year or more (Leigh *et al.*, 1982; Kinnaird, 1990; Ribeiro, 1991). The irregular fruiting of dicotyledons in tropical areas, coupled with the scattered dispersion of conspecifics (Leigh *et al.*, 1982), may make most dicotyledonous species unreliable cues for mapping foraging paths.

However, non-palm fruits constitute a significant portion of a peccary's diet (Bodmer, 1989). These are probably encountered incidentally as herds travel between palm stands.

The dry season is a time of fruit scarcity on Maracá (pers. obs.; Moskovits, 1985), as it is in other neotropical areas (Foster, 1982). Many frugivores may starve during this period (Foster, 1982). Since peccaries occur in large herds they rapidly deplete food patches (e.g. *buriti*). For example, the herd observed foraging in the dry season consumed most of the *buriti* fruit on the ground within hours of entering the patch. A peccary herd therefore probably increases its range in the dry season, since *buritizais* are widely dispersed throughout the forest. This hypothesis is supported by work performed by Ayres (1986) and Bodmer (1990), who noted that in the dry season peccaries expanded their range into seasonally flooded forests. In contrast, during the wet season a herd probably reduces its range since fruit availability is greater. This hypothesis is bolstered by the high density and great frequency with which peccaries were observed in eastern Maracá in the wet but not the dry season.

Peccaries incidentally buried *buriti* seeds after consuming the pulp. Seed burial can provide a good micro-habitat for germination, and reduce or eliminate predation on the buried seeds by insect larvae (Smythe, 1989). In addition, seeds buried by peccaries may avoid being eaten or carried by agoutis or pacas into unsuitable habitats. Since *buriti* palms occur only in well-defined patches, one would presume that *buritizais* are the best places for seeds to germinate. This relationship appears to be true for *bussú* (*Manicaria saccifera* Gaertn.), another Amazonian palm whose distribution is restricted to low seasonally inundated areas (Ribeiro, 1991).

In *terra firme* forests soil-furrowing by peccaries may benefit seeds falling onto rooted areas, since soil disturbance plays an important role in the establishment of tree species in tropical forests (Putz, 1983). For example, black palm (*Astrocaryum standleyanum* L.H. Bailey) seeds can germinate on the forest floor but they require burial for roots to penetrate the soil (Smythe, 1989).

## PECCARIES AND PALM DYNAMICS

Peccaries may be of critical importance to forest dynamics, because they are the only surviving neotropical terrestrial vertebrates to form huge herds which forage over immense areas. This foraging strategy results in peccaries consuming many post-dispersal seeds (Kiltie, 1981b), including those scatter-hoarded by agoutis and those dropped by arboreal frugivores (Kiltie, 1981b; Kiltie and Terborgh, 1983).

Seed predators can alter tree species composition, dispersion and density (e.g. Janzen, 1969, 1971, 1983; Hubbell, 1980; Smith, 1987). Peccaries feed on germinated *buriti* seeds which have avoided destruction by other seed-eaters, so they must therefore influence the community dynamics of the forest. These peccaries are also the only ungulates that eat the seeds of *Iriartea deltoidea* Ruiz & Pav., *Socratea exorrhiza* (Mart.) Wendl. and *buriti* palms (Kiltie, 1981c, 1982; Kiltie and Terborgh, 1983; Bodmer, 1989). They also eat the seeds of many other species of palms, including *Oenocarpus* sp., *Phytelephas* sp. and *Astrocaryum murumuru* Mart. var. *macrocalyx* (Kahn & Millán) Henderson (Kiltie and Terborgh, 1983; Bodmer, 1989).

The selective pressure exerted by herding animals on plants differs from that of solitary or small-group-forming seed predators (Janzen, 1974, 1976, 1986; Kiltie and Terborgh, 1983; Hallwachs, 1986). Herding animals may be responsible for the evolution of synchronized mast-fruiting in tropical plant species (Janzen, 1976). Theoretically, seed predators in large groups are more likely to locate and destroy most, if not all seeds produced by small isolated patches of trees (Janzen, 1976). Trees in large patches are more likely to swamp seed predators with more seeds than they can eat (Janzen, 1974, 1976). Interestingly, *inajá* occurs in large conspecific patches, as does *buriti*.

The extirpation of peccaries from Santa Rosa National Park in Costa Rica appears to have had repercussions on some plants. For example, Hallwachs (1986) found that the fruits of *Hymenaea courbaril* L. (Leguminosae) remained untouched for months beneath trees, until dispersed or eaten by agoutis. She suggests that if white-lipped peccaries still inhabited the park they would have quickly consumed most, if not all, of the fruits. Similarly, Smythe (1989) found that *Astrocaryum* seeds remained beneath parent trees for months or years on Barro Colorado Island (BCI), Panama, slowly being eaten by collared peccaries and gnawed open or scatter-hoarded by agoutis. If white-lipped peccaries still occurred on BCI they would probably quickly consume most seeds.

## HUNTING OF PECCARIES

In areas adjacent to Maracá, peccaries provided Amerindians with much of the meat they consumed. This is also true in other regions of Amazonia, both for indigenous and non-indigenous peoples (Dourojeanni, 1985; Redford and Robinson, 1987; Vickers, 1988). Unfortunately, extensive and intensive hunting by gold prospectors adjacent to the reserve threatens the peccary population. The species is probably extinct in El Salvador, and rapidly being extirpated throughout the neotropics (Sowls, 1984; Stearman, 1989). In fact the white-lipped peccary may be the most threatened large mammal of the neotropics (Smythe, 1987), yet there is no mention of it in the IUCN *Mammal Red Data Book* (Thornback and Jenkins 1982).

## CONCLUSION

Peccary densities correlated with changes in fruit production, which in turn were correlated with seasonal rainfall. Peccaries were abundant in the 1988 wet season when *inajá* masted and other fruits were abundant. In the 1991 dry season peccaries declined in abundance, perhaps because most tree species stopped fruit production. *Buriti* continued producing fruit in the dry season and these were eaten by peccaries, but *buritizais* have a dispersed distribution. This dispersion pattern, coupled with an extreme decline in the availability of other fruit species, may have forced peccaries to increase their foraging range.

The foraging range of peccaries probably decreases in the wet season when fruit is abundant. However, fruit production in tropical forests can vary greatly between years. In poor fruit production years (wet and dry season), peccaries probably increase their foraging range. They may even abandon their normal home range. Herds may have seasonal, yearly and possibly multi-year ranges. This hypothesis may explain peccary disappearance from areas where they were common for years, and the belief that peccary migrations have no pattern. Population decline resulting from outbreaks of disease is an alternative explanation.

Peccaries are rapidly disappearing, yet their ecology and natural history remains relatively unknown. Peccary extinction (or population extirpations) could have significant ecological repercussions for a tropical forest community, since seed predators exert selective pressures on plant populations. Since peccaries consume the seeds of many tree species, their extinction would alter the dynamics of tropical forests.

Extensive and intensive hunting by gold prospectors near the reserve, and in some areas of the island, poses a significant danger to the white-lipped peccary population of Maracá. Perhaps a more serious threat is the rapid progression of the agricultural frontier along the eastern edge of the reserve. If forest conversion continues, Maracá will eventually be isolated from the contiguous Amazon forest. Since we do not know peccary home range sizes, we cannot be sure that Maracá is sufficient in area to support a population. Peccary extirpation would have a serious impact on local indigenous peoples since they obtain much of their meat from this animal.

## ACKNOWLEDGEMENTS

I thank Jean Huffman for her encouragement, support and understanding; without her the study could not have been completed. John Eisenberg, Kent Redford and John Robinson provided moral and logistic support during critical moments in the project's development. Without Peter Feinsinger's criticisms and understanding I may not have attempted the study. Celso Morato de Carvalho (INPA Boa Vista) kindly provided essential space for the writing of the article. Thanks also to IBAMA for allowing me to work on Maracá, Dr Blos for his support, and Guttemberg Moreno de Oliveira (IBAMA) for his enlightened management of the Ecological Station. The project was funded by an Amazon Research and Training Grant from the Tinker Foundation, the Royal Geographical Society, Wildlife Conservation International, the World Wildlife Fund, and the Chicago Zoological Society.

## REFERENCES

Ayres, J.M. (1986). Uakaris and Amazonian flooded forest. Unpublished PhD thesis, University of Cambridge.
Bodmer, R.E. (1989). Frugivory in Amazon ungulates. Unpublished PhD Thesis, University of Cambridge.
Bodmer, R.E. (1990). Responses of ungulates to seasonal inundation in the Amazon flood plain. *Journal of Tropical Ecology*, 6, 191–201.
Caldecott, J. (1988). *Hunting and wildlife management in Sarawak*. IUCN, Gland, Switzerland.
Davis, D.E and Winstead, R.L. (1980). Estimating the numbers of wildlife populations. In: *Wildlife management techniques manual*, ed. S.D. Schemnits. Wildlife Society, Washington, DC.
Dourojeanni, M.J. (1985). Over-exploited and under-used animals in the Amazon region. In:

*Key environments: Amazonia*, eds G.T. Prance and T.E. Lovejoy, pp. 419–433. Pergamon, Oxford.

Eisenberg, J.G. (1981). *The mammalian radiations.* University of Chicago Press, Chicago.

Eisenberg, J.G. (1989). *Mammals of the Neotropics: Vol. 1., The Northern Neotropics.* University of Chicago Press, Chicago.

Foster, R.B. (1982). The seasonal rhythms of fruit fall on Barro Colorado Island. In: *The ecology of a tropical forest, seasonal rhythms and long-term changes*, eds E.G. Leigh Jr, A.S. Rand and D.M. Windsor, pp. 151–172. Smithsonian Institution Press, Washington, DC.

Hallwachs, W. (1986). Agoutis (*Dasyprocta punctata*): The inheritors of guapinol (*Hymenaea courbaril* : Leguminosae). In: *Frugivores and seed dispersal*, eds A. Estrada and T. Fleming, pp. 119–135. Dr W. Junk Publishers, Netherlands.

Hubbell, S.P. (1980). Seed predation and the coexistence of tree species in tropical forest. *Oikos*, 35, 214–229.

Janzen, D.H. (1969). Seed-eaters versus seed size, number, toxicity and dispersal. *Evolution*, 23, 1–27.

Janzen, D.H. (1971). Seed predation by animals. *Annual Review of Ecology and Systematics*, 2, 465–492.

Janzen, D.H. (1974). Tropical blackwater rivers, animals and mast fruiting by the Dipterocarpaceae. *Biotropica*, 6, 69–103.

Janzen, D.H. (1976). Why bamboo wait so long to flower. *Annual Review of Ecology and Systematics*, 7, 347–391.

Janzen, D.H. (1983). Food webs: who eats what, why, how, and with what effects in a tropical forest? In: *Tropical rainforest ecosystems. Structure and function*, ed. F.B. Golley. Elsevier Scientific, Amsterdam.

Janzen, D.H. (1986). Mice, big mammals and seeds: it matters who defecates where. In: *Frugivores and seed dispersal*, eds A. Estrada and T. Fleming, pp. 251–271. Dr W. Junk Publishers, Netherlands.

Jarman, P.J. (1974). The social organization of antelopes in relation to their ecology. *Behaviour*, 58, 215–267.

Kahn, F. and Castro, A. de (1985). The palm community in a forest of Central Amazonia, Brazil. *Biotropica*, 17, 210–216.

Kiltie, R.A. (1980). More on Amazon cultural ecology. *Current Anthropology*, 21, 541–546.

Kiltie, R.A. (1981a). Stomach contents of rainforest peccaries (*Tayassu tajacu* and *T. pecari*). *Biotropica*, 13, 234–236.

Kiltie, R.A. (1981b). Distribution of palm fruits on a rain forest floor: why white-lipped peccaries forage near objects. *Biotropica*, 13, 141–145.

Kiltie, R.A. (1981c). The function of interlocking canines in rain forest peccaries (Tayassuidae). *Journal of Mammalogy*, 62, 459–469.

Kiltie, R.A. (1982). Bite force as a basis for niche differentiation between rain forest peccaries (*Tayassu tajacu* and *T. pecari*). *Biotropica*, 14, 188–195.

Kiltie, R.A. and Terborgh, J. (1983). Observations on the behavior of rain forest peccaries in Peru: why do white-lipped peccaries form herds? *Zeitschrift für Tierpsychologie*, 62, 214–255.

Kinnaird, M.F. (1990). *Behavioral and demographic responses to habitat change by the Tana River crested Mangabey.* Unpublished PhD Dissertation, University of Florida, Gainsville, Florida.

Leigh, E.G. Jr, Rand, A.S. and Windsor, D.S. (1982). *The ecology of a tropical forest, seasonal rhythms and long-term changes.* Smithsonian Institution Press, Washington, DC.

Leuthold, W. (1977). *African ungulates: zoophysiology and ecology.* Springer-Verlag, New York.

Mayer, J.J. and Brandt, P.N. (1982). Identity, distribution and natural history of the peccaries. In: *Mammalian biology in South America: the Pymatuning symposia in ecology*, eds M.A. Mares and H.H. Genoways, pp. 433–455. University of Pittsburgh, Pittsburgh, USA.

Mayer, J.J. and Wetzel, R.M. (1987). *Tayassu pecari*. *Mammalian Species*, 293, 1–7.
Milliken, W. and Ratter, J.A. (1989). *The vegetation of the Ilha de Maracá*. RGS/Royal Botanic Garden Edinburgh.
Moore, H.E. (1973). The major groups of palms and their distributions. *Gentes Herbarium*, 11, 27–41.
Moore, H.E. (1977). Endangerment at the specific and generic levels in palms. In: *Extinction is forever: the state of threatened and endangered plants of the Americas*, eds G.T. Prance and T.S. Elias, pp. 267–282. New York Botanical Gardens, Bronx, New York.
Moskovits, D. (1985). *The behavior and ecology of the two Amazonian tortoises,* Geochelone carbonaria *and* Geochelone denticulata, *in Northwestern Brazil*. Unpublished PhD Thesis, University of Chicago, Illinois.
Putz, F.E. (1983). Tree fall pits and mounds, buried seeds, and the importance of soil disturbance to pioneer trees on Barro Colorado Island, Panama. *Ecology*, 64, 1069–1074.
Redford, P.H. and Robinson, J.G. (1987). The game of choice: patterns of indian and colonist hunting in the neotropics. *American Anthropologist*, 89, 650–667.
Ribeiro, A. de S. (1991). *Estrutura e dinâmica de uma população de bussu –* Manicaria martiana *Burret (Arecaceae) – em floresta úmida de terra firme na Amazônia Central, Manaus, Brasil*. INPA, Manaus, Brasil.
Robinson, J.G. and Redford, K.H. (1986). Body size, and population density of neotropical forest mammals. *American Naturalist*, 128, 665–680.
Smith, T.J. (1987). Seed predation in relation to tree dominance and distribution in mangrove forests. *Ecology*, 68, 266–273.
Smythe, N. (1986). Competition and resource partitioning in the guild of neotropical terrestrial frugivorous mammals. *Annual Review of Ecology and Systematics*, 17, 169–188.
Smythe, N. (1987). The importance of mammals in neotropical forest management. In: *Management of the forests of Tropical America: prospects and technologies*, ed. J.C.F. Colon. USDA Forest Service.
Smythe, N. (1989). Seed survival in the palm *Astrocaryum standleyanum*: evidence for dependence upon its seed dispersers. *Biotropica*, 21, 50–56.
Sowls, L.K. (1984). *The peccaries*. University of Arizona Press, Tucson.
Stearman, A.M. (1989). *The effects of settler incursion on fish and game resources of the Yugui, a native Amazonian society of Eastern Bolivia*. Unpublished manuscript, University of Central Florida.
Thornback, J. and Jenkins, A. (1982). *Mammal red data book*. IUCN, Gland, Switzerland.
Vickers, W.T. (1988). Game depletion hypothesis of Amazonian adaptation: data from a native community. *Science*, 239, 1521–1522.

# 9 The Bats of the Ilha de Maracá

FIF ROBINSON
*Whitby, UK*

## SUMMARY

Results of a field survey of the bats of the Ilha de Maracá are presented. Bats were captured at various sites around the island between July 1987 and February 1988. A total of 315 bats of 37 species were caught, using mist nets and a harp trap. A humane research policy was operated, with a minimum number of bat specimens being taken, and experiments were conducted with a new technique for making imprints of bats' dentition to reduce the need for taking specimens.

Results are combined with those of previous workers to produce a provisional species list for the island, including 49 species. The exceptionally high diversity of bats on Maracá, which may be shown by future studies to include as many as 60 species, may be partly due to the diversity of habitats on the island and the seasonal nature of the climate. Almost three times as many individual bats were caught at waterside sites as in closed forest, but the number of species recorded in the two habitats was the same (26). The Phyllostominae and the Emballonuridae were better represented in the forest than at waterside sites. The harp trap proved more efficient than mist nets in terms of captures per unit area, and was found to catch proportionally fewer phyllostomid bats and proportionally more small insectivorous species.

## RESUMO

São apresentados os resultados de um levantamento dos morcegos da Ilha de Maracá. Os morcegos foram capturados em várias localidades na ilha, entre julho de 1987 e fevereiro de 1988. Usando redes e uma 'armadilha de harpa', um total de 315 indivíduos de 37 espécies foram capturados. Foi adotada uma política de coleta do mínimo de exemplares, e experiências foram realizadas com uma nova técnica de impressão dental, para reduzir ao máximo a necessidade de matar.

Os resultados são reunidos com os de estudos anteriores para produzir uma lista provisória dos morcegos da ilha, incluindo 49 espécies. A diversidade excepcional de morcegos em Maracá, que futuros estudos podem mostrar a incluir tanto quanto 60 espécies, é possivelmente em parte o resultado da alta diversidade de habitats na ilha, e o caráter sazonal do clima. Maiores números de morcegos foram capturados na beira d'água do que na floresta fechada, mas o número de espécies registrado em cada um destes ambientes foi o mesmo (26). Os Phyllostominae e Emballonuridae foram melhor representados na floresta do que na beira d'água. A 'armadilha de harpa' se mostrou mais eficiente do que as redes em termos de capturas por metro quadrado, e capturou proporcionalmente menos morcegos das Phyllostomidae e mais das pequenas espécies insetívoras.

# INTRODUCTION

This survey of the bats of the Ilha de Maracá was carried out at various times between July 1987 and February 1988. The principal aims were to make an inventory of the bat species living on Maracá, and to investigate aspects of their ecology, particularly habitat preferences, vertical stratification within the forest, and diet. Much of the work was carried out around the Ecological Station at the eastern end of the island, but excursions were also made to a variety of sites in other parts of the island (see Figure Pr.1, p. xvii). This is a generalized account of the results of the survey. For more detailed information about the fieldwork carried out, and weights and measurements of the bats caught, refer to Robinson (1989a).

The Ilha de Maracá is 60 km long and up to 25 km wide, with a total area of about 1000 km$^2$. The majority of this area is undisturbed forest, though there are pockets of savanna at the eastern end of the island, some areas of permanent wetland, and a small amount of disturbed vegetation caused by the building of the Ecological Station. This survey was carried out almost entirely in undisturbed forest and in waterside habitats, which include the Causeway running across a marshy area between the Ecological Station and the river.

In addition to the survey, a subsidiary aim was to develop techniques for minimizing the necessity for collecting bat specimens. The three key elements of the minimum collection policy were: (i) only collecting a specimen for a known purpose; (ii) using accurate and verifiable techniques of field identification; and (iii) developing the 'chewit' system for making permanent imprints of the teeth of a live bat. Further details are given in the Discussion.

Prior to this survey the bats of Maracá were not well known. Initial work was carried out by Taddei and Reis (1980) who recorded 12 bats of eight species in the vicinity of the Ecological Station during a brief visit. Mok *et al.* (1982) caught 175 bats of 20 species during a week on the island. Unfortunately their paper only indicates two of the species which were caught on Maracá, and their data and specimens in INPA have now been lost.

In 1987–88 a parallel survey of the bats of Maracá was carried out by Rogério Gribel and colleagues from INPA. They kindly made available the results of their work during June and November of 1987, and I have incorporated species recorded by them into the provisional species list for the island (Table 9.1). I do not have available the results of subsequent visits in 1988.

# METHODS AND MATERIALS

Fieldwork was carried out during the periods 14 July to 10 August 1987, 27 November to 22 December 1987 and 26 January to 8 February 1988. During the period 22 July to 3 August 1987 I visited INPA in Manaus to discuss the work with Rogério Gribel and colleagues, and to examine their specimen collection and references. I am very grateful to Rogério for the invaluable assistance he gave me during this visit.

Two hundred and twenty-nine of the 315 bats caught during the survey were caught in mist nets. About 90% of netting effort was in mist nets strung between poles at ground level in the forest or at waterside sites, and 10% in the sub-canopy of the forest with nets strung up at heights of 3–10 m above ground level.

It was necessary to patrol mist nets constantly, as bats can struggle violently, causing serious entanglement of smaller bats or escape of larger species. A leather glove was worn on the left hand to avoid being bitten, while the right hand was kept free for disentanglement of the net. An adjustable 'Petzl' headtorch was used as a light source during extraction, while a miner's lamp with a lead-acid battery was used to give up to 12 hours of general illumination.

A further 85 bats were caught in a single 'harp trap' kindly lent to the project by Ben Gaskell. This consisted of a 2 × 2 m metal frame with two sets of very fine tungsten wires threaded vertically across it. Bats are caught by flying into the wires, which are too fine for most bats to detect, and sliding down them into a catching bag below. This trap was especially good for catching small insectivorous bats, and could be hauled into the sub-canopy without fear of entanglement with trees. Furthermore it was not necessary to patrol the harp trap constantly, since once caught the bats could sleep unharmed in the catching bag.

Some effort was also put into looking for bat roosts during the daytime. This met with little success, except for the unavoidable roosts of molossids in the roof of the Ecological Station and amongst cracked boulders of the river bed in the western half of the island.

As the survey was ecological (rather than taxonomic) in nature, I operated a minimum collection policy. I did not take specimens of species which I was confident had been reliably identified in the field. I tried only to take a single specimen of species which were more difficult to identify. In a few cases more than one specimen was taken if there was a possibility that two separate species were involved, or in cases of accidental injury. In total 29 specimens were collected from the 38 species which were recorded.

All bat specimens were donated to the collection of INPA in Manaus. Difficulties in identification of certain specimens will be resolved by consultation with experts in Brazil or abroad. The British Museum (Natural History) indicated that there were no species of bat which they specially required for their collection, though if specimens were provided they would be happy to identify them. As I am confident that all specimens can be reliably identified in Brazil, I considered it most appropriate that the specimens should remain in Amazonia.

When caught, each bat was weighed, measured and where possible identified in the field. As well as taking a standard set of measurements, careful note was made of features critical to the identification of each species or genus.

Wherever possible, bats were held in separate cotton bags so that their faeces could be collected for subsequent examination. Seeds were present in the faeces of some of the frugivorous species, but attempts to germinate them were unsuccessful. No attempt has been made to identify the insect remains from faeces. Ectoparasites were noted and/or collected from many of the bats, and donated to the entomology department of INPA.

As the identification of many species relies on the careful examination of the teeth, which often requires that the bat be killed, I experimented with making finely detailed imprints (called 'chewits') of the teeth, using silicone dental putty. For more details of this technique see Robinson (1989b).

## RESULTS AND DISCUSSION

Table 9.1 is a provisional bat species list for Maracá, including relevant collecting data. This is by no means definitive, as identification of some difficult species needs to be confirmed. The list includes species recorded by Rogério Gribel and colleagues in June and November 1987, but not in subsequent visits to the island in 1988. Species accounts are presented at the end of the paper.

### DIVERSITY OF THE MARACÁ BAT FAUNA

The bat fauna of Maracá is exceptionally diverse. The Shannon–Weaver species diversity index ($H'$) for all of the captures in this survey is 2.77, which is the highest value I have seen reported for any site in Latin America. The results of the survey by Rogério Gribel *et al.*, carried out in parallel with this work, yielded a similar figure. The number of species recorded from the island is currently at least 49, but new species were still encountered regularly towards the end of the survey and the total could well be around 60 species. In a typical night's netting one might catch as many as 15 bats of 10 different species.

This great variety of bats is a delight to the naturalist, every night's work providing new surprises. The bats of Maracá range from 3 g to 100 g in weight, and feed variously on insects, fruit, nectar and pollen, fish, frogs and blood. However, for the ecologist wishing to analyse the underlying structure of the bat community, this diversity creates many problems. Mist nets left in the forest will seldom catch the same bat species from one night to the next, and the number of bats of each species is generally so low that sample sizes are insufficient for valid statistical analysis. These problems are further complicated by local variation in the composition and structure of the forest, the phase of the moon, local seasonal migrations and the practical difficulties of standardizing capture effort. Consequently there are no straightforward answers to simple questions such as 'which bats live where?' A short survey such as this cannot therefore hope to provide much more than a rough sketch of the bat community, and to suggest lines for further enquiry. All conclusions are necessarily tentative.

Tropical rainforests are home to the most diverse bat faunas on the planet (Humphrey and Bonaccorso, 1979), largely due to their structural diversity and great variety of food resources. The causes of the exceptional diversity of the bats of Maracá are a matter for speculation, but contributory factors may include:

*Habitat diversity* Bats were caught throughout the island in a wide variety of sites in order to gain an overall picture of the bat fauna. The data used for calculating

**Table 9.1.** Provisional bat species list for the Ilha de Maracá, with field data

| Species | No. | Method | Location | Height | Sex | Voucher | Faeces | Ectoparasites | Other |
|---|---|---|---|---|---|---|---|---|---|
| **EMBALLONURIDAE** | | | | | | | | | |
| *Rhynchonycteris naso* Wied | 3 | M | Anta (R) | Low over water | M | 0 | | | |
| *Saccopteryx bilineata* (Temminck) | 10 | M(3); H(7) | FT; Ca | 1–3 m | | 1 | Insects | Eggs (acarine?) | |
| *Saccopteryx leptura* (Schreber) | 1 | H | FT | 15 m | F | 0 | | | RG |
| *Saccopteryx canescens* Thomas | 3 | M(1); H(2) | FT; Ca | 1–8 m | | 1 | | | |
| *Cormura brevirostris* (Wagner) | 1 | H | FT | | M | 1 | | | |
| **NOCTILIONIDAE** | | | | | | | | | |
| *Noctilio leporinus* (Linnaeus) | 1 | M | Anta (R) | Low over water | M | 0 | Fish scales; bone fragments | | |
| **MORMOOPIDAE** | | | | | | | | | |
| *Pteronotus parnelli* Gray | 5 | M(2); H(3) | FT; Ca | 1–7 m | | 0 | Insect remains (incl. beetle elytron) | Winged flies (2); mites (1) | RG,T+R |
| *Pteronotus personatus* (Wagner) | 51 | H | FT; Ca | 1–3 m | | 0 | Insect remains | Ticks (17); mites (10); winged flies (2) | RG |
| *Pteronotus gymnonotus* (Natterer) | 4 | H | FT; Ca | 5–7 m | | | | | |
| **PHYLLOSTOMIDAE** | | | | | | | | | |
| *Micronycteris megalotis* (Gray) | 3 | M | FT | 0–2 m | | 0 | Fruit pulp; hairs | | RG |
| *Micronycteris sylvestris* (Thomas) | — | | | | | | | | RG |
| *Micronycteris nicefori* Sanborn | 1 | M | FT | 0–2 m | | 1 | | | RG |
| *Micronycteris hirsuta* (Peters) | — | | | | | | | | RG,M |
| *Micronycteris minuta* (Gervais) | 1 | M | Ca | 0–2 m | | 0 | | | RG |
| *Tonatia bidens* (Spix) | 1 | M | FT | 10–12 m | | 1 | | | RG |
| *Tonatia silvicola* (D'Orbigny) | — | | | | | | | | RG |
| *Tonatia brasiliense* (Parets) | 2 | M | Ca; Abacaxi (R) | Low | | 1 | | | RG |
| *Mimon crenulatum* (E. Geoffroy) | 4 | M(2); H(2) | FT; Anta (F) | | | 0 | | Louse (1); mites (25) | RG |
| *Phyllostomus discolor* (Wagner) | — | | | | | | | | RG |
| *Phyllostomus elongatus* (E. Geoffroy) | 9 | M(7); H(3) | FT; Ca; Anta (F/R) | 0–2 m | | 0 | Insect remains | | RG |
| *Phyllostomus hastatus* (Pallas) | 9 | | Ca; FT | | | 0 | Seeds; insect remains | | RG |
| *Phylloderma stenops* Peters | — | | | | | | | | RG |
| *Trachops cirrhosus* (Spix) | 5 | M | Fumaça (R); Anta (F); FT | 0–1 m | | 0 | | Winged flies (2); mites (1) | |
| *Lonchorhina aurita* Tomes | 4 | M | Abacaxi (R); Fumaça (R) | | | 1 | | | |
| *Glossophaga soricina* (Pallas) | 13 | M | Fumaça (R); Abacaxi (R); Anta (F); FT | 0–8 m | | 0 | Fruit (small seeds) | | RG |
| *Lonchophylla* cf. *thomasi* Allen | — | | | | | | | | RG,M |
| *Choeroniscus* cf. *minor* (Peters) | 2 | M | Anta (F) | | | 0 | | | |

*continues overleaf*

**Table 9.1.** (*continued*)

| Species | No. | Method | Location | Height | Sex | Voucher | Faeces | Ectoparasites | Other |
|---|---|---|---|---|---|---|---|---|---|
| *Carollia perspicillata* (Linnaeus) | 55 | M(54); H(1) | Fumaça (R); Abacaxi (R); Filhotes (R); Anta (F/R); Ca; FT | 0–2 m (8 m) | | 3 | Fruit pulp; seeds (*Piper*?) | Winged flies; ticks; mites; acarine eggs | RG,T+R |
| *Rhinophylla* cf. *pumilio* Peters | 2 | | | | | | | | RG |
| *Sturnira lilium* (E. Geoffroy) | 1 | M | Ca | 0–2 m | | 1 | Seeds; resin/gum | | RG |
| *Sturnira tildae* de la Torre | 3 | | Anta (R) | | | | Fruit; seeds | | RG,T+R |
| *Uroderma bilobatum* Peters | 6 | | FT; Abacaxi (R); Anta (R) | | | 1 | | | T+R |
| *Uroderma magnirostrum* Davis | – | | | | | | | | RG |
| *Artibeus lituratus* (Olfers) | 10 | | | | | 2 | Seeds; fruit pulp; resin/gum | | RG |
| *Artibeus jamaicensis* Leach | 14 | | | | | 1 | Seeds; fruit pulp; resin/gum | | RG |
| *Artibeus fuliginosus* Gray | – | | | | | | | | T+R |
| *Artibeus concolor* Peters | 2 | | Ca; Anta (R) | | | 2 | | | RG |
| *Artibeus* cf. *anderseni* Osgood | 4 | M | Ca; FT; Anta (R) | | | 2 | | | RG |
| *Vampyrops* sp. | 2 | | FT; Anta (F) | 0–2 m | M/F | 1 | | | RG |
| *Vampyressa* sp. | – | | | | | | | | RG |
| *Mesophylla macconnelli* Thomas | 2 | | Anta (R); Abacaxi (R) | | | 1 | Fruit pulp; black fibres | | RG,T+R |
| *Chiroderma villosum* Peters (*Stenodermine* sp.) | 1 | M | FT | 9 m | | 1 | | | RG,T+R |
| *Desmodus rotundus* (E. Geoffroy) | 3 | M | Anta (F/R) | | | 0 | | | |
| **VESPERTILIONIDAE** | | | | | | | | | |
| *Myotis* cf. *nigricans* (Schinz) | 17 | M(7); H(9); Hand (1) | FT; Ca; Abacaxi (R); Anta (R); Dan/Steve (R); Station | | | | | Wingless flies (2); mites (1); lice (2) | RG |
| *Eptesicus* sp. | 1 | | | | | | | | |
| **MOLOSSIDAE** | | | | | | | | | |
| *Tadarida laticaudata* (E. Geoffroy) | 53 | M | Dan/Steve (R); Abacaxi (R); Anta (R) | | | 1 | | Wingless flies; ticks; mites | |
| *Molossus molossus* (Pallas) | 2 | M | Station | | | 1 | | | RG,T+R |

Method: M = mist net (no. caught); H = harp trap.
Location: FT = forest track and environs (near Station), Ca = Causeway. Other names refer to camps around the island (F = forest; R = riverside). See Figure Pr.1.
Other (records): RG = Rogério Gribel and team from INPA, to 11 November 1987 (unpublished); T+R = Taddei and Reis (1980); M = Mok *et al.* (1982).

the diversity index have been pooled from different areas of primary forest (both at ground level and in the sub-canopy), from rocky rivers, quiet creeks and swampy areas. Each part of the island habitat mosaic supports different bats, and when the data from different areas are pooled the resultant diversity index is greater than that of each area taken on its own. For example, for all mist net captures in primary forest sites $H' = 2.45$ and for all waterside sites $H' = 2.25$, but when pooled these data yield $H' = 2.63$.

*Mixed sampling techniques* The apparent diversity of the bat fauna depends on the sampling methods used. Most calculations of $H'$ are based on samples taken in mist nets, which consistently under-record aerial insectivores of the families Emballonuridae, Mormoopidae, Vespertilionidae and Molossidae (bats which are able to detect mist nets). This survey used both mist nets and a harp trap, the latter strung with very fine tungsten wires and very successful at catching these insectivorous species, thus increasing the overall diversity of the sample.

*Forest structure* The marked dry season at Maracá seems to be responsible for the forest's relatively open structure. The canopy tends to be less dense and the sub-canopy and understorey better developed than in the forests of Central Amazonia. There is an interesting possibility that this could create a distribution of food resources and open space which favours a greater variety of foraging strategies than would be found in a forest with a denser canopy structure. In particular this might apply to insectivorous species, which need enough space to manoeuvre whilst chasing insects.

## DIFFERENCES BETWEEN HABITATS

These data do not permit a fine-scale analysis of the bats of different habitat types, since there are many confounding variables and the data collection was not systematic. Capture sites have been grouped broadly into 'forest' and 'waterside'.

The single most notable difference between forest and waterside sites was in the rate of capture of bats (see Table 9.2). The total number of captures per unit catching effort in mist nets was more than three times higher at waterside sites than in forest: 171 bats were caught at waterside sites during 4395 metre hours (mh) of netting effort (= 8790 $m^2$ hours, a standard mist net being 2 m high), i.e. 3.89 bats per 100 mh, whereas 4508 mh of netting in forest yielded only 56 bats (1.24 per 100 mh). Similarly the rate of capture in the harp trap was considerably higher at waterside sites, although the data have been skewed by one exceptional night's trapping on the Causeway (discussed below). The greater abundance of bats at these waterside habitats could be due to the influence of open water on food resources or flight activity, or it may simply be that the sites are on the edge of the forest, where the resources of two habitat types are available.

Although there was a greater abundance of bats in waterside habitats than in the forest, a lower species diversity was found in those localities (see Table 9.3). It should be noted, however, that the waterside harp trap value has been skewed by the capture of 38 *Pteronotus personatus* on the same night. Looking at the relative

**Table 9.2.** Capture rate of bats in mist nets in different habitats

|  | Total bats caught | Effort (mh) | Mean bats/100 mh |
|---|---|---|---|
| Waterside | 171 | 4395 | 3.89 |
| Forest | 56 | 4508 | 1.24 |

**Table 9.3.** Shannon–Weaver diversity index ($H'$) for bat captures in different habitats on the Ilha de Maracá

| Habitat | Mist net captures | Harp trap captures |
|---|---|---|
| Waterside | 2.25 | 0.65 |
| Forest | 2.45 | 1.91 |

abundance of species in the two habitats, it can be seen that whereas the numbers of species recorded are equal (26), several abundant species (notably *Tadarida laticaudata*, *Carollia perspicillata*, and the larger *Artibeus* spp.) dominate the waterside catch, greatly reducing the diversity index. By contrast, in the forest the majority of species recorded were only represented by one or two examples, and no species was greatly abundant.

The species composition of the catch also varied between habitats. Of the total of 36 confirmed separate species in this survey, 26 were recorded in each habitat. Sixteen of these species were recorded in both, 10 species only in the forest and 10 only at waterside sites. Sample sizes, however, are small, so that although these results may reflect genuine ecological differences between species, they may also be simply due to sampling error.

The species recorded only in forest sites were (number of captures): *Saccopteryx leptura* (1); *Cormura brevirostris* (1); *Micronycteris megalotis* (3); *Micronycteris nicefori* (1); *Tonatia bidens* (1); *Mimon crenulatum* (4); *Choeroniscus* cf. *minor* (2); *Rhinophylla* cf. *pumilio* (2); *Vampyrops* sp. (2); *Chiroderma villosum* (1). Species only recorded in waterside sites were: *Rhynchonycteris naso* (3); *Noctilio leporinus* (1); *Micronycteris minuta* (1); *Tonatia brasiliense* (2); *Lonchorhina aurita* (4); *Sturnira lilium* (1); *Sturnira tildae* (3); *Artibeus concolor* (2); *Mesophylla macconnelli* (2); *Tadarida laticaudata* (53).

Two groups of bats seem to be better represented in the forest than in waterside habitats: the Phyllostominae and Emballonuridae. Bats of the phyllostomid subfamily Phyllostominae are generally broad-winged and highly manoeuvrable, adapted to gleaning food from the surface of plants or the ground. They represent 25.0% of mist net captures in the forest (14 of 56), but only 10.2% by the waterside (17 of 166). The emballonurids are also highly manoeuvrable, feeding on insects caught in flight. One species, *Rhynchonycteris naso*, typically lives and feeds over water and was not caught in the forest. Of the remaining four emballonurid species 12 out of 15 were taken in the forest, although 9 of these 12 were caught in the harp trap, so the difference may simply reflect greater harp-trapping effort in the forest (120.5 hours) than in waterside sites (24 hours).

Table 9.4. Capture rates of mist nets versus harp trap

| Technique | Total bats caught | Effort (m²h) | Rate (bats m$^{-2}$h$^{-1}$) |
|---|---|---|---|
| Mist nets | 229 | 18 966 | 0.012 |
| Harp trap | 83 | 578 | 0.144 (0.073*) |

* Excluding one exceptionally high catch.

Table 9.5. Comparison of taxonomic composition of bats caught by mist nets and harp trap on the Ilha de Maracá

| % of catch | Mist net | Harp trap |
|---|---|---|
| Phyllostomids | 68.3% | 7.3% |
| Aerial insectivores* | 31.3% | 92.7% |

* Emballonuridae, Mormoopidae, Vespertilionidae.

Of the bats which occur in waterside habitats but not in the forest there are no groups which stand out, though some specialist species are obvious. These include the fishing bat *Noctilio leporinus*, the high-speed aerial insectivore *Tadarida laticaudata* which lives in colonies in cracked riverine rocks, and the previously mentioned *Rhynchonycteris naso*. Large *Artibeus* species are also notably more abundant at the waterside than beneath the forest canopy.

## MIST NETS VERSUS HARP TRAPS AS ANALYTICAL TOOLS

For this survey, bats were caught in mist nets and a single harp trap. The harp trap consisted of an aluminium frame 2 × 2 m, strung with very fine tungsten wires (see Methods and Materials), which bats detected less easily than mist nets. Comparison of the capture rates of the nets and the trap (see Table 9.4) shows that in total the nets caught 229 bats in 9483 metre hours (i.e. 18 966 m²h) of operation (0.012 bats per square metre per hour), whereas the trap caught 83 bats during 144.5 hours of operation (0.144 bats m$^{-2}$h$^{-1}$) – almost 12 times the rate of the nets. If the single night on which the trap caught 44 bats is excluded as atypical, its capture rate still works out at 0.073 bats m$^{-2}$h$^{-1}$, which is six times the rate of the nets. If only forest captures are taken into account, it again proves to be 12 times as effective at catching bats as is the equivalent area of net (0.075 bats m$^{-2}$h$^{-1}$ and 0.0062 bats m$^{-2}$h$^{-1}$ respectively).

The higher catch rate of the harp trap could in part be explained by the fact that often only part of a mist net will be able to catch bats, with much of its length being obscured by vegetation, whereas the harp trap will always be set in the centre of a likely flight path for bats. Even more remarkable, however, is the difference in the species composition of the catch: 68.3% of the mist net catch consisted of phyllostomid bats, with 31.3% aerial insectivores of the families Emballonuridae, Mormoopidae, Vespertilionidae and Molossidae. Of the harp trap's captures only 7.3% were phyllostomids, whereas 92.7% were insectivores from these four families. Hence the harp trap is less efficient than mist nets at capturing phyllostomid bats, and all of the extra catch rate is accounted for by the aerial insectivores. These bats are sufficiently sensitive and manouevrable to avoid capture in a mist net, but only with great difficulty do they detect the fine tungsten wires of the harp trap.

It is tempting to speculate that since the harp trap is capable of capturing a wider variety of bats, its samples are more truly representative, and it would be a superior tool for quantitative analysis of bat faunas. However, it seems to catch a disproportionately high selection of small insectivores. Larger non-insectivorous species are less likely to be caught for two reasons. First, due to the tension of the wires larger species are more likely to bounce off the trap without being caught. Secondly, aerial insectivores have to fly more or less continuously to feed themselves (unlike, say, frugivores or nectarivores), so their greater catch rate may reflect greater flight activity rather than relative abundance of bats in the forest.

For ease of operation, mist nets are in some ways far superior to harp traps. They are cheap, relatively easy to obtain, and portable. A small rucksack can carry many hundreds of square metres of net, whereas a single 4 $m^2$ harp trap needs two people to carry it, and large numbers could not be used without great expense and an established study area. Harp traps would be superior for many aspects of intensive ecological study, especially if insectivorous species are being studied, but extensive survey work is far more convenient with mist nets. Neither technique is capable of producing an unbiased picture of the composition of a bat community, and data from these sources should be treated with caution.

## HUMANE RESEARCH – MINIMUM COLLECTION POLICY AND 'CHEWITS'

While normally associated with finding alternatives to the abuse of laboratory animals, the concept of humane research could well be applied to many aspects of ecological fieldwork. For some considerable time the moral justification for killing birds and large mammals for ecological research has been questioned, and only highly selective killing is now practised. This is not the case, however, with the study of neotropical bats. There are a number of reasons for this, of which the principal is that many bat species cannot be reliably identified in the field, and the taxonomic status of some groups is in a constant state of turmoil. Secondly, the majority of the work on these animals is still carried out by researchers from abroad, whose field collections sustain their investigations after their return to their country of origin. Thirdly, bats still carry with them the image of being vermin, particularly in the United States where fear of bat-transmitted rabies is high, so there are very few people prepared to stand up and defend them against ill-treatment.

Many of the scientific papers written about South American bats involve the killing of hundreds, or even thousands, of individuals. While many of these deaths are necessary for achieving the objectives of a research programme, much of the killing is automatic and indiscriminate. Howell (1983) cites the case of a routine study by a vertebrate ecology course which killed 217 *Glossophaga soricina* for studies of their dietary habits. Only 38 of the bats' stomachs contained any food, and the taxonomically identifiable parts of the diet are the same in the stomach or the faeces. He concludes: 'Since bats are prone to defecate upon being handled, condensed samples of the diet can be obtained without killing the animals. The hundreds of alcohol-preserved specimens on deposit with the Universidad de Costa Rica might be used for projects, thus conserving dwindling bat populations.'

At a time when ecological disruption of natural ecosystems is at a peak, I believe that it is essential for ecologists and taxonomists to set an example, and only to kill bats (or other animals) when it is necessary. A first principle to work by would be only to kill discriminately, not indiscriminately. This would mean only ever killing a bat for a known purpose which could not be achieved without killing it. Current practice, however, is often to kill all bats caught in case they might be useful: to shoot first and ask questions later.

However, even within a framework of discriminate killing, it is still difficult to avoid taking quite large numbers of specimens. This is usually due to the difficulties of accurate and verifiable field identification. If a bat has been identified incorrectly and then released, the data collected are worse than useless – they can be dangerously misleading. Moreover, if the taxonomy of a particular group is reviewed, many old species identifications must be reinterpreted by reference to preserved specimens. Consequently there is an understandable caution about releasing bats.

Nevertheless, there has so far been little effort to reduce the necessity for taking specimens, and some simple steps could save the lives of many bats. Most importantly, keys should be made available for use by fieldworkers. Those keys which are currently available are nearly all made by museum workers for museum workers, and refer mainly to cranial and dental characteristics which cannot be seen in the field. In many cases there are perfectly good external characteristics which could be substituted without prejudicing the accuracy of the keys. The description of the genus *Carollia* in Nowak and Paradiso (1983) provides a classic example of this 'skull and teeth'-centred view of bats: 'This genus differs from *Rhinophylla* in that the lower molars are of a different form from the lower premolars, and in the presence of a tail.'

Many bat species can be identified easily and unambiguously in the field by reference to external characters alone. Keys could easily be devised which lead the fieldworker either to a positive identification, or to a problem which renders field identification unreliable if a specimen is not taken. The worker will then be able to make an informed decision as to whether or not to kill the bat.

For this survey I operated a 'minimum collection policy'. I did not take specimens of species which I was confident had been reliably identified in the field, using unambiguous absolute characters (presence or absence of a feature) or clear-cut relative characters (size of a particular feature). I tried only to take a single specimen of species which were more difficult to identify. In a few cases more than one specimen was taken either due to confusion between very similar species, or due to accidental injury or death. In total 29 specimens were collected from the 38 species recorded. All of these specimens were left in Brazil.

## Chewits

The single biggest obstacle to field identification of tropical bats is the need for careful examination of minute details of the teeth of many species. Even if the required details can be seen, which is not easy on a live wriggling bat in the dark, they cannot be shown to a sceptical third party (using conventional methods) unless the bat is killed. In order to help remove this technological barrier, I have experimented with a new system for making a permanent imprint of the teeth of a live bat. The imprint, called a 'chewit', is made by gently pressing a thin layer of a silicone dental putty against the bat's teeth until set. The putty sets through mixture with a catalyst shortly before application to the teeth, and the imprint can be kept permanently. It is also possible to make a plaster model of the teeth, using the chewit as a mould. This permits examination of the fine details of a bat's teeth without needing to kill it, and allows duplicates to be sent to fellow bat workers without losing the original chewit. Full details of how to use the technique have been published in Robinson (1989b).

If chewits becomes a widely used technique, then it will be possible to incorporate dental criteria (as recorded on the chewit) into field keys in positions which would otherwise be dead ends, in both senses of the words.

## SPECIES ACCOUNTS

In the discussion of families and genera which follows, measurements are all in millimetres and weights in grams. FA represents length of forearm. The time of capture is standardized throughout the survey, the change of the clocks for Brazilian Summer Time having been ignored for this purpose.

### FAMILY EMBALLONURIDAE

Representatives of this family are found throughout the tropics, though the three genera recorded on Maracá are restricted to the neotropics. All species are insectivorous, usually taking small insects in flight. The five species recorded on Maracá are all small to medium in size (3.5–10 g). Of the 16 species which have been recorded in South America (Koopman, 1982), 13 have been recorded in the Brazilian Amazon (Mok *et al.*, 1982).

#### Genus *Rhynchonycteris*

The single species, *Rhynchonycteris naso*, is found throughout northern Brazil (Koopman, 1982). It is very easy to identify in the field, due to the sharply bicoloured fur which imparts a grizzled appearance, the absence of wing sacs, a very long calcar (>tibia) and the presence of whitish tufts of fur on the forearm. Bats of this species, which are easily identified when at rest, were frequently disturbed from their daytime roosting sites of logs and rocks overhanging the water by boats travelling up the Furo de Maracá. At the Corredeira dos Filhotes a group of 14 of these bats regularly roosted on a large boulder overhanging a stream. They formed a loose group, with a regular spacing of 15–20 cm between each bat. They could be clearly seen through binoculars with their bodies lifted away from the rock surface and their pointed noses lifted upwards. In this characteristic posture, together with their grizzled fur, they achieve an excellent camouflage, resembling patches of lichens.

**Genus *Saccopteryx***

Four species of this genus have been recorded in the Brazilian Amazon (Mok *et al.*, 1982), of which three appear to be present on Maracá: *S. bilineata*, *S. leptura* and *S. canescens*. *Saccopteryx bilineata* is the largest species, distinguishable from others of the genus on size alone. Although no adult ectoparasites were recorded on any of the individuals caught, all of the February 1988 captures were moderately to densely infested with bright orange objects *c*. 0.5 mm in diameter attached to unfurred areas of the skin, especially on the base of the tail membrane, the wings and the bare skin of the face. These are thought to have been acarine eggs (Lucille Anthony, INPA, pers. comm.)

The dorsal fur of the single *Saccopteryx leptura*, a female, was uniform brown from the base to the tip. While the external measurements of this species overlap considerably with those of the slightly smaller *S. canescens*, and differences in coloration are not consistent, they can be separated on the width across the third molars M3–M3 (Husson, 1962; Davis, 1976). Although no voucher specimen was taken, I believe that the chewit of this bat in conjunction with field notes provides a reliable identification. Dental measurements of the specimen identified as *Saccopteryx canescens* all accord with the diagnostic characters of that species rather than *S. leptura*.

**Genus *Cormura***

The single species, *Cormura brevirostris*, is found throughout the northern and western parts of the Amazon basin (Koopman, 1982; Mok *et al.*, 1982). It can be distinguished from *Saccopteryx* by the absence of pale lines on the dorsal fur, and from that genus and *Peropteryx* by the shape and position of the wing sac, and the position of attachment of the wing membrane onto the leg or the foot.

FAMILY NOCTILIONIDAE

This family contains two species of the single genus *Noctilio*. These are the incredible fishing bats, endemic to the neotropics. The larger species *Noctilio leporinus* feeds primarily on a diet of fresh fish, gaffed out of the water with its giant claws. Both species are found in the Brazilian Amazon (Koopman, 1982; Mok *et al.*, 1982).

**Genus *Noctilio***

The appearance of *Noctilio leporinus* is so distinctive that it could not be confused with any other bat except the smaller *N. albiventris*, which can be distinguished on forearm length alone. A feature of this bat which I have not seen referred to in the literature is that the short velvety fur and the wing and tail membranes are covered with a fine sweetly scented oil, presumably to waterproof them. This highly distinctive and slightly sickly smell is very powerful and tenacious; the cloth bag used to hold the fishing bat could be identified by smell alone more than a week later. Several months after the Maracá Project, in the far southwest of the Amazon basin in lowland Bolivia, I was able to detect *Noctilio* in flight around the plaza of a town from their powerful scent alone, prior to actually seeing them.

## FAMILY MORMOOPIDAE

Formerly considered a sub-family of the Phyllostomidae, this family contains two genera (*Pteronotus* and *Mormoops*) which are restricted to the neotropics and southern United States. They are characterized by the presence of plate-like outgrowths around the mouth, and are exclusively insectivorous.

### Genus *Pteronotus*

Four of the six species have been recorded in the Brazilian Amazon (Mok *et al.*, 1982; R. Gribel, pers. comm.), and three of these have been recorded on Maracá: *Pteronotus parnelli, P. personatus* and *P. gymnonotus*.

*Pteronotus parnelli* is the largest bat of the genus, and is easily recognizable by the presence of fur on the dorsal side of the body between the wings, and its FA of >54. Although data were not collected systematically on flight activity through the night, it would appear that this species regularly feeds late at night or early in the morning, as all three harp trap captures were made between 2230 hrs and dawn, when capture rates for most species are very low. *P. personatus* could only be confused with the previous species, *P. parnelli*, but is clearly distinguishable on size alone. Thirty-eight were caught on the Causeway in a single night (7–8 February 1988). On this occasion the harp trap was the centre of a great turmoil of activity for several hours, with many bats already in the catching bag and others flying around or into the trap, apparently attracted by the calls of the others.

There was a very marked sexual segregation in these captures along the Causeway, with 36 females but only two males caught. The significance of this gathering of females is unknown; it may have only been temporary as the captures were made in a single night. One of the females appeared to be pregnant. By contrast, in the forest there was a more or less even sex ratio (seven females, five males), and even a suggestion of pairing. On three occasions a male and a female were caught together in the trap, which suggests that the bats could have been foraging together, though this could not be proven without further work.

In the forest the harp trap was set from sunset to sunrise (1800–0600 hrs) each night, yet 11 of the 12 *P. personatus* were caught between 2030 and 2330 hrs, indicating a late peak of flight activity. On 7 February 1988 the trap was not set on the Causeway until 2200 hrs, but bats were caught immediately, and activity continued until 0500 hrs (14 bats 2200–2300 hrs; 8 bats 2300–0100 hrs; 13 bats 0100–0300 hrs; 3 bats 0300–0500 hrs; 0 bats 0500–0600 hrs).

The colour of *P. personatus* was highly variable, from a dull reddish-brown to a vivid cinnamon. About 10 of the bats had patches of brown across the back. As the colour change from brown to red is thought to represent ageing of the fur, these patches may reflect moult patterns (Hill and Smith, 1984).

*Pteronotus gymnonotus* is quite similar to the previous two bats of this genus, except that its wing membranes join on the midline of the back, giving it the remarkable appearance of having a naked back. The only other species of the genus to share this feature is *P. davyi*, which has a shorter forearm (some overlap) and a relatively longer tail. This species had not been recorded in the Brazilian Amazon prior to 1982 (Mok *et al.*, 1982), though it is present in adjacent parts of Venezuela and Guyana (Koopman, 1982). However, Gribel (pers. comm.) has recorded it in the area in recent years, so these are not the first records. The first three bats were captured between 2230 hrs and dawn, and the last between 0015 and 0300 hrs, again suggesting a late peak of activity, as for the previous members of this genus.

## FAMILY PHYLLOSTOMIDAE

The Phyllostomidae or leaf-nosed bats are by far the largest and commonest family of bats in tropical South America. They are endemic to the neotropics and southern United States, and in the absence of many Old World bats they have radiated into a wide range of trophic niches. They are very important in the functioning of tropical forest ecosystems, particularly through the dispersal of seeds and the pollination of flowers. The family is divided into six well-defined and easily recognizable sub-families. Thirty-four of the 47 bat species recorded on Maracá belong to this family (see Table 9.1).

## SUB-FAMILY PHYLLOSTOMINAE

### Genus *Micronycteris*

These long-eared bats are small to medium in size, and are primarily insectivorous but may also eat a variety of fruits (Gardner, 1977). Six of the 10 known species have been recorded in the Brazilian Amazon (Mok *et al.*, 1982). Five species have been recorded on Maracá, of which three were recorded in this survey.

*Micronycteris megalotis* can be recognized by its small size, presence of a low band of skin without a prominent notch connecting the bases of the ears, and a calcar which is longer than the foot. All three bats were caught in the first hour after sunset, which could indicate an early peak of activity. *M. nicefori* can be recognized by its larger size, the absence of an interauricular band and its very short calcar (<½ hind foot). *M. minuta* is a very small species which can be distinguished from *M. megalotis* by the presence of a high interauricular band with a deep V-shaped notch, and a calcar shorter than the foot. The single specimen carried a particularly heavy ectoparasite infestation, with a louse and about 25 mites on its body.

### Genus *Tonatia*

This genus contains eight species of large-eared bats, which can be distinguished from other Phyllostomine genera, except for *Mimon*, by the presence of just one pair of lower incisors. In *Mimon* the noseleaf is about twice as long as broad, whereas in *Tonatia* it is much shorter than this. Four of the eight species have been recorded from the Brazilian Amazon (Mok *et al.*, 1982; Gribel, pers. comm.). Three species have been recorded on Maracá, of which two were recorded in this survey.

*Tonatia bidens* can be distinguished by its large size, relatively small ears, and broad lower canines. *T. brasiliense* can be distinguished from others of the genus likely to occur in the area by its small size alone.

### Genus *Mimon*

The two species in this genus, together with *Tonatia*, are characterized by the possession of a single pair of lower incisors. They can easily be distinguished from *Tonatia* by the greater development of the noseleaf, which is about twice as long as broad. Only *Mimon crenulatum* has been recorded in Amazonia (Mok *et al.*, 1982), though *M. bennetti* occurs in adjacent regions (Koopman, 1982). *Mimon crenulatum* can be separated from *M. bennetti* by its shorter forearm, crenulate margin of the noseleaf, and attachment of the wing membrane to the outer toe rather than the side of the ankle.

### Genus *Phyllostomus*

The nominate genus of the family Phyllostomidae and sub-family Phyllostominae contains four species, three of which have been recorded in the Brazilian Amazon (Mok *et al.*, 1982). Members of this genus are large (FA>55), with two pairs of lower incisors and a tail which is enclosed in, but not running to the end of, the tail membrane. Three *Phyllostomus* species have been recorded on Maracá, of which two were found in this survey.

*Phyllostomus hastatus* is the largest species of the genus, and can be distinguished on size alone (FA>75) from the other species. The eight bats of this species caught on the Causeway were taken in mist nets set at ground level close to a group of *Cecropia* trees, on whose fruits the bats were feeding. Several were covered in hundreds of sticky circular seeds, adhering to the fur and membranes. Faeces collected from these bats had varied contents: one sample contained only seeds of about 2 mm diameter; a second contained large seeds *c.* 6 × 4 mm; a third contained a mixture of these two types of seeds; while the faeces of another bat consisted entirely of insect remains. This indicates that these bats are variable and opportunistic in their choice of food.

*Phyllostomus elongatus* can be distinguished from others of the genus on forearm and tibia length, combined with the length of the calcar, which is longer than the foot. All but two specimens were caught in the first three hours of darkness, which could indicate an early peak of feeding activity.

### Genus *Trachops*

*Trachops cirrhosus*, the only species of this genus, is found throughout the Amazon region (Koopman, 1982; Mok *et al.*, 1982). These bats are well known for eating small vertebrates such as frogs and mice, as well as insects and possibly some fruit (Gardner, 1977). They are easy to distinguish from other phyllostomine bats by the presence of conspicuous wart-like protuberances around the mouth, and a finely toothed margin to the noseleaf. All were caught in the lower half of mist nets, suggesting that they were flying very close to the ground while searching for food. The third metacarpal of this species is considerably shorter than the forearm, giving a broadly rounded wing tip, which would be an advantage while flying low over water or the ground.

### Genus *Lonchorhina*

The genus contains three species of which one, *Lonchorhina aurita*, has been recorded from the Brazilian Amazon (Mok *et al.*, 1982). This is the most widespread species of the genus. The other two species have been recorded from adjacent parts of Venezuela and Colombia. All species have a very long sharply-pointed noseleaf (as long as the ears). The specimens accord with measurements of *L. aurita* in Husson (1962), so I have provisionally placed them under this species. However, I am not yet aware of the distinctions between this species, *L. orinocensis* and *L. marinkellei*, and need to compare descriptions of these species with the Maracá specimens before identification can be certain.

## SUB-FAMILY GLOSSOPHAGINAE

### Genus *Glossophaga*

Of the five species in this genus, only *Glossophaga soricina* has been recorded from the Brazilian Amazon (Mok *et al.*, 1982), although *G. longirostris* occurs in adjacent Colombia, Venezuela and Guyana (Koopman, 1982). The snout is elongated, as in other members of the sub-family Glossophaginae, but not as markedly as in some other genera.

*Glossophaga soricina* can be distinguished from other genera by the presence of lower incisors, and by upper incisors which are similar in size, forming an even row. *G. soricina* can be distinguished from *G. longirostris* by its less elongate rostrum and smaller size (although there is some overlap). Those captured at the Fumaça rapids were all caught in the early evening, flying onto the island from an islet in the river as though dispersing into the forest from their roost.

### Genus *Choeroniscus*

This genus contains four species of which one, *Choeroniscus minor*, has been recorded in the Brazilian Amazon (Mok *et al.*, 1982), but another two are likely to occur in the area (Koopman, 1982). The genus has an exceptionally long muzzle, and can be recognized by the absence of lower incisors and the spacing of the upper incisors. Measurements and descriptions of the specimen taken are in accord with those of *C. minor* as given in Husson (1962). However, this identification still needs to be confirmed.

## SUB-FAMILY CAROLLINAE

### Genus *Carollia*

This genus contains four species, of which three have been recorded from the Brazilian Amazon (Mok *et al.*, 1982). They can be distinguished from members of the Phyllostominae by the presence on the lower lip of a central wart flanked by rows of smaller warts, and from the Stenodermatinae by the presence of an external tail.

*Carollia perspicillata* is the most widespread and frequently caught bat species in tropical forests of South and Central America. It is notoriously difficult to separate from other species of the genus, but can be distinguished in the field from *C. castanea* and *C. brevicauda* on length of the tibia, which is never greater than 16 in these two species (Pine, 1972). The tibia length was checked on 35 of the 55 captures, and confirmed their identification as *C. perspicillata*, although there is a possibility that the other 20 captures could include other *Carollia* species. On the whole the Maracá specimens are quite small for the species, which could be due to the absence of others of the genus (see Pine, 1972).

### Genus *Rhinophylla*

This genus contains three species, of which two have been recorded from the Brazilian Amazon (Mok *et al.*, 1982). They can be distinguished from the members of the Phyllostominae by the absence of an external tail, and from the Stenodermatinae by the presence on the lower lip of a central wart flanked on each side by a larger elongate wart. As far as I am aware, the forearm lengths of the specimens ascribed to *Rhinophylla pumilio* preclude the possibility that they are the slightly smaller *R. fischerae*. However, until confirmation of this point has been made, this identification remains provisional.

## SUB-FAMILY STENODERMATINAE

### Genus *Sturnira*

This genus contains 11 species, of which two have been recorded in the Brazilian Amazon (Mok *et al.*, 1982). They can easily be distinguished from all other phyllostomid bats by the absence of a tail and tail membrane, combined with the short, broad jaw typical of the frugivorous Stenodermatinae.

*Sturnira lilium*, the most widespread species of the genus in lowland Amazonia, can be separated from *S. tildae* on forearm length and dental characteristics (Davis, 1980). The teeth of the single bat assigned to this species agreed in every detail with the descriptions of Davis (1980) and Husson (1962), but it had an FA of 44.6 whereas Davis states that it is 'seldom as long as 41'. R. Gribel (pers. comm.) reports regularly finding oversized *S. lilium* in the Brazilian Amazon, and Husson (1962) reports a specimen thought to be of this species with an FA of 45.7. Albuja (1982) also reports that Ecuadorian specimens have an FA of 41–44.3.

*Sturnira tildae* can be separated from other *Sturnira* species on the characteristics of the lower incisors, lingual cusps of the upper molars and forearm length (Davis, 1980).

### Genus *Uroderma*

This genus contains two species, both of which have been recorded at Maracá (Taddei and Reis, 1980). They can be distinguished from other Stenodermatine bats by the form and size of the upper incisors, combined with the presence of a narrow white line in the middle of the lower back. *Uroderma bilobatum* can be distinguished from *U. magnirostrum* by the presence of a depression in the rostral profile, clear facial stripes and whitish edging to the ears, all of which are reduced or absent in the latter species (Davis, 1968).

### Genus *Artibeus*

This genus contains about 14 species, of which six have been recorded from the Brazilian Amazon (Mok *et al.*, 1982). They can be distinguished from other Stenodermatine bats by the form and size of their upper incisors and the absence of a white line down the middle of the back. However, there is great variation within some species and the classification of the genus remains somewhat confused, rendering field identification of many species very difficult.

The large *Artibeus* complex (*lituratus/jamaicensis/planirostris/fuliginosus*) contains from two to four species. *A. lituratus* and *A. jamaicensis* are well established, but there is much dispute over the status of the last two names (either as separate species or as subspecies of either of the first two (Koopman, 1978; Jones and Carter, 1979)). Furthermore, acceptance of *A. planirostris* and *A. fuliginosus* as species is based on the work of Patten (1971), an unpublished PhD thesis which is not available to me. In the absence of a thorough review of the confusion, I remain confused. It is certain that *A. lituratus* and *A. jamaicensis* are present on the island, and Taddei and Reis (1980) reported the presence of *A. fuliginosus*, the first record for Brazil.

Field identification was based principally on coloration, *A. jamaicensis* being much darker and with reduced or no facial stripes and a very dark or black tongue. However, many specimens intermediate in coloration were observed, and a scatter-diagram of forearm length against weight shows no discontinuity between the very large pale *A. lituratus* and the small dark *A. jamaicensis*.

*Artibeus concolor* is a medium-sized *Artibeus* species which can be distinguished by its size and the absence of facial stripes. *Artibeus* cf. *anderseni* is a small *Artibeus* which can easily be confused with *A. cinereus*. The identification is provisional, based on the absence of rear lower molars (Koopman, 1978), and still needs to be confirmed by examination of the specimens. The two voucher specimens taken could possibly represent separate species.

**Genus *Vampyrops***

This genus contains eight small to medium-sized species, four of which have been recorded in the Brazilian Amazon (Mok *et al.*, 1982). They can be distinguished by the form of their upper incisors, the presence of a white mid-dorsal streak and the presence of a fringe of hairs on the rear edge of the wing membrane. The specimens of *Vampyrops* caught are not yet identified to species.

**Genus *Mesophylla***

The single species *M. macconnelli* is found throughout the Amazon basin (Koopman, 1982; Mok *et al.*, 1982). It can be recognized by its small size, dull coloration and dental characteristics. Two were caught, both at riverside sites: one at Abacaxi on the Furo de Santa Rosa; the other in a net across a quiet tree-lined channel close to Anta Camp at the far west of the island.

**Genus *Chiroderma***

This genus contains five species, of which two have been recorded from the Brazilian Amazon (Mok *et al.*, 1982). They differ from other stenodermatines in having a heavy, deep rostrum, and a broad rounded noseleaf. *Chiroderma villosum* can be distinguished from *C. trinitatum*, the only other species likely to occur at Maracá, by its larger size, the absence of a mid-dorsal stripe, and dental characteristics (Goodwin and Greenhall, 1961).

SUB-FAMILY DESMODONTINAE

**Genus *Desmodus***

*Desmodus rotundus*, the common vampire bat, is the only species of the genus. It is the most widely distributed bat in South America, and is found throughout the Amazon basin. It can be distinguished from the other two genera of vampires by its pointed ears, longer thumb with a distinct basal pad, naked interfemoral membrane, and dental features. The specimens were taken many kilometres from *fazendas* (farms) with domestic animals, so these bats must have been subsisting on the blood of forest mammals such as peccaries or tapirs.

FAMILY VESPERTILIONIDAE

This is the largest and most widespread family of bats, found throughout the tropical and temperate regions of the world. All species are insectivorous. Of the 28 species found in South America (Koopman, 1982), eight have been recorded in the Brazilian Amazon (Mok *et al.*, 1982).

### Genus *Myotis*

This is the most widespread mammal genus of the world, excluding *Homo*. LaVal (1973) reviewed the neotropical species of the genus, but the classification of South American species is still in question. The differences between species are very slight, so field identification of many species is very difficult. Three species have been recorded from the Brazilian Amazon (Mok *et al.*, 1982).

Of the 17 *Myotis* bats captured, most were probably *M. nigricans*, judging by the length of the dorsal fur (3.5–4 mm), the absence of a fringe of hairs on the tail membrane, and the weakly bicoloured fur. However, there is a possibility that some were *M. albescens*, including one which was considerably heavier than other specimens (7 g) and had fur with strongly contrasting golden tips. For the time being I shall pool all of the data, pending the identification of the two voucher specimens and those of Rogério Gribel and colleagues.

### Genus *Eptesicus*

A single specimen of this genus was dissected out from the stomach of a tree snake collected by Mark O'Shea. The bat was partly digested, but still in quite good condition. It was identified as *Eptesicus* from its upper incisors, protruding last vertebra of the tail and tail membrane without dorsal fur. The snake was an Amazon tree boa, *Corallus enydris enydris*, which was captured at night on low bushes in the cleared area around the Station. Snakes of this genus are known to take bats frequently, though they will also take lizards, frogs and other small mammals. The detection and capture of bats flying in darkness is achieved with the aid of heat-sensitive pits located on the labia.

## FAMILY MOLOSSIDAE

The Molossidae or free-tailed bats are another widespread insectivorous family. Of the 27 species found in South America (Koopman, 1982), 10 have been recorded in the Brazilian Amazon (Mok *et al.*, 1982), of which two have been recorded on Maracá.

### Genus *Tadarida*

There are 50 species of this genus worldwide but only six occur in South America (Koopman, 1982), and *T. laticaudata* is the only species to have been recorded from the Brazilian Amazon (Mok *et al.*, 1982). Bats of this genus can be recognized by the deep vertical grooves or wrinkles on the upper lip and the separation of the upper incisors. *Tadarida laticaudata* is separated from others of the genus which could occur on Maracá by the presence of two (as opposed to three) pairs of lower incisors, and an FA of 40–50 (Vizotto and Taddei, 1973).

Bats of this species were abundant in the upper reaches of the Furo de Maracá and Furo de Santa Rosa, where the river is broad and fast-flowing with many protruding rocks. The bats live in colonies in the cracks in these rocks, and were easy to detect by their powerful musty odour (typical of molossids) and the high-pitched squeaking which was heard at all times of day. All 53 captures came from rocky sites in or at the edge of the river. The 19 bats caught in a riverside net at Anta Camp came from the edge of a huge flock of these bats (estimated at several thousand) which flew downstream together at dusk. None were caught amongst vegetation of any kind, nor in the slow-moving tree-lined channels of the river.

**Genus *Molossus***

Of the seven species of this genus, two have been recorded from the Brazilian Amazon (Mok *et al.*, 1982) and two occur in adjacent parts of Venezuela and the Guianas (Koopman, 1982). They can be distinguished from other genera by the absence of vertical grooves on the upper lip, ears which join at the base, a circular antitragus and the presence of one pair of lower incisors and one pair of upper premolars (Vizotto and Taddei, 1973).

A large colony of *Molossus molossus* lived in the buildings of the Station. They can be distinguished from *M. ater* by the shorter forearm (FA 34–41). As *Molossus* often live in mixed roosts, there is a possibility that other species also lived there. The bats had taken up residence in the roofs of all four buildings, and were generally unpopular. They made a lot of noise as they ran around in the roof, their powerful odour pervaded the air, and the dried and crumbled faeces constantly rained through cracks between the ceiling tiles onto the clothes and belongings of people below. This last problem, however, is the only one which is at all serious and could be cured very simply by laying polythene sheets in the roof to hold the faeces. Exclusion of the bats would be extremely difficult and, I believe, undesirable.

## REFERENCES

Albuja, L. (1982). *Murciélagos del Ecuador*. Escuela Politécnica Nacional, Quito, Ecuador.
Davis, W.B. (1968). Review of the genus *Uroderma* (Chiroptera). *Journal of Mammalogy*, 49 (4), 676–698.
Davis, W.B. (1976). Notes on the bats *Saccopteryx canescens* Thomas and *Micronycteris hirsuta* (Peters). *Journal of Mammalogy*, 57, 604–607.
Davis, W.B. (1980). New *Sturnira* (Chiroptera: Phyllostomidae) from Central and South America, with key to currently recognized species. *Occasional Papers of the Museum of Texas Technical University*, No. 70.
Gardner, A.L. (1977). Feeding habits. In: *Biology of bats of the New World family Phyllostomatidae. Part II*, eds R.J. Baker, J.K. Jones Jr and D.C. Carter, pp. 293–350. Special Publications of the Museum of Texas Technical University.
Goodwin, G.G. and Greenhall, A.M. (1961). A review of the bats of Trinidad and Tobago. *Bulletin American Museum Natural History*, 122, 187–302.
Hill, J.E. and Smith, J.D. (1984). *Bats, a natural history*. British Museum of Natural History Publications, London.

Howell, D.J. (1983). *Glossophaga soricina* (Murciélago Lengualarga, Nectar Bat). In: *Costa Rican natural history*, ed. D.H. Janzen, pp. 472–474. University of Chicago Press.

Humphrey, S.R. and Bonaccorso, F.J. (1979). *Population and community ecology*. In: *Biology of bats of the New World family Phyllostomatidae*, eds R.J. Baker, J.K. Jones Jr and D.C. Carter, pp. 409–441. Special Publications of the Museum of Texas Technical University, No. 16.

Husson, A.M. (1962). The bats of Suriname. *Zoologische Verhandelingen, Leiden*, 58, 1–282.

Jones, J.K. Jr and Carter, D.C. (1979). Systematic and distributional notes. In: *Biology of bats of the New World family Phyllostomatidae*, eds R.J. Baker, J.K. Jones Jr and D.C. Carter, pp. 7–11. Special Publications of the Museum of Texas Technical University, No. 16.

Koopman, K.F. (1978). Zoogeography of Peruvian bats with special emphasis on the role of the Andes. *American Museum Novitates*, No. 2651.

Koopman, K.F. (1982). Biogeography of the bats of South America. In: *Mammalian biology in South America*, eds M.A. Mares and H.H. Genoways, pp. 273–302. Pymatuning Symposia in Ecology No. 6.

LaVal, R.K. (1973). A revision of the Neotropical bats of the genus *Myotis*. *Los Angeles County Natural History Museum Scientific Bulletin*, No. 15.

Mok, W.Y., Wilson, D.E., Lawrence, A.L. and Luizão, R.C.C. (1982). Lista atualizada de quirópteros da Amazônia Brasileira. *Acta Amazonica*, 12 (4), 817–823.

Nowak, R.M. and Paradiso, J.L. (1983). *Walker's mammals of the world*, 4th edition. John Hopkins University Press.

Patten, D.R. (1971). A review of the large species of *Artibeus* (Chiroptera: Phyllostomatidae) from western South America. Unpublished PhD dissertation, Texas A. & M. University.

Pine, R.H. (1972). *The bats of the genus Carollia*. Texas Agricultural Experimental Station Technical Monograph, No. 8.

Robinson, F.J. (1989a). Bats of Ilha de Maracá, Roraima, Brazil. Interim Report, January 1989. In: Ratter, J.A. and Milliken, W., eds, *Maracá Rainforest Project Preliminary Report. Mammals (Part 1)*. Royal Botanic Garden Edinburgh.

Robinson, F.J. (1989b). Dental, palate and tongue imprints of bats: A new field technique. *Journal of Zoology, London*, 219, 681–684.

Taddei, V.A. and Reis, N.R. (1980). Notas sobre alguns morcegos da Ilha de Maracá, Território Federal de Roraima (Mammalia, Chiroptera). *Acta Amazonica*, 10 (2), 363–368.

Vizotto, L.D. and Taddei, V.A. (1973). Chave para determinação de quirópteros brasileiros. *Boletim Ciências São José do Rio Preto, São Paulo*, 1, 1–72.

# 10 Small Mammals of the Ilha de Maracá

ADRIAN A. BARNETT[1] AND ALÉXIA C. DA CUNHA[2]

[1]*The Roehampton Institute, London, UK;* [2]*Rio de Janeiro, Brazil*

## SUMMARY

Fieldwork was carried out on the small mammals of the Ilha de Maracá during both the wet and the dry seasons. Ant infestation of traps prevented successful live-trapping, and overall trap success was very low. Ten habitat types were studied. Eleven species of small mammals were trapped and six more were recorded in other ways, giving a total of 17 species: *Agouti paca, Cavia aperea, Couendou prehensilis, Dactylomys dactylinus, Dasyprocta agouti, Holochilus brasiliensis, Marmosa murina, Mesomys hispidus, Monodelphis brevicaudata, Nectomys squamipes, Oryzomys concolor, O. delicatus, O. megacephalus, Proechimys guyannensis, Rhipidomys mastacalis, Sciurus pyrrhonotus* and *Zygodontomys* cf. *lasiurus*. A further six species of marsupials and nine species of rodents may also occur on the island. Habitat preferences of the species trapped do not appear to differ greatly from those previously observed elsewhere. Existing knowledge about each species is summarized.

## RESUMO

Um levantamento de pequenos mamíferos foi realizado na Ilha de Maracá, nas estações seca e chuvosa. Dez tipos de ambientes foram estudados. A infestação das armadilhas por formigas impediu a captura de animais vivos, e o sucesso geral de captura foi muito baixo. Onze espécies foram capturadas e mais seis foram registradas por outros meios, dando um total de 17 espécies: *Agouti paca, Cavia aperea, Couendou prehensilis, Dactylomys dactylinus, Dasyprocta agouti, Holochilus brasiliensis, Marmosa murina, Mesomys hispidus, Monodelphis brevicaudata, Nectomys squamipes, Oryzomys concolor, O. delicatus, O. megacephalus, Proechimys guyannensis, Rhipidomys mastacalis, Sciurus pyrrhonotus* e *Zygodontomys* cf. *lasiurus*. Além destas, é possível que seis espécies de marsupiais e nove de roedores (julgando pela distribuição) possam ocorrer na ilha. Preferências ambientais das espécies capturadas parecem ser pouco diferentes do que já foi observado em outros lugares. Dá-se um resumo do conhecimento existente sobre cada espécie.

## INTRODUCTION

### THE STATUS OF SMALL MAMMAL RESEARCH IN RORAIMA STATE

This paper is the result of small mammal studies conducted at the SEMA Ecological Station on the Ilha de Maracá between 25 November 1987 and 22 April 1988 (dry season), and 1 October and 9 October 1988 (wet season).

*Maracá: The Biodiversity and Environment of an Amazonian Rainforest.*
Edited by William Milliken and James A. Ratter. © 1998 John Wiley & Sons Ltd.

The fauna of Roraima State was, until recently, one of the least known within the Amazon basin. Even groups such as primates, relatively well known elsewhere in Brazil, had been little studied there. The situation is similar for small mammals. Despite work in the Venezuelan *llanos* (e.g. Handley, 1976), and in such habitat types as Brazilian *cerrado* (e.g. Alho *et al.*, 1986; Mares *et al.*, 1986), the *caatinga* (e.g. Streilein, 1981) and Amazonian forest (e.g. Peres, 1968; Carvalho and Toccheton, 1969), we have been unable to locate any previous publications dealing specifically with the small mammals of Roraima State.

Tate (1931a, b) mentions the region, and collections have been made on Mount Roraima in the far northeast of the State (Anthony, 1929), but nevertheless the area's mammals seem to be very poorly known. Moojen (1948), in his extensive review of the genus *Proechimys* (Echimyidae) gives only one collection locality in the area, in contrast to the many locations given for Amazonas and Pará. A more recent review of the genus by Patton (1987) also shows this bias in collecting sites. Honacki *et al.* (1982) list 106 species of marsupials and rodents in 34 genera as occurring within Brazil, of which 68 had their types taken there. But although 13 federal divisions are mentioned, Roraima State is not amongst them, despite a good representation from the northern states of Amapá, Amazonas and Pará.

Alho (1981), in his overview of Brazilian rodent research, tabulated species by habitat and political unit. Of the 182 species listed, 31 are recorded for Amazonas State and 31 from Pará State, but only two species (*Proechimys guyannensis* and *Coendou prehensilis*) are listed for Roraima State. The INPA/RGS/SEMA research initiative (Projeto Maracá) was thus particularly valuable in providing an opportunity to study the biology of a poorly known mammal fauna in a largely undisturbed location.

## STUDY SITES

The majority of the collections were made at the eastern end of the Ilha de Maracá, in the vicinity of the Ecological Station (Figure Pr.1, p. xvii). Ten habitat types were trapped:

- *Aningal* – *Montrichardia linifera* Schott (Araceae)-dominated vegetation complex in a fringing band around a seasonal lake near the Ecological Station. The lake dried out completely by the end of the dry season. The vegetation and lake were surrounded by wet grassland.

- *Dry grassland* – Natural grassland including Gramineae and Cyperaceae (*Andropogon*, *Scleria*, etc.), with scattered shrubby Melastomataceae (including *Comolia*), Malvaceae (including *Hibiscus furcellatus* Desr.), and Leguminosae. The more open parts included occasional small trees of *Curatella americana* L. and *Byrsonima crassifolia* (L.) Kunth (see Plate 1b). Grass was growing to approximately 1.5 m in height, and there was no permanent water present.

- *Wet grassland* – Similar to the above, but beside a small seasonal lake. In the late dry season the lake dried up completely, and in the wet season the water was up to 2 m deep with a well-developed community of emergent plants (including *Canna*, *Sagittaria* and *Thalia*). The water surface was covered by Nymphaeaceae and other floating aquatics (including *Cabomba* and *Utricularia*). The soil was wet underfoot for most of the dry season, and partially inundated during the wet season. Milliken and Ratter (Chapter 5) have used the terms 'seasonally flooded grassland' for the grassy part of this habitat, and '*vazante*' for the soak.

- *Camp clearing* – Area adjacent to the Ecological Station. The grass was kept below 10 cm by regular cutting. A few shrubs and bushes and occasional scattered trees were present.

- *Rough grassland* – The same site as the Station clearing, after six months without cutting. The vegetation was 1–1.5 m high, with *Cenchrus echinatus* L. (Gramineae) abundant. Melastomataceous shrubs, *Mimosa* spp., and other low leguminous shrubs were common. The lower part of the trapping grid possessed much sparser vegetation, with fewer shrubs and some bare soil showing through.

- *Uninhabited building* – An old, semi-derelict wooden hut (Casa de Santa Rosa, see Figure Pr.1, p. xvii) in a partly overgrown clearing adjacent to riverside forest. Oranges and other fruit trees had been planted in the clearing.

- *Inhabited building* – Traps were placed on the ground in and around the kitchen, dormitories and laboratories of the Ecological Station, and off the ground on the wooden ornamental trellis surrounding it.

- *Riverside forest* – Forest adjacent to the Rio Uraricoera. The dominant canopy trees included *Ceiba pentandra* (L.) Gaertn. and *Enterolobium cyclocarpum* (Jacq.) Griseb. The palm *Maximiliana maripa* (Correa) Drude was locally very common, as was the spiny *Xylosma intermedium* (Seemann) Tr. & Pl. close to the river. Vines (including *Entada* and *Uncaria*) were very abundant.

- *Terra firme forest* – Tall mixed forest with a canopy to 35 m, and emergents to 40 m. Typical emergents included *Ceiba*, *Enterolobium*, *Hymenaea* and *Jacaranda*; the main canopy included *Alseis*, *Lecythis*, *Licania*, *Pradosia* and *Sloanea*. Palms, principally *Astrocaryum aculeatum* G.F.W. Mey. and *Maximiliana maripa*, were present in the understorey together with trees of *Apeiba*, *Casearia*, *Duguetia*, *Guatteria* and *Himatanthus*. Lower understorey included *Abuta*, *Neea*, *Phenakospermum* and the small palm *Astrocaryum gynacanthum* Mart. The ground layer was locally dense, with coverings of *Aechmaea*, *Calathea*, *Costus*, *Heliconia*, *Monotagma* and *Olyra latifolia* L., and *Renealmia* were also characteristic ground components. Lianas were common, but the epiphytic flora very poorly developed.

- *Riverine islands* – Areas effectively composed of large vegetated bars of white sand, above water when visited (middle of the dry season), but probably inundated in the wet season (see Plate 2b). There was a very sparse herb layer, with terrestrial orchids locally abundant on the higher ground, and little understorey vegetation. The tree canopy reached approximately 15 m, with very low species diversity.

For more detailed descriptions of the vegetation of the Ilha de Maracá (including the above habitat types) see Milliken and Ratter (Chapter 5).

## METHODS

### TRAPPING TECHNIQUES

Wooden 'Nipper' snap-traps and metal 'Longworth' live-traps were placed in groups of 5–50, in representative parts of the 10 habitat types. Traps were generally placed in pairs and located at sites of likely small mammal activity. These included runs, holes, piles of feeding debris and beside fallen logs (see Barnett and Dutton, 1995).

Baited Longworth traps placed in forest habitats quickly attracted heavy infestations of ants. It would have been cruel to expose confined animals to these, so live-trapping was restricted to the relatively ant-free grassland zones. This limited the number of traps available for forest work to approximately 250 snap-traps. In order to maximize capture rates, traps in forest areas were moved every few days to new areas (following recommendations in Delany, 1974). Following a suggestion from Dr Joe Anderson, areas were ground-baited for two nights prior to trap introduction with a paste of oats, soya oil and water. Upon introduction, traps were pre-baited for two nights, the first with additional ground bait and the second without. On the third night the traps were set baited with paste and a selected additive. The additive was normally banana for terrestrial traps, mango for arboreal traps and fish for waterside traps. Traps were set on a 'hair trigger' to maximize their effectiveness.

The vegetation was often so dense that it was impossible to see one trap from the next, and indeed to find the traps without disturbing the vegetation – which can significantly lower capture rates (Delany, 1974; Barnett and Dutton, 1995). To avoid this the technique of 'festooning' was used, whereby a continuous length of thin white cord is deployed to link all the traps in an area, and in addition each trap was tagged with orange marker tape. Traps were also firmly anchored to prevent removal by small carnivores (see Barnett and Dutton, 1995).

Traps were checked two to three times a day, between 0600 and 0800 hrs and at dusk (approx. 1800 hrs), with an additional check about midday (depending on the number, location and success of the traps being operated). Tree-trapping was included in 10 of the forest grids, up to 3 m above the ground. In addition, three lines of pit-fall traps with associated drift fences (originally set up for herpetological collection) were used to capture small mammals. Two were in primary forest, and one was at the savanna–forest boundary.

## SPECIMEN PROCESSING

Flat-skin specimens were prepared following Corbet (1968). Standard measurements were taken to the nearest millimetre, and weights to the nearest gram. Gut length was also measured in selected specimens. Stomach, caecum and faecal pellets were collected for dietary analysis.

Weight, number and length of embryos were recorded where present. Weight and length of testes were recorded for reproductively active males. Skulls were removed from selected specimens and cleaned on site. Others, from specimens preserved in alcohol, were removed and cleaned at the Federal University of Paraíba (UFPb). Ectoparasites were collected and preserved in alchohol. Information on this collection and others made at Maracá is presented by Linardi *et al.* (1991). Skin and skull collections were prepared so that they could be divided between Brazilian and British institutes. Unicates remain in Brazil, in the collection of the Zoological Museum, INPA, Manaus. Other specimens are lodged in the Natural History Museum, London.

# RESULTS

A total of 73 small mammals of 10 species, from three families in two orders, were trapped by the authors in 3951 trap nights. Dry season work yielded 61 animals of 11 species in 3578 trap nights (a mean trap success of 1.7%), over nine habitat types. Wet season work yielded 12 animals of three species in 373 trap nights in one habitat type (a mean trap success of 3.2%). The combined average trap success was 1.84%. All animals were caught in snap-traps. In addition to trapping, seven species from seven families were recorded visually, aurally or by finding their dead bodies (see species accounts). An overall list of recorded species and numbers of individuals per habitat type is given in Table 10.1, and a provisional key to the known small mammal species is given in Appendix 2, p. 447. A list of the other non-volant mammals known to occur on the Ilha de Maracá is provided in Appendix 3, p. 449.

## TRAPPING SUCCESS

The failure of drift fences to catch animals was an unsolved mystery. M.J. Coe (pers. comm.) has used this technique with great success in Kenya and Liberia, while M. O'Shea (pers. comm.) found capture rates of small mammals from drift fences exceeded those from standard mammal traps in the lowland forests of Cameroon, and the technique has been widely used elsewhere (Barnett and Dutton, 1995).

Using standard trap-based techniques, several authors have reported very high rates of capture success from tropical small mammal communities, e.g. 34% in African savanna (Cheeseman and Delany, 1979) and 46% in the *paramo* grasslands of Ecuador (Barnett, 1997). However, those for Brazil and adjacent areas are generally much lower, e.g. 2.4% in *cerrado* gallery forest (Fonseca and Redford, 1984). Some trap rates have been very low indeed: August (1983), in a two-year study of small mammals of Venezuelan *llanos*, attained an average trap success of

Table 10.1. Species records by habitat

| Habitat | Trap nights | Species | No. | Trap success (%) |
|---|---|---|---|---|
| Dry grassland | 459 | *Holochilus brasiliensis* | 1 | |
| | | *Oryzomys delicatus* | 5 | |
| | | Total | 6 | 1.3 |
| Wet grassland | 693 | *Holochilus brasiliensis* | 5 | |
| | | *Oryzomys delicatus* | 5 | |
| | | *Zygodontomys* cf. *lasiurus* | 1 | |
| | | Total | 11 | 1.6 |
| Rough grassland | 265 | *Zygodontomys* cf. *lasiurus* | 10 | |
| | | *Proechimys guyannensis* | 1 | |
| | | Total | 11 | 4.1 |
| Camp clearing | 120 | No captures | 0 | 0 |
| Uninhabited building | 139 | *Oryzomys concolor* | 1 | |
| | | *Proechimys guyannensis* | 3 | |
| | | *Rhipidomys mastacalis* | 7 | |
| | | Total | 11 | 7.9 |
| Inhabited building | 200 | *Zygodontomys* cf. *lasiurus* | 3 | 1.5 |
| *Aningal* | 120 | *Oryzomys delicatus* | 3 | 2.5 |
| Riverside forest | 428 | *Nectomys squamipes* | 2 | |
| | | *Proechimys guyannensis* | 3 | |
| | | Total | 5 | 1.16 |
| *Terra firme* forest | 1393 | *Marmosa murina* | 1 | |
| | | *Monodelphis brevicaudata* | 3 | |
| | | *Proechimys guyannensis* | 19 | |
| | | Total | 23 | 1.65 |
| Riverine islands | 100 | No captures | 0 | 0 |

approximately 1%. Streilein (1981), working in *caatinga* forest, averaged 0.78% trap success and cites the work of the AGGEU (a Brazilian public health agency), whose professional trappers caught one animal in 29 250 trap nights (a trap success of 0.0003%). According to Fittkau and Klinge (1973) 'the rarity of rodents (in central Amazonian forests) can be established from the fact that collections for blood parasite studies by zoology staff of INPA had to be suspended because of the difficulty of getting animals'.

Our mean value of 1.84% falls somewhere between these extremes, but is still much lower than was initially expected. That this value is not an artefact of our techniques is suggested by the fact that other workers have also obtained notably low trap success values on Maracá. While collecting for ectoparasite studies in February and March 1988, a team from UFMG caught 26 rodents in 2920 trap nights, a trap success of 0.89% (P. Linardi, pers. comm.). There are five possible explanations for these low trapping successes: trap disturbance (which included insects, peccaries, deer, ground doves, lizards, giant cockroaches and heavy rain), effects of the baiting programme (which included pre-baiting), bait type, seasonal

Table 10.2. Records of pregnant females over the dry season

| Species | Nov | Dec | Jan | Feb | Mar | Apr |
|---|---|---|---|---|---|---|
| *Oryzomys concolor* | | | | 1 | | |
| *Oryzomys delicatus* | | | | 1 | 1 | 1 |
| *Proechimys guyannensis* | | | 3 | 2 | | |
| *Rhipidomys mastacalis* | | | | 1 | | |
| *Zygodontomys* cf. *lasiurus* | | | | 1 | | |

changes in populations of animals (see Table 10.2), including possible low proportions of juveniles and overall population declines during the dry season, and an intrinsically low number of small mammals in the Guiana Shield region (see Emmons, 1984). It is probable that all of these played a role to some extent. However, we consider the most important factor to have been the strongly seasonal availability of fruits and seeds and consequent competition with other terrestrial frugivores during the dry season resource low, and the resulting effects on the small mammal populations. Bait removal, particularly by ants, was an important secondary factor, reducing trap attractiveness to an already apparently sparse small mammal fauna.

## SPECIES RECORDS

### MARSUPIALS

**Family Didelphidae**

*Marmosa murina* (Linnaeus 1758) – Murine mouse opossum, *gambazinha*

One animal, an adult male, was caught on the ground in *terra firme* forest. A female was also found by the Station caretaker in a riverside water-pump in the second week of September 1988. There was a nest of grass and she had six babies at the teat.

Despite an apparently wide geographical range (see Streilein, 1981), little appears to be known about this species. Eisenberg and Redford (1979) collected it in Guatopo National Park, Venezuela, and it has been recorded for Pará State by Carvalho and Toccheton (1969). Food consists of fruits and insects (Fleming, 1971; August, 1983). Most *Marmosa* species appear to be arboreal or scansorial. August (1983) caught 84% of his *M. robinsoni* in trees. O'Connell (1981) trapped 59% of her collection of the same species in trees, and likewise 50% of the larger *M. cinerea*. The studies of both August and O'Connell were carried out in Venezuela. There appears to be no published information on reproduction in *M. murina*.

*Monodelphis brevicaudata* (Erxleben 1777) – Red-legged short-tailed opossum, *catita*

Three animals of this widely distributed terrestrial marsupial were recorded. All were taken in forest, two beside logs in dense undergrowth. All our animals were males.

O'Connell (1981) reports that *M. brevicaudata* is nocturnal, and gives the diet as rodents, insects, fruits, seeds and carrion. Her two-year capture–mark–recapture study showed that animals moved a mean distance of 61 m (for males) and 30 m (for females) between captures. Density was about 1 per hectare (*llanos* vegetation, Venezuela). At Maracá two animals were caught in the same area of forest within four days of each other. O'Connell (1981) captured lactating females in May, August and November, and females with young in July and August. The rainy season at her study site is May–November. Streilein (1982) believes this species to be seasonally polyoestrus.

## RODENTS

### Family Cricetidae

*Holochilus brasiliensis* (Desmarest 1819) – Marsh rat

Seven individuals were trapped. A single wet season animal, an adult male, was caught in rough grassland. Five of the dry season animals were caught in wet grassland beside a small body of shallow water. One was taken in dry grassland. Most were taken in large runs through grass, which had probably been made by large caviomorph rodents (probably *Cavia aperea*). In the dry season we obtained two large males with scrotal testes, one small male with abdominal testes, three small females and four juveniles.

*H. brasiliensis* is a widely distributed species with a complex taxonomy. Gardner and Patton (1976) analysed the karyotype and found this single 'species' to be a series of closely allied cytological forms. Honacki *et al.* (1982) review the taxonomy. Carvalho and Toccheton (1969) report that it prefers moist habitats, mainly near streams, and that it is nocturnal and feeds on grass stems. They imply that it is difficult to trap. Twigg (1965) concurs on habitat and feeding preferences, and notes that signs of its presence include mounds of damaged plant tissues from its nocturnal feeding. He reports a mean litter size of 3.47 from 30 litters (range 2–5). In the *cerrado* biome Fonseca and Redford (1984) found this species to be restricted to gallery forests.

*Nectomys squamipes* (Brants 1827) – Water rat, *rato d'agua, rato paca*

Two animals were trapped, both in riverside forest in the dry season. A non-reproductive female was caught emerging from a hole at the base of a tree some 3 m from the river, while a male with scrotal testes was taken on 21 April 1988 in palm and bamboo-dominated riverside forest some 5 m from the river, on open forest floor with little herbaceous cover.

Mares *et al.* (1986) found that in the *cerrado* this species appears to be restricted to gallery forest. Alho *et al.* (1986) reported a positive correlation between captures and the density of ferns, and a negative one with the density of bamboo in *cerrado* gallery forest. They attributed this to a preference by this species for mesic environments. It is said to be a good swimmer: 93% of their captures were in flooded forest. Density was estimated at 1.2–3.4 animals per hectare. Home range is between 2200 and 21 000 m². Captive animals were observed to catch fish and small snails. Deitz (1983) found a nest of this species on the ground in grass.

*Oryzomys concolor* (Wagner 1845) – Rice rat

One specimen, a pregnant female, was taken in February 1988 (dry season), in the gap between the wall and the roof of an uninhabited wooden building in a small overgrown clearing in riverine forest. Traps had also been set in the surrounding forest.

Mares *et al.* (1986) trapped *O. concolor* in gallery forest trees. It comprised 8% of their total catch. All the specimens from Alho's study (Alho *et al.*, 1986) were in gallery forest trees, but it was reported that individuals also explored the wetter parts of the forest floor. Fleming (1971) caught *O. concolor* in Panama. He records the species as uncommon there, having a localized distribution (Fleming, 1970). Specimens were taken in scrubby vegetation, grasslands and in forest clearings, and he considered it to be semi-arboreal. Hershkovitz (1960) recorded the species as usually nesting in shrubby vegetation near streams.

*Taxonomic note*: This species has been called *Oecomys concolor* by Gardner and Patton (1976) and by Emmons (1997). Carvalho and Toccheton (1969) report that it is difficult to separate in the field from *Oryzomys bicolor* (Tomes, 1860), although it is bigger.

*Oryzomys delicatus* (J.A. Allen and Chapman 1897) – Rice rat

Thirteen specimens of this house-mouse-sized species were captured, making this our second most commonly trapped species (17.8% of total catch over all habitats) after *Proechimys*. Five were trapped in dry grassland, three in wet grassland and three in *aningal* (i.e. clumps of the aroid *Montrichardia linifera* on watersides). It was the commonest rodent in dry season grassland, where it comprised 65% of wet season trapping. Specimens were all taken on the ground. On the grassland grids this was normally in runs between grass stems. On the *aningal* grid individuals were trapped between the roots of the *Montrichardia*, and on the dry bare soil between living plant stems around the margin of a seasonal lake. Specimens were captured throughout the dry season. Five adult males had scrotal testes, and two taken later in the season had abdominal testes. Two adult-sized females trapped early in the dry season were not in reproductive condition. Two taken towards the end of the dry season were both pregnant. Juveniles (1♂, 1♀) were obtained early in the dry season.

The species has been recorded (as *O. microtus* J.A. Allen, 1916) from Pará by Carvalho and Toccheton (1969). They state that it is not very common (in collections), but in the field it is frequent and is easily recognized by its small size and characteristic yellowish back and very pale belly. Husson (1978) gives the habitat in Surinam as 'grass savannas'. Here it may reach considerable densities and become harmful to agriculture. Fleming (1970) recorded this species (as *O. fulvescens* Saussure, 1860) from grasslands, scrubby forest and forest clearings in Panama. He recorded the species as commonly seen by night, hopping in frog-like fashion in the roadside grasses. It has been recorded as an important item of the prey of the barn owl (*Tyto alba hellmayri*). Pregnant females have been reported in June, and juveniles in June, October and December.

*Taxonomic note*: Eisenberg (1990) regards *O. delicatus* and *O. fulvescens* as two separate species and states that the former 'apparently replaces *O. fulvescens* in the Guianas'. *O. fulvescens* is recorded as having a more reddish back, white (as opposed to grey) underparts and a tail which is noticeably bicoloured.

*Oryzomys megacephalus* (Fischer 1814) (= *O. capito* Olfers, 1818) – Rice rat

One individual was trapped at a hole in the roots of a big forest tree. The trap site was in primary forest near the bank of a dry stream. *Astrocaryum* palms were abundant on this grid. When found, the specimen was badly eaten by ants, so that little information could be obtained from the corpse.

This common, nocturnal, forest-dwelling species is widely distributed. Fleming (1971) records it from Panama, and Mares *et al.* (1986) found it in the gallery forests of the Brazilian *cerrado* area. There are also records from Costa Rica and northeastern Argentina (Nowak, 1991). Alho *et al.* (1986) reported a home range of 500–2200 $m^2$, and noted that the species is easily captured. Fleming (1971) obtained peak densities of 3.2–4.3 animals per hectare. These densities fluctuate seasonally. The species is omnivorous and commonly collected under logs and rocks in Panamanian forests. Everard and Tikasingh (1973) found the Trinidadian subspecies to be the most abundant of the ground-dwelling small mammals in their grids, comprising 28.6% of the catch. They noted that it climbed to a height of 3 m, and was often caught in arboreal traps. Carvalho and Toccheton (1969) describe this species as the most common animal in the Brazilian collections they examined.

*Taxonomic note*: Gardner and Patton (1976) have synonymized this species with *O. goeldi* Thomas, 1892 and *O. perenensis* J.A. Allen, 1901. Other authors (e.g. Peres, 1964) have treated *O. goeldi* as a sub-species of *O. capito*. The use of *megacephalus* for this species follows A. Langguth (pers. comm.) – see also Honacki *et al.* (1982).

*Rhipidomys mastacalis* (Lund 1840) – Climbing rat

Seven individuals were taken in the same hut as *O. concolor*. All individuals were taken in traps set above the ground, except for one found dead on the ground. None were taken in other parts of the hut, nor in the nearby riverine forest, although traps were set in both. Specimens were all captured in one 13-day period in February 1988. Of the five adults, three were males with scrotal testes, one was a non-reproductive female and one a pregnant female. Two juveniles (1♂, 1♀) were also taken.

Mares *et al.* (1986) recorded this species from gallery forest in *cerrado*, but at low density. Fonseca and Redford (1984) found *R. mastacalis* to be one of the commonest rodents in this habitat (39% of captures), and commented that previous studies had not included extensive use of arboreal traps. They trapped this species only in the forest, and never in adjacent grassland. In the forest it occurred only in swampy areas.

Handley (1976) caught all of his Venezuelan *R. mastacalis* in moist areas. Specimens were taken on logs on the ground and also on lianas. Alho *et al.* (1986) record this species as nocturnal and arboreal, and note a preference for more mesic parts of gallery forest where it moves along the vines on which it also nests. Eisenberg and Thorington (1973) gave densities of 187 animals per $km^2$ of forest in Venezuela. Alho (1981) gives the species range in Brazil as being from Belém to Rio de Janeiro, and notes that it lives in trees in the summer and on the ground in the winter.

*Zygodontomys* cf. *lasiurus* (J.A. Allen 1897)

Four animals of this normally abundant terrestrial omnivore were trapped during the dry season. One male with scrotal testes was trapped in wet grassland, while a pregnant female and a juvenile were caught in the Ecological Station. Another juvenile, blind and with a large tick embedded in its tail, was discovered crossing the lawn in front of the Station. Several other small (presumably juvenile) animals were observed in and around the Ecological Station at various times. Ten animals were trapped during the wet season, nine in dense secondary growth behind the Station. Six of these were adult: four males with scrotal testes, one pregnant female and one non-reproductive female. Two juvenile males were also taken. A third juvenile was badly mauled by ants and the sex could not be determined. The tenth animal was a pregnant female taken in the kitchen of the Ecological Station. This species comprised 83.3% of the total wet season catch.

Alho *et al.* (1986) report *Bolomys* (= *Zygodontomys*) *lasiurus* to be the most abundant small mammal in *cerrado* vegetation. It comprised 60% of their captures. It is considered to be a generalist and has been reported from human habitations in rural areas. In *cerrado*, densities range from 8 to 14 animals per hectare, with a home range of 300–4000 $m^2$. Populations are at their height in *cerrado* during the dry season, when grass seed (a major food source) is most abundant. The species is mostly nocturnal, and begins its activities in the late afternoon (a specimen from our study was trapped at dusk within five minutes of a trap being set). Karimi *et al.* (1976) analysed several thousand female *Z. lasiurus* from the *caatinga*, and found pregnant females in April, May and June. Mean litter size was six (range 1–13). Alho (1981) notes that *Z. lasiurus* increases in population density at the onset of the rainy season.

Seeds constituted 82% of the diet by volume, but some individuals from the *campo* had 70–90% insects in their stomachs. Over 50% of captures occurred between sunrise and sunset. Alho (1981) considered this species to be the most abundant and widely distributed small rodent in central Brazil.

*Taxonomic note*: The taxonomy of this species is extremely complex (Honacki *et al.*, 1982; Voss, 1991). The skull and body measurements of the Maracá *Zygodontomys* lie outside the recognized range of variation of *Z. lasiurus* (A. Langguth, pers. comm.). However, given the enigmatic geographic patterns of morphological variation reported for *Zygodontomys* by Voss *et al.* (1990), it would be premature to consider erecting a new taxon for the Maracá specimens.

**Family Echimyidae**

*Dactylomys dactylinus* (Desmarest 1817) – Amazon bamboo rat, *toró*

The characteristic calls of this species were heard many times at night from brakes and bamboo stands near the river – its reported preferred habitat (Lavel, 1976; Emmons, 1981, 1984). However, 428 trap nights in this habitat failed to secure a specimen. A local rancher, a long-term resident, also described both the call and the animal itself, including such characteristic features as the furring at the base of the tail. He considered it quite common in riverside vegetation, where it moves about at night in the branches of trees. It lives in holes in tree trunks some distance from the ground.

Information on this species appears to be rather sparse (see Lavel, 1976). Although it is an echimyid, the fur lacks spines and bristles (Eisenberg, 1990; Nowak, 1991). Nowak records it as having been found in dense vegetation near water; it is nocturnal and arboreal, has a vegetarian diet and may be capable of caecal fermentation of masticated vegetation (Eisenberg, 1990). Though included in their key to Brazilian mammals by Peterson and Pine (1982), it is mentioned neither by Pine (1973) for Belém, nor by Husson (1978) for Surinam. Emmons (1984) believed this species to be absent from the Guiana region.

*Mesomys hispidus* (Desmarest 1817) – Spiny tree rat

A fragment of skin, bearing the characteristic heavy spines with light tips and deep dorsal grooves, was found as part of a carnivore kill. It was identified by comparison with specimens at INPA and UFPb by Dr A. Langguth.

The species appears to be strictly arboreal. This is certainly suggested by the proportionately broad hind foot, though there appears to be very little actual ecological information on the species (see Eisenberg, 1990). Handley (1976) collected specimens in trees and near streams or moist places. The genus is very little known.

*Proechimys guyannensis* (E. Geoffroy 1803) – Spiny rat, *sauiá*, *rato-de-espinho*

Twenty-four animals were trapped in the dry season, and one was taken in the wet season. It was the most abundant species trapped, constituting 33.7% of the total catch. All the dry season specimens were taken in forest, with the exception of three animals taken in a wooden hut. Three of the forest-trapped animals came from the riverside forest, and 12 from *terra firme* forest. Two specimens, collected for ectoparasite studies, were donated by Professor Célio Valle, UFMG. The single wet season specimen was taken in rough dense herbage near the Ecological Station. The majority of specimens were trapped alongside logs or in holes in tree roots or dead logs. One animal was caught in an arboreal trap, set 2 m from the ground on a thick, living branch – it was our only arboreal capture.

Fleming (1971) and Everard and Tikasingh (1973) both record botfly larvae (Diptera: Cuterbridae) from *Proechimys*. We did not find any. Lice were very commonly collected from the Maracá *Proechimys*, but were absent from all other small mammals trapped. Five of our specimens were juveniles (two males, three females), nine were adult-sized males (eight with scrotal testes and one with abdominal testes), while all five adult dry season females were pregnant. The single wet season animal was also a pregnant female. No reproductive data are available for two males and one female. Ant damage rendered three other animals unsexable.

Autotomy of the tail by *Proechimys* appears to be quite common (see comments in Husson, 1978). Three of our specimens displayed this phenomenon. Husson records that abscission occurs at the fifth caudal vertebra, but one of our specimens had no external tail at all.

This genus is considered to be nocturnal, terrestrial (or occasionally scansorial), and omnivorous (with a strong tendency to frugivory). It is frequently abundant, comprising 40.9% of the small mammals trapped in a secondary forest in Trinidad (Everard and Tikasingh, 1973). The same authors reported it to be abundant in seasonal marsh forest. Though members of the genus appear to breed all year round (see Nowak, 1991), there is generally a peak in density during the season of greatest food abundance. At such times high densities can be attained; Fleming (1971) records densities for *P. semispinosus* (Tomes, 1860) of up to 9.2 animals per hectare. Everard and Tikasingh (1973) cite data from Belém where home ranges of 1.4 ha were reported.

*Taxonomic note*: The systematics of the genus are notoriously complex. Patton (1987), in his review of the genus, published photographs of the palate of several species of *Proechimys*. These allow species to be distinguished on the basis of how far the mesopterygoid fossa extends in relation to the upper molar row. On this basis, and others, all our specimens are *P. guyannensis*. However, Patton (1987) notes that *P. cuvieri* Petter, 1978 and *P. guyannensis* can be sympatric. Studies by Dr P. Linardi (UFMG) of the ectoparasites collected from *Proechimys* at Maracá indicate that there may be two species on the island, and that the second may well be *P. cuvieri*.

**Family Dasyproctidae**

*Dasyprocta agouti* (Linnaeus 1758) (= *D. leporina*) – Red-rumped agouti, *cutia*

Individuals of this species were regularly sighted by project members in *terra firme* forest. The species was not trapped as it is too large for the trap types used.

*D. agouti* has been recorded from the *llanos* and Guatopo National Park in Venezuela by Eisenberg and Redford (1979), and Alho (1981) described its distribution as including all of Brazil. The species feeds particularly on fallen fruits and seeds, and is attracted to the sounds of such objects falling to the forest floor (Smythe, 1978). Animals generally live in pairs in year-round home ranges of 1–2 ha (Smythe, 1978). Burrows are often located on hillsides or by stream beds. Well-worn paths may radiate from these (Fleming, 1970). Litter size is generally two. Smythe (1978) reported that *Dasyprocta* on Barro Colorado Island, Panama, breed all year round, but that only those young born in the fruiting season survived to maturity.

## Family Agoutidae

*Agouti paca* (Linnaeus 1766) – Paca

Individuals of this species were observed infrequently in *terra firme* forest. As the species is nocturnal this should not be taken as an indication of the relative abundance against the diurnal *D. agouti*.

This species has been recorded by Eisenberg and Redford (1979) from the Venezuelan *llanos* and from Guatopo National Park, and is widely distributed throughout Amazonia. Enders (1935) reported that it eats a wide variety of fruits and nuts and also roots. Nowak (1991) notes that individuals generally have several burrows, that it is solitary, and that it calls less frequently than *Dasyprocta*. The spoor of *Dasyprocta* and *Agouti* may be distinguished by the presence of three toes on the hind foot of *Dasyprocta* and five on that of *Agouti* (see Emmons, 1997).

## Family Erethizontidae

*Coendou prehensilis* (Linnaeus 1758) – Brazilian porcupine, *ouriço*, *porco-espinho*

Spines of this species were picked up from the forest floor from the remains of a carnivore kill. They were identified by comparison with specimens in the collection of INPA, Manaus.

*Coendou* is nocturnal, arboreal and has diurnal rest sites 6–10 m above the ground. The location of these is reported to be changed every night. Such behaviour could make the species difficult to see, and probably accounts for the lack of observations (see also Eisenberg and Redford, 1979; Emmons, 1984, in this context). Alho (1981) recorded the species in *cerrado* forest and in scrub. Eisenberg and Redford (1979) found the species in the Venezuelan *llanos*.

## Family Sciuridae

*Sciurus pyrrhonotus* (Allen 1915)

Specimens of a large, black-furred squirrel were occasionally observed high in the canopy in *terra firme* forest, but no specimens were secured. Biogeographical data point to this species.

There appears to be very little information on the ecology of this species. However, the taxonomic position is very confused (see Honacki *et al.*, 1982), so that the data on the other 'species' referred to below may well be of relevance to the animals that live on Maracá.

*Sciurus pyrrhonotus* is a relatively large animal (head and body 300 mm; tail 300 mm) and is generally placed in a group commonly known as giant neotropical squirrels. Members of this species-group seem to specialize in eating hard seeds and palm nuts. Smythe (1978) reports that 60% of the diet of the Panamanian species *S. granatensis* Humboldt, 1811 consisted of hard palm nuts. Enders (1935), Kiltie

(1981) and Kiltie and Terborgh (1983) believed that the activities of such large squirrels were important in providing fallen fruit upon which agouti and peccary largely depend.

**Family Caviidae**

*Cavia aperea* (Erxleben 1777) – Guinea pig

Several specimens were trapped by the ectoparasite study group from UFMG, headed by Dr C. Valle. All records came from the seasonally flooded grassland, early in the dry season. Extensive trapping by the authors later in the year failed to procure a specimen, despite the presence of numerous runs in the dense tall grass.

*Cavia* is herbivorous, terrestrial and mostly nocturnal, and can achieve high local densities (Eisenberg, 1990). The species is probably native to the area, and not part of the recently escaped feral population (but see Eisenberg, 1990). The use of *C. aperea* runs and tunnels through vegetation by other species of small mammals has also been reported by Bilencia *et al.* (1995).

# GENERAL DISCUSSION

The habitat preferences of the species recorded from Maracá appear to differ little, in the broad sense, from those previously reported elsewhere. This is unsurprising, since fine-tuning to the immediate environment generally occurs via dietary and reproductive phenomena (see Cody and Diamond, 1975; Wilson, 1980). Barnett and Cunha (1994) provide an analysis of these patterns. There is a considerable need for continuous long-term monitoring of small mammal populations (with live trapping) at Maracá, to give an insight into the demographic processes occurring.

Whilst composition of the small mammal fauna is in accordance with the predictions made on biogeographical grounds, two points stand out: firstly the low numbers of *Oryzomys capito* (only one animal captured); and secondly the failure to trap any arboreal species. *O. capito* is reported as very common wherever it is found (see species records). It is not a habitat specialist, and has been reported both north (Panama, Venezuela), south (Brazilian *cerrado*) and east (Surinam, Manaus, Belém) of Maracá. Consequently, the Maracá populations can be regarded as marginal neither in a biogeographical nor in an ecological sense. The low numbers recorded may be a product of trapping techniques and intensity, or may reflect the effects of seasonality.

The failure to obtain arboreal species could be attributed to any of the factors dealt with in the discussion of low terrestrial trapping rates. However, a contributing factor may have been the very low density of epiphytes at Maracá (see Milliken and Ratter, Chapter 5), resulting from the high seasonality of the climate. This would have reduced both the diversity of available fruits and seeds and, perhaps more importantly, the number of insects. M. O'Shea (pers. comm.) found a similarly low diversity of arboreal species amongst the herpetofauna.

In passing it should be noted that the normal human commensals *Mus* and *Rattus* do not yet seem to have reached the island. This may account for the presence of *Zygodontomys* in the Ecological Station.

## COMPARISON OF THE SMALL MAMMAL FAUNA OF MARACÁ WITH OTHER NEOTROPICAL SITES

### GENERAL OBSERVATIONS

The Ilha de Maracá lies adjacent to the Gran Sabana refugium of Steyermark (1979). The Workshop 90 group identified the region as one meriting 'highest priority' in their ranking of Amazonian sites for conservation (Prance, 1990a, b). However, this would appear to be mainly on account of its high level of botanical endemism (see Maguire, 1970; Tryon, 1972; Steyermark and Dunsterville, 1980), rather than for its mammalian fauna which has a low percentage of endemics (Emmons, 1990; Eisenberg, 1997). Roraima State itself has had few mammalian studies. Emmons (1997) has 23 question marks showing uncertainty in mammalian distributions between the Amazon and the northern borders of Brazil. Twelve of these refer to rodents and marsupials. Of these, seven are within Roraima State. This clearly impedes the comparison of the small mammal fauna of Maracá with that of other sites.

The recorded small mammal fauna of the Ilha de Maracá comprises 17 species in three habitat groups. The forest group comprises 12 species: one arboreal marsupial (*Marmosa murina*), one terrestrial marsupial (*Monodelphis brevicaudata*), one arboreal cricetid rodent (*Rhipidomys mastacalis*), one scansorial forest cricetid (*Oryzomys concolor*), one terrestrial cricetid (*O. megacephalus*), one (possibly two) terrestrial echimyid rodents (*Proechimys guyannensis* and, possibly, *P. cuvieri*), two arboreal echimyids (*Dactylomys dactylinus* and *Mesomys hispidus*), a porcupine (*Coendou prehensilis*), a squirrel (*Sciurus pyrrhonotus*), and two larger terrestrial rodents (*Dasyprocta agouti* and *Agouti paca*). The grassland group comprises two cricetids (*Oryzomys delicatus* and *Zygodontomys* cf. *lasiurus*) and one caviomorph (*Cavia aperea*). The third group, of semi-aquatic small mammals, comprises two rodents (*Holochilus brasiliensis* and *Nectomys squamipes*).

In addition, the following species may also be present in forests: two arboreal marsupials (*Caluromys philander* (Linnaeus, 1758) and *Marmosa emilae* Thomas, 1909), one scansorial marsupial (*Lutreolina crassicaudata* Thomas, 1910), two terrestrial marsupials (*Didelphis marsupialis* (Linnaeus, 1758) and *Metachirus nudicaudatus* (E. Geoffroy, 1803)), two scansorial cricetids (*Oryzomys bicolor* (Tomes, 1860) and *O. macconnelli* Thomas, 1910), two terrestrial cricetids (*Akodon urichi* J.A. Allen & Chapman, 1897 and *Neacomys guianae* Thomas, 1905), one arboreal echimyid (*Makalata armatus* (I. Geoffroy, 1830)), and two squirrels (*Sciurus aestuans* Linnaeus, 1776 and *Sciurillus pusillus* (Desmarest, 1817)). The semi-aquatic marsupial *Chironectes minimus* (Zimmerman, 1780) is also a candidate. *Sigmodon alstoni* (Thomas, 1881) and *Akodon urichi* may occur in grasslands. If all these possible species were present, in addition to those definitely recorded, the total number of small mammals in all habitats on Maracá would be 32. This compares favourably with survey results from Cocha Cashu, Peru (33 spp.; Janson and Emmons, 1990) and from Manaus (25 spp.; Malcolm, 1990).

The relatively high number of species (potentially) occurring on Maracá given its position on the margins of the Amazon basin is probably due to its broad habitat range, i.e. the inclusion of both forest and savanna habitats. The other two sites with which it has been compared are composed solely of forest (Foster, 1990; Prance, 1990c), which is undoubtedly a significant factor. This phenomenon of higher species diversities at ecotones has commonly been recorded for tropical ecosystems (e.g. Happold, 1974; Jeffrey, 1977; Barnett et al., 1996). However, the situation is not entirely simple on Maracá, as species otherwise restricted to forest may sometimes be found among grasslands, living in gallery forests (Mares et al., 1986) or in shrubby areas (Handley, 1976). Only two species have been recorded from the area of grassland sampled on Maracá. August (1973) records 12 small mammal species from the grasslands and open woodlands of the Venezuelan *llanos*, while Alho et al. (1986) found 13 species of small mammals in grassland and 12 species in gallery forests (with an overlap of five species) in *cerrado*, the Brazilian equivalent. In a similar study, Deitz (1983) trapped 10 species in the same habitat types. Given the comparative paucity of records from grasslands on Maracá, it would be interesting to carry out a more thorough inventory of the small mammal fauna of this habitat.

## PATTERNS OF COMMUNITY COMPOSITION

Emmons (1984) compared the numbers, densities and compositions of mammal communities at seven moist evergreen forest sites in Amazonian Peru, Ecuador and Brazil. She reported that *Proechimys* was the most abundant small mammal at all sites. This was also true at Maracá. Her nearest site to the Ilha de Maracá was the Minimum Critical Size of Ecosystems (CMSE) project near Manaus. This is some 700 km SSE of the present study site. Emmons found a strong relationship between soil type and mammal density, reporting that mammal densities were highest on white-water alluvial soils and lowest on sandy, Guiana Shield weathered tertiary sediments. Thus, at Cocha Cashu in Peru (an alluvial site) there were 6.4 times as many small mammals as at the CMSE site (a site of Guiana Shield sandy sediments). This included 2.4 times as many opossums and 21.3 times as many small rodents. This was reflected in the trapping success. Mean trap success at Cocha Cashu was 6.9%, and 0.75% at the CMSE site. The Ilha de Maracá shares the same soil type as the CMSE site, which may help to explain the very low trap success (mean of 1.84%) attained by the present study.

Emmons also observed that the composition of the small mammal community changed as one moved to the rim of the Amazon basin. Not only did numbers of individuals fall, but the communities had fewer species and the frequency of these (in terms of both biomass and individuals) was lower in rim communities. Emmons identified three types of species absent from the Guayanan fauna: species at the small end of the size range for their taxon or guild, species from locally polytypic genera, and those belonging to monotypic genera that are everywhere rare.

These observations may help to explain some of the patterns seen in the Maracá fauna. Thus, *Oryzomys capito* (reported by Emmons to be rare at the CMSE site) was trapped only once at Maracá. Emmons regards the CMSE case as an example of the atrophy of a feeding guild in the more marginal basin rim habitat. This may also explain the apparent absence from Maracá of *Neacomys guianae*, a very small primarily insectivorous mouse. Emmons' observation on species of monotypic genera is difficult to verify. Species previously considered rare are often found upon subsequent analysis to be quite common if the right habitat is investigated (see Barnett, 1991a, b). The absence from the Maracá record of such species as *Chironectes minimus* is therefore not conclusive (especially given its biogeography – see Emmons, 1997). In some cases our work has amplified known distribution patterns; for example the Maracá records for *Dactylomys dactylinus* represent a range extension of approximately 200 km. One does not know how complete are the records of the Maracá forest fauna: with one exception all the expected-but-unrecorded species are arboreal or scansorial and their apparent absence may simply be due to the technical difficulties of trapping in tropical trees (see Malcolm, 1991; Barnett and Dutton, 1995; Voss and Emmons, 1996).

## SUGGESTIONS FOR FUTURE WORK

The INPA/RGS/SEMA research initiative at Maracá has provided a substantial corpus of biological knowledge for what is, in Amazonian terms, a fairly small area. The following suggestions are designed to establish a long-term programme of research on small mammals.

- Long-term simultaneous monitoring of small mammal populations in Maracá's diverse habitat types, using, if possible, live traps and capture–mark–recapture methods (CMR).
- Quantified study of the availability of seeds and fruits used by small mammals, and of the seasonal variation in the availability of such foods.
- Quantified studies of food competition between small mammals and larger mammalian frugivores (e.g. peccaries, deer and tapir).
- Determination of densities and home ranges of small mammal species by CMR.
- A programme of intensive arboreal trapping.
- Study of interactions between primates (especially *Cebus* and *Saimiri*) and arboreal insectivorous small mammals.
- A programme of intensive trapping in gallery forests.
- Testing ant-proof baits.
- The use of larger traps (e.g. 'Shermans') in both arboreal and terrestrial work to catch larger species.

## ACKNOWLEDGEMENTS

The authors would like to thank Ângelo dos Santos and Fernando Paulo Teles (INPA) for their help in carrying out this project, Steve Bowles and Jill Thompson (loan of equipment),

William Milliken (technical assistance), Joe Anderson (technical advice), Robert Miller (auxiliary data collection) and Mark O'Shea for his role in obtaining equipment indispensable for the continuation of the work, and for his help with the fieldwork.

Identification of specimens was made possible by the kindness and knowledge of Alfredo Langguth (Federal University of Paraíba), Martin Perry (Natural History Museum, London) and Sueli Marques (Museu Paraense Emílio Goeldi). Célio Valle (Federal University of Minais Gerais, UFMG) loaned specimens, and gave helpful advice on capture methodology. We are also indebted to Sue Branford and Isabella Barber in London, Ning Labbish Chao, Paulo Petry and Jorge Portugal in INPA (Manaus), and Paulo Ribeiro in Rio de Janeiro, all of whom advised and helped us with word-processing.

We are grateful to Pedro Linardi for use of unpublished data on trap success rates on Maracá. Finally, we would like to express our deepest thanks to Nélia Tamanini for her continuing kindness.

# REFERENCES

Alho, C.A.H. (1981). Brazilian rodents: their habitats and habits. In: *Mammalian biology in South America*, eds M.A. Mares and H.H. Genoways, pp. 143–166. Special Publications Series, Pymatuning Laboratory, University of Pittsburgh.

Alho, C.A.H., Pereira, L.A. and Paula, A.C. (1986). Patterns of habitat utilization by small mammal populations in the *cerrado* biome of central Brazil. *Mammalia*, 50, 447–460.

Anthony, H.E. (1929). Two new genera of rodents from South America. *American Museum Novitates*, 383.

August, P. (1983). The role of habitat complexity and heterogeneity in structuring tropical mammal communities. *Ecology*, 64, 1495–1507.

Barnett, A.A. (1991a). The wildlife detectives. In: *Wildlife fact file yearbook 1992*, ed. S. Francis, pp. 20–25. International Masters Publishing, London.

Barnett, A.A. (1991b). Records of the grey-bellied shrew opossum, *Caenolestes caniventer*, and Tate's shrew opossum, *Caenolestes tatei* (Caenolestidae, Marsupialia), from Ecuadorian montane forests. *Mammalia*, 55, 443–445.

Barnett, A.A. (1997). The ecology and natural history of a fishing mouse *Chibchanomys* sp. nov. (Ichthyomyini: Muridae) from the Andes of Southern Ecuador. *Zeitschrift für Saugetierkunde*, 62, 43–52.

Barnett, A.A. and Cunha, A.C. da (1994). Notes on the small mammals of Ilha de Maracá, Roraima State, Brazil. *Mammalia*, 58, 131–137.

Barnett, A.A. and Dutton, J. (1995). *Small mammals (excluding bats)*. Expedition Field Techniques Series No. 1. Royal Geographical Society/Expedition Advisory Centre, London.

Barnett, A.A., Prangley, M., Hayman, P.V., Diawara, D. and Koman, J. (1996). A survey of the mammals of the Kounounkan massif, south-western Guinea. *Journal of African Zoology*, 110, 235–240.

Bilencia, D.N., Cittadino, E.A. and Kravetz, F.O. (1995). Influencia de la atividad de *Cavia aperea* sobre la estrutura del habitat y la distribucion de *Akodon azurae* y *Oryzomys flavescens* (Rodentia: Caviidae, Muridae) en bordes de cultivos de la region Pampeana (Argentina). *Iheringia* (Sér. Zool.), 79, 67–75.

Carvalho, C.T. de and Toccheton, A.J. (1969). Mamíferos do Nordeste do Pará, Brasil. *Revista Biologia Tropical*, 15, 215–226.

Cheeseman, C.L. and Delany, M.J. (1979). The population dynamics of small rodents in a tropical African grassland. *Journal of Zoology, London*, 188, 451–475.

Cody, M.L. and Diamond, J.M. (eds) (1975). *Ecology and evolution of communities*. Harvard University Press, Cambridge, Massachusetts.

Corbet, G. (1968). *British Museum (Natural History) guides for collectors, No. 1: mammals (excluding marine mammals)*. BMNH, London.

Deitz, J.M. (1983). Notes on the natural history of some small mammals in central Brazil. *Journal of Mammology*, 64, 521–523.
Delany, M.J. (1974). *The ecology of small mammals*. Institute of Biology – Studies in Biology, No. 51. Edward Arnold, London.
Eisenberg, J.F. (1990). *Mammals of the Neotropics: Volume 1 – The Northern Neotropics; Panama, Colombia, Venezuela, Guyana, Suriname, French Guiana*. University of Chicago Press.
Eisenberg, J.F. and Redford, K.H. (1979). A biogeographic analysis of the mammalian fauna of Venezuela. In: *Vertebrate ecology in the Northern Neotropics*, ed. J.F. Eisenberg, pp. 31–36. Smithsonian Institution, Washington, DC.
Eisenberg, J.F. and Thorington, J. (1973). A preliminary analysis of a neotropical mammal fauna. *Biotropica*, 5, 150–161.
Emmons, L.H. (1981). Morphological, ecological and behavioural adaptations for arboreal browsing in *Dactylomys dactylinus* (Rodentia, Echimyidae). *Journal of Mammology*, 62, 183–189.
Emmons, L.H. (1984). Geographic variation in densities and diversities of non-flying mammals in Amazônia. *Biotropica*, 16, 210–222.
Emmons, L.H. (1997). *Neotropical rainforest mammals: A field guide*. Second edition. University of Chicago Press.
Enders, R.K. (1935). Mammalian life histories for Barro Colorado. *Bulletin of the Museum of Comparative Zoology, Harvard*, 78, 385–502.
Everard, C.O.R. and Tikasingh, E.S. (1973). Ecology of the rodents *Proechimys guyannensis trinitatis* and *Oryzomys capito velutinus* on Trinidad. *Journal of Mammology*, 54, 875–886.
Fittkau, E.J. and Klinge, H. (1973). On biomass and trophic structure of the central Amazonian rainforest ecosystem. *Biotropica*, 5, 2–14.
Fleming, T.H. (1970). Notes on the rodent faunas of two Panamanian forests. *Journal of Mammology*, 42, 439–455.
Fleming, T.H. (1971). Population ecology of three species of neotropical rodents. *Miscellaneous Publications of the Museum of Zoology, Michigan*, 143, 1–77.
Fonseca, H. and Redford, K.H. (1984). The mammals of IBGE's Ecological Reserve, Brasília, and an analysis of the role of gallery forest in increasing diversity. *Revista Brasileira Biologia*, 44, 517–523.
Foster, R.B. (1990). The floristic composition of the Rio Manu floodplain forest. In: *Four Neotropical forests*, ed. A.H. Gentry, pp. 99–111. Yale University Press, New Haven.
Gardner, A.L. and Patton, J.L. (1976). Karyotypic variation in Oryzomine rodents (Cricetinae) with comments on chromosomal evolution in the Neotropical Cricetine complex. *Occasional Papers of the Museum of Zoology, Louisiana State University*, 49, 1–48.
Handley, C.O. (1976). Mammals of the Smithsonian Venezuelan Project. *Brigham Young University Bulletin, Biology Series*, 20, 1–89.
Happold, D.C.D. (1974). The small rodents of the forest–savanna–farmland association near Ibadan, Nigeria, with observations on reproductive biology. *Revue Zoologie Africaine*, 88, 814–837.
Hershkovitz, P. (1960). Mammals of northern Colombia, preliminary report No. 8: Arboreal rice rats, a systematic revision of the sub-genus *Oecomys*, genus *Oryzomys*. *Proceedings of the US National Museum*, 110, 513–568.
Honacki, J.H., Kinman, K.E. and Koeppl, J.W. (1982). *Mammal species of the world: a taxonomic and geographical reference*. Allen Press Inc. and the Association of Systematic Collections, Collins, Kansas.
Husson, A.M. (1978). *Mammals of Suriname*. Zoologische Monographieen, Rijkmuseum van Natuurlijke Historie, No. 2.
Janson, C.H. and Emmons, L.H. (1990). Ecological structure of the nonflying mammal community at Cocha Cashu Biological Station, Manu National Park, Peru. In: *Four Neotropical forests*, ed. A.H. Gentry, pp. 314–338. Yale University Press, New Haven.

Jeffrey, S.M. (1977). Rodent ecology and land use in western Ghana. *Journal of Applied Ecology*, 14, 741–755.
Karimi, Y., Rodrigues de Almeida, C. and Petter, F. (1976). Notes sur les rongeurs du Nord-est du Brasil. *Mammalia*, 40, 257–266.
Kiltie, R.A. (1981). Distribution of palm fruits on a rainforest floor: why white-lipped peccaries forage near objects. *Biotropica*, 13, 141–145.
Kiltie, R.A. and Terborgh, J. (1983). Observations on the behaviour of rain forest peccaries in Peru: why do white-lipped peccaries form herds? *Zeitschrift für Tierpsychologie*, 62, 241–255.
Lavel, R.K. (1976). Voice and habitat of *Dactylomys dacrylinus* (Rodentia: Echimyidae) in Ecuador. *Journal of Mammology*, 57, 402–404.
Linardi, P.M., Botelho, J.M., Rafael, J.A., Valle, C.M.C., Cunha, A.C. da and Machado, P.A.R. (1991). Ectoparasitos de pequenos mamíferos da Ilha de Maracá, Roraima, Brasil. I. Ectoparasitofauna, registros geográficos e de hospedeiros. *Acta Amazonica*, 21, 131–140.
Maguire, B. (1970). On the flora of the Guayana Highland. *Biotropica*, 2, 85–100.
Malcolm, J.R. (1990). Estimation of mammalian densities in continuous forest north of Manaus. In: *Four Neotropical forests*, ed. A.H. Gentry, pp. 339–357. Yale University Press, New Haven.
Malcolm, J.R. (1991). Comparative abundances of neotropical small mammals by trap height. *Journal of Mammalogy*, 72, 188–191.
Mares, M.A., Ernest, K.A. and Gettinger, D.D. (1986). Small mammal community structure and composition in the *cerrado* province of Brazil. *Journal of Tropical Ecology*, 2, 289–300.
Moojen, J. (1948). Speciation in the Brazilian spiny rats, of the genus *Proechimys* (Rodentia: Family Echimyidae). *University of Kansas Publications, Museum of Natural History*, 1, 301–406.
Nowak, R.H. (1991). *Walker's mammals of the world*, 5th edition. Johns Hopkins University Press, Baltimore.
O'Connell, M.A. (1981). Population biology of North and South American grassland rodents: a comparative review. In: *Mammalian biology in South America*, eds M.A. Mares and H.H. Genoways, pp. 167–185. Special Publications Series, Pymatuning Laboratory, University of Pittsburgh.
Patton, J.L. (1987). Species groups of spiny rats, genus *Proechimys* (Rodentia: Echimyidae). *Fieldiana Zoology, Chicago, New Series*, 39, 305–345.
Peres, D. de A.F. (1964). Mamíferos colecionados na região de Rio Negro (Amazonas, Brasil). *Boletim do Museu Paraense Emílio Goeldi* (*Zoologia*), 42, 1–23.
Peres, D. de A.F. (1968). Tipos de mamíferos recentes no Museu Nacional, Rio de Janeiro. *Arquivos del Museo Nacional, Rio de Janeiro*, 53, 161–193.
Peterson, N.E. and Pine, R.H. (1982). Chave para identificação de mamíferos da região Amazônica Brasiliera com exceção dos quirópteros e primatas. *Acta Amazonica*, 12, 465–482.
Pine, R.H. (1973). Mammals (exclusive of bats) of Belém, Pará, Brazil. *Acta Amazonica*, 3, 47–69.
Prance, G.T. (1990a). Consensus for conservation. *Nature*, 345, 384.
Prance, G.T. (1990b). Future of the Amazonian rainforest. *Futures*, November 1990, 891–903.
Prance, G.T. (1990c). The floristic composition of the forests of Central Amazonian Brazil. In: *Four Neotropical forests*, ed. A.H. Gentry, pp. 112–140. Yale University Press, New Haven.
Smythe, N. (1978). The natural history of the Central American agouti (*Dasyprocta punctata*). *Smithsonian Contributions to Zoology*, No. 257.
Steyermark, J.A. (1979). Plant refuge and dispersal centres in Venezuela: their relict and endemic element. In: *Tropical botany*, eds K. Larsen and L.B. Holm-Nielsen, pp. 185–221. Academic Press, New York.
Steyermark, J.A. and Dunsterville, G.C.K. (1980). The lowland floral element on the summit

of Cerro Guaiquinima and other cerros of the Guayana highland of Venezuela. *Journal of Biogeography*, 7, 285–303.

Streilein, K.E. (1981). Behavior, ecology and distribution of South American marsupials. In: *Mammalian biology in South America*, eds M.A. Mares and H.H. Genoways, pp. 231–250. Special Publications Series, Pymatuning Laboratory, University of Pittsburgh.

Streilein, K.E. (1982). Reproductive biology of marsupials. *Annals of the Carnegie Museum*, 51, 251–269.

Tate, G.H.H. (1931a). Brief diagnoses of twenty-six apparently new forms of *Marmosa* (Marsupiala) from South America. *American Museum Novitates*, 403, 1–14.

Tate, G.H.H. (1931b). Random observations on habits of South American mammals. *Journal of Mammology*, 12, 248–256.

Tryon, R. (1972). Endemic areas and geographic speciation in tropical American ferns. *Biotropica*, 4, 121–131.

Twigg, G.I. (1965). Studies on *Holochilus sciureus sciureus*, a Cricetine rodent from the coastal region of British Guiana. *Proceedings of the Zoological Society, London*, 145, 263–283.

Voss, R.S. (1991). An introduction to the Neotropical murid genus *Zygodontomys*. *Bulletin of the American Museum of Natural History*, 206, 414–432.

Voss, R.S. and Emmons, L.H. (1996). Mammalian diversity in Neotropical lowland forests: a preliminary assessment. *Bulletin of the American Museum of Natural History*, 230, 1–115.

Voss, R.S., Marcus, L.F. and Escalente, P.P. (1990). Morphological evolution in murid rodents: conservative patterns of craniometric covariance and their ontogenic basis in the Neotropical genus *Zygodontomys*. *Evolution*, 44, 1568–1587.

Wilson, E.O. (1980). *Sociobiology*. Harvard University Press, Cambridge, Massachusetts.

# 11 Birds of the Ilha de Maracá

JOSÉ MARIA C. DA SILVA
*Museu Paraense Emílio Goeldi, Belém, Brazil*

## SUMMARY

The Ilha de Maracá is an ecological reserve in the north of the Brazilian Amazon. As a result of the efforts of a number of professional and amateur field ornithologists it is possible to write a first synthesis of the avifauna of this interesting conservation unit. So far 442 species of birds have been recorded for the island. These are principally distributed in forest (293 spp.) and savanna (149 spp.). Maracá's avian richness supports the hypothesis that diversity is greater at the margin of the Amazon basin than at its centre. The ornithological data also support the botanists' hypothesis that the *terra firme* forest at the eastern end of Maracá appears to be in a phase of expansion perhaps after some catastrophic event. Four mixed flock types were recorded on Maracá. Two were in seasonally flooded grassland (seed-eaters and fish-eaters) and two in forest habitats (insectivores in the understorey, and insectivores and insectivore–frugivores in the canopy). There are three migrant categories on Maracá: migrant species or subspecies from the northern hemisphere (15); migrant species or subspecies from the southern hemisphere (2); and altitudinal migrants (1). A biogeographic analysis of the forest birds, using toucans, aracaris and woodcreepers, showed that Maracá's forest avifauna is clearly in a contact zone between the Guianan and Upper Amazon (Napo and Imeri) areas of endemism. Maracá's savanna avifauna is more closely related to those of the Venezuelan and Colombian savannas than to those of Central Brazil. A discussion of the importance of considering biogeographic transition areas in the planning of conservation in the Amazon is presented.

## RESUMO

A Ilha de Maracá é uma estação ecológica localizada no norte da Amazônia brasileira. Como uma consequência do esforço de vários ornitólogos amadores e profissionais é possivel escrever uma primeira síntese da avifauna desta interessante unidade de conservação. Até agora, 442 espécies de aves foram registradas para Maracá. Elas estão distribuidas principalmente nas florestas (293 spp.) e savanas (149 spp.). A riqueza de espécies de aves em Maracá apoia a hipótese de que a riqueza de aves na Amazônia é maior nas margens do que no centro desta região. Os dados ornitológicos também apoiam a hipótese dos botânicos de que a floresta de *terra firme* no leste de Maracá parece estar em fase de expansão, talvez depois de algum evento catastrófico. Quatro tipos de bandos mistos foram registrados em Maracá. Dois em um campo sazonalmente alagado (granívoros e piscívoros) e dois em hábitats florestais (insetívoros no sub-bosque e insetívoros mais insetívoros–frugívoros na copa). Há três categorias de migrantes em Maracá: espécies ou subespécies migrantes do hemisfério norte (15); espécies ou subespécies migrantes do sul da América do Sul (2); e migrantes altitudinais (1). Uma análise biogeográfica das aves de florestas usando tucanos, araçaris e

*Maracá: The Biodiversity and Environment of an Amazonian Rainforest.*
Edited by William Milliken and James A. Ratter. © 1998 John Wiley & Sons Ltd.

arapaçús, mostra que a avifauna das florestas de Maracá é claramente uma transição entre as avifaunas das áreas de endemismo das Guianas e do Alto Amazonas (Napo e Imeri). A avifauna das savanas de Maracá é mais relacionada à avifauna das savanas de Venezuela e Colômbia do que das savanas do Brasil Central. Uma discussão sobre a importância de se considerar áreas de transição no planejamento da conservação na Amazônia é apresentada.

## INTRODUCTION

The aim of this chapter is to present an overview of the composition, ecology and biogeography of the avifauna of the Ilha de Maracá. It is a product of the combined efforts of various professional and amateur ornithologists who have visited the reserve since its creation in 1978. As in any such composite work, it has been necessary to make critical analyses of the constituent data sets, and to adjust them accordingly. There is clearly a need for further ornithological research on the Ilha de Maracá, and this article reflects that need. It was written, above all, to lend scientific support to all those who recognize the importance of Maracá for the conservation of the extraordinary biological resources of Brazilian Amazonia.

### ORNITHOLOGICAL INVESTIGATIONS ON THE ILHA DE MARACÁ

Until recently the avifauna of the region of Roraima State in which Maracá is situated was little known. The first ornithological observations made in close proximity to the island were those of Dr A. Hamilton Rice's seventh Amazonian expedition, which explored the upper Rio Branco and the Rio Uraricoera in 1924. One of the expedition's members, Dr George C. Shattuck, prepared a list of 55 species of birds observed in the region (Pinto, 1966).

Moskovits *et al.* (1985) presented the first major study of the avifauna of the Ilha de Maracá. This paper, the result of observations made between January 1980 and November 1982, comprises a list of the birds (with scientific names and with common names in English and Portuguese) recorded on and in the immediate vicinity of the island, together with data on the relative abundance of each species and the habitats in which they were recorded. In addition to the authors of the article (D. Moskovits, J. Fitzpatrick and D. Willard), D.C. Oren and G.M. Russel also assisted with the study.

Mike Hopkins was on Maracá during March 1979 and June 1986, and although his principal interest was the collection of botanical material, he also recorded several bird species which were new records for the Ilha de Maracá.

Under the aegis of the Maracá Rainforest Project (February 1987 to March 1988), various people made studies of the island's avifauna. A team from the Museu Paraense Emílio Goeldi (MPEG), composed of J.M.C. da Silva, M.S. Brigida and R. Pereira, worked on Maracá from 12 February to 11 March 1987. Studies were made of the behaviour of several bird species and of the avifaunal composition of the savannas and forests in the vicinity of the Ecological Station. In addition, a reference collection of the birds of the island was prepared, and is presently deposited in the MPEG. Some of the principal scientific results of this study are presented in Silva and Oren (1990).

D.F. Stotz, currently at the Field Museum of Natural History, Chicago, studied the birds of Maracá between 22 and 24 September 1987. His observations were made in the vicinity of the Ecological Station. A. Whittaker and M. Cohn-Haft, from the World Wildlife Fund PDBFF (Manaus), visited the island between 17 and 24 December 1987. Cohn-Haft returned to Manaus after a few days, but Whittaker continued his observations and was the only ornithologist to visit the forests of the western extremities of the reserve. In addition, the following carried out ornithological observation during the Maracá Rainforest Project: D. Scott, J. Searight and R.B. Cavalcanti.

## THE DATA BASE – DEFINITIONS AND LIMITATIONS

SPECIES LIST

To prepare a final inventory of the bird species of Maracá (Appendix 4, p. 451), the lists produced by Moskovits *et al.* (1985) and the new records made during the Maracá Rainforest Project were combined, and each of the species in this overall list was then subjected to critical analysis.

Moskovits *et al.* (1985) had recorded 377 species of birds (excluding *Chaetura*), and not the 386 that was stated in the text of the original article. Adding to these the new records of Hopkins and the Maracá Rainforest Project scientists (69 spp.), one arrives at a total of 446. Given that the majority of the records stem from observations made by different people with different levels of experience of Amazonian avifauna, it was necessary to evaluate the level of reliability of each record.

The means used to refine the preliminary list was primarily that of checking which of the species had previously been recorded for Roraima State. In order to do this, I used the excellent summary of the ornithology of Roraima by Pinto (1966), and the following works: Novaes (1965), Phelps (1973), Dickerman and Phelps (1982) and Silva and Willis (1987). In addition, I used my own data collected in the colony of Apiaú, municipality of Mucajaí, in March 1990, and the specimens collected in Roraima in the 1960s by Emílio Dente and J. Hidasi, currently deposited in the MPEG.

As a result of this investigation, I concluded that 342 (76.7%) of the 446 species of birds recorded for Maracá had already been collected in Roraima, thus making them reliable records. It was therefore left to evaluate the situation of the remaining 104 species. Based on the known geographical distributions and ecological requirements of each of these, the majority (87.5%) would be expected to occur in Roraima. The records of 14 species, however, presented problems. Of these, three were confirmed (*Brotogeris chrysopterus*, *Pteroglossus pluricinctus* and *Euphonia violacea*); four were excluded (*Brotogeris cyanoptera*, *Pteroglossus aracari*, *Electron platyrhynchum* and *Euphonia laniirostris*); three had their records included in the final species list but require confirmation (*Coccyzus pumilus*, *Pipra coronata* and

*Tyrannus dominicensis*); and four identifications were corrected (*Celeus undatus, Celeus elegans, Tyranneutes virescens* and *Basileuterus mesoleuca*). As a consequence, the final list of species recorded for Maracá now totals 442. The problematic cases are discussed below.

### *Brotogeris chrysopterus/cyanoptera*

Moskovits *et al.* (1985) recorded both of these species for Maracá. The two are phylogenetically related and were combined by Haffer (1987) into a superspecies. In the north of the Amazon the two species appear to replace each other geographically, whereas in the south there is a hybridization zone along the Beni river in Bolivia (Gyldenstope, 1945). Neither had previously been recorded in Roraima. In the distribution map of the two species that was presented by Haffer (1987), Roraima was included hypothetically as an area of occurrence of *B. chrysopterus*, and during Projeto Maracá only that species was recorded (J.M.C. Silva, D. Stotz, A. Whittaker and M. Cohn-Haft). As a result of this evidence, I prefer to exclude *B. cyanoptera* from the Maracá species list.

### *Coccyzus pumilus*

The dwarf Cuckoo was recorded visually on Maracá by A. Whittaker (Whittaker, 1995). It is a bird which occupies a broad spectrum of habitats including gallery forests, savannas and disturbed areas in humid forest regions (Hilty and Brown, 1986). The species had already been recorded for the limitrophe region of Venezuela (Bolivar and Amazonas) with Brazil (Meyer de Schauensee and Phelps, 1978), and its occurrence within Roraima is to be expected. I have included this species in the final list of the birds of Maracá, but since it is the first record for Brazilian territory it needs to be confirmed by other observers or by specimens.

### *Electron platyrhynchum*

Moskovits *et al.* (1985) recorded this species for Maracá. To the east of the Andes the Broad-billed Motmot is known principally from the regions south of the Solimões and Amazon rivers (up to Santa Cruz and Cochabamba in Bolivia), and from Caquetá in Colombia (Meyer de Schauensee, 1982). This being the case, its occurrence on Maracá would constitute a very great range extension. As the record was a visual one and no other researcher has since observed this bird on Maracá or in Roraima, I have preferred to exclude it pending more concrete evidence for its existence in the region.

### *Pteroglossus pluricinctus/aracari*

Moskovits *et al.* (1985) recorded both of these species for Maracá. They both belong to the same superspecies (Haffer, 1974), and appear to be mutually exclusive geographically. On the distribution map of the two species presented by Haffer (1974), Roraima is left blank due to the lack of specimens. On Maracá and at

Colônia Apiaú I observed and collected only *P. pluricinctus*, and M. Cohn-Haft and A. Whittaker also observed only this species on Maracá. Based on these records, I have decided to exclude *P. aracari* from the Maracá species list.

*Celeus grammicus/undatus*

Moskovits *et al.* (1985) recorded *C. undatus* for Maracá, whereas I observed *C. grammicus*. Haffer (1974) considers them members of the same superspecies. In his map there are no collection points for Roraima, but he inferred that *C. undatus* must occur in the region. At Colônia do Apiaú I observed and collected only *C. grammicus*, but no observer recorded both of the species for Maracá. Given that they are very similar and can easily be confused in the field, I have decided to retain only *C. grammicus* in the list of the birds of Maracá.

*Celeus jumana/elegans*

Moskovits *et al.* (1985) recorded *C. elegans* for Maracá, whereas I collected and observed only *C. jumana*. Short (1972, 1982) states that the two are co-specific, basing this on some supposed hybrids collected in Venezuela. *C. elegans* is known from Serra da Lua, which is approximately 150 km south of Maracá (close to Boa Vista). Since the Maracá specimens show no tendency towards the typical characters of *C. elegans*, Silva and Oren (1990) have suggested that *elegans* and *jumana* should be considered distinct species, and that a detailed study should be carried out between Maracá and Serra da Lua to examine the nature of the contact between them. As a result, I have kept only *C. jumana* in the list of Maracá's bird species.

*Pipra coronata*

A. Whittaker recorded *P. coronata* for Maracá, a taxon which forms a superspecies with *P. serena*, *P. exquisita*, *P. nattereri*, *P. iris* and *P. vilasboasi* (Snow, 1979). Of this group, only *P. serena* has been recorded for Roraima State, and only on Mount Roraima (Snow, 1979). However, the subspecies collected there (*P. s. suavissima*) is endemic to the highland tepuis (Mayr and Phelps, 1967) and is not known to occur below 500 m altitude (Meyer de Schauensee and Phelps, 1978). *P. coronata* occurs in the State of Amazonas, in the frontier zone between Venezuela and Roraima, and is known to occur at low altitudes (Meyer de Schauensee and Phelps, 1978). Thus its record for Maracá is not discrepant with the species' known geographical and ecological distribution. I have therefore kept *P. coronata* in the final list of the birds of Maracá, drawing attention to the necessity for more data (more observations or collections) to corroborate the record.

*Tyranneutes stolzmanni/virescens*

Moskovits *et al.* (1985) recorded *T. virescens* for Maracá, but I collected and observed only *T. stolzmanni*. D. Stotz also recorded only the latter species there. *T.*

*stolzmanni* and *T. virescens* are phylogenetically related and replace each other geographically, forming a superspecies (Snow, 1979). Because of that, I have preferred to retain only *T. stolzmanni* in the list.

### *Tyrannus dominicensis*

Moskovits *et al.* (1985) recorded this species for Maracá, that being its first record for Brazil. During the Maracá Rainforest Project there were no sightings of the Grey Kingbird. Since this species is a migrant from the northern hemisphere and has been recorded in a variety of habitats in the regions neighbouring Roraima, its recording for this State does not come as a surprise. As in the case of the Dwarf Cuckoo, I have kept *T. dominicensis* in the list of the birds of Maracá, awaiting, however, more concrete evidence of its occurrence in Roraima.

### *Basileuterus rivularis*

Moskovits *et al.* (1985) treated this species as *B. mesoleuca*. However, the species which has been collected in Roraima is *B. rivularis* (Hellmayr, 1935; Pinto, 1966).

### *Euphonia violacea/laniirostris*

Moskovits *et al.* (1985) recorded both these species on Maracá, but I observed and collected only *E. violacea*, as did D. Stotz. The two species are phylogenetically related and replace each other geographically, forming a superspecies (Haffer, 1975). Up to now only *E. violacea* has previously been collected in Roraima (Pinto, 1966), and since the two species are similar in appearance and easily confused in the field, I have included only *E. violacea* in the final list.

## ECOLOGICAL INFORMATION

Moskovits *et al.* (1985) divided the avifauna of Maracá into four principal habitat types: forest, dry savanna, wet savanna and river margins. All of the subsequent observers, with the exception of J.M.C. da Silva and D.F. Stotz, used the same classification system in their reports. Given that at the time the botany of the Ilha de Maracá had not been adequately studied, this system could be considered adequate for a first approximation.

Milliken and Ratter (1989 and Chapter 5) recognized 13 different vegetation types on Maracá, but with the data that have been collected for birds it would be impossible to approach the same level of detail. Thus I have adjusted the botanical classification to suit the available information on the ecological distribution of the avifauna of Maracá. Two principal habitat groups are recognized, the forests and the savannas, both of which are subdivided. The forests include five distinct types: *terra firme* forest (including both the eastern and western ends of the island and forest dominated by *Peltogyne gracilipes*); riverine forest; *buritizal* and associated damp forest; *campina* (*caatinga*, including all types of low forest); and *caapoeira* (secondary vegetation). Amongst the savannas I have recognized only the two major types proposed by the botanists: *Curatella americana/Byrsonima crassifolia* savanna and seasonally flooded savanna. I have classified the avifauna into the following dietary groups: small insects, large insects and other invertebrates, fish, terrestrial vertebrates, carrion, fruits and/or large seeds, small seeds, nectar, and leaves.

To allocate habitat types to the bird species (see Appendix 4, p. 451), I have used the data presented by Moskovits *et al.* (1985), those submitted to the Royal Geographical Society by some of the Maracá Rainforest Project scientists (D. Stotz, A. Whittaker and M. Cohn-Haft) and, principally, my own observations from Maracá, from the forests and savannas around Boa Vista, and from the forests of Colônia do Apiaú. For the allocation of dietary groups, I have referred to my own field observations and, when those are scanty, the information presented by Schubart *et al.* (1965) and Sick (1985).

## COMPOSITION OF THE AVIFAUNA

Birds are generally divided into two main taxonomic groups: the non-passerines and the passerines. The non-passerines are represented on Maracá by 18 orders and 42 families, and include all of the birds listed in Appendix 4 between Tinamidae and Picidae. The passerines comprise a single order and are represented on Maracá by 20 families. Among these, two distinct groups can be recognized on the basis of the syrinx (song-organ): those with a structurally simple syrinx (sub-oscine) and those with a complex syrinx (oscine). There are six sub-oscine families on Maracá (from Dendrocolaptidae to Tyrannidae), and 14 oscine families (Hirundinidae to Fringillidae). I have further sub-classified each of these groups into forest birds and savanna birds (Table 11.1). The non-passerines comprise approximately half of the bird species found on Maracá (two more than the passerines), but are represented by more species (82) in the savanna areas than are the passerines (67). Conversely in the forests there are more passerine species (153) than there are non-passerine (140).

One of the interesting relationships to be investigated is between the sub-oscines and the oscines. The ratio amongst the resident passerine species of Maracá is 1.69 sub-oscines/oscines for the forest habitats, and 0.56 for the savanna habitats. These figures illustrate a fact well known to neotropical ornithologists, that sub-oscines are largely found in forest environments, whereas oscines predominate in more open areas (Novaes, 1973; Willis, 1976; Slud, 1976). Slud (1976), when he mapped these ratios of sub-oscine/oscine passerines throughout South America, commented that the ratio found in Roraima was very low in comparison with the surrounding regions. Considering Maracá as a whole, i.e. the forests and savannas together, our overall ratio of 1.22 corroborates this observation. However, if one considers only the values obtained for the forest habitats of Maracá, these ratios are found to be very similar to those recorded for other regions of Amazonia where savannas are absent or insignificant.

Table 11.1. Ecological distribution of the principal groups of bird species on the Ilha de Maracá. Numbers in parentheses refer to migratory species

|  | Forest | Savannas | Total |
|---|---|---|---|
| Non-passeriformes | 139 (+1) | 76 (+6) | 215 (+7) |
| Passeriformes: |  |  |  |
| Sub-oscines | 93 (+0) | 22 (+2) | 115 (+2) |
| Oscines | 55 (+5) | 39 (+4) | 94 (+9) |
| Total | 287 (+6) | 137 (+12) | 424 (+18) |

## SPECIES RICHNESS AND ECOLOGICAL DISTRIBUTION

The total number of bird species recorded for Maracá (442), when compared with the data presented by Haffer (1990, Figure 2), strongly supports the prediction that avian species richness is greater at the margin of the Amazon basin than at its centre. At the margin, the regions closest to the Andes in turn exhibit greater diversity than those regions bordering the Guianan and Brazilian Shields. Haffer (1990) proposes that this pattern correlates with the geochemical divisions of Amazonia presented by Fittkau (1982), wherein the richness of mineral nutrients (and possibly ecological diversity) diminishes in the following direction: from (i) regions close to the Andes (Western Peripheric Region) to (ii) regions close to the Guiana Shield (Northern Peripheric Region) to (iii) the Brazilian Shield (Southern Peripheric Region), and finally to (iv) Central Amazonia.

If it is environmental heterogeneity that is responsible for Maracá's high avian species diversity, it is necessary to investigate the contribution of each of the component habitats to the island's overall diversity.

Although I have recognized five forest habitats on Maracá, which together contribute 293 species to the overall list, there are insufficient data to compare the avifaunal relationships between each of them. In general, the different forest types of a given area of Amazonia, if they are continuous, exhibit great similarity in the bird species they contain (Silva and Constantino, 1988). Those species which are common to all the forest types do not affect local diversity of the avifauna, and it is thus necessary to discuss those which are restricted to a single habitat type.

The *terra firme* forest at the eastern end of the island appears still to be in a phase of expansion (cf. Milliken and Ratter, 1989 and Chapter 5), or has undergone a recent catastrophic event (e.g. fire, cf. Miller and Proctor, Chapter 6). Some of the faunal and vegetational data support this hypothesis:

- The high density of clearings recorded, 10.8–14.4/ha (Moskovits, 1988), which is one of the highest that have been recorded in South America (cf. Almeida, 1989, Table 2).
- The depauperate avifauna of the understorey; using 20 mist nets it took longer (960 net-hours vs. 600 net-hours) to catch 100 birds, and the number of species caught (25 vs. 46) was lower, than in a low vine forest at Carajás, Pará.

- The presence in the understorey, in relatively high densities, of certain species commonly associated with riverine habitats in Roraima (e.g. *Myrmoborus leucophrys*, *Ramphotrigon ruficauda*, *Turdus fumigatus*, etc.).
- The greater abundance of a species of tortoise (*Geochelone carbonaria*) commonly associated with savannas or forests close to grasslands, than another syntopic species (*G. denticulata*) typical of forest environments (Moskovits, 1988).

The *terra firme* forests of the western end of Maracá are more diverse than those of the east, and exhibit a closer floristic similarity to those of central Amazonia (Milliken and Ratter, 1989 and Chapter 5). They also contain certain bird species (recorded by A. Whittaker), such as *Xiphorhynchus pardalotus*, *Pithys albifrons* and *Pipra pipra*, which were not encountered at the eastern end of the island but are the most common species of the understorey at Apiaú (Silva, pers. obs.) and Manaus (Stotz and Bierregard, 1989). Several other understorey bird species which are common at Apiaú (e.g. *Philydor ruficaudatus*, *Automolus rubiginosus*, *A. ochrolaemus*, *Sclerurus mexicanus*, *Cymbilaymus lineatus*, *Thamnomanes caesius*, *Myrmotherula longipennis*, *Myrmoborus myotherinus*, *Hylophylax punctulata* and *Myrmornis torquata*) appear to be inexplicably absent from Maracá. However, I suspect that these species must occur in the forest at the western end of the island, and that an adequate ornithological survey of that area would add substantially to the list of forest bird species.

The riverine forests of the island, although they are only narrow strips, support many characteristic species which are not found in the *terra firme* forests (e.g. *Opisthocomus hoazin*, *Aratinga solstitialis*, *Amazona amazonica*, *Crotophaga major*, *Celeus flavus*, *Sakesphorus canadensis*, *Hypocnemoides melanopogon*, *Cephalopterus ornatus* and *Ochthornis littoralis*). There are also species such as *Piaya cayana* which live preferentially in riverine forest only when there is a congener of the same size in the *terra firme* forest (*P. melanogaster*), as is the case on Maracá. In the forests of Carajás, Santarém and Belém, where *P. melanogaster* is absent, *P. cayana* is found in both *terra firme* and riverine (*várzea*) forest.

The *buritizais* (*Mauritia flexuosa* palm stands) and associated flooded forests support many characteristic bird species such as *Ara ararauna*, *A. severa*, *A. manilata*, *A. nobilis*, *Reinarda squamata* and *Icterus chrysocephalus* which rely on the *buriti* palm for feeding and nesting, but which also utilize other forest habitats.

The areas of secondary vegetation are poor in restricted species; those species which are found in high densities in this habitat (e.g. *Manacus manacus*, *Turdus leucomelas*, *Euphonia violacea*, *Thraupis episcopus*, *T. palmarum*, *Ramphocelus carbo* and *Saltator maximus*) tend to live in low densities in the *terra firme* forest margin and in riverine forests. These species include in their diets many small fruits of the plants typical of the first stages of forest succession (such as *Cecropia* spp.) and actively disperse the seeds of forest margin species into the secondary vegetation, thus accelerating succession after human disturbance (Silva *et al.*, 1996).

The *campina* or *caatinga* (low forest on sandy soil) vegetation was little studied ornithologically. However, since this habitat type appears sparsely distributed on the island, it seems unlikely that there are many bird species restricted to its environs. Of the 21 species known to be restricted to this type of vegetation in Amazonia (Oren, 1981), only two (*Aratinga pertinax* and *Elaenia ruficeps*) have so far been recorded on Maracá.

The transition between the forests and the savannas on Maracá affords considerable expanses of marginal habitat, which are useful as much for the forest bird species as for the savanna species, but also add significantly to the species richness of Maracá by supporting characteristic species such as *Piaya minuta*, *Dryocopus lineatus*, *Xiphorhynchus picus*, *Taraba major*, *Thamnophilus doliatus*, *Myrmeciza atrothorax*, etc. I have estimated that about 29% of Maracá's forest bird species were recorded in these ecotonal strips, and that if one includes the species which live preferentially in the canopy but generally move to the forest edge to forage, the percentage of species relying upon this habitat reaches 43%.

The savannas of Maracá, although occupying only 0.5% of the island's area (Milliken and Ratter, 1989 and Chapter 5), support 33.7% of its bird species. The seasonally flooded grasslands are richer than the *Curatella americana/Byrsonima crassifolia* savannas (107 vs. 87 species), with 45 species common to both.

## MIXED FLOCKS

The formation of multispecific bird associations is a common phenomenon in the neotropics, and is easily observed in a number of habitats from open grassy savannas to dense tropical forests. These associations may be classified into two groups: aggregations or mixed flocks. An aggregation is the temporary meeting of bird species in response to an environmental factor, such as fruit production or termite flights, whereas mixed flocks are 'groupings whose cohesion is dependent on members' responses to one another' (Powell, 1985). The adaptive significance of these mixed flocks is the subject of controversy (see review in Powell, 1985), and may be due to a number of causes of which increased foraging efficiency and decreased predation risk stand out.

Four types of mixed flocks were studied during my stay on Maracá: two in the seasonally flooded grassland (seed-eaters and fish-eaters) and two in the forest (insectivores in the understorey, insectivores and insectivore–frugivores in the canopy). Four species of seed-eaters (*Volatinia jacarina*, *Sporophila plumbea*, *S. intermedia* and *S. minuta*) formed large groups of 50–80 individuals in the flooded grassland. These groups were dominated by *V. jacarina* and *S. minuta* and moved as a cohesive whole amongst the patches of seeding grasses. As I made few observations, I do not have data on any factor (e.g. alarm-call systems in one of the species), other than mutual exploitation of a single resource, which might be influencing the cohesion of this type of mixed flock.

Five species of herons (*Casmerodius albus*, *Egretta thula*, *Florida caerulea*, *Butorides striatus* and *Pilherodius pileatus*) formed mixed flocks of 20–50 individuals whilst foraging for fish in the small lagoons and streams spread through the flooded grassland. A study in Florida (Caldwell, 1981) has demonstrated that the species aggregate around *E. thula*, which is the most aggressive species and appears to be capable of indicating the best fishing sites. This also seems to be the case on

Maracá, but studies must nevertheless be made to examine this hypothesis in more detail.

Mixed flocks of insectivores were observed on eight occasions in the understorey of the *terra firme* forest at the eastern end of Maracá. They were made up of 15 regular species (observed on more than 25% of the sightings) and six accessory species (Table 11.2). The most frequently encountered species were *Thamnomanes ardesiacus* (the group leader), *Sittasomus griseicapillus*, *Xenops minutus*, *Thamnophilus punctatus*, *Myrmotherula axillaris*, *M. menetriesii* and *Myiobius barbatus*. Mixed flocks of insectivores and insectivore–frugivores were observed on 10 occasions in the canopy, for which 43 species were recorded of which 32 appear to be regulars (Table 11.2). The most frequent members of this type of flock (more than eight records) were *Herpsilochmus rufimarginatus*, *Sittasomus griseicapillus*, *Pygiptila stellaris*, *Myrmotherula menetriesii*, *Tyranneutes stolzmanni*, *Tolmomyias sulphurescens*, *T. flaviventris*, *Myiopagis gaimardii*, *Microbates collaris*, *Vireo olivaceus*, *Hylophilus pectoralis*, *H. muscicapinus*, *Tachyphonus luctuosus* and *Hemithraupis guira*.

The two types of mixed flocks shared nine species in common, and on a few occasions, in the lower regions of the forest, they would meet. These forest flocks were made up of many species represented by few individuals, in contrast to those of the savannas where few species accumulated in large numbers (Table 11.2). The geographical variation in the composition of mixed flocks through Amazonia is also a topic worthy of investigation. In the forest at Apiaú the mixed flock of understorey insectivores contained fewer regular species (20), and in addition to *Thamnomanes ardesiacus*, *T. caesius* is also a nuclear species. In the mixed flocks of the canopy on Maracá and at Apiaú, the species of *Herpsilochmus* encountered was *H. rufimarginatus*, and no species of *Terenura* was recorded. *Terenura* is behaviourally similar to *Herpsilochmus* and is known to participate actively in the mixed insectivorous flocks of tropical regions (Munn, 1985; Teixeira, 1987). At Manaus, *Herpsilochmus dorsimaculatus* and *Terenura spodioptila* were both recorded (Stotz and Bierregard, 1989), whereas at Cosha Cashu (Peru) no record was made of *Herpsilochmus*, and *T. humeralis* was a common component of the mixed flocks of the forest canopy (Munn, 1985). In the east of Pará, where *Terenura* is absent, one of the most characteristic antwrens in the mixed flocks is *H. rufimarginatus*. It is probable that the two genera replace each other in certain parts of Amazonia. To test this hypothesis, however, further information on the composition of mixed flocks of the forest canopy would be required from other parts of the Amazon, as would data on the natural history of the two genera.

## MIGRANTS

There are three recognizable classes of migrants in the avifauna of Maracá: migrants from the northern hemisphere, migrants from the southern hemisphere, and altitudinal migrants. There are 15 winter migrants from the north, which use a range of habitats on Maracá: (i) the flooded grasslands (*Pandion haliaetus*, *Tringa solitaria*, *T. flavipes*, *T. melanoleuca*, *Actitis macularia* and *Riparia riparia*), (ii) the

**Table 11.2.** Species participating in the mixed flocks of the forest canopy and understorey on the Ilha de Maracá in January 1987

| Species | Understorey flocks | Canopy flocks |
|---|---|---|
| *Piaya melanogaster* | – | 2 (2)* |
| *Galbula galbula* | 3 (1.6) | – |
| *Monasa atra* | – | 3 (2) |
| *Capito niger* | – | 1 (1) |
| *Picumnus exilis* | – | 2 (2) |
| *Piculus flavigula* | – | 3 (1) |
| *Veniliornis cassini* | – | 4 (2) |
| *Dendrocincla fuliginosa* | – | 1 (2) |
| *Deconychura longicauda* | 2 (2) | – |
| *Sittasomus griseicapillus* | 5 (2) | 8 (2.2) |
| *Glyphorynchus spirurus* | – | 1 (1) |
| *Xiphorhynchus guttatus* | 3 (2) | 2 (1) |
| *Lepidocolaptes albolineatus* | – | 6 (2) |
| *Automolus rufipileatus* | 1 (2) | – |
| *Xenops minutus* | 6 (1) | – |
| *Thamnophilus aethiops* | 4 (2) | – |
| *Thamnophilus punctatus* | 6 (2.5) | – |
| *Pygiptila stellaris* | – | 8 (1.2) |
| *Thamnomanes ardesiacus* | 8 (2) | – |
| *Myrmotherula brachyura* | – | 6 (2) |
| *Myrmotherula guttata* | 6 (2) | 5 (2) |
| *Myrmotherula axillaris* | 7 (2) | 3 (2) |
| *Myrmotherula menetriesii* | 5 (2) | 8 (2) |
| *Herpsilochmus rufimarginatus* | – | 10 (2.6) |
| *Hypocnemis cantator* | 3 (2) | – |
| *Pachyramphus polychopterus* | – | 2 (1) |
| *Tityra inquisitor* | – | 3 (1) |
| *Tyranneutes stolzmanni* | – | 8 (1) |
| *Myiodynastes maculatus* | – | 3 (1) |
| *Atilla spadiceus* | – | 2 (1) |
| *Lathotriccus euleri* | 3 (1) | – |
| *Myiobius barbatus* | 5 (2) | – |
| *Tolmomyias sulphurescens* | – | 9 (2) |
| *Tolmomyias flaviventris* | – | 9 (2) |
| *Myiopagis gaimardii* | – | 9 (2) |
| *Mionectes oleagineus* | 2 (1) | – |
| *Thryothorus leucotis* | 1 (1) | – |
| *Microbates collaris* | 3 (2) | 8 (2) |
| *Ramphocaenus melanurus* | 2 (2) | 6 (2) |
| *Vireo olivaceus* | – | 8 (2) |
| *Hylophilus pectoralis* | – | 9 (2.5) |
| *Hylophilus muscicapinus* | – | 10 (2) |
| *Dendroica striata* | – | 5 (1) |
| *Cyanerpes caeruleus* | – | 3 (2) |
| *Cyanerpes cyaneus* | – | 6 (2) |
| *Chlorophanes spiza* | – | 3 (2) |
| *Dacnis cayana* | – | 4 (2) |
| *Euphonia xanthogaster* | – | 6 (2) |
| *Euphonia violacea* | – | 1 (2) |
| *Euphonia chrysopasta* | – | 2 (1) |

Table 11.2. (*continued*)

| Species | Understorey flocks | Canopy flocks |
|---|---|---|
| *Tangara mexicana* | – | 3 (2) |
| *Tachyphonus cristatus* | 2 (1) | 5 (2) |
| *Tachyphonus luctuosus* | 3 (2) | 8 (2.5) |
| *Hemithraupis guira* | – | 8 (2) |
| *Hemithraupis flavicollis* | – | 1 (1) |

* Numbers without parentheses represent the number of sightings of the species (out of 10 flock sightings for the canopy and eight for the understorey), and numbers within parentheses represent the average number of individuals sighted.

*Curatella/Byrsonima* savannas (*Tyrannus dominicensis* and *Setophaga ruticilla*), (iii) both types of savanna (*Chordeiles minor, Tyrannus savana monachus* and *Hirundo rustica*), or (iv) the various types of forest (*Catharus fuscescens, C. minimus, Dendroica petechia* and *D. striata*). The majority of these species are recorded only between August and September and between April and May, and their numbers are highly variable. Two special cases – *Pandion haliaetus* and *Tyrannus savana* – are worthy of discussion.

*Pandion haliaetus* migrates when young to South America, and remains there for one or two years until sexually mature. It returns to the northern hemisphere to breed, and from then on returns annually to the south during the northern winter (Sick, 1985).

*Tyrannus savana* poses an interesting question; four distinct subspecies of this conspicuous flycatcher are recognized (Zimmer, 1937) of which two, *T. s. monachus* and *T. s. savana*, have been recorded for Roraima. The migratory movements of *T. s. monachus* are little known, but it probably reproduces between southern Mexico and Colombia and migrates to Venezuela, Surinam and Brazilian Amazonia (Traylor, 1979; Hilty and Brown, 1986). *T. s. savana* reproduces south of the 15°S parallel, and migrates northwards between January and September (Silva, pers. obs.). The two subspecies were first recorded for Roraima by Pinto (1966), and although only *monachus* has been collected on Maracá (18 February 1987) I believe that *savana* must also occur there. We would thus have on Maracá the meeting of two migrant populations of the same species, and the interaction between them would constitute an interesting topic of research for a resident ornithologist on the island.

Only *T. s. savana* (whose presence on the island still requires confirmation), *Chaetura andrei meridionalis* and *Sporophila lineola* could be considered as migrants from the southern hemisphere. *C. a. meridionalis* is known from Piauí State to northern Argentina, Paraguay and Rio Grande do Sul, migrating to the extreme north of South America and to southern Central America (Pinto, 1978). It was recorded for Roraima by Pinto (1966), and must be the population recorded on Maracá, although a smaller population of *C. a. andrei* is found in Venezuela (Guarico, Sucre and northern Bolivar states) (Meyer de Schauensee and Phelps, 1978). The latter subspecies may also occur in Roraima, but specimens and observations are necessary to evaluate this hypothesis. The individuals of *Sporophila lineola* which arrive in Roraima are, judging by the dates of their sightings (February), probably part of the population which breeds in southeastern Brazil, southern Bolivia, Paraguay and northern Argentina, and which migrates to Amazonia from March/April until January/February (Silva, 1995a).

Moskovits *et al.* (1985) suggested that *Brotogeris chrysopterus*, *Turdus olivater* and *Piranga flava* are altitudinal migrants. This is in fact true only of *T. olivater*, which is recognizably a species of high-altitude forests (Meyer de Schauensee and Phelps, 1978). *P. flava* includes a subspecies endemic to the tepuis (*P. f. haemala*) and one of the lowland regions of northern Amazonia (*P. f. macconnelli*), and only the latter has been recorded for lowland Roraima (Pinto, 1966). It is indeed probable that *P. f. haemala* migrates to the Maracá region, but actual evidence for that movement is currently lacking. *Brotogeris chrysopterus* is commonly encountered in lowland Amazonia throughout the year (Silva, pers. obs.) and is clearly not an altitudinal migrant. Its seasonality on Maracá, as observed by Moskovits *et al.* (1985), may be a response to fruit production of the plants on which it feeds.

## BIOGEOGRAPHY

### FOREST BIRDS

The biogeography of the forest birds of Amazonia has recently been dominated by a single principal theme: the model of biological diversification in the tropics, or, as it is inadequately labelled (Vanzolini, 1981), the refuge theory. It postulates that the current areas of continuous forest were extensively broken up by the expansion of open formations (*cerrados* or *caatingas*) in the dry periods associated with the ice ages. During the wetter climatic phases in the interglacial periods, the forests would have expanded again. As a consequence, various types of interactions were established between periodically isolated and united bird populations, resulting in diverse patterns of geographical distribution and differentiation (see the discussion in Haffer, 1974; Simpson and Haffer, 1978; Brown and Ab'Saber, 1979; Prance, 1982; Whitmore and Prance, 1987).

As regards the areas of endemism (generally associated with refugia) recognized for the Amazonian forest avifauna (Haffer, 1978, 1985), the Ilha de Maracá is located between the Guiana area (to the east) and the Napo and Imeri areas (to the west). Given this position, and assuming that the forests are still in a phase of expansion, one can predict that the avifauna of Maracá's forests will include a mixture of the characteristic elements of those areas of endemism.

An analysis of the ramphastids (toucans and araçaris) of Maracá, using Haffer's (1974) excellent synthesis, supports this hypothesis well. The avifauna of Maracá was found to include elements with centres of distribution in the Guianas (*Pteroglossus aracari*), elements from the upper Amazon region, which includes the Napo and Imeri areas of endemism (*P. pluricinctus* and *P. flavirostris*), and elements whose Guianan and Napo–Imeri populations have hybridized in the Maracá region (*Ramphastos vitellinus culminatus* × *R. v. vitellinus* and *R. t. tucanus* × *R. t. cuvieri*).

Analyses of the other groups show a clear tendency for the predominance of species and subspecies whose principal distribution is within the Guianan region, although in some cases extending into the northeast and southeast of Venezuela (Silva, unpubl.). In the family Dendrocolaptidae (woodcreepers), for example, there exists good taxonomic and distributional information for 10 of the 14 species recorded on Maracá, which allows us to discern some clear patterns. Well-differentiated species or subspecies were recorded: (i) with wide distributions throughout Amazonia (*Sittasomus griseicapillus amazonicus*), (ii) with wide distributions in northern Amazonia (centred in the Guianan region) to the west of the lower Rio Negro (*Deconychura l. longicauda*, *Dendrexetastes r. rufigula*, *Hylexetastes perrotii*, *Dendrocolaptes c. certhia*, *D. p. picumnus* and *Xiphorhynchus guttatus polystictus*), (iii) with distributions restricted to the Rio Negro–Rio Branco interfluve and the extreme south of Venezuela (*Dendrocincla fuliginosa phaeochroa* and *Lepidocolaptes albolineatus duidae*), and (iv) with a centre of distribution in Upper Amazonia (*Dendrocincla merula bartletti*). Maracá's specimens of the last species were identified by D.C. Oren (in Silva and Oren, 1990) as *D. m. merula*. However, this was a misidentification based on a comparison with few specimens. I have now carefully re-examined these specimens, and without doubt they belong to *D. m. bartletti*, an Upper Amazonian subspecies.

## SAVANNA BIRDS

In the context of the model of biological diversification in the tropics, the Amazonian open vegetation types play an important role: they are interpreted as relicts of a biome that was more widespread in the past (Haffer, 1974; Prance, 1982; Silva, 1995b). The biogeographical relations between the isolated Amazonian savannas may offer important evidence for the determination of possible expansion cycles of these formations during the ice ages. A recent biogeographical analysis of the Amazonian savanna avifauna indicates that the avifauna of the savannas of Roraima, in contrast to the other savannas north of the Amazon such as Amapá and Trombetas-Paru (Pires, 1973), are more influenced by the avifauna of the *llanos* of Colombia and Venezuela than by that of the *cerrado* of Central Brazil (Silva, 1995b).

In addition to widespread South American species, the savanna avifauna of Maracá is made up of the following components (species marked with an asterisk possess disjunct vicariants in *cerrado* and *caatinga* areas): (i) those distributed principally in the *llanos* of Colombia–Venezuela, occasionally occurring in the Guianas (*Athene cunicularia minor\**, *Xenopsaris albinucha minor\**, *Arremonops conirostris* and *Sporophila intermedia*), (ii) those restricted to one or more savanna patches located north of the Amazon (*Burhinus bistriatus*, *Colinus cristatus*, *Polytmus guainumbi guainumbi\**, *Falco sparverius isabellinus\**, *Picumnus spilogaster spilogaster*, *Furnarius leucopus leucopus\**, *Euphonia finschii*, *Sporophila plumbea whiteleyana\** and *Campylorhynchus griseus griseus*), and (iii) those restricted to Amazonian savannas (*Buteo albicaudatus colonus\**, *Chordeiles rupestris rupestris*, *Inezia subflava*, *Gymnomystax mexicanus* and *Agelaius icterocephalus icterocephalus*).

# CONSERVATION – THE IMPORTANCE OF TRANSITION AREAS

The choice of priority areas for the establishment of conservation units in Amazonia has frequently been the subject of intense debate. The proposal adopted by official Brazilian papers on this question (Wetterberg *et al.*, 1976; Wetterberg and Pádua, 1978) is based on the model of biological diversification in the tropics, and considers as areas of prior importance 'those which more than one author has considered as Pleistocene refugia'.

Various criticisms have been made of this proposal (Vanzolini, 1980, 1986), particularly that by this approach the transition areas, with their special biological characteristics, are considered as of little importance or as merely secondary issues. Brown (1979) has shown, however, that the regions with greatest genetic diversity of certain tropical forest butterfly groups (Heliconiini and Ithomiinae) are located between the areas of endemism, or rather between the supposed 'refugia'. These areas of diversity thus correspond to contact zones between biotas rather than the central regions of endemism. This suggests that a more effective conservation strategy would be to preserve both the centres ('pure genetic heritage') and the peripheries ('more diverse genetic heritage') of the areas of endemism (Brown, 1979; Vanzolini, 1980).

As shown, the Maracá area contains a mixture of the avifauna of adjacent areas of endemism, and sufficient environmental heterogeneity to maintain this diversity. Thus its importance as a 'laboratory' for ecological–evolutionary study is very great. The region has also been recognized as an area of endemism for certain groups of tropical butterflies (Brown, 1979). It is therefore also important for the conservation of the 'pure' genetic heritage of, at least, certain types of organisms.

Maracá still supports populations of three raptor species which were recently added to the list of Brazilian birds threatened with extinction (Portaria No 1.522, 19 December 1989): *Accipiter poliogaster*, *Harpia harpyja* and *Morphnus guianensis*. Two birds endemic to the basin of the Rio Branco (*Poecilurus kollari* and *Cercomacra carbonaria*), also included in this list, have yet to be registered for Maracá. Both are birds of the gallery forests in more extensive Roraiman savannas (Silva, pers. obs.). This habitat is poorly represented on the Ilha de Maracá, but could be preserved by the creation of other complementary reserves. One possibility would be to conserve the marginal forests of the Rio Branco and its affluents, as well as to create a corridor of small reserves linking Maracá with the Caracaraí Ecological Station to conserve tracts of the original Roraima savannas. The latter would maintain an ecosystem about which we still know very little and which, thanks to the rapid expansion of human activity in the region, is suffering grave and perhaps irrevocable damage.

# ACKNOWLEDGEMENTS

I thank David C. Oren, Andrew Whittaker, William Milliken, James A. Ratter and M. Cohn-Haft for comments and help with some data. My field assistants in Maracá, Manoel

Santa Brígida and the late Rosemiro Pereira, were extremely dedicated to their work. Field and museum studies were funded by the Royal Geographical Society, the John D. and Catherine T. MacArthur Foundation, WWF-US and Conselho Brasileiro de Desenvolvimento Científico e Tecnológico (CNPq).

# REFERENCES

Almeida, S.S. de (1989). *Clareiras naturais na Amazônia Central: abundância, distribuição, estrutura e aspectos da colonização vegetal*. Dissertação de Mestrado, Instituto Nacional de Pesquisas da Amazônia/Fundação Universidade do Amazonas.

Brown, K.S. Jr (1979). *Ecologia geográfica e evolução nas florestas tropicais*. Tese de Livre-Docência, Universidade Estadual de Campinas.

Brown, K.S. Jr and Ab'Saber, A. (1979). Ice-age refuges and evolution in the tropics: correlation of paleoclimatological, geomorphological and pedological data with modern biological endemism. *Paleoclimas*, 5, 1–30.

Caldwell, G.S. (1981). Attraction to tropical mixed-species heron flocks: proximate mechanisms and consequences. *Behaviour, Ecology and Sociobiology*, 8, 99–103.

Dickerman, R.W. and Phelps, W.H. Jr (1982). An annotated list of the birds of Cerro Urutani on the border of Estado Bolivar, Venezuela, and Territorio Roraima, Brazil. *American Museum Novitates*, 2732, 1–20.

Fittkau, E.J. (1982). Struktur, Funktion und Diversitat zentralamazonischer Okosysteme. *Arch. Hydrobiol.*, 95, 29–45.

Gyldenstope, N. (1945). A contribution to the ornithology of northern Bolivia. *Kungl. Svenska. Vetenskapsakademien – Akad.* (Ser. 3), 23, 1–300.

Haffer, J. (1974). Avian speciation in Tropical South America. *Publication of the Nuttall Ornithological Club*, 14, 1–390.

Haffer, J. (1975). Avifauna of northwestern Colombia, South America. *Bonner Zool. Beitr.*, 28, 48–76.

Haffer, J. (1978). Distribution of Amazon forest birds. *Bonner Zool. Beitr.*, 29, 38–78.

Haffer, J. (1985). Avian zoogeography of the neotropical lowlands. In: *Neotropical ornithology*, eds P. Buckley, M.S. Foster, E.S. Morton, R.S. Ridgely and F.G. Buckley, pp. 113–146. Ornithological Monographs No. 36.

Haffer, J. (1987). Biogeography of Neotropical birds. In: *Biogeography and Quaternary history in Tropical America*, eds T.C. Whitmore and G.T. Prance, pp. 105–150. Oxford Science Publications.

Haffer, J. (1990). Avian species richness in tropical South America. *Studies on Neotropical Fauna and Environment*, 25 (3), 157–183.

Hellmayr, C.E. (1935). Catalogue of the birds of South America. *Field Museum of Natural History, Zoological Series*, 13 (7), 1–541.

Hilty, S.L. and Brown, W.L. (1986). *A guide to the birds of Colombia*. Princeton University Press.

Mayr, E. and Phelps, W.H. Jr (1967). The origin of the bird fauna of the south Venezuelan highlands. *Bulletin of American Museum of Natural History*, 136 (5), 269–328.

Meyer de Schauensee, R.M. (1982). *A guide to the birds of South America*. The Academy of Natural Sciences of Philadelphia.

Meyer de Schauensee, R.M. and Phelps, W.H. Jr (1978). *A guide to the birds of Venezuela*. Princeton University Press.

Milliken, W. and Ratter, J.A. (1989). *The vegetation of the Ilha de Maracá*. Royal Botanic Garden Edinburgh.

Moskovits, D.K. (1988). Sexual dimorphism and population estimates of the two Amazonian tortoises (*Geochelone carbonaria* and *G. denticulata*) in northwestern Brazil. *Herpetologica*, 44 (2), 209–217.

Moskovits, D., Fitzpatrick, J.W. and Willard, D.E. (1985). Lista preliminar das aves da

Estação Ecológica de Maracá, Território de Roraima, Brasil e áreas adjacentes. *Papéis Avulsos de Zoologia, São Paulo*, 36, 51-68.

Munn, C.A. (1985). Permanent canopy and understorey flocks in Amazonia: species composition and population density. In: *Neotropical ornithology*, eds P. Buckley, M.S. Foster, E.S. Morton, R.S. Ridgely and F.G. Buckley, pp. 683-712. Ornithological Monographs No. 36.

Novaes, F.C. (1965). Notas sobre algumas aves de Serra Parima, Território de Roraima (Brasil). *Boletim do Museu Paraense Emílio Goeldi, Zool*, (n.s.), 54, 1-10.

Novaes, F.C. (1973). Aves de uma vegetação secundária na foz do Amazonas. *Publicações Avulsas do Museu Goeldi*, 21.

Oren, D.C. (1981). *Zoogeographic analysis of the white sand campina avifauna of Amazonia*. PhD Thesis, Harvard University.

Phelps, W.H. Jr (1973). Adiciones a las listas de aves de Sur America, Brasil y Venezuela y notas sobre aves venezolanas. *Boletim da Sociedad Venezoelana de Ciências Naturales*, 30, 23-40.

Pinto, O.M.O. (1966). Estudo crítico e catálogo remissivo das aves do Território Federal de Roraima. *Cadernos da Amazônia, Manaus*, 8, 1-176.

Pinto, O.M.O. (1978). *Novo catálogo das aves do Brasil I*. Editora Gráfica dos Tribunais, São Paulo.

Pires, J.M. (1973). Tipos de vegetação da Amazônia. *Publicações Avulsas do Museu Goeldi*, 20, 179-202.

Powell, G.V.N. (1985). Sociobiology and adaptative significance of interspecific foraging flocks in the Neotropics. In: *Neotropical ornithology*, eds P. Buckley, M.S. Foster, E.S. Morton, R.S. Ridgely and F.G. Buckley, pp. 713-756. Ornithological Monographs No. 36.

Prance, G.T. (ed.) (1982). *Biological diversification in the tropics*. Columbia University Press.

Schubart, O., Aquirre A.C. and Sick, H. (1965). Contribuição para o conhecimento da alimentação das aves brasileiras. *Arquivos de Zoologia de São Paulo*, 12, 95-249.

Short, L.L. (1972). Relationships among the four species of the superspecies *Celeus elegans* (Aves, Picidae). *American Museum Novitates*, 2487, 1-26.

Short, L.L. (1982). *Woodpeckers of the world*. Delaware Museum of Natural History.

Sick, H. (1985). *Ornitologia Brasileira, uma introdução*. Editora Universidade de Brasília, 2 vols.

Silva, J.M.C. (1995a). Seasonal distribution of the Lined Seedeater *Sporophila lineola*. *Bulletin of the British Ornithologists' Club*, 115, 14-21.

Silva, J.M.C. (1995b). Biogeographic analysis of the South American avifauna. *Steenstrupia*, 21, 49-67.

Silva, J.M.C. and Constantino, R. (1988). Aves de um trecho de mata no baixo rio Guamá: uma reanálise: riqueza, raridade, diversidade, similaridade e preferências ecológicas. *Boletim do Museu Paraense Emílio Goeldi, sér. zool.*, 4 (2), 201-210.

Silva, J.M.C. and Oren, D.C. (1990). Resultados de uma excursão ornitológica à ilha de Maracá, Roraima, Brasil. *Goeldiana Zoologia, Belém*, 5, 1-8.

Silva, J.M.C. and Willis, E.O. (1986). Notas sobre a distribuição de quatro espécies de aves da Amazônia brasileira. *Boletim do Museu Paraense Emílio Goeldi, sér. zool.*, 2, 151-158.

Silva, J.M.C, Uhl, C. and Murray, G. (1996). Plant succession, landscape management, and the ecology of frugivorous birds in abandoned pastures. *Conservation Biology*, 10, 491-503.

Simpson, B.B. and Haffer, J. (1978). Speciation patterns in the Amazon forest biota. *Annual Review of Ecology and Systematics*, 9, 497-518.

Slud, P. (1976). Geographic and climatic relationships of avifaunas with reference to comparative distribution in the Neotropics. *Smithsonian Contributions to Zoology*, 212.

Snow, D. (1979). Family Pipridae. In: *Check-list of birds of the world*, ed. M.A. Traylor Jr, pp. 245-280. Museum of Comparative Zoology, Cambridge, Massachusetts.

Stotz, D.F. and Bierregaard, R.A. Jr (1989). The birds of the Fazendas Porto Alegre, Esteio and Dimona, north of Manaus, Amazonas, Brazil. *Revista Brasileira de Biologia*, 29 (3), 861-872.

Teixeira, D.M. (1987). Notas sobre *Terenura sicki* Teixeira and Gonzaga, 1983 (Aves, Formicariidae). *Boletim do Museu Paraense Emílio Goeldi, sér. zool.*, 3 (2), 241–251.

Traylor, M.A. Jr (1979). *Check-list of birds of the world, 7*. Museum of Comparative Zoology, Cambridge, Massachusetts.

Vanzolini, P.E. (1980). Questões ecológicas ligadas à conservação da natureza no Brasil. *Biogeografia*, 16, 1–22.

Vanzolini, P.E. (1981). A quasi-historical approach to the natural history of the differentiation of reptiles in tropical geographical isolates. *Papeis Avulsos de Zoologia*, 29, 111–119.

Vanzolini, P.E. (1986). *Levantamento Herpetológico da área do estado de Rondônia sob a influência da rodovia BR-364*. Programa Polonoroeste, Relatório de Pesquisa, No. 1, 1–50.

Wetterberg, G.G. and Pádua, M.T.J. (1978). *Preservação da natureza na Amazônia brasileira. Situação em 1978*. PRODEPEF (série técnica), 13, 1–44.

Wetterberg, G.G., Pádua, M.T.J., Castro, C.S. and Vasconcellos, J.M. (1976). *Uma análise de prioridade em conservação da natureza na Amazônia*. PRODEPEF (série técnica), 8, 1–62.

Whitmore, T.C. and Prance, G.T. (eds) (1987). *Biogeography and Quaternary history in Tropical America*. Clarendon Press, Oxford.

Whittaker, A. (1995). First report of *Coccyzus pumilus* for Brazil (Cuculiformes: Cuculidae). *Ararajuba*, 3, 82–83.

Willis, E.O. (1976). Effects of a cold wave on an Amazonia avifauna in the upper Paraguay drainage, Western Mato Grosso, and suggestions on oscine–suboscines relationships. *Acta Amazonica*, 6 (3), 379–394.

Zimmer, J.T. (1937). Studies of Peruvian birds, XXVII. *American Museum Novitates*, 962, 1–28.

# 12 The Reptilian Herpetofauna of the Ilha de Maracá

**MARK T. O'SHEA**
*Wolverhampton, UK*

## SUMMARY

The herpetological survey of the Ilha de Maracá was conducted over a 10 month period from June 1987 to March 1988. During this period the known reptile fauna was raised by 46%, from 45 to 66 species. In addition, two further subspecies (of species already known from Maracá) were recorded. The reptilian herpetofauna of Maracá is now known to comprise at least seven turtles, one caiman, two amphisbaenians, 22 lizards and 34 snakes.

Maracá's reptile fauna was compared statistically with those of eight other neotropical sites (six Amazonian and two lower Central American), using the methodology of Duellman (1990) in an effort to determine its 'coefficient of biogeographical resemblance' (CBR) to the other sites, and to predict its species richness. Not surprisingly, Maracá demonstrated greatest similarity to the Amazon basin sites to the south.

However, it was apparent from the data that Maracá has been incompletely sampled and that a further 20 species may yet be recorded if its species richness is compatible with those of the other sites studied. Maracá is already the type locality for two lizard species with restricted ranges, and others amongst the specimens collected may prove to represent new taxa in the light of further research.

## RESUMO

Um levantamento herpetológico da Ilha de Maracá foi realizado durante um período de 10 meses entre junho de 1987 e março de 1988. Durante este período, a fauna conhecida de répteis aumentou em 46%, de 45 para 66 espécies. Em adição, duas subespécies novas (de espécies já conhecidas na ilha) foram registradas. A herpetofauna réptil de Maracá compreende pelo menos sete tartarugas, um jacaré, dois anfisbenídeos, 22 lagartos e 34 serpentes.

Comparações estatísticas foram feitas entre a herpetofauna réptil da Ilha de Maracá e a de oito outros lugares neotropicais (seis na Amazônia e dois na América Central), usando a metodologia de Duellman (1990) para determinar semelhança e prognosticar a diversidade total. Como esperado, Maracá mostrou o maior índice de semelhança com os outros sítios Amazônicos ao sul.

Entretanto, ficou aparente que a Ilha de Maracá tinha sido incompletamente estudada, e que, se a diversidade for comparável com a dos outros sítios, ainda podem existir umas 20 espécies para serem registradas.

Maracá já representa a localidade do tipo de duas espécies de distribuição restrita, e, entre o material coletado, estudos futuros poderiam revelar mais taxa novos.

## INTRODUCTION

Prior to this study, herpetological collections had been made on the Ilha de Maracá by Celso Morato de Carvalho and Márcio Martins from INPA, resulting in an impressive inventory of known species. Debra Moskovits (1985, and Chapter 13) had also conducted detailed studies of the tortoise species of the island. The author spent seven months on Maracá, from June to November 1987 and February to March 1988. During the intervening two months, when it was necessary for the author to return to the UK, observations, data collection and some specimen collecting were continued by Aléxia Celeste da Cunha. As a result of all studies, a total in excess of 100 species of amphibians and reptiles are now known to occur on the island.

This chapter provides an overview of the reptile fauna of the island, incorporating earlier records made by other workers, and discusses the data in the wider Latin American context. A detailed account of the fauna, including keys and species descriptions, has already been produced in a report for the Royal Geographical Society and an overall description of the herpetofauna (including amphibians) has been published (O'Shea, 1989).

## METHODS

The purpose of the survey was to investigate the herpetofauna in each of the diverse habitats represented on the Ilha de Maracá (see Figure Pr.1, p. xvii for locations of cited collecting localities), using a variety of techniques ranging from the construction of drift-fences, for sampling the secretive subsurface herpetofauna, to the use of crossbows and climbing ropes for access to the canopy fauna.

Two 40 m drift-fences were constructed out of lengths of synthetic tent fabric, specially designed for the purpose, and a third was manufactured *in situ* out of black polythene sheeting. They were set up so that each fence sampled several habitats, e.g. running from the savanna through the transition vegetation into the forest or up an incline from a damp area onto drier ground. Fences were never higher than 45 cm, enabling large mammals to jump over them and thereby reducing the chances of accidental damage. The lower edges were buried to a depth of 10 cm in the ground to deter excavation. Each drift-fence straddled 21 plastic bucket pitfall traps, which were sunk into the ground every 2 m with their lips flush with the soil surface. Fences were checked every morning.

Canopy access was achieved by the use of crossbows, which fired blunt-headed bolts attached to braided nylon fishing line. The line issued from a fixed-spool fishing reel attached to the stock of the bow, and was sent over a carefully selected branch at a height of 30 m or so. Nylon monofilament line, which possesses a 'memory' and thus creates drag, is unacceptable for canopy access using crossbows. Once the bolt had returned to the ground an 8 mm cord was hauled over the branch, followed by a kernmantel caving/climbing rope. When this rope had in turn been passed over the selected branch, load-tested for safety and anchored securely to a tree at ground level, the canopy was accessed using standard ascenders and descenders (as used in single-rope caving techniques). Although on this occasion canopy access proved of limited value herpetologically, it was an extremely useful means of setting up high-level bat and butterfly traps which provided some interesting collections.

Most other collecting took the form of active searching of macro- and micro-habitats in the forest and on the savanna. Many captures were purely opportunistic, particularly those of the snakes, and many specimens were brought in by colleagues working on other sub-projects. Nocturnal collecting was accomplished with the use of head-torches, mainly in the vicinity of the borrow-pits alongside the causeway. Herpetological field techniques are still evolving, and have some way to go before they reach the sophisticated standards employed in the capture of birds, bats, other mammals, fish and invertebrates. Apart from the commoner frogs and lizards, encounters with many species (e.g. snakes) occurred only infrequently. New collecting methods are currently being devised and refined, and some were tested during the present study. Pitfall traps, for example, tend to be ineffective for collecting agile snakes, so on Maracá 'pipe-traps' leading to secure half-buried holding boxes were incorporated in the drift-fences. However, in this case they were not effective. The herpetological field techniques employed in this study are discussed in more detail elsewhere (O'Shea, 1992, 1996).

With conservation as a prime motive, many specimens were returned alive to their environment when the relevant data concerning date and time of capture, macro- and micro habitat, locality, activity, length, sex and necessary scale counts had been recorded. In addition, the author maintained a photographic record of almost every species of amphibian and reptile collected. Those specimens which were preserved (mainly small lizards and frogs requiring further museum examination) will be lodged in Brazilian collections (INPA Manaus and Museum of Zoology of the University of São Paulo), and partly in the British Museum (Natural History), London.

## OVERVIEW OF THE REPTILIAN HERPETOFAUNA OF THE ISLAND

### FOREST

The most immediately obvious reptile in the dry, sandy *terra firme* forest at the eastern end of the island is the diminutive leaf-litter gecko *Coleodactylus septentrionalis*, which rarely exceeds 40 mm in total length. This tiny lizard, brown with a series of paired white dorsal spots and occasionally a pair of broken white dorsolateral stripes, is found everywhere in the dry forest. Its range halts abruptly at the damper leaf-litter found in wet hollows and alongside *igarapés* (streams), where it is replaced by the small, drab and rather inconspicuous semi-aquatic microteiid lizards *Leposoma percarinatum* and *Neusticurus racenisi*. Another slightly larger microteiid, also commonly sighted in the dry *terra firme* forest leaf-litter, is the dark-brown-flanked and light-brown-backed *Gymnophthalmus underwoodi*, which reaches 60–70 mm. This species, together with *L. percarinatum*, is believed to be parthenogenetic, existing in female-only asexual populations (C. Cole, pers. comm.). A further microteiid species reported from Maracá, but one which the author found to be uncommon, was *Cercosaura ocellata ocellata*, with its startling lateral blue eye-markings reminiscent of a diminutive European eyed-lizard.

These small leaf-litter lizards (in particular *C. septentrionalis*), together with the numerous tiny forest-floor leptodactylid frogs (e.g. *Adenomera hylaedactyla*, the *Leptodactylus wagneri* complex and *Physalaemus ephippifer*), form a major part of the diet of the fast-moving, diurnal terrestrial forest colubrids. These snakes include the common striped whipsnake *Mastigodryas boddaerti boddaerti*, the semi-aquatic *Liophis (Leimadophis) poecilogyrus*, and the aggressive rear-fanged green loras *Philodryas viridissimus* and *P. olfersii*. Larger snakes probably prey on the common ameiva lizard *Ameiva ameiva*, which occurs in considerable numbers in tree-fall sunspots, along trails and in the Ecological Station clearing. Such a broad range of lizard sizes, from the minuscule juvenile *C. septentrionalis* (less than 15 mm long) to adult ameivas (in excess of 300 mm total length), would provide suitable prey for a wide variety of saurophagous forest snakes.

Drift-fences were used to capture nocturnal or semi-fossorial species. Daily checks of their pitfall traps yielded the commoner leaf-litter frogs, lizards, small colubrid snakes and the leptotyphlopid blindsnakes *Leptotyphlops dimidiatus*, *L. septemstriatus* and the large white-headed and -tailed typhlopid blindsnake *Typhlops reticulatus*. Although *L. dimidiatus* and another species, *L. macrolepis*, were already known to occur on Maracá, the other two blindsnakes constituted first records for the reserve. Both species of amphisbaenid (the black and white *Amphisbaena fuliginosa* and the large white *A. alba*) have been collected from Maracá, and several specimens of *A. fuliginosa* were taken in drift-fence traps. In addition, a juvenile amphisbaenid which was found under attack by ants may constitute a new species.

Chelonians are well represented in the forests of Maracá, both by terrestrial and aquatic species. Two forest tortoises are present on the island, apparently occurring sympatrically: the common red-footed *Geochelone carbonaria*, and the less frequently encountered yellow-footed *G. denticulata* (which possesses a serrated carapace and is therefore easily distinguished from its congener). Juveniles were found in August, and adults over 350 mm in length and weighing 5 kg were captured throughout the year. These species are discussed in detail by Moskovits (Chapter 13). The *igarapés*, damp *baixada* forest and seasonally flooded *vazante* vegetation (see Milliken and Ratter, Chapter 5) contain several species of aquatic turtles, including the spotted-legged terrapin *Rhinoclemmys punctularia punctularia*, the side-neck turtle *Platemys platycephala* and the scorpion mud turtle *Kinosternon scorpioides*. The curious leaf-like matá-matá, *Chelus fimbriatus*, has also been collected from a forest *igarapé* on the island.

Two small caiman were sighted by project members in a *buritizal* (*Mauritia flexuosa* palm swamp) close to the Ecological Station. It is possible that these were specimens of smooth-fronted caiman, either *Palaeosuchus trigonatus* or *P. palpebrosus*, since these diminutive species are known to inhabit forest *igarapés* in Amazonia. Neither species, however, has yet been positively identified on Maracá.

During the dry season much of the *buritizal* dried out, and by searching underneath fallen palm fronds the author collected the geckoes *Coleodactylus septentrionalis* and *Hemidactylus palaichthus*, and the black-spotted mabuya *Mabuya nigropunctata*. Small colubrid snakes were also found beneath these fronds, including the minute semi-fossorial brown *Atractus trilineatus*, and the tiny rear-fanged black-headed *Tantilla melanocephala*. A species from the complex genus *Thamnodynastes*, comprising small semi-fossorial, rear-fanged snakes with vertically elliptical pupils, and the forest racer colubrid *Drymoluber dichrous* have also been collected from Maracá by earlier workers, but these species eluded the author.

In addition to the terrestrial forest lizards, Maracá was home to several arboreal species. By day, light grey juvenile and female specimens of the gecko *Gonatodes humeralis* were visible on fallen tree trunks, on cut stumps in forest regeneration study plots, and on young saplings up to 1 m above the ground. The more vividly patterned male, with his green-speckled body and yellow-chevron-marked head, was observed much less frequently. These lizards were rarely encountered on the ground, and never above 1.5 m from the forest floor. On the trunks of the larger trees the diurnal iguanid *Plica plica* would regularly be seen, most commonly in pairs but occasionally in threes, preying on tree-ants. The larger geckoes such as the house-gecko *Hemidactylus palaichthus* and the turnip-tailed gecko *Thecadactylus rapicauda* could generally be encountered by night. The former, however, was also occasionally sighted during the day. These nocturnal lizards often spend the day sleeping inside tree-holes, or within the lattice-work of lianas and strangler-figs which shroud many of the larger forest trees.

Two particularly interesting records for arboreal iguanids were reported during the project. Two eggs collected from leaf-litter in a tree buttress were incubated and hatched to reveal Maracá's first *Anolis ortonii*, the only other anole reported from the island being *Anolis auratus*, which is fairly common within the Station compound. An extremely attractive and heavily gravid female iguanid was also collected on the edge of one of the large cleared forest regeneration study plots. It was transversely banded with emerald-green and yellow-green with fine intervening black lines and was identified as the long-legged iguanid *Polychrus marmoratus*. The lizard was photographed and released after laying its eggs, but unfortunately these proved to be infertile. The rare and secretive short-headed grey-brown iguanid *Uranoscodon superciliosus*, so named because of the raised superciliary scales around the eye, was also collected on a couple of occasions from terrestrial or semi-arboreal habitats along dark, narrow, drying-out tree-shrouded *igarapés*.

A series of huge jumbled rock outcrops occur within the forest less than 1 km inland from the Cachoeira de Fumaça on the Furo de Santa Rosa. Heavily clothed in lianas, ferns and broad-leaved epiphytes, and incised with a maze of fissures, cracks and caves, these rocks provide an ideal home for colonies of bats and lizards. The iguanid *Plica plica* was frequently seen basking on the steep rock faces, and the dart-poison frog, *Dendrobates leucomelas*, was found calling from horizontal fissures.

In addition to the 10 species of snakes reported above as occurring within the forests of Maracá, many others were encountered and captured. The tree boa *Corallus hortulanus hortulanus* was a particularly common and extremely variable nocturnal species, and the 11 specimens collected exhibited five different colour-morphs or patterns. These varied from a black-blotched form with a red zig-zag vertebral stripe to all-grey, yellow, orange or red animals. These nocturnal snakes were often observed along the main trail from the Ecological Station, where they captured bats on the wing. The brilliant red–orange, black and blue Brazilian rainbow boa *Epicrates cenchria cenchria* was also recorded on Maracá, and a large 1.75 m pair was collected under fallen *Maximiliana maripa* palm fronds on the forest floor, close to the Station.

A number of arboreal colubrid snakes inhabit the forest, particularly in the secondary growth-fringed clearings and in the transition zone bordering savanna. Amongst the species now known to occur on Maracá are the diurnal brown vine-snake *Oxybelis aeneus*, and the parrot-snake *Leptophis ahaetulla ahaetulla*. Nocturnally active tree-dwelling colubrids included the cat-eyed snake *Leptodeira annulata* and Catesby's slug-eater *Dipsas catesbyi*.

Both true and false coral snakes inhabit the island: *Pseudoboa neuwiedii* does not really resemble a coral snake but is often referred to as a 'cobra-coral falsa' by the Brazilians. The adult is entirely dried-blood brown with an even darker brown head, but the juvenile is bright salmon-pink with a black head and nape and a broad white band around the temporal region of the head. More than resembling a coral snake, the juvenile *P. neuwiedii* is almost identical to the juvenile of the 'mussurana' *Clelia clelia*, from which it can only be distinguished by the presence of single (rather than paired) subcaudal scales. A much more convincing false coral snake is the common tricoloured *Erythrolamprus aesculapii aesculapii*, which inhabits the same leaf-litter and rotten fallen logs as does the true coral. It also preys upon the same small snakes and lizards as does Maracá's true forest coral snake *Micrurus lemniscatus lemniscatus*. However, the possession of a much larger eye, a yellow snout and a loreal scale easily distinguishes *Erythrolamprus aesculapii* from *Micrurus lemniscatus*.

One of the most surprising snakes to be captured in the forest was the Mt Roraima 'cascavel' rattlesnake *Crotalus durissus ruruima*. Although small and medium-sized specimens were fairly common on the savannas, several large adults were observed up to 0.5 km into the forest and a large pair, both over 1.5 m in length, was collected from the main trail in the forest behind the Ecological Station. It was surprising how easily these large snakes could be overlooked, since they gave no warning rattle. Specimens were also observed in the forest near the isolated Santa Rosa savanna, although the author and other workers failed to find any rattlesnakes on that small savanna itself. Occasionally, specimens of the common lancehead, *Bothrops atrox*, were encountered in the forest, one specimen being observed devouring an *Ameiva ameiva* alongside a trail through a damp swampy area.

The forest at the western end of the island is floristically significantly different from that at the eastern end where much of the study was based (see Milliken and Ratter, 1989 and Chapter 5). The occasional forays that were made to the western forests produced some very interesting species. A juvenile of the yellow-striped, forest *igarapé*-dwelling teiid lizard *Kentropyx calcarata* was collected there, at that time not only a first record for Maracá but also a first for the Territory of Roraima (now State of Roraima). The aquatic fish-eating colubrid *Helicops angulatus*, another Maracá first record, and a specimen of a rapid-moving olive-brown racer snake from the genus *Drymobius*, possibly *D. rhombifer* or a new species (O'Shea and Stimson, 1993), were also collected. The occurrence of these species suggests that the area deserves considerable further investigation, and the author is of the opinion that more intensive studies there would probably uncover the bushmaster, *Lachesis muta*, which was not encountered at the drier eastern end of the island.

## SAVANNA

The savannas (*campos*) constitute only a tiny fraction of Maracá's total surface, and are situated almost entirely in the far southeastern corner of the island. The isolated semi-permanent or permanent pools and the man-made lagoons in the causeway borrow-pits provide further aquatic habitats for amphibians, aquatic reptiles and wading birds. Several lizards were found to be fairly common around the lagoons, on the causeway or along the forest–savanna boundary, but few were sighted on the scorched savanna where they would have been easy prey for the many predatory birds. The teiid *Kentropyx striata*, however, was frequently seen either running through the grass or climbing in low vegetation. Superficially this species resembles *Ameiva ameiva* with its yellow stripes on a green background, but the presence of keeled scales easily distinguishes it from the larger common ameiva. The typical open situation-loving whiptail, *Cnemidophorus lemniscatus*, was also reported from the savannas of Maracá (Martins, pers. comm.).

The ability of *K. striata* to climb also separates it ecologically from its larger relative, since ameivas move along the ground away from rising water levels whilst *Kentropyx* escapes by climbing into the littoral vegetation. Both lizards were found in close proximity to each other, especially in the forest–savanna transition zone. The brown, smooth-scaled skink *Mabuya nigropunctata* was also observed in the leaf-litter along the causeway, and in the forest–savanna edge. A small juvenile *Mabuya* species collected in the same habitat may prove to be new, since its head scalation differed from that of other specimens collected. The tegu *Tupinambis teguixin* was also sighted occasionally, either running through the long grass along the edge of the lagoon or rushing across the causeway. These large lizards obviously use small mammal runs, as several were captured by the mammalogists in their trap-lines (Barnett and da Cunha, Chapter 10). The tortoise *Geochelone carbonaria* was observed moving surprisingly rapidly across the savanna grassland during the early part of the dry season.

The lagoon and the neighbouring pools on the seasonally flooded savanna (see Plate 4b) were home to a substantial population of the spectacled caiman, *Caiman crocodilus*. During the wet season only juveniles and sub-adults were visible in the water, the adults having dispersed across the savanna. However, as the dry season approached and the water level began to diminish, the adult caiman began to return to the lagoon. At the height of the dry season more than a dozen adults could be seen lying, gaping to keep cool, on the banks or in the shallow water, whilst others sheltered under the limited vegetation cover.

Snakes were less in evidence on the savannas than in the forest, and like the lizards the smaller species seemed only to occur in the vicinity of the causeway and the lagoon. A specimen of the nocturnal rear-fanged false coral snake *Oxyrhopus petola* was collected on the causeway, and the coral snake *Micrurus lemniscatus diutius* has also been reported from that locality (in leaf-litter) by Celso Morato de Carvalho. A live juvenile anaconda, *Eunectes murinus gigas*, was discovered in a wheel-rut rain-puddle less than five minutes after a vehicle had passed. A few treesnakes inhabited the causeway trees, mainly *Chironius* spp., but the occasional thunder-and-lightning or tiger ratsnake *Spilotes pullatus pullatus* was also sighted. The lagoon itself was inhabited by larger anacondas which preyed on young caiman and waterbirds, and several specimens 3–4 m long were sighted.

The largest savanna snakes recorded were the 'cascavel' rattlesnake *Crotalus durissus ruruima* (of which the largest specimens were actually found in the forest) and the 'cribo' *Drymarchon corais corais*. The largest cribo captured, a 2 m male, was taken by the author on the isolated savanna at Santa Rosa, but other specimens were seen on the savannas and within the forest nearby. These large ophiophagous snakes were not very common, and had not previously been reported from the island. The small striped snake *Liophis* (*Lygophis*) *lineatus* was reported once, also from the Santa Rosa savanna. This snake was extremely common on the savannas between Maracá and Boa Vista, as was the rattlesnake, so it would have been surprising if it were not present on the island itself. The reason why more snakes were reported from the small Santa Rosa savanna than on the more extensive savanna patches below the Station was almost certainly due to the higher density of scientists working at Santa Rosa, where several of the longer-running botanical, pedological and hydrological studies were carried out.

The Ecological Station and the out-station buildings also provided niches for a number of species, mostly savanna dwellers but also a few forest inhabitants. The most obvious reptile species in any tropical human dwelling is the house gecko, in this case *Hemidactylus palaichthus*, and the presence of this lizard provided easy meals for the island's commonest snake, the cat-eyed snake *Leptodeira annulata*. These snakes could be found on most evenings hunting geckoes on the verandah of the Station, and the author captured 18 without much effort. A population of bats (*Molossus molossus*) in the roofs of these buildings also tempted the occasional tree boa, *Corallus hortulanus*, into venturing amongst the Station buildings. A tree snake, *Chironius carinatus*, was found in a tree behind the kitchen, and juveniles of both the rattlesnake and the common lancehead were found on the ground directly outside the dormitories where they were probably hunting the ameivas which scuttle about in the station clearing. A common lancehead was found in the out-station building at the Casa de Santa Rosa.

The largest snake encountered within the Ecological Station compound itself was a 2 m common boa, *Boa constrictor constrictor*, captured in the centre of the camp after a heavy rainstorm. This specimen had obviously struck at a porcupine, since it had five quills embedded in the roof of its mouth, with one penetrating the top of its snout. These spines were removed and the boa was released.

Possibly the most important specimen collected in the compound was a juvenile rainbow boa, *Epicrates cenchria*. This snake was quite unlike the large and typical Brazilian *Epicrates cenchria cenchria* captured in the forest, as it was brown dorsally with a series of cream ocelli and a broken cream dorsolateral stripe. It was released, as a larger specimen (to act as a study voucher) was found dead on the road between the island and Boa Vista. The author is not convinced that these two specimens, obviously of a savanna taxon, represent the Guianan *Epicrates cenchria maurus*, from which they seem to differ considerably, and although that name is tentatively applied here, further study of the preserved specimen is planned. There is also a record of an identical specimen being collected on the savanna between Maracá and Boa Vista in the 1970s (da Cunha and Nascimento, 1980).

A number of species are present on the station which may have been introduced by man, perhaps in supplies or building materials. These include the anolis lizard *Anolis auratus*, found by the author on the introduced bushes outside the dormitories. The latter has also been reported by Celso Morato de Carvalho as present on the causeway (the supply-route to the station). A blue-tailed skink of the genus *Mabuya*, with a yellow-striped brown dorsum, appeared to be confined to the timber framework of the verandah. It inhabited holes in the timber made by carpenter bees, and was extremely difficult to catch. Whether these skinks, described by Rebouças-Spieker and Vanzolini (1990) as a new species, *M. carvalhoi*, are indigenous or introduced has yet to be determined. Specimens were not observed in any other habitats on the island, suggesting that they may have been introduced with building materials. So far the species is only known from one other locality, the Rio Catrimani in western Roraima State (Yanomami Indian territory). However, *Mabuya* is not an oviparous genus so the possibility of their having arrived from elsewhere as eggs, deposited inside holes in construction timber, must be discounted.

## RIVERS

Many of the species found in the riverine forest were the same as those found on the savanna or in the inland forests. Minute leptodactylids such as *Pseudopaludicola pusillus* and the tiny gecko *Coleodactylus septentrionalis* hid in the leaf-litter, and snakes of the genera *Chironius* and *Spilotes* basked stretched out across the branches of riverside shrubs and trees. The unusual yellow, black and white coral snake *Micrurus hemprichii*, collected one morning on the opposite bank of the Furo de Santa Rosa, constituted the first record of this species for the Maracá environs.

The rocks along the river's edge were inhabited by the dorsally flattened, brown or grey iguanid *Tropidurus hispidus*. These rapid-moving lizards could be seen during the heat of the day dashing out from dark rock crevices to grab an insect before retreating to cover. *Tropidurus* was also recorded inhabiting totally different habitats such as the deserted house at the Casa de Maracá out-station and the many ramshackle wooden bridges on the road to Boa Vista.

Green iguanas, *Iguana iguana iguana*, and small unidentified turtles dived into the river from rocks and from the overhanging branches of drowned trees at the approach of an outboard motor. Large river turtles rested and basked on rocks in mid-stream, and it was a considerable time before one was cornered and positively identified as *Podocnemis unifilis*. Other species of this genus were probably also present, but getting close enough to capture or identify them was not a simple matter. Larger anacondas were occasionally seen on the river banks, and the Maiongong Indians told of giant individuals further up the river.

It is clear that even after the completion of the INPA/RGS/SEMA Maracá Rainforest Project (Projeto Maracá), the herpetofauna of the island was still incompletely sampled. This conclusion is supported by the statistical analysis which follows. The wary turtles, elusive forest creek caiman, disappearing snake tails and secretive lizards, not to mention the inhabitants of the floristically different and herpetologically under-explored western end of the island, all entice the field worker to return to this exciting and largely unspoiled rainforest island.

## COMPOSITION AND RELATIONSHIPS OF THE REPTILE HERPETOFAUNA OF THE ILHA DE MARACÁ

Sixty-six species of reptiles (55 genera) representing 20 families currently comprise the known reptilian herpetofauna of the Ilha de Maracá (see Appendix 5, p. 462). The structure and composition of the Maracá reptile fauna is here compared with data from other lowland Central and South American rainforests. Data have been obtained from the following localities and literature sources:

1. Estación Biológica La Selva, Costa Rica (Guyer, 1990).
2. Barro Colorado Island and adjacent mainland, Panama (Rand and Myers, 1990).
3. Santa Cecilia, Ecuador (Duellman, 1978, 1990).
4. Cocha Cashu, Parque Nacional Manu, Peru (Rodriguez and Cadle, 1990).
5. Iquitos, Peru (Dixon and Soini, 1986) – confined to a 30 km radius of Iquitos.
6. Minimum Critical Size of Ecosystems Project, Manaus, Amazonas, Brazil (Zimmerman and Rodrigues, 1990) – confined to a 15 km radius of Cabo Frio, Gavião.
7. Carajás, Pará, Brazil (da Cunha *et al.*, 1985) – confined to a 45 km radius of Sierra Norte.
8. Belém, Pará, Brazil (Crump, 1971; da Cunha and Nascimento, 1978) – confined to a 40 km radius of Belém.
9. Ilha de Maracá, Roraima, Brazil (O'Shea).

No attempt will be made here to describe the floristic composition of the sites, their relief, geology or climatology, as these data are available in the original papers and the botany and aspects of the ecology of the island are dealt with elsewhere in this volume. However, it should be realized that in contrast to the majority of the above localities, which comprise mostly lowland rainforest, the Ilha de Maracá does contain a small area of natural savanna and therefore exhibits herpetofaunal elements associated with open formations.

In the interests of uniformity I have adopted the analytical methodology and statistical techniques used by Duellman (1990) in his comparison of two Central American and three South American herpetofaunas (nos 1, 2, 3, 4 and 6 listed above). Only the data pertaining to the reptiles from these five localities have been included in this analysis, as the amphibians are dealt with by Martins (Chapter 14).

Some of the values for these sites may vary slightly from those quoted by Duellman as there appear to be a few minor typographic errors in the original data and calculations (i.e. Tables 24.1 and 24.2).

To complement the data from the five localities analysed by Duellman and the Maracá material presented in his volume, further reptile faunal data were obtained from literature sources for three additional Amazonian localities (listed above):

### Iquitos, Peru

Dixon and Soini (1986) carried out an intensive long-term survey of the reptile fauna of the Iquitos region. However, many of their collecting localities were distributed over an area of 150 km radius from Iquitos and since the purpose of this analysis was to compare the reptile fauna of a small, localized area in Roraima with similarily localized and well-defined sites elsewhere in Latin America, the overall area surveyed by Dixon and Soini was considered too extensive. It was therefore decided to restrict the Iquitos data to those species recorded within 30 km of Iquitos. This area incorporated five of Dixon and Soini's main collecting sites: 1 (Centro Union), 3 (Iquitos), 5 (Mishana), 6 (Moropon), and 8 (Paraiso). The delimitation of the study area did not appear to affect the overall analysis adversely since it was only necessary to exclude two species of lizards and three of snakes, and the authors had themselves expressed some doubts as to the validity of the original collecting site data for these species.

### Belém, Brazil

Data were obtained from two primary sources: Crump (1971 – lizards[1]) and da Cunha and Nascimento (1978 – snakes). Records of amphisbaenian, chelonian and crocodilian species occurring in the vicinity of Belém were obtained from a number of scattered literature sources. Whilst Crump's sampling sites were all located close to Belém, those of da Cunha and Nascimento were distributed over a wide area and it was again felt necessary to restrict further analysis to those species recorded within 40 km of Belém. Most of the snakes recorded by da Cunha and Nascimento were collected from sites 1–15 within the delimited area but eight species were only reported from sites 16–37, outside the area, and these have been omitted.

### Carajás, Brazil

Data obtained from da Cunha et al. (1985) relate to an area within a 45 km radius of Sierra Norte, Pará. No data were available for crocodilians.

Duellman (1990) analysed the species composition, historical faunal assemblages and resource use of the five herpetofaunas. Following the same methods, data from the nine reptile faunas have been assessed here in an attempt to ascertain their relationships to the Ilha de Maracá reptile fauna.

Table 12.1. Mean species richness. Percentages given in parentheses

|  | Crocodilia | Chelonia | Sauria | Amphisbaenia | Ophidia | Total Reptilia |
|---|---|---|---|---|---|---|
| 5 sites | 2 (2.5) | 4 (5.0) | 24 (30.0) | 1 (1.3) | 49 (61.3) | 80 |
| 9 sites | 2 (2.3) | 5 (5.8) | 25 (28.7) | 1 (1.2) | 54 (62.1) | 87 |
| Maracá | 1 (1.5) | 7 (10.6) | 22 (33.3) | 2 (3.0) | 34 (51.1) | 66 |

## SPECIES COMPOSITION AND RICHNESS

Duellman's data for the five neotropical sites comprise 240 reptile species.[2] When these data were expanded to include the reptiles of Maracá and the three additional Amazonian sites, the total reptile species represented rose to 307. The complete checklists, with details of species occurrence, were too large for inclusion here. The species present at the nine sites varied from 54 to 138 (mean 87) and the low apparent diversities recorded for Manu, Maracá, and possibly also Carajás suggest that the herpetofaunas of these localities are either species-poor or are currently incompletely known.

From data obtained from the five original sites, Duellman proposed that the average reptilian assemblage for a neotropical lowland rainforest comprises two crocodilians, four turtles, 24 lizards, a single amphisbaenian and 49 snakes – a total of 80 reptile species. Expanding the data set to include the nine sites considered here, the average species diversity is fractionally higher (Table 12.1), with two crocodilians, five turtles, 25 lizards, a single amphisbaenian and 54 snakes comprising a total reptile fauna of 87 species. The recorded Maracá fauna contains one crocodilian, seven turtles, 22 lizards, two amphisbaenians and 34 snakes, totalling 66 reptile species.

The high turtle species count reflects Maracá's riverine character and the numerous smaller temporary and permanent water-courses in the study area. The close proximity of the open formations also allows for the presence of both species of *Geochelone*. The low snake diversity reinforces the belief that the inventory for the reserve is far from complete even though the lizard count is close to the mean figures recorded from the other sites. Whilst an intensive survey of an area should identify the majority of lizard species in a relatively short time, the elusive habits of most snakes will often result in incomplete checklists with 'first records' for some species often being delayed until many years after the initiation of a survey. Collecting techniques will also affect numbers of species recorded and the use of drift-fences greatly enhanced the recorded species diversity for fossorial and semi-fossorial snakes, lizards and amphisbaenians on Maracá.

## COEFFICIENT OF BIOGEOGRAPHICAL RESEMBLANCE (CBR)

Duellman (1990) defined the coefficient of biogeographical resemblance as $CBR = 2C/(N_1 + N_2)$ where: $C$ = the number of species in common to both areas; $N_1$ =

number of species in area 1 and $N_2$ = number of species in area 2, based on an algorithm by Pirlot (1956). The CBR was also used here to compare the nine sites (Tables 12.2a–e). Whilst the CBRs for the two Central American sites (0.462–0.667) and seven South American localities (0.291–0.789) are in most cases moderately high, those between the two areas are low (<0.269) demonstrating a considerable degree of differentiation between Central and South American reptile herpetofaunas. The CBRs for snakes between the two areas were higher than those for lizards and other reptiles, since many snakes exhibit extremely wide distributions. One lizard (*Thecadactylus rapicauda*) and one snake (*Leptophis ahaetulla*) were recorded from all nine sites whilst one crocodilian (*Caiman crocodilus*), a single lizard (*Ameiva ameiva*) and six snake species (*Boa constrictor*, *Epicrates cenchria*, *Drymarchon corais*, *Imantodes cenchoa*, *Leptodeira annulata* and *Lachesis muta*) were reported from eight sites. Considering only the seven Amazonian locations, one turtle (*Kinosternon scorpioides*), five lizards (*Gonatodes humeralis*, *Thecadactylus rapicauda*, *Mabuya bistriata–nigropunctata*,[3] *Ameiva ameiva* and *Tupinambis teguixin*) and seven snakes (*Corallus hortulanus*, *Epicrates cenchria*, *Dipsas catesbyi*, *Helicops angulatus*, *Leptophis ahaetulla*, *Micrurus lemniscatus* and *Bothrops atrox*) occurred at all sites whilst *Caiman crocodilus*, two turtles (*Platemys platycephala* and *Geochelone denticulata*), two lizards (*Anolis fuscoauratus* and *A. ortonii*) and a further nine snake species (*Boa constrictor*, *Drymarchon corais*, *Drymoluber dichrous*, *Erythrolamprus aesculapii*, *Imantodes cenchoa*, *Leptodeira annulata*, *Tripanurgos compressus*, *Xenodon severus* and *Lachesis muta*) were recorded from six of the seven locations. Wide-ranging snake species constitute 70% of the reptile species occurring at eight or more of the nine neotropical sites, and 59% of those species occurring at six or more of the seven Amazonian localities. Several of the widely distributed species present at the other six Amazonian sites were not recorded from Maracá, but may eventually be collected there also.

Many of the species recorded from Maracá were not reported from any of the other Amazonian sites. Seven lizards (32%) and six snakes (18%), representing 32% of the squamate fauna and 20% of the total reptiles, were not even reported from the three closest Amazonian sites at Manaus, Carajás and Belém. Duellman (1990) demonstrated that no correlations existed between the CBRs for his five neotropical sites and the distances separating them. Similar comparisons were carried out in the present investigation for the seven Amazonian sites, but confined to squamate reptiles for which complete sets of data were available (Table 12.3). Again no correlations between distance and CBR were in evidence but the differences between the squamate reptile fauna of Maracá and the other Amazonian sites was greatly emphasized, particularly in the comparison between Maracá and Manaus. Although separated by less than 800 km, the couplet was placed 18th out of 21 in the table. Couplets comparing Maracá with other Amazonian localities occupied the five lowest positions in the table, although Maracá and Carajás appeared 11th, suggesting a much closer relationship than exists with any other site. The low CBRs between Maracá and other Amazonian sites are further emphasized by the high CBRs existing between other, much more widely separated, couplets e.g. Belém and Iquitos separated by 2760 km (positioned 2nd), and Santa Cecilia and Belém, separated by 3025 km (positioned 7th).

**Table 12.2.** Species diversity, species in common (in bold) and coefficients of biogeographial resemblance (in italics) for reptiles from nine Latin American localities

(a) *Crocodilians and turtles*

|  | La Selva | Barro Colorado | Santa Cecilia | Manu | Iquitos | Manaus | Carajás | Belém | Maracá |
|---|---|---|---|---|---|---|---|---|---|
| La Selva | **5** | *0.667* | *0.154* | *0.167* | *0.111* | *0.200* | *0* | *0.200* | *0.154* |
| B. Color. | 4 | **7** | *0.133* | *0.143* | *0.100* | *0.167* | *0* | *0.167* | *0.133* |
| S. Cecilia | 1 | 1 | **8** | *0.533* | *0.571* | *0.769* | *0.462* | *0.462* | *0.625* |
| Manu | 1 | 1 | 4 | **7** | *0.700* | *0.667* | *0.500* | *0.500* | *0.667* |
| Iquitos | 1 | 1 | 6 | 7 | **13** | *0.556* | *0.333* | *0.444* | *0.571* |
| Manaus | 1 | 1 | 5 | 4 | 5 | **5** | *0.400* | *0.600* | *0.615* |
| Carajás | 0 | 0 | 3 | 3 | 3 | 2 | **5** | *0.400* | *0.615* |
| Belém | 1 | 1 | 3 | 3 | 4 | 3 | 2 | **5** | *0.615* |
| Maracá | 1 | 1 | 5 | 5 | 6 | 4 | 4 | 4 | **8** |

Closest resemblance to Maracá – Manu (0.667); Santa Cecilia (0.625); Manaus, Carajás and Belém (0.615).

(b) *Lizards and amphisbaenians*

|  | La Selva | Barro Colorado | Santa Cecilia | Manu | Iquitos | Manaus | Carajás | Belém | Maracá |
|---|---|---|---|---|---|---|---|---|---|
| La Selva | **25** | *0.462* | *0.071* | *0.049* | *0.062* | *0.080* | *0.080* | *0.080* | *0.042* |
| B. Color. | 12 | **27** | *0.103* | *0.093* | *0.119* | *0.154* | *0.154* | *0.115* | *0.196* |
| S. Cecilia | 1 | 3 | **31** | *0.426* | *0.789* | *0.393* | *0.464* | *0.429* | *0.291* |
| Manu | 1 | 2 | 10 | **16** | *0.429* | *0.439* | *0.488* | *0.390* | *0.300* |
| Iquitos | 2 | 4 | 28 | 12 | **40** | *0.431* | *0.523* | *0.492* | *0.375* |
| Manaus | 2 | 4 | 11 | 9 | 14 | **25** | *0.640* | *0.520* | *0.490* |
| Carajás | 2 | 4 | 13 | 10 | 17 | 16 | **25** | *0.680* | *0.653* |
| Belém | 2 | 3 | 12 | 8 | 16 | 13 | 17 | **25** | *0.531* |
| Maracá | 2 | 5 | 8 | 6 | 12 | 12 | 16 | 13 | **24** |

Closest resemblance to Maracá – Carajás (0.653); Belém (0.531); Manaus (0.490); Iquitos (0.375).

(c) *Snakes*

|  | La Selva | Barro Colorado | Santa Cecilia | Manu | Iquitos | Manaus | Carajás | Belém | Maracá |
|---|---|---|---|---|---|---|---|---|---|
| La Selva | **56** | *0.621* | *0.165* | *0.161* | *0.213* | *0.224* | *0.233* | *0.203* | *0.200* |
| B. Color. | 32 | **47** | *0.200* | *0.231* | *0.242* | *0.262* | *0.234* | *0.269* | *0.247* |
| S. Cecilia | 9 | 10 | **53** | *0.619* | *0.667* | *0.584* | *0.560* | *0.640* | *0.345* |
| Manu | 7 | 9 | 26 | **31** | *0.483* | *0.484* | *0.487* | *0.505* | *0.338* |
| Iquitos | 15 | 16 | 46 | 28 | **85** | *0.648* | *0.561* | *0.713* | *0.370* |
| Manaus | 13 | 14 | 33 | 22 | 47 | **60** | *0.561* | *0.667* | *0.426* |
| Carajás | 12 | 11 | 28 | 19 | 37 | 30 | **47** | *0.622* | *0.444* |
| Belém | 13 | 16 | 40 | 26 | 56 | 44 | 37 | **72** | *0.415* |
| Maracá | 9 | 10 | 15 | 11 | 22 | 20 | 18 | 22 | **34** |

Closest resemblance to Maracá – Carajás (0.444); Manaus (0.426); Belém (0.415); Iquitos (0.370).

**Table 12.2.** (*continued*)

(d) *All squamates*

|  | La Selva | Barro Colorado | Santa Cecilia | Manu | Iquitos | Manaus | Carajás | Belém | Maracá |
|---|---|---|---|---|---|---|---|---|---|
| La Selva | **81** | 0.568 | 0.121 | 0.125 | 0.165 | 0.181 | 0.183 | 0.169 | 0.158 |
| B. Color. | 44 | **74** | 0.165 | 0.182 | 0.201 | 0.226 | 0.178 | 0.222 | 0.227 |
| S. Cecilia | 10 | 13 | **84** | 0.550 | 0.708 | 0.521 | 0.526 | 0.575 | 0.324 |
| Manu | 8 | 11 | 36 | **47** | 0.465 | 0.470 | 0.487 | 0.472 | 0.324 |
| Iquitos | 17 | 20 | 74 | 40 | **125** | 0.581 | 0.548 | 0.649 | 0.372 |
| Manaus | 15 | 18 | 44 | 31 | 61 | **85** | 0.586 | 0.626 | 0.448 |
| Carajás | 14 | 13 | 41 | 29 | 54 | 46 | **72** | 0.647 | 0.523 |
| Belém | 15 | 19 | 52 | 34 | 72 | 57 | 54 | **97** | 0.452 |
| Maracá | 11 | 15 | 23 | 17 | 34 | 32 | 34 | 35 | **58** |

Closest resemblance to Maracá – Carajás (0.523); Belém (0.452); Manaus (0.448); Iquitos (0.372).

(e) *All reptiles*

|  | La Selva | Barro Colorado | Santa Cecilia | Manu | Iquitos | Manaus | Carajás | Belém | Maracá |
|---|---|---|---|---|---|---|---|---|---|
| La Selva | **86** | 0.575 | 0.124 | 0.129 | 0.161 | 0.182 | 0.172 | 0.170 | 0.158 |
| B. Color. | 48 | **81** | 0.162 | 0.178 | 0.192 | 0.222 | 0.190 | 0.240 | 0.218 |
| S. Cecilia | 11 | 14 | **92** | 0.548 | 0.696 | 0.538 | 0.521 | 0.526 | 0.354 |
| Manu | 9 | 12 | 40 | **54** | 0.490 | 0.486 | 0.489 | 0.449 | 0.367 |
| Iquitos | 18 | 21 | 80 | 47 | **138** | 0.579 | 0.530 | 0.633 | 0.392 |
| Manaus | 16 | 19 | 49 | 35 | 66 | **90** | 0.575 | 0.625 | 0.462 |
| Carajás | 14 | 15 | 44 | 32 | 57 | 48 | **77** | 0.626 | 0.531 |
| Belém | 16 | 22 | 51 | 35 | 76 | 60 | 56 | **102** | 0.464 |
| Maracá | 12 | 16 | 28 | 22 | 40 | 36 | 38 | 39 | **66** |

Closest resemblance to Maracá – Carajás (0.531); Belém (0.464); Manaus (0.462).

There are several possible reasons for the disparity existing between Maracá and other Amazonian sites:

1. The reserve has not been completely sampled and several widespread species recorded from other localities may yet be collected, thereby increasing Maracá's CBR with other sites. However, Manu is even less well known, yet it scores considerably higher values in Table 12.3.
2. The herpetofauna of Maracá includes a large proportion of open-formation species and savanna relicts. The open-formation dwellers, including *Cnemidophorus lemniscatus*, *Kentropyx striata*, *Tropidurus hispidus* and *Liophis lineatus*, were only found on Maracá's natural savannas and in forest edge situations. The savanna relicts were found within the forest but only in fairly close proximity to open habitats. Examples include *Geochelone carbonaria*, *Hemidactylus palaichthus*, *Leptotyphlops dimidiatus*, *Philodryas olfersii*, *Pseudoboa neuwiedii* and *Crotalus durissus*. Most of the other Amazonian sites, with the

Table 12.3. CBR and percentage similarity of squamates between Amazonian sites in relation to approximate distance between sites (with Maracá comparisons in bold)

| | Combined total species | Species in common (CBR) | Percentage species in common | Approx. distance in km |
|---|---|---|---|---|
| Iquitos – Santa Cecilia | 135 | 74 (0.708) | 54.8 | 470 |
| Iquitos – Belém | 150 | 72 (0.649) | 48.0 | 2760 |
| Carajás – Belém | 115 | 54 (0.647) | 46.6 | 550 |
| Manaus – Belém | 125 | 57 (0.626) | 45.6 | 1340 |
| Manaus – Carajás | 111 | 46 (0.586) | 41.4 | 1135 |
| Manaus – Iquitos | 149 | 61 (0.581) | 40.9 | 1450 |
| Belém – Santa Cecilia | 129 | 52 (0.575) | 40.3 | 3025 |
| Manu – Santa Cecilia | 95 | 36 (0.550) | 37.9 | 1370 |
| Iquitos – Carajás | 143 | 54 (0.548) | 37.8 | 2520 |
| Carajás – Santa Cecilia | 115 | 41 (0.526) | 35.7 | 2840 |
| **Maracá – Carajás** | **96** | **34 (0.523)** | **35.4** | **1615** |
| Manaus – Santa Cecilia | 125 | 44 (0.521) | 35.2 | 1730 |
| Carajás – Manu | 90 | 29 (0.487) | 32.2 | 2330 |
| Belém – Manu | 110 | 34 (0.472) | 30.9 | 2760 |
| Manaus – Manu | 101 | 31 (0.470) | 30.7 | 1575 |
| Iquitos – Manu | 132 | 40 (0.465) | 30.3 | 945 |
| **Maracá – Belém** | **120** | **35 (0.452)** | **29.2** | **1575** |
| **Maracá – Manaus** | **111** | **32 (0.448)** | **28.8** | **710** |
| **Maracá – Iquitos** | **149** | **34 (0.372)** | **22.8** | **1500** |
| **Maracá – Santa Cecilia** | **119** | **23 (0.324)** | **19.3** | **1615** |
| **Maracá – Manu** | **88** | **17 (0.324)** | **19.3** | **1970** |

exception of Carajás, comprise areas of primary forest, secondary forest or man-made cleared areas but lack natural savannas. Areas of natural *campo* in the Carajás study area allow for the presence of several open-formation species in common with Maracá and increase the CBR of the couplet.

3. Maracá is situated in the transition zone between the Amazonian and Guianan sub-regions and exhibits Guianan herpetofaunal elements which are not reported further south. Such species include *Hemidactylus palaichthus*, *Anolis auratus*, *Gymnophthalmus underwoodi*, *Neusticurus racenisi*, *Leptotyphlops dimidiatus* and *Atractus trilineatus*.

4. Maracá is the type locality for two lizard species with extremely limited distributions. *Coleodactylus septentrionalis* also occurs in a few scattered Venezuelan and Guianan localities but *Mabuya carvalhoi* is a Roraima endemic known only from Maracá and the Rio Catrimani 240 km to the southwest.

## HISTORICAL FAUNAL ASSEMBLAGES

Savage (1982) defined four historical generic units in order to determine the origins of the Central American herpetofauna. Duellman (1990) used three of Savage's units in his analysis of five neotropical herpetofaunas whilst noting that Savage's

Young Northern Unit was not represented in the faunas he surveyed. However, elements of this unit are present in the herpetofauna of Maracá. The faunal assemblages, following Savage (1982) and Duellman (1990), are:

*The Young Northern Unit* of temperate North American origin, represented in South America by savanna-dwelling species. Since this unit contains no true tropical forest species in South and Lower Central America, it was not represented in the data compiled by Duellman (1990) but was weakly represented in the herpetofauna of Maracá (e.g. *Crotalus durissus* and *Cnemidophorus lemniscatus*), Carajás and Belém.

*The Old Northern Unit* was originally derived from subtropical and warm temperate North American taxa although the Central and South American species, with the exception of the widely distributed *Drymarchon corais*, are believed to have evolved *in situ* during the Tertiary and Quaternary. The main components of the South American herpetofauna from this unit are snakes of the subfamily Colubrinae (Colubridae).

*The Middle American Unit* contains those tropical taxa which developed in Central America north of the Panamanian Portal and only moved into South America after the land-bridge was established between South and Middle America. Typical elements of the Middle American Unit present in South America include the *Norops* group of the genus *Anolis* (Polychrotidae) and xenodontine snakes of genera *Dipsas*, *Leptodeira* and *Imantodes* (Colubridae).

*The South American Unit* contains those species of reptiles which arose in tropical and subtropical South America in isolation from Central America during the Cenozoic. Some of the taxa may have originally entered from Central America during an earlier land-bridge, but still other groups may be of Gondwanan origin. Xenodontine colubrid snakes are a dominant part of the South American Unit, as are most of the teiid lizards, geckoes, and cheliid and pelomedusid turtles.

The reptile herpetofauna of Maracá is most strongly influenced by the South American Unit (Table 12.4) which accounts for 43 species (65%) of the total 66 species. The Middle American Unit is represented by six widely distributed tropical species of the genera *Anolis*, *Iguana*, *Boa*, *Dipsas* and *Leptodeira*. The colubrine colubrid snakes form the backbone of the Old Northern Unit's contribution but other taxa represented include the scincids and the kinosternid and emydid turtles. The Young Northern Unit is only represented by the open-formation species, *Cnemidophorus lemniscatus* and *Crotalus durissus*. The origin of the typhlopid blindsnake *Typhlops reticulatus* has not been determined and *Hemidactylus palaichthus*, although a Guianan endemic, probably originated from West African *H. brooki* stock (Kluge, 1969). Both *T. reticulata* and *H. palaichthus* have been omitted from this analysis of historical origins.

**Table 12.4**. Historical faunal assemblages for the herpetofauna of the Ilha de Maracá (adapted from Savage 1982; figures in parentheses indicate species numbers when more than one)

| Young Northern Unit | Old Northern Unit | Middle American Unit | South American Unit |
|---|---|---|---|
| | | | Caiman |
| | Rhinoclemmys | | Chelus |
| | Kinosternon | | Platemys |
| | | | Podocnemis |
| | | | Geochelone (2) |
| Cnemidophorus | Mabuya (2) | Anolis (2) | Coleodactylus |
| | | Iguana | Gonatodes |
| | | | Thecadactylus |
| | | | Polychrus |
| | | | Plica |
| | | | Tropidurus |
| | | | Uranoscodon |
| | | | Ameiva |
| | | | Kentropyx (2) |
| | | | Tupinambis |
| | | | Cercosaura |
| | | | Gymnophthalmus |
| | | | Leposoma |
| | | | Neusticurus |
| | | | Amphisbaena (2) |
| Crotalus | Chironius (2) | Boa | Leptotyphlops (3) |
| | Drymarchon | Dipsas | Corallus |
| | Drymobius | Leptodeira | Epicrates |
| | Drymoluber | | Eunectes |
| | Leptophis | | Atractus |
| | Mastigodryas | | Erythrolamprus |
| | Oxybelis | | Helicops |
| | Spilotes | | Liophis (2) |
| | Tantilla | | Oxyrhopus |
| | | | Philodryas (2) |
| | | | Pseudoboa |
| | | | Thamnodynastes |
| | | | Micrurus (2) |
| | | | Bothrops |
| 2 species | 14 species | 6 species | 42 species |

Maracá's historical assemblages can be compared with those compiled for the other eight localities. The reptile faunas of Lower Central America (La Selva and Barro Colorado Island – Table 12.5) are strongly Middle American in content but both Old Northern and South American Units are also well represented. Even between these two locations separated by a mere 540 km, a shift from Old Northern and Middle American composition towards South American species can be seen.

From Panama into South America this shift is continued both southwards through the Upper Amazon region of Ecuador and Peru and to the southeast across

**Table 12.5.** Historical faunal assemblages (figures in parentheses are percentages of total reptile fauna for the sites)

|  | Young Northern Unit | Old Northern Unit | Middle American Unit | South American Unit | Total reptile species |
|---|---|---|---|---|---|
| Lower Central American sites |  |  |  |  |  |
| La Selva | 0 | 29 (33.7) | 39 (45.3) | 18 (20.9) | 86 |
| Barro Colorado | 0 | 21 (25.9) | 30 (37.0) | 30 (37.0) | 81 |
| Upper Amazonian sites |  |  |  |  |  |
| Santa Cecilia | 0 | 14 (15.2) | 12 (13.0) | 66 (71.7) | 92 |
| Iquitos | 0 | 20 (14.4) | 14 (10.1) | 100 (71.9) | 139* |
| Manu | 0 | 7 (13.0) | 6 (11.1) | 41 (75.9) | 54 |
| Amazon basin sites |  |  |  |  |  |
| Maracá | 2 (3.0) | 14 (21.2) | 6 (9.1) | 42 (63.6) | 66* |
| Manaus | 0 | 18 (20.0) | 10 (11.1) | 62 (68.9) | 90 |
| Carajás | 1 (1.3) | 20 (26.0) | 11 (14.3) | 45 (58.4) | 77 |
| Belém | 1 (1.0) | 19 (18.6) | 12 (11.8) | 68 (66.7) | 102* |

* Due to difficulties over determining their origins the typhlopid blindsnakes have been omitted from these tables as have the geckoes of genus *Hemidactylus*, which arose from West African stock.

the Amazon basin states of Amazonas, Roraima and Pará in Brazil (Table 12.5). The decrease in Old Northern and Middle American elements in favour of South American species is most evident in the Upper Amazonian sites from north to south, but whilst it is still apparent in the Amazon basin sites, from north and west to southeast and east, there do appear to be a few discrepancies.

The Old Northern Unit species, presumably colubrine snakes, comprise a considerably larger portion of the total reptile fauna at the Amazon basin sites than the Upper Amazon locations. Unlike the Lower Central American and Upper Amazon sites, all of the Amazon basin sites, with the exception of Manaus, are situated relatively near to areas of natural savanna and also exhibit elements of the savanna-dwelling Young Northern Unit either in the form of savanna-dwelling taxa (e.g. *Cnemidophorus*) or forest-invading savanna relicts (e.g. *Crotalus*).

Carajás registers much higher percentages for both Old Northern and Middle American elements than do Belém, Manaus or Maracá, despite its position south of the Amazon and further from Central America than most of the other sites. However, the actual numbers of species recorded representing the Old Northern and Middle American Units at Carajás fit the trend more closely than the other Amazonian sites, the high percentages being the result of fewer South American Unit species producing a lower total number of species for the site.

Maracá fits the pattern for decreasing percentages of Young and Old Northern Units and a gradual increase in South American Unit species, but it scores slightly lower than the other three Amazon basin sites for Middle American Unit species. This could be an artefact of incomplete sampling of groups such as anoline lizards or of xenodontine snakes such as *Imantodes* or *Dipsas*.

## DISTRIBUTION PATTERNS OF REPTILES FROM THE ILHA DE MARACÁ

The reptile fauna of Maracá comprises species exhibiting a considerable degree of variation in distribution patterns (Table 12.6). Forty-eight per cent (32 species) are wide ranging, occurring either further south to Argentina or west to Panama with 29% (19 species) extending further into Central America and two species (3.0%), *Drymarchon corais* and *Oxybelis aeneus*, recorded within the USA. Most of the Maracá species with distributions beyond Panama are members of the Old and Young Northern and Middle American Units, but six species (*Caiman crocodilus*, *Geochelone carbonaria*, *Thecadactylus rapicauda*, *Corallus hortulanus*, *Epicrates cenchria* (*maurus*) and *Oxyrhopus petola*) are South American Unit species which have invaded Central America, three as far north as Mexico.

Of those reptile species found on Maracá with entirely South American distributions, 22 (33% of the total Maracá reptile fauna) are distributed through northern South America between Colombia and Bolivia. A further 12 species (18%) are confined to the Guianan sub-region as defined by Hoogmoed (1979).

This sub-region contains the three Guianas, eastern Venezuela and northern Brazil (Roraima, Amazonas, Pará and Amapá north of the Amazon and Negro). The colubrine snakes *Chironius exoletus*, *Drymoluber dichrous* and *Mastigodryas boddaerti* which occur in the northern South American region and the turtle *Rhinoclemmys punctularia* from the Guianan sub-region are members of the Old Northern Unit with distributions confined to South America. *Mabuya carvalhoi*, a Roraima endemic, is also an Old Northern Unit species and *Dipsas catesbyi* is a South American representative of a Middle American genus.

## RESOURCE AND HABITAT PARTITIONING

Duellman (1990) looked at herpetofaunal resource partitioning at the five neotropical sites in terms of habitat, time and diet. Data have only been compiled here for squamate reptiles, and since Maracá possesses areas of natural savanna in addition to extensive forests – and hence exhibits open-formation elements in its herpetofauna – it will be necessary to expand the categories of Duellman to the following (Table 12.7):

- Macrohabitat – forest, open, and edge situations
- Habitat usage – terrestrial (including fossorial, aquatic, etc.) or arboreal
- Diel activity – diurnal or nocturnal (including crepuscular)
- Preferred diet

Any analysis of macrohabitat preferences is flawed by the fact that reptiles, especially large snakes, will move around and may be encountered in widely differing habitats. This was illustrated on Maracá by the occurrence of large specimens of *Crotalus durissus*, a savanna relict, in the forest. The squamate fauna of Maracá, although that of primarily a forested island, does demonstrate a high level of savanna influence (Table 12.8), with 12% of species confined to open habitats and a

**Table 12.6.** Distribution patterns for the reptiles of the Ilha de Maracá*

| Widespread South American (Panama to Argentina) | Northern South America (Amazonian countries) | Guianan region (Guianas, Roraima and Venezuela) |
|---|---|---|
| *Caiman crocodilus*** | | |
| *Kinosternon scorpioides*** | *Chelus fimbriatus* | *Rhinoclemmys punctularia* |
| *Geochelone carbonaria*** | *Platemys platycephala* | |
| | *Podocnemis unifilis* | |
| | *Geochelone denticulata* | |
| *Thecadactylus rapicauda*** | *Gonatodes humeralis* | *Coleodactylus septentrionalis* |
| *Anolis auratus* | *Polychrus marmoratus* | *Hemidactylus palaichthus* |
| *Anolis ortonii* | *Plica plica* | *Uranoscodon superciliosus* |
| *Iguana iguana*** | *Tropidurus hispidus* | *Mabuya carvalhoi*† |
| *Mabuya nigropunctata*‡ | *Kentropyx calcarata* | *Kentropyx striata* |
| *Ameiva ameiva* | *Tupinambis teguixin* | *Gymnophthalmus underwoodi* |
| *Cnemidophorus lemniscatus*** | *Cercosaura ocellata* | *Leposoma percarinatum* |
| | | *Neusticurus racenisi* |
| *Amphisbaena alba* | | |
| *Amphisbaena fuliginosa* | | |
| *Leptotyphlops macrolepis* | *Typhlops reticulatus* | *Leptotyphlops dimidiatus* |
| *Boa constrictor*** | *Eunectes murinus* | *Leptotyphlops septemstriatus* |
| *Corallus hortulanus*** | *Chironius exoletus* | *Atractus trilineatus* |
| *Epicrates cenchria*** | *Dipsas catesbyi* | |
| *Chironius carinatus*** | *Drymoluber dichrous* | |
| *Drymarchon corais*** | *Helicops angulatus* | |
| *Drymobius rhombifer*** | *Mastigodryas boddaerti* | |
| *Erythrolamprus aesculapii* | *Micrurus hemprichii* | |
| *Leptodeira annulata*** | *Micrurus lemniscatus* | |
| *Leptophis ahaetulla*** | *Bothrops atrox* | |
| *Liophis lineatus* | | |
| *Liophis poecilogyrus* | | |
| *Oxybelis aenus*** | | |
| *Oxyrhopus petola*** | | |
| *Philodryas olfersii* | | |
| *Philodryas viridissimus* | | |
| *Pseudoboa neuwiedii* | | |
| *Spilotes pullatus*** | | |
| *Tantilla melanocephala*** | | |
| *Crotalus durissus*** | | |
| 32 species | 21 species | 12 species |

\* *Thamnodynastes* sp. is omitted due to incomplete identification.
\*\* Widespread species occurring into Central America beyond Panama.
† *Mabuya carvalhoi* is endemic to Roraima, Brazil.
‡ Forest-dwelling Amazonian *Mabuya bistriata* = *M. nigropunctata* (Avila-Pires, 1995).

further 7% occurring only in open and edge situations. In both cases these are mostly lizards, which tend to be more localized in their distributions than the larger snakes which comprise the majority of the wide-ranging, all-habitat species. Several families, the teiids, tropidurids and colubrids, contain forest and open habitat species and all four boids occur widely in forest and open situations. Overall the majority of lizards and amphisbaenians on Maracá occurred in open and edge situations whilst the majority of snakes present were forest dwellers.

**Table 12.7.** Resource partitioning of squamate reptiles on the Ilha de Maracá: amp = amphisbaenians; amn = annelids; ant = ants; art = arthropods; bat = bats; bir = birds; cae = caecilians; cai = caiman; fis = fish; fre = frog eggs; frl = frog larvae; fro = frogs; gas = gastropods; liz = lizards; mam = mammals; sna = snakes; ter = termites; veg = vegetation

(a) *Lizards and amphisbaenians*

| | Macrohabitat | Habitat usage | Diel activity | Preferred diet |
|---|---|---|---|---|
| *Coleodactylus septentrionalis* | forest | terrestrial | diurnal | art |
| *Gonatodes humeralis* | forest/edge | terrestrial/arboreal | diurnal | art |
| *Hemidactylus palaichthus* | open/edge | arboreal | nocturnal | art |
| *Thecadactylus rapicauda* | forest/edge/open | arboreal | nocturnal | art |
| *Anolis auratus* | open | arboreal | diurnal | art |
| *A. ortonii* | forest/edge | arboreal | diurnal | art |
| *Polychrus marmoratus* | open/edge | arboreal | diurnal | art, veg |
| *Iguana iguana* | forest/edge | arboreal | diurnal | art, veg |
| *Plica plica* | forest/edge | arboreal | diurnal | ant, art |
| *Tropidurus hispidus* | open | terrestrial/arboreal | diurnal | art, liz, veg |
| *Uranoscodon superciliosus* | forest/edge | terrestrial/arboreal | diurnal | art, amn |
| *Mabuya nigropunctata* | forest/edge/open | terrestrial/semi-arboreal | diurnal | art |
| *M. carvalhoi* | open | arboreal | diurnal | ? art |
| *Ameiva ameiva* | open/edge | terrestrial | diurnal | art, liz |
| *Cnemidophorus lemniscatus* | open | terrestrial | diurnal | art |
| *Kentropyx calcarata* | forest/edge | terrestrial | diurnal | art |
| *K. striata* | open | terrestrial/arboreal | diurnal | art |
| *Tupinambis teguixin* | forest/edge/open | terrestrial | diurnal | art, liz, mam, veg |
| *Cercosaura ocellata* | forest/edge | terrestrial/semi-arboreal | diurnal | art |
| *Gymnophthalmus underwoodi* | forest/edge/open | terrestrial | diurnal | art |
| *Leposoma percarinatum* | forest | terrestrial | diurnal | art |
| *Neusticurus racenisi* | forest | terrestrial/semi-aquatic | diurnal | art, fis, frl |
| *Amphisbaena alba* | forest/edge/open | fossorial | ? both* | art, amn, liz |
| *A. fuliginosa* | forest/edge/open | fossorial | ? both* | art, amn |

* Amphisbaenians were collected in drift-fences above ground after rainy nights but being largely fossorial they may be active subterraneally both by day and night.

(b) *Snakes*

| | Macrohabitat | Habitat usage | Diel activity | Preferred diet |
|---|---|---|---|---|
| *Leptotyphlops dimidiatus* | forest/edge/open | fossorial | ? both* | ter, ant |
| *L. macrolepis* | forest | fossorial | ? both* | ter, ant |
| *L. septemstriatus* | forest | fossorial | ? both* | ter, ant |
| *Typhlops reticulatus* | forest | fossorial | ? both* | ter, ant, amm, amp |
| *Boa constrictor* | forest/edge/open | terrestrial/arboreal | both | liz, bir, mam |
| *Corallus hortulanus* | forest/edge/open | arboreal | nocturnal | liz, bir, bat, mam |
| *Epicrates cenchria* | forest/edge/open | terrestrial/arboreal | both | bir, mam |
| *Eunectes murinus* | forest/edge/open | aquatic | both | bir, mam, cai |
| *Atractus trilineatus* | forest | semi-fossorial | nocturnal | ter, art, ann |
| *Choronius carinatus* | forest/edge/open | terrestrial/arboreal | diurnal | fro, liz, bir, mam |
| *C. exoletus* | forest | terrestrial/arboreal | diurnal | fro, liz, bir, mam |
| *Dipsas catesbyi* | forest | arboreal | nocturnal | gas |
| *Drymarchon corais* | open/edge/forest | terrestrial | diurnal | liz, sna, mam |
| *Drymobius rhombifer* | forest | terrestrial | diurnal | liz |
| *Drymoluber dichrous* | forest | terrestrial/arboreal | diurnal | fro, liz |
| *Erythrolampus aesculapii* | forest | terrestrial | diurnal | fro, liz, sna, mam |
| *Helicops angulatus* | forest/edge/open | aquatic | both | fis, fro, frl, liz |
| *Leptodeira annulata* | forest/edge/open | terrestrial/arboreal | nocturanl | fro, fre, liz |

*continues overleaf*

**Table 12.7.** (*continued*)

| | Macrohabitat | Habitat usage | Diel activity | Preferred diet |
|---|---|---|---|---|
| *Leptophis ahaetulla* | forest/edge/open | arboreal | diurnal | fro, bir |
| *Liophis lineatus* | open | terrestrial | diurnal | fro |
| *L. poecilogyrus* | forest/edge/open | terrestrial | diurnal | fro |
| *Mastigodryas boddaerti* | forest/edge/open | terrestrial/arboreal | diurnal | fro, liz, mam |
| *Oxybelis aeneus* | forest/edge/open | arboreal | diurnal | fro, liz |
| *Oxyrhopus petola* | forest/edge/open | terrestrial/semi-fossorial | both | liz, mam |
| *Philodryas olfersii* | open/edge | terrestrial/arboreal | diurnal | fro, liz, bir, mam |
| *P. viridissimus* | forest | terrestrial/arboreal | diurnal | fro, liz, bir, mam |
| *Pseudoboa neuwiedii* | forest/edge/open | terrestrial/semi-fossorial | diurnal | liz |
| *Spilotes pullatus* | forest/edge/open | terrestrial/arboreal | diurnal | fro, liz, sna, bir, mam |
| *Tantilla melanocephalus* | forest | terrestrial/semi-fossorial | diurnal | art |
| *Thamnodynastes* sp. | open | terrestrial/semi-fossorial | nocturnal | fro |
| *Micrurus hemprichii* | forest | terrestrial/semi-fossorial | nocturnal | amp |
| *M. lemniscatus* | forest/edge/open | terrestrial/semi-fossorial | both | cae, amp, sna |
| *Bothrops atrox* | forest/edge/open | terrestrial/arboreal | both | fro, liz, mam |
| *Crotalus durissus* | open/edge/forest | terrestrial | both | liz, mam |

* Leptotyphlopids and typhlopids were collected in drift-fences above ground after rainy nights but being largely fossorial they may be active subterraneally both by day and night.

Table 12.8. Macrohabitat. Percentages given in parentheses

|  | Lizards and amphisbaenians | Snakes | All squamates |
|---|---|---|---|
| Forest only | 3 (12.5) | 12 (35.3) | 15 (25.9) |
| Open only | 5 (20.8) | 2 (5.9) | 7 (12.1) |
| Forest and edge only | 7 (29.1) | 0 | 7 (12.1) |
| Open and edge only | 3 (12.5) | 1 (2.9) | 4 (6.9) |
| Forest, edge and open | 6 (25.0) | 19 (55.9) | 25 (43.1) |
| Total forest | 16 (66.7) | 31 (91.2) | 47 (81.0) |
| Total open and edge | 21 (87.5) | 22 (64.7) | 43 (74.1) |

Table 12.9. Habitat usage. Percentages given in parentheses

|  | Lizards and amphisbaenians | Snakes | All squamates |
|---|---|---|---|
| Terrestrial only* | 10 (41.7) | 19 (55.9) | 29 (50.0) |
| Arboreal only | 8 (33.3) | 4 (11.8) | 12 (20.7) |
| Both | 6 (25.0) | 11 (32.4) | 17 (29.3) |
| Total terrestrial | 16 (66.7) | 30 (88.2) | 46 (79.3) |
| Total arboreal | 14 (58.3) | 15 (44.1) | 29 (50.0) |

* Including fossorial and aquatic species.

The terrestrial and arboreal lizards on Maracá were fairly well matched (Table 12.9) but the majority of snakes were either partially or entirely terrestrial in habit, only four truly arboreal snake species being recorded. Amongst the lizards, the teiids, gymnophthalmids and scincids dominated the terrestrial component whilst the other families were primarily the arboreal representatives. The strong influence of the terrestrial snakes, especially the leptotyphlopids, single typhlopid, elapids and viperids, resulted in a primarily terrestrial squamate herpetofauna.

Only two of the lizards, both geckoes, are truly nocturnal whereas 18 snakes are either entirely or partially nocturnal (Table 12.10), the majority of both snakes and lizards being diurnal in their habits. There does not appear to be any correlation between diel activity cycles and habitat usage other than the fact that both species of truly nocturnal lizards are arboreal.

Two species of amphisbaenians and four species of blindsnakes are included as both diurnal and nocturnal, since although they were only collected in drift-fences following nights of heavy rain their fossorial habits would appear to make diel cycles irrelevant in determining their activity patterns. With the exception of nocturnal arboreal snakes, and also diurnal terrestrial snakes which score slightly lower than Barro Colorado Island, the percentage values for diel activity and habitat usage for lizards and snakes on Maracá (Table 12.11) are consistently slightly higher than those quoted by Duellman (1990) for his five neotropical sites. This may be because six species of lizards (27%) and 11 species of snakes (32.4%) were reported here as inhabiting both terrestrial and arboreal niches, whilst 12 species of snakes were recorded as both diurnal and nocturnal. Duellman agrees

**Table 12.10.** Diel activity. Percentages given in parentheses

|  | Lizards and amphisbaenians | Snakes | All squamates |
|---|---|---|---|
| Diurnal only | 20 (83.3) | 16 (47.1) | 36 (62.1) |
| Nocturnal only | 2 (8.3) | 6 (17.6) | 8 (13.8) |
| Both | 2 (8.3)* | 12 (35.3) | 14 (24.1) |
| Total diurnal | 24 (91.7) | 28 (82.4) | 50 (86.2) |
| Total nocturnal | 4 (16.6) | 18 (52.9) | 22 (37.9) |

\* Including blindsnakes and amphisbaenians.

**Table 12.11.** Diel activity and habitat usage in squamates. Percentages given in parentheses

|  | Diurnal terrestrial | Diurnal arboreal | Nocturnal terrestrial | Nocturnal arboreal |
|---|---|---|---|---|
| Lizards | 14 (63.6) | 12 (54.5) | 0 | 2 (9.1) |
| Amphisbaenids | 2 (100.0) | 0 | 2 (100.0) | 0 |
| Snakes | 26 (76.5) | 12 (35.3) | 15 (44.1) | 6 (17.6) |
| All squamates | 42 (72.4) | 24 (41.4) | 17 (29.3) | 8 (13.8) |

with the habitat usage of most of the dual niche snakes. However, a number of the Maracá lizards exhibiting edge or open habitat preferences are peculiar to the Guianan region and are therefore absent from his data, i.e. *Kentropyx striata*, *Cercosaura ocellata*, *Tropidurus hispidus* and *Mabuya carvalhoi*. Other species exhibit more wide-ranging habitat use than reported by Duellman, i.e. *Gonatodes humeralis* and *Uranscodon superciliosus*, both of which were encountered on the ground. Duellman does not quote a diel activity for the leptotyphlopids, although literature sources suggest both diurnal and nocturnal subterranean activity.

It is interesting to note that all of the New and Old Northern Unit elements of the squamate fauna of Maracá (Table 12.4) are diurnal, i.e. colubrine colubrids, scincids and the Young Northern Unit *Crotalus* and *Cnemidophorus*. Most of the South American Unit xenodontine colubrids, with the exception of *Philodryas* spp., are entirely terrestrial, whilst the Middle American xenodontines, *Dipsas* and *Leptodeira*, are nocturnal and arboreal. This is also true of the five sites analysed by Duellman. Three snakes, *Boa constrictor*, *Epicrates cenchria* and *Bothrops atrox*, occur by day and night both terrestrially and arboreally, in the latter case usually as juveniles.

Most lizards and amphisbaenians prey on arthropods (Table 12.12) although larger species such as adult teiids will feed on smaller lizards, and the tegu, *Tupinambis teguixin*, which achieves a snout–vent length in excess of 300 mm, also preys on small mammals. Although adult green iguanas are primarily herbivorous, the other species recorded as feeding on vegetation only occasionally eat flower heads or fallen fruit. Some dietary specialization is exhibited within the lizards by

**Table 12.12.** Dietary components of the reptile fauna of Maracá, and the number of species (percentages of total in parentheses)

|  | Lizards and amphisbaenians | Snakes | All squamates |
|---|---|---|---|
| Vegetation | 4 (16.7) | 0 | 4 (6.9) |
| Arthropods | 24 (100) | 2 (5.9) | 26 (44.8) |
| Ants or termites | 1 (4.2) | 5 (14.7) | 6 (10.3) |
| Earthworms | 3 (12.5) | 2 (5.9) | 5 (8.6) |
| Gastropods | 0 | 1 (2.9) | 1 (1.7) |
| Fish | 1 (4.2) | 1 (2.9) | 2 (3.4) |
| Frogs | 0 | 16 (47.1) | 16 (27.6) |
| Frog eggs | 0 | 1 (2.9) | 1 (1.7) |
| Frog larvae | 1 (4.2) | 1 (2.9) | 1 (1.7) |
| Caecilians | 0 | 1 (2.9) | 1 (1.7) |
| Lizards | 3 (12.5) | 19 (55.9) | 22 (37.9) |
| Amphisbaenians | 0 | 3 (8.8) | 3 (5.2) |
| Snakes | 0 | 4 (11.8) | 4 (6.9) |
| Caiman | 0 | 1 (2.9) | 3 (5.2) |
| Birds | 0 | 10 (29.4) | 10 (17.2) |
| Bats | 0 | 1 (2.9) | 1 (1.7) |
| Mammals | 1 (4.2) | 15 (44.1) | 16 (27.6) |
| Total invertebrates | 24 (100) | 7 (20.6) | 31 (53.4) |
| Total poikilotherms | 6 (25.0) | 28 (82.4) | 24 (41.4) |
| Total homoiotherms | 1 (4.2) | 16 (47.1) | 17 (29.3) |
| Total vertebrates | 6 (25.0) | 28 (82.4) | 24 (41.4) |

*Plica plica* which preys largely on tree ants and beetles (Dixon and Soini, 1986), and the semi-aquatic *Neusticurus* spp. which will take fish and the larvae of frogs (Obst et al., 1988).

The snakes are entirely carnivorous and most species prey on vertebrates, only the fossorial blindsnakes and diminutive leaf-litter dwelling colubrids *Atractus trilineatus* and *Tantilla melanocephala* feeding on arthropods and annelids. However, one invertebrate feeder, the nocturnally arboreal *Dipsas catesbyi*, is a specialized gastropod feeder (Dixon and Soini, 1986; Duellman, 1990). Most of the snakes (82%) feed on poikilothermic animals, with frogs and lizards the usual prey, but there are again some specializations. The cat-eyed snake *Leptodeira annulata* has been observed feeding on arboreal frog spawn in bromeliads (Mitchell, 1986), although not on Maracá, and the watersnake *Helicops angulatus* readily feeds on fish and tadpoles. Semi-fossorial snakes such as the true elapid coral snakes *Micrurus* spp. and the colubrid false coral snake *Erythrolamprus aesculapii* prey primarily on amphisbaenians and also occasionally caecilians and small snakes. Larger diurnal snake species such as the terrestrial *Drymarchon corais* and the arboreal *Spilotes pullatus* include snakes in their otherwise primarily mammalian diets. A large number of snakes (16) feed on homoiothermic animals: 10 species (29%) on birds and 15 species (44%) on mammals ranging from small rodents to capybara in the case of the anaconda *Eunectes murinus*. *Eunectes* is the closest to a total homoiothermic feeder since it preys almost entirely on mammals and

Table 12.13. Activity and habitat usage of frog-, lizard- and mammal-eating snakes. Percentages given in parentheses

|  | Diurnal terrestrial | Diurnal arboreal | Nocturnal terrestrial | Nocturnal arboreal |
|---|---|---|---|---|
| Frogs | 8 (28.6) | 0 | 16 (57.1) | 10 (35.7) |
| Lizards | 14 (63.6) | 12 (54.5) | 0 | 2 (9.1) |
| Frog*-eating snakes – 16 | 10 (62.5) | 10 (62.5) | 3 (18.8) | 2 (12.5) |
| Lizard†-eating snakes – 19 | 12 (63.2) | 8 (42.1) | 5 (26.3)‡ | 4 (21.1) |
| Mammal-eating snakes – 14§ | 11 (78.6) | 9 (64.3) | 5 (35.7) | 4 (28.6) |

\* Martins (Chapter 14) reports 28 species of anurans for Maracá.
† Amphisbaenians are excluded.
‡ Terrestrial lizard-eating snakes active during the night may prey on sleeping diurnal lizards or other groups of prey.
§ Aquatic *Eunectes murinus* excluded.

waterbirds, but adults also occasionally kill caiman (pers. obs. and Ross, 1989). *Eunectes* is also, rarely, cannibalistic (O'Shea, 1994). One of the most specialized homoiothermic feeders is the nocturnally arboreal tree boa *Corallus hortulanus* which on Maracá preys largely on bats detected by its heat-sensitive labial pits. A number of viperids and boids which are homoiothermic in their prey preferences when adult exhibit saurophagous predatory behaviour as juveniles, i.e. *Boa constrictor*, *Bothrops atrox* and *Crotalus durissus*.

The most frequent dietary components of the snakes of Maracá are lizards (19 snake species – 55.9%) and frogs (16 snake species – 47.1%) (Table 12.12). These are also the most common prey items in the diets of snakes at Duellman's sites, who reports that 9.4–16.1% of snakes prey on cylindrical vertebrates (caecilians, amphisbaenians and snakes). Six species of snakes (17.6%) on Maracá prey on these three groups.

Duellman (1990) made a number of statements regarding reptilian guilds which are relevant to the herpetofauna of Maracá and which should be included here. With the exception of the large and fairly omnivorous and/or herbivorous *Tupinambis* and *Iguana*, most of the lizards, whether terrestrial or arboreal, diurnal or nocturnal, are insect generalists, although *Plica plica* could be said to be something of an ant specialist. Most of the arboreal species are heliophilic and will bask in sunspots. The exceptions to this rule are the geckoes. The arboreal iguanids are sit-and-wait strategists whilst the geckoes are active foragers. *Mabuya carvalhoi* appeared to be an active forager on the Ecological Station verandah and it seems likely that *Kentropyx striata*, when climbing, adopts a similar habit in the edge situations. Most of the terrestrial microteiids, *Gymnophthalmus, Leposoma, Neusticurus*, etc. are heliophobic, whilst terrestrial tropidurids (*Tropidurus*), scincids (*Mabuya*), and macroteiids (*Ameiva* and *Cnemidophorus*) are heliophilic. Duellman reports that terrestrial *Mabuya* are not active foragers, unlike the other terrestrial heliophilics.

Duellman recognized three guilds of anuran-eating snakes: (i) diurnal terrestrial, (ii) nocturnal terrestrial, and (iii) nocturnal arboreal. Most of the anuran-eating snake guild are terrestrial and diurnal. Species of snakes which will feed on frogs but which are active arboreally during the day (Table 12.13) may either be preying on sleeping nocturnal treefrogs or taking other prey.

There are also three guilds of lizard-eating snakes on Maracá: (i) diurnal terrestrial, (ii) nocturnal arboreal, and (iii) diurnal arboreal. The nocturnal terrestrial guild is not represented as Maracá has no terrestrially active nocturnal lizards. Since most lizards are diurnal some nocturnal saurophagous snakes may actively seek sleeping lizards.

Duellman (1990) describes five guilds of mammal-eating snakes which comprise the four diel activity/habitat usage couplets mentioned above, and a fifth guild devoted to the aquatic anaconda *Eunectes*. The close relationships reported by Duellman between the two terrestrial mammal-eating guilds, and to a lesser degree between the two arboreal guilds, were not evident in the data from Maracá (Table 12.13) although close relationships can be seen between the two diurnal guilds and the two nocturnal guilds. Duellman also reports that most bird-eating snakes are diurnal and arboreal, and this is true of Maracá's avian predator snakes with the exception of the nocturnal *Corallus hortulanus*, primarily a bat specialist which may also prey on sleeping birds, and the aquatic *Eunectes murinus* which will take diurnal wading birds.

Minor snake guilds also reported by Duellman which are represented on Maracá include the nocturnal arboreal gastropod-feeder *Dipsas catesbyi*, both nocturnal and diurnal terrestrial cylindrical caecilian, amphisbaenian and snake-eaters such as *Micrurus* and *Erythrolamprus*, and the earthworm specialist, *Atractus trilineatus*.

Arnold (1972) reported that numbers of sympatric snakes, lizards and frogs increased at lower altitudes and that there were correlations between the numbers of predatory snake species and the numbers of prey species. Duellman considered the numbers of lizards and frogs in relation to the numbers of sympatric snakes preying upon them for his five neotropical sites and found that there were no significant correlations. Sites exhibiting high potential prey diversity demonstrated low predator species richness and *vice versa*. Maracá also failed to show any correlation between numbers of predator and prey species. With 19 species of lizard-eating snakes to 22 species of lizards Maracá demonstrated similar species richness to La Selva and Manaus (20 snakes to 24 lizards and 19 snakes to 24 lizards respectively). Although Maracá has 16 species of anuran-eating snakes the anuran fauna appears fairly depauperate, consisting of only 28 species (Martins, Chapter 14). In contrast, Barro Colorado Island exhibits 15 species of anuran-eating snakes to 50 species of frogs, and Santa Cecilia possesses 19 species of anurophagous snakes to 86 anuran species.

Prey species abundance does not appear to be a limiting factor in the reptilian diversity of Maracá and the fairly wide variation of habitat types: forest, open, edge and aquatic, provides for the colonization of numerous niches not available to species in other sites surveyed, thus increasing the resources to be partitioned. Given that the herpetofauna of Maracá has yet to be fully sampled, it must be considered likely that it is fairly rich by lowland Amazonian standards, exhibiting elements of both forest and savanna habitats and Amazonian and Guianan origins (Dixon, 1979, p. 230). A more complete survey of the western end of the island would greatly increase our understanding of the diversity of the reserve, result in the collection of several expected species, and probably also uncover a few surprises.

## ACKNOWLEDGEMENTS

Thanks for project and equipment funding are due to the Royal Geographical Society, the Royal Society and the Percy Sladen Memorial Fund. For assistance with identification of problematic snake specimens I am grateful to Andrew Stimson and Colin McCarthy (BMNH, London) and Celso Morato de Carvalho (INPA-Roraima). Persons and companies who assisted with equipment sponsorship include: Barnett International Ltd and Redmond O'Hanlon (crossbows); Brian Sheen and Clothtec Ltd (drift-fences); R. Taylor and Sons (snake grabs); Jessops Ltd and Kodak (UK) Ltd (trade price photographic equipment and film); Ever Ready (UK) Ltd (batteries); White Mountain, The Rope Shop and Fenwicks Ltd (climbing ropes, etc.).

Finally I am indebted to Aléxia Celeste da Cunha, William Milliken, Duncan Scott, Rogério Gribel Neto and all the other participants in the Maracá Rainforest Project who spared time and effort from their own projects to capture specimens for me.

## NOTES

1. Crump (1971) records the reptile fauna of Belém as comprising a single crocodilian, three turtles, three amphisbaenians, 24 lizards and 44 snakes. Only 23 lizards are actually named in the paper and judging by the data provided in Tables 1 and 2 (pp. 18–19) this would appear to be the true number of lizards present.
2. Duellman actually includes 239 reptile species in his checklist but it seems unlikely that the single unspecified *Liophis* species recorded for Manu and Santa Cecilia are the same species.
3. Although Duellman (1990) and other authors list *Mabuya bistriata* for their localities, Avila-Pires (1995) confines this species to open habitats in major Amazonian river valleys. She applies the name *M. nigropunctata* to more widespread forest *Mabuya* in this complex. Certain diverse study sites may possess both species, e.g. Belém.

## REFERENCES

Arnold, S.J. (1972). Species densities of predators and their prey. *Amer. Nat.*, 106, 206–236.
Avila-Pires, T.C.S. (1995). Lizards of Brazilian Amazonia (Reptilia: Squamata). *Zool. Verh.*, 299, 1–706.
Crump, M.L. (1971). Quantitative analysis of the ecological distribution of a tropical herpetofauna. *Occ. Pap. Mus. Nat. Hist. Kansas*, 3, 1–62.
da Cunha, O.R. and Nascimento, F.P. (1978). Ofídios da Amazônia X. As cobras da região leste do Pará. *Publ. Avul. do Mus. Para. Emílio Goeldi*, 31.
da Cunha, O.R. and Nascimento, F.P. (1980). Ofídios da Amazônia XI. Ofídios de Roraima e notas sobre *Erythrolampus bauperthuisii* Dumeríl, Bibron & Dumeril, 1854, sinônimo de *Erythrolampus aesculapii aesculapii* (Linnaeus, 1758). *Bol. Mus. Para. Emílio Goeldi*, 102, 1–21.
da Cunha, O.R., Nascimento, F.P. and Avila-Pires, T.C.S. de (1985). Os répteis da área de Carajás, Pará, Brasil (Testudines e Squamata). *Publ. Avul. do Mus. Para. Emílio Goeldi*, 40, 11–92.
Dixon, J.R. (1979). Origin and distribution of reptiles in lowland tropical rainforests of South

America. In: *The South American herpetofauna: Its origin, evolution, and dispersal*, ed. W.E. Duellman, pp. 217–240. Mus. Kansas Nat. Hist. Monogr. 7.

Dixon, J.R. and Soini, P. (1986). *The reptiles of the Upper Amazon Basin, Iquitos region, Peru*, 2nd edition. Milwaukee Public Museum.

Duellman, W.E. (1978). The biology of an equatorial herpetofauna in Amazonian Ecuador. *Misc. Publ. Mus. Nat. Hist. Univ. Kansas*, 65, 1–352.

Duellman, W.E. (1990). Herpetofaunas in neotropical rainforests: comparative composition, history, and resource use. In: *Four Neotropical rainforests*, ed. A. Gentry, pp. 455–505. Yale University Press, New Haven.

Estes, R., Queiroz, K. and Gauthier, J. (1988). Phylogenetic relationships within Squamata. In: *Phylogenetic relationships of the lizard families – Essays commemorating Charles L. Camp.*, ed. R. Estes and G. Pregill, pp. 119–281. Stanford University Press, California.

Frost, D.R. and Etheridge, R. (1989). A phylogenetic analysis and taxonomy of iguanine lizards (Reptilia: Squamata). *Misc. Publ. Mus. Nat. Hist. Univ. Kansas*, 81, 1–65.

Guyer, C. (1990). The herpetofauna of La Selva, Costa Rica. In: *Four Neotropical Rainforests*, ed. A. Gentry, pp. 371–385. Yale University Press, New Haven.

Hoogmoed, M.S. (1979). The herpetofauna of the Guianan region. In: *The South American herpetofauna: Its origin, evolution, and dispersal*, ed. W.E. Duellman, pp. 241–279. Mus. Kansas Nat. Hist. Monogr. 7.

Kluge, A.G. (1969). The evolution and geographical origin of the New World *Hemidactylus mabouia–brooki* Complex (Gekkonidae, Sauria). *Misc. Publ. Mus. Zool. Univ. Mich.*, 138.

McDiarmid, R.W., Toure, T. and Savage, J.M. (1996). The proper name of the neotropical tree boa often referred to as *Corallus enydris* (Serpentes: Boidae). *J. Herpetol.*, 30 (3), 320–326.

Milliken, W. and Ratter, J.A. (1989). *The vegetation of the Ilha de Maracá*. Royal Botanic Garden Edinburgh.

Mitchell, A.W. (1986). *The enchanted canopy*. Collins, London.

Moskovits, D.K. (1985). The behavior and ecology of two Amazonian tortoises, *Geochelone carbonaria* and *Geochelone denticulata*, in northwestern Brazil. PhD Dissertation, University of Chicago.

Obst, F.J., Richter, K. and Jacob, U. (1988). *The completely illustrated atlas of reptiles and amphibians for the terrarium*. TFH Publications, Neptune City, New Jersey.

O'Shea, M. (1989). The herpetofauna of the Ilha de Maracá, State of Roraima, Northern Brazil. *Reptiles: Proceedings of the 1988 UK Herpetological Society symposium on captive breeding*, pp. 51–72. British Herpetological Society.

O'Shea, M. (1992). *Expedition field techniques: reptiles and amphibians*. Expedition Advisory Centre, London.

O'Shea, M. (1994). *Eunectes murinus gigas* (Northern green anaconda) cannibalism. *Herpetol. Review*, 25 (3), 124.

O'Shea, M. (1996). Amphibians and reptiles. In: *Biodiversity assessment: A field guide to good practice. Field Manual 2: Data and specimen collection of animals*, ed. N. Stork and J. Davies, pp. 61–66. HMSO, London.

O'Shea, M. and Stimson, A.F. (1993). An aberrant specimen of *Drymobius rhombifer* (Colubridae: Colubrinae): a new generic record for Brazil. *Herpetological Journal*, 3, 70–71.

Peters, J.A., Orejas-Miranda, B., Donoso-Barros, R. amd Vanzolini, P.E. (1986). *Catalogue of the Neotropical Squamata. Part 1. Snakes, Part 2. Lizards and amphisbaenians*, 2nd edition. Smithsonian Institution, Washington, DC.

Pirlot, P.L. (1956). Les formes européenes du genre *Hipparion*. *Inst. Geol. Barcelona, Mem. y Commun.*, 14, 1–150.

Rand, A.S. and Myers, C.D. (1990). The herpetofauna of Barro Colorado Island, Panama, an ecological summary. In: *Four Neotropical rainforests*, ed. A. Gentry, pp. 386–409. Yale University Press, New Haven.

Rebouças-Spieker, R. and Vanzolini, P.E. (1990). *Mabuya carvalhoi*, espécie nova do estado de Roraima, Brasil (Sauria, Scincidae). *Rev. Brasil. Biol.*, 50 (2), 377–386.

Rodriguez, L.B. and Cadle, J.E. (1990). A preliminary overview of the herpetofauna of Cocha Cashu, Manu National Park, Peru. In: *Four Neotropical rainforests*, ed. A. Gentry, pp. 410–425. Yale University Press, New Haven.

Ross, C.A. (1989). *Crocodiles and alligators*. Merehurst Press, London.

Savage, J.M. (1982). The enigma of the Central American herpetofauna: dispersal or vicariance? *Ann. Mo. Bot. Gard.*, 69, 464–547.

Zimmerman, B.L. and Rodrigues, M.T. (1990). Frogs, snakes, and lizards of the INPA–WWF Reserves near Manaus, Brazil. In: *Four Neotropical rainforests*, ed. A. Gentry, pp. 426–454. Yale University Press, New Haven.

# 13 Population and Ecology of the Tortoises *Geochelone carbonaria* and *G. denticulata* on the Ilha de Maracá[1]

**DEBRA MOSKOVITS**
*Field Museum of Natural History, Chicago, USA*

## SUMMARY

The results of detailed studies of the population and feeding ecology of *Geochelone carbonaria* and *G. dentata*, the two tortoise species occurring on the Ilha de Maracá, are presented, and the nutrient composition of their diets is examined. *Geochelone carbonaria*, with a density of slightly over one individual per hectare, was more abundant than *G. dentata* in the study area. Little difference was found between the two species in terms of habitat use or sex ratio, although males of *G. denticulata* showed a greater preference for dense forest than did females of the same species, and inactive females of *G. carbonaria* spent more time in moist habitats during the dry season than in the wet season. However, marked individual variation was found in habitat use regardless of species, sex, or season.

The diets of both species were similar, and diverse in composition, fruits being the most important food item and flowers the second most important. Fruit consumption was found to be highest in the wet season and flower consumption highest in the dry season, reflecting availability. However, distinct preferences were shown for certain foods, e.g. the fruits of *Bagassa guianensis* and the flowers of *Cochlospermum orinocense*. Food sources predominating in the diet (i.e. abundant in the forest) were generally found to be nutrient-poor, whereas those for which distinct preference was shown were generally nutrient-rich.

## RESUMO

São apresentados os resultados de estudos de população e ecologia alimentar de *Geochelone carbonaria* e *G. dentata*, as duas espécies de jabotis que ocorrem na Ilha de Maracá, e a composição de nutrientes nas suas dietas é examinada. *Geochelone carbonaria* foi mais abundante do que *G. dentata* na área de estudo, com pouco mais de um indivíduo por hectare. Poucas diferenças entre as duas espécies foram evidentes no uso de habitat ou no proporções dos sexos, embora os machos de *G. denticulata* mostrassem mais preferência pela floresta densa do que as fêmeas da mesma espécie, e as fêmeas inativas de *G. carbonaria* passaram mais tempo em ambientes úmidos durante a estação seca do que nas chuvas. Porém, foi observada uma forte variação individual no uso de habitat, independentemente de espécie, sexo, ou estação.

As dietas das espécies foram semelhantes, e diversas em composição, frutos sendo o elemento mais importante e flores, o segundo mais importante. Consumo de frutos foi maior na estação chuvosa, e de flores na estação seca, refletindo disponibilidade.

---

*Maracá: The Biodiversity and Environment of an Amazonian Rainforest.*
Edited by William Milliken and James A. Ratter. © 1998 John Wiley & Sons Ltd.

Porém, mostraram preferências distintas por certos alimentos, por exemplo, frutos de *Bagassa guianensis* e flores de *Cochlospermum orinocense*. As fontes de alimento predominantes nas dietas (i.e., abundantes na floresta) geralmente se mostraram pobres em nutrientes, enquanto alimentos preferidos foram relativamente ricos.

## INTRODUCTION

The two Amazonian tortoises *Geochelone carbonaria* (Spix, 1824) and *G. denticulata* (Linnaeus, 1766), as well as the other terrestrial chelonians that inhabit the humid, relatively benign environment of tropical rainforests, are very poorly known. Aside from Castaño-Mora and Lugo-Rugeles' (1981) study of the Amazonian *Geochelone* species in captivity in Colombia, there are only a few brief reports of the two species (Snedigar and Rokosky, 1950; Medem, 1962; Legler, 1963; Auffenberg, 1965; Moll and Tucker, 1976; Fretey, 1977; Medem *et al.*, 1979; Castaño-Mora, 1985), most of which were summarized by Pritchard and Trebbau (1984). Basic information, including size and structure of their populations, reproductive behaviour, sexual size, dimorphism and geographic variation, is lacking for these tortoises in their natural habitats.

Following a preliminary survey in 1980, detailed studies were carried out on the populations of *G. carbonaria* and *G. denticulata* on the Ilha de Maracá, where the two species are syntopic (Moskovits, 1985). The studies included population censuses, behavioural (including feeding) observations and scat analysis. Nutrient composition of some of the items regularly eaten, preferred or rejected by the tortoises was determined.

## STUDY AREA

Observations were carried out in the vicinity of the Ecological Station at the eastern end of the Ilha de Maracá. The 570 ha study area was roughly composed of 85% forest (13.1% dominated by *Maximiliana maripa* palms; 5.3% by vine tangles), 9.4% dense edge vegetation (3–5 m tall walls of razor-grass (*Scleria*), bromeliads and *Heliconia*, with a few shrubs), 4% swamps and *Mauritia flexuosa* palm stands, and 1.6% open grassy areas, some of which were seasonally flooded. Both species of *Geochelone* used all of the different habitat types during at least part of the year. *Mauritia* palm stands fruited asynchronously throughout the forest, providing year-round food for the tortoises. These habitats are described in detail by Milliken and Ratter (Chapter 5).

Treefalls were common, ranging in density from 10.8 to 14.4 ha$^{-1}$ in the forest habitats, and from 0.9 to 7.6 ha$^{-1}$ in the open grass and edge habitats (Moskovits, 1985). These clusters of debris were frequently used by the tortoises during active and inactive periods.

The temperature range on Maracá (20.5–36.5°C in the forest shade) was also favourable for the tortoises, being well within the extremes tolerated by turtles (Brattstrom, 1965). Monthly temperature means oscillated slightly during the period of this study, ranging from 25 to 28°C. The rains on Maracá are sharply seasonal

(RADAMBRASIL, 1975), and most of the precipitation (76% of the 2385 mm annual total during the study) fell between April and August. Even during the height of the dry season, however, the tortoises found refuge in deep, humid burrows and in dense, moist treefalls.

# METHODS

## CAPTURES AND MARKING CODE

Searches for tortoises consisted of walks along the forest trails (see Figure 23.2, p. 436) throughout the daylight hours (0600–1830 hrs). Times in the field and individual routes taken were recorded, and the entire study area was covered every 3–4 days. When favoured tortoise fruits were in season, the trails surrounding the fruiting trees were covered at short intervals. Active tortoises could be spotted up to 8–10 m away from a trail, depending on the vegetation; their shuffling could be heard more than 15 m away, especially in the brittle leaf-litter of the dry season.

A small triangular file was used to make deep notches on the marginal scutes of tortoises. A slightly modified version of the marking code adopted by Schwartz and Schwartz (1974) was followed.

## SPECIES AND SEXES

All tortoises encountered in the study area or in its immediate vicinity were marked and identified to species, and the sex was determined. Although the two species of *Geochelone* are very similar, they are reliably distinguishable in the field, and sex determination is relatively simple (Castaño-Mora and Lugo-Rugeles, 1981; Pritchard and Trebbau, 1984; Moskovits, 1985). However, while examining specimens in museums, it was not possible to sex tortoises <28 cm curved carapace length (CCL) reliably; 28 cm was therefore defined as the cut-off measure between 'juveniles' and 'adults' in this study.

## POPULATION ESTIMATES

The Petersen method (Seber, 1973; Caughley, 1977; Southwood, 1978) was used to estimate the size of both populations of *Geochelone* in the study area. The total study period was divided into two roughly equivalent segments (Table 13.1), consisting of approximately equal numbers of wet (mating) and dry (non-mating) months. Animals recaptured within the period segment in which they were marked ('repeats') were not counted as 'recaptures'.

The assumptions necessary for the Petersen estimates to be valid, and the various associated problems, were discussed by Seber (1973), Poole (1974), Caughley (1977) and Morgan and Bourn (1981). The combination of few marked and few recaptured animals results in substantially negatively biased Petersen estimates, so the estimates are presented here only as indicators of the magnitude of the *Geochelone* population at the study site, and as means of comparison with similar data in future studies.

## FEEDING ECOLOGY

Between March 1981 and November 1982, 227 individuals (285 captures) were marked and measured. Of these, 56 (20 male and 25 female *G. carbonaria*; eight male and three female *G. denticulata*) were radiotracked or trailed for a total of 1442 tortoise-days (Moskovits and Kiester, 1987). Location and activity of the tortoises were checked at least once daily (0600–1830 hrs), and active tortoises were followed until lost from sight. Indirect evidence was obtained from trailed routes (e.g. partially eaten fruits and dug-up soil). Feeding observations (direct and indirect) were recorded on 132 occasions (95 for *G. carbonaria*, 37 for *G. denticulata*): 110 in the dry season (August–March) and 22 in the wet season (April–July). Samples of foods were collected for identification, and, when possible, larger samples were collected for nutrient analyses (see below).

Tortoise scats found along routes of trailed individuals or in the proximity of tortoises encountered, or collected while keeping tortoises overnight for processing, were individually picked and searched for recognizable items. Most seeds recovered from the 43 scats examined were matched with those of fruits present at the time in the forest. Other seeds and animal parts were identified in museums and herbaria. Due to wide variation in degree of food digestion, retrieved items were not quantified. The few items digested beyond recognition were not further analysed. Several of the seeds recovered were planted to test for viability.

### Estimates of fruit availability

Fruit availability was estimated by the spatial and temporal frequency with which fruits were encountered along two census trails (each of 2 km). In the 25 samples taken over a period of 18 months in the field (one sample per three weeks), 5 m wide strips were surveyed to the left or right of the designated trails (chosen at random with respect to habitat or microhabitat types). The side of the trail surveyed was switched every 50 m. Each tree encountered in fruit was recorded, and the total number of fruits on the tree and the proportion apparently ripe were estimated. Specimens of plants that we collected were sent to the herbarium at the Instituto Nacional de Pesquisas da Amazônia (INPA), in Manaus, for identification. The presence of flowers along these trails was noted, but quantities were not estimated.

### Enclosure observations

Observations were made on eight tortoises (three male, three females of *Geochelone carbonaria*; one male, one female of *G. denticulata*) held captive for up to 20 days during both wet and dry seasons. The 400 $m^2$ outdoor enclosure had red clay soil and was at the savanna–forest ecotone. Captive tortoises were fed large quantities of the fruits and flowers available in the field, and fresh water was supplied daily.

Items consistently refused by the tortoises and not observed in the diet of free-ranging individuals were considered 'non-foods'. Preference for different foods was easier to distinguish in the enclosure, where tortoises chose certain foods and ate them first or exclusively. Preference levels were subjectively categorized as: items refused or parts of foods not eaten (non-foods); items eaten often or in moderation (i.e. in amounts smaller or proportional to their abundance in the study area); and items highly preferred (i.e. eaten immediately when available, or presented in captivity, seen to be eaten simultaneously by several tortoises in the field, or found in higher concentration in tortoise scats than in the forest).

**Analyses of nutrient composition and fermentability**

Thirty-three food and non-food items were collected in sufficient quantity to be analysed for nutritional composition and fermentability. Samples were sun-dried in the field and subsequently re-dried to constant mass in the laboratory. They were then ground through a 1 mm screen and subjected to a number of analyses, the methodological details of which are summarized in Moskovits and Bjorndal (1990).

# RESULTS

## CAPTURES

A total of 227 tortoises were marked in the study area during a total of approximately 2850 daylight hours in the field. Of these, 44 tortoises were recaptured at least once (285 captures). Fifteen (7%) of the 189 marked *Geochelone carbonaria* were 'juveniles'. Of the adults, 98 were males and 76 were females (Table 13.1), and the resulting sex ratio (1.3:1) is not significantly different from 1:1 (G-test, $\chi^2$ = 3.32, df = 1, $P > 0.05$). Far fewer individuals of *G. denticulata* were marked (38 representing <17% of all tortoises encountered), and they included only one juvenile. The adult sex ratio for *G. denticulata* (1.5:1) was also not significantly different from 1:1, but the sample size was small, with 23 males and 15 females.

The Petersen estimate (adjusted for small numbers of recaptures; $N^*$ in Seber, 1973) for the adult population of *Geochelone carbonaria* in the 570 ha study area was 598, with 95% confidence limits (Seber, 1973) of 392 and 1176. For adults of *G. denticulata*, the adjusted population estimate was 114, with limits of 55 and 636.

## PREDATION AND MORTALITY

The only obvious source of *Geochelone* mortality on Maracá was observed at the egg stage. Although the only lizard observed destroying *Geochelone* nests was a medium-sized *Tupinambis* (probably *T. teguixin*), a number of other lizards, birds and mammals are potential egg and hatchling predators. Since no intact nests were located, it would not be possible to estimate the proportion that failed, but given the number of destroyed nests encountered it seems likely that the failure rate is similar to those recorded for other turtle populations (e.g. Burger, 1977; Swingland and Coe, 1979; Thompson, 1983), whose averages ranged from 75 to 96%.

Hatchling mortality of *Geochelone* in rainforests must be very high, given the vast number of potential predators. These include large lizards and snakes, ground-walking birds (e.g. *Crax*, *Penelope*, *Neomorphus*, *Aramides* and *Psophia*), forest

Table 13.1. Frequency of capture and recapture of adult (CCL > 28 cm) tortoises, broken into two periods for the Petersen estimate. Period 1 = July–September 1980 + March–October 1981; Period 2 = November 1981–March 1982 + May–October 1982. 'Repeats' are individuals recaptured within the period marked

|  | Period | No. captured | | No. marked (repeats) | | No. recaptured | |
| --- | --- | --- | --- | --- | --- | --- | --- |
|  |  | Male | Female | Male | Female | Male | Female |
| *Geochelone carbonaria* | 1 | 71 | 46 | 62 (9) | 38 (8) | – | – |
|  | 2 | 55 | 52 | 36 (11) | 38 (8) | 8 | 6 |
| *Geochelone denticulata* | 1 | 13 | 10 | 12 (1) | 9 (1) | – | – |
|  | 2 | 17 | 6 | 11 (3) | 6 (0) | 3 | 0 |

falcons (*Micrastur*), smaller birds (e.g. *Momotus*) and a number of mammals (e.g. opossum, *Didelphis*; tayra, *Eira*; fox, *Cerdocyon*; peccaries, *Tayassu*; and several cat species).

On reaching a certain size and shell rigidity (probably after two to three years), the tortoises become immune to most non-human predators. One exception to this is the jaguar *Panthera onca*, which is capable of opening the carapace of adult tortoises (Louise Emmons, pers. comm.), which form one of the principal prey items of this animal in the Manu National Park, Peru (Robin Foster, pers. comm). Only two shells of dead tortoises were found during the study, both of which were of adult males. The larger of these (*Geochelone carbonaria*) showed no signs of predator attack, and the smaller (*G. denticulata*) had obviously been killed by humans. In certain areas of Amazonia, human hunting pressure on *Geochelone* is high, but at the time of study this was not the case. However, with the relatively recent influx of *garimpeiros* (mineral prospectors) into the region of the Ilha de Maracá, humans may now play a significant part in tortoise mortality in certain parts of the island.

Around Maracá, hunters claim that the grey-winged trumpeters, *Psophia crepitans* – medium-sized ground birds which travel and forage in groups and are commonly encountered on the island – occasionally eat tortoises. They are said to surround their victim and collectively peck it to death. White-lipped peccaries *Tayassu pecari*, however, although at times very abundant in the study area (see Fragoso, Chapter 8), appeared to pose no threat to the tortoises which were sometimes encountered within foraging groups of over a hundred of these animals. Likewise the army ants *Eciton* spp., whose activities are discussed by Benton (Chapter 16), do not appear to interact with tortoises. A large swarm raid was once observed moving over a resting *Geochelone*, climbing over its limbs, carapace and head, but apparently inflicting no bites.

Malaria, shell fungi and ectoparasites must affect the *Geochelone* population in rainforests more seriously than any of the agents mentioned above. Tortoises were often observed surrounded by clouds of mosquitoes, particularly during the rainy season, and were visibly agitated by their biting. Ticks were also present in abundance, particularly on *G. denticulata* which were commonly found with over 100 ticks attached to the neck, bases of the limbs and the carapace. It is not yet known whether these ticks are vectors for diseases, but the loss of blood and infections often created at the attachment sites almost certainly affected the health of the host.

## HABITAT USE

According to most accounts, *Geochelone carbonaria* is associated with open country formations such as *cerrado* (scrub savannas), gallery forests, forest islands or forests bordered by savanna (Williams, 1960; Fretey, 1977; Medem *et al.*, 1979; Castaño-Mora and Lugo-Rugeles, 1981; Defler, 1983). *G. denticulata*, on the other hand, has been reported only from humid rainforests (Williams, 1960; Medem, 1962; Castaño-Mora and Lugo-Rugeles, 1981). Auffenberg (1971) suggests that the two species coexisted in the rainforests during the Miocene, and that *G. carbonaria* invaded the newly emerging savannas during the Quaternary period of forest expansions and contractions. In their present distributions, the two species are known to overlap where rainforests and savannas meet in Venezuela (Medem *et al.*, 1979), northwest Brazil (present study), Surinam (Mittermeier, pers. comm.), French Guiana (Fretey, 1977) and northeast Brazil (Balée, pers. comm.). Mittermeier recalls finding *G. carbonaria* and *G. denticulata* in different microhabitats in his Surinam site, with the former concentrating in the higher, drier areas and the latter in the lower, more humid environments.

Observations made during the present study indicated little difference in habitat use between the two species. *Geochelone denticulata* was not seen strictly to avoid open areas; individuals were often seen entering dry seasonal lakes or foraging beyond the savanna boundary, occasionally for periods in excess of three weeks. The characteristics and areas of the habitat types studied are summarized in Table 13.2. The proportion of time spent by tortoises in each of these showed significant departure from the estimated composition of the study area (see Table 13.3). Both sexes of *G. carbonaria* frequented vine tangles, edge and swamp habitats more often than would be expected, dense understorey forests less often, and open forests, *Mauritia* palm stands and seasonal lakes as often as one would expect from the distribution of these habitats. The magnitude of these deviations was generally much greater for the females, especially in their use of vine tangles, swamps and edge habitats. In *G. denticulata* there were distinct habitat use differences between the two sexes. Although both sexes avoided dense forests, the females were most frequently found in edge habitats (which the males slightly avoided), whereas males showed a preference for dense understorey forests (used less often than expected by the females).

The most salient feature in habitat use, however, is the marked individual variation regardless of species, sex or time of year (see Moskovits and Kiester, 1987). Indeed it is probably this individual variation which accounts for the recorded habitat use differences between the male and female *Geochelone denticulata*. Seasonal effects were significant only for inactive *G. carbonaria* females, which used the moist habitats (swamps, *Mauritia* stands and seasonal lakes) more often during the dry (nesting) season, and the dense understorey forests, vine tangles and edge habitats during the wet (mating) season.

Table 13.2. Habitat types in the study area*: their characteristics and potential use by tortoises

| | Characteristics | | | Potential effect on tortoises | | | Area (ha) (% of site) | Debris clusters per ha |
|---|---|---|---|---|---|---|---|---|
| Habitat type | Understorey | Canopy | | Food | Shelter | Humidity | | |
| High-ground forest, open understorey | 0–60% of ground with vegetation (2 layers) | Continuous 25–40 m; some emergents | | Fruits and flowers | Debris clusters and burrows | Dry to moist | 12.5 (9.0) | 13.2 |
| High-ground forest, shrubby understorey | 60–100% of ground with vegetation (2–3 layers) | Broken; 25–35 m; few emergents | | Fruits and flowers | Debris clusters and burrows | Dry to moist | 33.5 (24.2) | 10.8 |
| Floodable forest, sparse understorey | Absent | Open; few emergents | | ? | Many treefalls | Moist/dry to wet | 0.3 (0.2) | 20.0 |
| High-ground forest with vines | 40–70% vines | Continuous to broken; 25–40 m; some emergents | | Fruits, flowers, vine fruits | Debris clusters and burrows | Dry to moist | 33.5 (24.6) | 14.3 |
| Dense vine forest | 70–100% vines or lianas | Dense, continuous; 15–20 m; no emergents | | Vines and liana fruits | Thick vegetation; debris clusters | Dry to moist | 7.8 (5.6) | 14.4 |

| Habitat | | | | | | | |
|---|---|---|---|---|---|---|---|
| Palm forest | Sparse to shrubby | 15–25 m; few emergents; 65–100% palms | Few fruits (?) | Frond falls | Dry to moist | 18.0 (13.0) | 12.4 |
| Forest with bromeliads | Dense and very low | 25–40 m; some emergents | Bromeliad and other fruits | Bromeliads; debris clusters | Dry to moist/wet | 12.4 (9.0) | 12.4 |
| Edge | Dense mat; low, spiny vegetation | Very sparse; few trees or shrubs up to 18 m tall | Bromeliad and *Heliconia* fruits, vines | Dense vegetation; tunnels in sand | Dry/moist to moist/wet | 13.0 (6.1) | 7.6 |
| Swamp | Sparse; shrubs, dense bromeliads and razor-grass | Sparse; 18–25 m; 0–50% *Mauritia* palms | *Mauritia* and bromeliad fruits | Debris clusters; grass clumps | Dry/moist to moist/wet | 2.7 (1.9) | 4.8 |
| *Mauritia* palm stand | Very sparse | 18–25 m; 50–100% *Mauritia* palms | *Mauritia* fruits and flowers | Frond falls | Moist to flooded | 2.7 (1.9) | 1.9 |
| Seasonal grassy ponds | Dense vegetation up to 3 m tall | Absent | Grasses | Clusters of vegetation | Dry to flooded | 2.2 (1.6) | 0.9 |

\* Based on a 138.5 ha census (24.4% of the study area).

272

**Table 13.3.** Tortoise use (percentage) of the different habitats during activity and inactivity*

| | Geochelone carbonaria | | | | | | | | Geochelone denticulata | | | | Individuals | | | | | | | | | | |
|---|---|---|---|---|---|---|---|---|---|---|---|---|---|---|---|---|---|---|---|---|---|---|---|
| | Female | | | | Male | | | | Male | | | | Geochelone carbonaria | | | | | | | | Geochelone denticulata | | |
| | Mating season | | Non-mating season | | Mating season | | Non-mating season | | Mating season | | Non-mating season | | Non-mating season | | | Inactive | | Mating season | | Inactive | | Male | | Female |
| Habitat type** | A† | I† | A | I | A | I | A | I | A | I | A | I | Active | | | | | | | | | M§ | N§ | M | N |
| | | | | | | | | | | | | | 109‡ | 119 | 141 | 16 | 109 | 119 | 10 | 216 | 700 | 4 | 132 104 | 217 129 |
| Dense forest | 27 | 30 | 21 | 22 | 47 | 45 | 28 | 26 | 8 | 30 | 31 | 20 | 50 | 9 | 10 | 22 | 50 | 0 | 11 | 49 | 4 | 65 | 26 | 19 | 9 |
| Open forest | 34 | 33 | 39 | 40 | 40 | 30 | 44 | 43 | 50 | 51 | 62 | 50 | 23 | 39 | 50 | 56 | 6 | 64 | 58 | 38 | 0 | 6 | 68 | 31 | 9 |
| Vine tangles | 16 | 17 | 14 | 11 | 0 | 11 | 8 | 11 | 8 | 9 | 9 | 15 | 27 | 0 | 10 | 22 | 44 | 0 | 11 | 10 | 0 | 6 | 0 | 13 | 14 |
| Edge | 16 | 17 | 17 | 13 | 7 | 10 | 16 | 14 | 0 | 2 | 0 | 7 | 0 | 44 | 20 | 0 | 0 | 25 | 16 | 3 | 96 | 12 | 5 | 19 | 46 |
| Swamp | 5 | 2 | 4 | 8 | 7 | 4 | 0 | 1 | 33 | 7 | 0 | 7 | 5 | 0 | 10 | 0 | 0 | 2 | 5 | 0 | 0 | 12 | 0 | 13 | 5 |
| Mauritia palms | 0 | 1 | 0 | 3 | 0 | 0 | 0 | 2 | 0 | 0 | 0 | 0 | 0 | 0 | 0 | 0 | 0 | 0 | 0 | 0 | 0 | 0 | 0 | 0 | 5 |
| Grassy pond/savanna | 2 | 0 | 4 | 3 | 0 | 0 | 4 | 0 | 30 | 0 | 0 | 0 | 0 | 9 | 0 | 0 | 0 | 9 | 0 | 0 | 0 | 0 | 0 | 0 | 14 |
| N (samples) | 44 | 174 | 98 | 159 | 15 | 71 | 25 | 102 | 12 | 43 | 13 | 46 | 22 | 23 | 20 | 18 | 16 | 45 | 19 | 39 | 25 | 17 | 19 | 16 | 22 |

* Based on the point samples of radiotracked and trailed tortoises.
** As in Table 13.2.
† A = active; I = inactive.
‡ Individual tortoise identification numbers.
§ M = mating season; N = non-mating season.

Most of the mating aggregations observed during the study occurred in the vicinity of fruiting trees of *Genipa americana*. These trees may have been particularly attractive since their peak fruit production period coincided with the April/May tortoise mating peak; their ripe fruit are large and fleshy and their strong sweet smell can be detected (at least by humans) from more than 50 m away. Not all fruiting *Genipa* trees attracted large numbers of tortoises, however, and the presence of a nearby swamp or of adequate sleeping shelters may have affected their attractiveness.

No obvious habitat, microhabitat or soil preferences were found for the nesting sites of these tortoises, although none were ever found in swamp habitats. Castaño-Mora and Lugo-Rugeles (1981) report that slightly humid, open areas were favoured for egg-laying, and local hunters around Maracá claim that *Geochelone carbonaria* tend to lay eggs at the forest edge. Although only one nest was found in the forest–savanna ecotone during the present study, there was a definite increase in captures of the females of this species during nesting months, possibly reflecting a preference for this habitat.

## USE OF SHELTERS

*Geochelone carbonaria* and *G. denticulata* do not excavate their own burrows, but occupy available treefalls, hollow logs or holes. The retreats observed during this study were divided into four categories:

- Burrows – including hollow logs, and holes 0.5–3 m deep formed by upturned tree roots or burrowing mammals such as armadillos (*Dasypus* spp.) and agoutis (*Dasyprocta agouti*).
- Dense vegetation cover – including dense understorey bush, palm frond debris or thick bromeliad stands.
- Debris clusters – formed by tree, branch or vine-falls.
- Open shelters – including slight ground depressions, large buttressed trees, roots or lianas – where the tortoise is firmly lodged but completely exposed.

All tortoises showed a marked preference for dense protected retreats during both seasons, debris clusters being the most frequently used by all but the female *Geochelone denticulata*. Individual tortoises showed distinct behavioural differences, generally using only one or two types of shelter. Nevertheless, two shelter-use patterns did emerge:

- No *G. denticulata* were seen in burrows of any kind during the survey.
- Males of *G. carbonaria* showed distinct seasonal differences in the types of shelters used.

*Geochelone denticulata* is reported to occur mostly in the more humid rainforests with hard clay soils (see above), where the presence of holes may not be as abundant as in the sandy soils of Maracá. Possibly the densities of armadillos, agoutis and other small burrowing mammals is lower in these areas. It is hard to imagine, however, why the tortoises would not be able immediately to exploit these retreats wherever they are available. Regarding the seasonal shift in behaviour among male *G. carbonaria*, the harsher environmental conditions during the dry season, together with the tendency for tortoises to take longer rest bouts during those months, could be expected to drive tortoises towards the more protected retreats. Stable temperatures and high humidity are most likely to be found in deep burrows, and yet the behavioural shift observed in these animals was quite the opposite (i.e. away from deep burrows and towards dense vegetation). The fact that neither female *G. carbonaria* nor male *G. denticulata* exhibited any seasonal change in shelter-use suggests that the dry season, at least during the period of study, was not sufficiently harsh to necessitate a change in behaviour. The shift observed in male *G. carbonaria* may therefore reflect no more than individual variation (see habitat use), and the relatively low number of individuals sampled.

## SEASONAL DIET AND FOOD PREFERENCES

The diverse diet of *Geochelone carbonaria* and *G. denticulata* on Maracá consisted of various vegetative and reproductive plant parts (grasses, leaves, vines, roots, bark, fruits and flowers), fungi (several gilled and woody mushrooms), animal matter (vertebrate carrion, insects and snails), soil, sand and pebbles (Table 13.4). The proportions of these various food items in the diet differed markedly when based on foraging observations (FO) or on scat examination (SC) (Table 13.5). Although fruits ranked highest in both categories, flowers, which ranked second highest in foraging observations, were never detected in the visual examination of scats. In contrast, animal matter (both vertebrate and invertebrate) was recovered from nearly half of the scats examined, but was observed eaten only once in the field (one male *G. carbonaria* feeding on an agouti carcass). Vegetative plant parts (grasses, leaves, leaf-litter, bark) were also detected more easily in scats than in field observations.

Diet composition did not differ significantly between species (Table 13.5; $P > 0.1$ for the nine food categories in FO; SC not tested). Within species, the only sex difference detected was the higher consumption of fruit by male than by female *G. carbonaria* (significant for FO only, $\chi^2 = 5.5$, df = 1, $P = 0.02$).

Seasonal differences, testable for *G. carbonaria* only, were pronounced in the consumption of fruit and flowers. Fruit consumption was considerably higher in the wet than in the dry season ($\chi^2 = 5.4$, df = 1, $P = 0.02$ for both FO and SC), while flowers were observed eaten only in the dry season. Clearly these differences are related to availability.

Fruits (27 identified and at least four unidentified species) were the major item consumed. They comprised nearly half of the feeding observations for both species and were recovered from over 60% of all scats examined (Table 13.5). Up to five different species of fruits were retrieved from a single faecal sample, and up to 34 seeds of *Spondias mombin* (one seed per fruit) and 40 seeds of *Duguetia* (multi-seeded fruits) were recovered simultaneously. Of six species retrieved from scats and planted, four germinated successfully.

**Table 13.4.** Items recorded in the diets of *Geochelone carbonaria* and *G. denticulata*

- **Fruits** – *Spondias mombin* L. (Anacardiaceae), *Anacardium giganteum* Hancock ex Engl. (Anacardiaceae), *Annona* sp. 1 (Annonaceae), *Annona* sp. 2 (Annonaceae), *Duguetia* sp. (Annonaceae), *Philodendron* sp. (Araceae), Unidentified (Bromeliaceae), *Trattinnickia rhoifolia* Willd. (Burseraceae), *Licania kunthiana* Hook.f. (Chrysobalanaceae), Unidentified (Lecythidaceae), *Myriaspora egenensis* DC. (Melastomataceae), *Bagassa guianensis* Aubl. (Moraceae), *Brosimum* sp. (Moraceae), *Ficus* sp. (Moraceae), *Mauritia flexuosa* L.f. (Palmae), *Desmoncus polyacanthos* Mart. (Palmae), *Passiflora coccinea* Aubl. (Passifloraceae), *Passiflora vespertilio* L. (Passifloraceae), *Duroia eriopila* L.f. (Rubiaceae), *Genipa americana* L. (Rubiaceae), *Guettarda spruceana* Steyerm. (Rubiaceae), *Posoqueria latifolia* (Rudge) R. & S. (Rubiaceae), *Pradosia surinamensis* (Eyma) Penn. (Sapotaceae), *Pouteria surumuensis* Baehni (Sapotaceae), *Pouteria hispida* Eyma agg. (Sapotaceae), *Ecclinusa guianensis* Eyma (Sapotaceae), *Clavija lancifolia* Desf. (Theophrastaceae)
- **Flowers** – *Jacaranda copaia* ssp. *spectabilis* (Mart. ex DC.) A. Gentry (Bignoniaceae), *Cochlospermum orinocense* (Kunth) Steud. (Cochlospermaceae), *Mauritia flexuosa* (Palmae)
- **Miscellaneous** – mushrooms (several species); *Maximiliana maripa* (Correa) Drude (Palmae) frond base; unidentified grasses, living leaves, vine stems, roots; sand, soil, pebbles, tortoise scat
- **Animal matter** – snails (at least two species); ants (several species); termites; beetles (assorted body parts recovered); butterflies (wings recovered); Euglossine bees; snakes (ventral scales recovered); lizards; birds (leg of *Psarocolius viridis* recovered); carcasses of agouti, peccary and deer

In the rainy season, when fruiting peaked in both volume and number of species (Figure 13.1), fruits made up over 70% of the feeding observations and were present in 100% of the scats examined. During this season, large fruits (such as *Duguetia* sp., *Pouteria surumuensis*, *Pouteria hispida*) or smaller fruits of large-canopy trees (*Spondias mombin*, *Pradosia surinamensis*) were widely available (present in 43% of the blocks censused).

The rich, oily fruit of the *Mauritia flexuosa* palm was available in the forest and was consumed by the tortoises throughout the year (Table 13.6). Tortoises often ate the fruits of two species, *Bagassa guianensis* (Moraceae) and *Myriaspora egenensis* (Melastomataceae) which were rare in the study area. Only one tree of *Myriaspora* and three trees of *Bagassa* were found in the study track, and neither was present along the fruit census trails. Groups of two to five individuals fed simultaneously under *Bagassa* trees bearing ripe fruit, even though individuals of *Geochelone* are generally solitary animals. Tortoises also aggregated under *Genipa americana* trees bearing ripe fruit. Fruits not seen eaten by tortoises, and considered unlikely items in their diet (type '3' in Figure 13.1), were generally non-fleshy and wind-dispersed, or too large and hard for tortoises to consume.

Flowers constituted close to 30% of the feeding observations during the dry season for both species of *Geochelone*. The flowering peak on Maracá occurred in the middle to late dry season (Figure 13.1), when the fruits in the forest were generally small or present singly or in small clusters. Three species comprised over 72% of the flower feeding observations, and the bright yellow *Cochlospermum orinocense* made up 30% of all flowers observed to be eaten. Yet the visually very similar *Allamanda nobilis* T. Moore (Apocynaceae) was never seen being eaten in

**Table 13.5.** Foods in the diet of *Geochelone carbonaria* (C) and *G. denticulata* (D), grouped by sex (M = male, F = female, J = juvenile) and season (D = dry, W = wet). Percentage of foraging observations (FO) is presented first, followed by % occurrence in scats (SC, in parentheses). The two highest entries in each row are in bold for FO and in italic for SC

(a) Grouped by species (Sp) and sex (Sx)

| Sp | Sx | n | Fruit | Flower | LV | DV | Fungi | Soil | Vertebrates | Invertebr. | Animal parts (any) |
|---|---|---|---|---|---|---|---|---|---|---|---|
| C | M | 22 (16) | **68.2** (*81.3*) | **13.6** (0) | 9.1 (*81.3*) | 0.0 (*43.7*) | 4.6 (18.7) | 0.0 (*43.7*) | 4.6 (25.0) | 0.0 (31.3) | 4.6 (*43.7*) |
| C | F | 70 (17) | **40.2** (*70.6*) | **27.1** (0) | 21.4 (*64.7*) | 2.9 (33.5) | 4.3 (17.7) | 4.3 (41.2) | 0.0 (*52.9*) | 0.0 (5.9) | 0.0 (*52.9*) |
| C | J | 3 (5) | **66.7** (60.0) | 0.0 (0) | 0.0 (60.0) | **33.3** (60.0) | 0.0 (0.0) | 0.0 (*100.0*) | 0.0 (*80.0*) | 0.0 (0.0) | 0.0 (*80.0*) |
| C |   | 95 (38) | **47.4** (*73.7*) | **23.2** (0) | 17.7 (*71.1*) | 3.2 (42.1) | 4.2 (15.8) | 3.2 (50.0) | 1.0 (44.7) | 0.0 (15.8) | 1.0 (*52.6*) |
| D |   | 37 (5) | **46.0** (60.0) | **29.7** (0) | 8.1 (*80.0*) | 5.4 (*80.0*) | 0.0 (*40.0*) | 10.8 (20.0) | 0.0 (20.0) | 0.0 (0.0) | 0.0 (20.0) |

(b) Grouped by species (Sp) and season (Sn)

| Sp | Sn | n | Fruit | Flower | LV | DV | Fungi | Soil | Vertebrates | Invertebr. | Animal parts (any) |
|---|---|---|---|---|---|---|---|---|---|---|---|
| C | D | 76 (27) | **40.8** (63.0) | **29.0** (0) | 15.8 (66.7) | 4.0 (44.4) | 5.2 (11.1) | 2.6 (*51.9*) | 0.0 (48.2) | 1.3 (11.1) | 1.3 (*55.6*) |
| C | W | 20 (11) | **70.0** (*100.0*) | 0.0 (0) | **25.0** (*81.8*) | 0.0 (36.4) | 0.0 (27.3) | 5.0 (45.5) | 0.0 (36.4) | 0.0 (27.3) | 0.0 (45.5) |
| D | D | 36 (3) | **41.7** (33.3) | **30.6** (0) | 8.3 (*100.0*) | 5.6 (*66.6*) | 0.0 (33.3) | 11.1 (0.0) | 0.0 (0.0) | 0.0 (0.0) | 0.0 (0.0) |
| D | W | 2 (2) | **100.0** (*100.0*) | 0.0 (0) | 0.0 (50.0) | 0.0 (*100.0*) | 0.0 (*50.0*) | 0.0 (*50.0*) | 0.0 (*50.0*) | 0.0 (0.0) | 0.0 (*50.0*) |

*n* = number of observations or scats (pooled over all individuals); LV = live vegetative plant parts (leaves, stems, roots); DV = dead leaves (leaf-litter) and bark. Soil includes sand and pebbles.

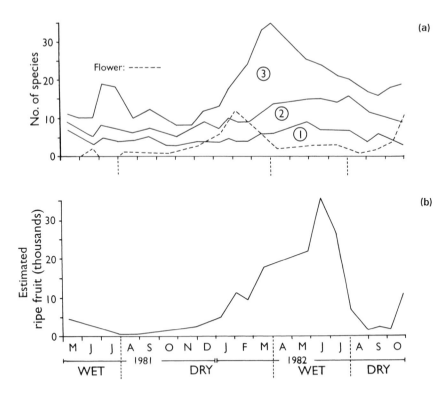

Figure 13.1. Phenology of fruiting and flowering trees at the eastern end of the Ilha de Maracá. (a) Species availability of fruit trees. Fruits are grouped as eaten (1), possibly eaten (2), or not eaten (3) by *Geochelone*. (b) Estimated number of ripe fruits (total over all trees censused) present per sample

the field and was rejected in captivity (Table 13.5). The small and fragrant, orange–yellow flowers of *Mauritia* palms were available over much of the year (Table 13.6), but were observed eaten only during the dry season.

Several species of mushrooms were eaten readily by the tortoises. Often the fungi were recovered seemingly intact from tortoise scats. Mushrooms were considerably more conspicuous in the forest during the rainy season, but tortoises were observed eating them only during the dry season (Table 13.5). No seasonal difference was evident from the scat data. Also consumed year-round were live grasses, leaves, stems, vines, roots, dead leaves, bark, soil, sand and pebbles. Scat reingestion was observed once in the enclosure.

## FEEDING BEHAVIOUR

When feeding in patches of dense fruit or flowers, tortoises did not systematically eat the first item encountered. Instead, they zig-zagged back and forth, holding their heads low and sniffing loudly. They often took only a few bites of a fruit or flower

Table 13.6. Phenology of the important species of fruits and flowers eaten by *Geochelone* tortoises

| Family | Species | May | Jun | Jul | Aug | Sep | Oct | Nov | Dec | Jan | Feb | Mar | Apr | May | Jun | Jul | Aug | Sep | Oct |
|---|---|---|---|---|---|---|---|---|---|---|---|---|---|---|---|---|---|---|---|
| **FRUITS** | | | | | | | | | | | | | | | | | | | |
| Anacardiaceae | *Spondias mombin* | | | = | = | = | = | | | | | | | | | | | | |
| Annonaceae | *Annona* sp. 1 | | | = | | | | | | | | | | | | | | | |
| | *Annona* sp. 2 | | | = | = | = | = | | | | | | | | | | | | |
| Araceae | *Duguetia* sp. | = | = | = | = | = | = | | | | = | = | = | = | = | = | = | = | |
| | *Philodendron* sp. | | | | | | | | | | | ÷ | ÷ | | | | | | |
| | ?? | | | | | | | | = | = | | | | | | | | | |
| Burseraceae | *Trattinnickia rhoifolia* | | | | = | = | = | | | | | | | | | | | | |
| Chrysobalanaceae | *Licania kunthiana* | | = | | | | | | | | | = | | = | = | = | | | |
| | ?? | | | | | | | | | | = | = | | | | | | | |
| Melastomataceae | *Myriaspora egensis* | | | | | | | | | | = | = | = | = | = | = | = | = | |
| Moraceae | *Bagassa guianensis* | | = | = | = | | | | | | = | = | = | | | | | | |
| | *Brosimum* sp. | | | | | | | | | | = | = | | | | | | | |
| | *Ficus* sp. | | | | | | = | = | = | = | = | = | = | | | | | | |
| Passifloraceae | *Passiflora coccinea* | = | = | = | = | = | = | = | = | = | = | = | = | = | = | = | = | = | = |
| | *Passiflora vespertilio* | | | | | | | | | | | = | = | | | | | | |
| Palmae | *Mauritia flexuosa* | | | | | | | | | | | | = | = | = | = | = | = | = |
| Rubiaceae | *Duroia eriopila* | = | = | ÷ | ÷ | ÷ | ÷ | ÷ | ÷ | ÷ | | | | | = | ÷ | ÷ | ÷ | ÷ |
| Sapotaceae | *Genipa americana* | = | = | = | ÷ | ÷ | = | | | | = | = | = | = | = | = | = | = | |
| | *Guettarda argentea* | | | | | | | | | | = | = | = | = | = | = | = | = | |
| | *Pradosia surinamensis* | = | = | = | | | | | | | | | | | | | | | |
| | *Pouteria suramuensis* | = | = | | | | | | | | | | | | | | | | |
| | *Pouteria hispida* | = | = | | | | | | | | | | | | | | | | |
| | *Ecclinusa guianensis* | | | | | | | | = | = | = | = | = | = | | | | | |
| Theophrastaceae | *Clavija lancifolia* | | | | | | | | | | | | = | = | | | | | |
| **FLOWERS** | | | | | | | | | | | | | | | | | | | |
| Bignoniaceae | *Jacaranda copaia* | | | | | | | | = | = | = | | | | | | | | |
| Cochlospermaceae | *Cochlospermum orinocense* | | | | | | | = | = | = | = | | | | | | | | |
| Palmae | *Mauritia flexuosa* | | | | | | | = | = | = | | | | | = | | = | = | |

= Fruits ripe
÷ Fruits probably unripe

that they later returned to eat whole, and when eating leaves, they generally bit and rejected several before swallowing one.

The same pick-and-choose behaviour was displayed in the enclosure, even when the tortoises were presented non-natural foods (e.g. banana, papaya). Captive tortoises often ate two or more items simultaneously, taking a few bites from each and switching frequently. Individual preferences varied among the tortoises tested, but all ate meat first when presented with a variety of items. Large live grubs were consumed immediately when presented to one captive individual.

The tortoises swallowed most of their foods whole in the field, only occasionally using their forelimbs to help tear larger items. However, in captivity, they manipulated the foods in their mouths until seeds or hard items were ejected. They did not swallow the seeds or bones as they did in the field. Both species showed keen discriminating capabilities in captivity, coming directly to a favoured item from as far as 20 m away and learning to avoid permanent obstacles (such as a dividing fence) along the way.

## NUTRIENT ANALYSES

Nutrient analyses were performed on 12 rejected, 18 regularly eaten, and three preferred items. Of the variables considered, median values differed significantly for abundance, fermentability, cell-wall composition (cellulose, lignin and cutin) concentration, nitrogen (N) concentration, phosphorus (P) concentration, and calcium (Ca):P ratios (Figure 13.2). The foods regularly eaten seem 'poor' in quality, with high cell-wall concentrations (especially cutin), low fermentability, and low N, P and total mineral concentrations. They were relatively abundant, and high in Ca concentration with high Ca:P ratios. In contrast, preferred foods had low Ca concentrations and Ca:P ratios, low cell-wall concentrations, high fermentability, and high N, P and total mineral concentrations (Figure 13.2). Surprisingly, items preferred and rejected were often similar in nutrient composition. This may be, in part, because regularly eaten foods analysed were mostly fruits (16 of 18 items), while those preferred or rejected were mostly flowers (2 of 3 and 9 of 12 respectively).

## DISCUSSION

### SPECIES RATIO

Reports on natural populations of the two Amazonian tortoises suggest that *Geochelone carbonaria* typically occurs in moist savannas or in forested areas adjacent to grasslands, while *G. denticulata* is restricted to rainforests (Williams, 1960; Medem, 1962; Fretey, 1977; Castaño-Mora and Lugo-Rugeles, 1981; Defler, 1983; Pritchard and Trebbau, 1984). Sympatry has been observed in areas of ecotone, where rainforest and savanna meet (Pritchard and Trebbau, 1984), and at Maracá the two species were syntopic.

Over five times more *Geochelone carbonaria* than *G. denticulata*, however, were marked in the rainforest study site. This large difference in apparent population

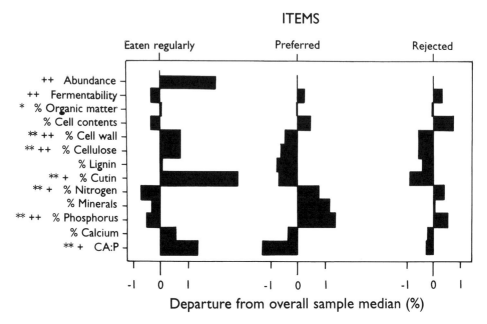

Figure 13.2. Nutrient composition profiles for food items preferred, accepted or rejected by *Geochelone* tortoises. Levels of significance in comparison of medians: Wilcoxon scores (rank sums), * = $P<0.05$, ** = $P<0.01$; median scores (number of points above the median), + = $P<0.05$, ++ = $P<0.01$

sizes, and the unexpected species ratio, remained unexplained at the end of the two-year study. There was no evidence that one species was more conspicuous or easier to catch than the other; on the contrary, the two species displayed very similar behaviour and did not even exhibit strong differences in habitat preferences, often being found side by side (Moskovits, 1985). It is possible that the individuals of *G. denticulata* encountered were peripheral to a main population centred further inland, in the deeper rainforest.

## POPULATION DENSITIES

No quantitative comparisons are possible for the densities recorded on Maracá, since there are no previous studies of the population size of *Geochelone* tortoises in South America. However, based on qualitative information collected during a period of residence in the lowland forest of Manú, eastern Peru (see Terborgh (1983) for a description of the site and its habitats), the *G. denticulata* density of that area (which has clayey soils) was much lower than the densities of *G. denticulata* and *G. carbonaria* on the eastern end of Maracá (which has sandy soils). Sandy soils may be important for nest building, and they may also affect the densities of small burrowing mammals (such as agouti and armadillo) whose burrows are used by the tortoises for shelter. Furthermore, Maracá's relatively open-canopy forest allows more sunlight to reach the ground than does Manú's closed canopy. More light on the forest floor not only provides greater opportunity for sunning (which may be important for juvenile tortoises), but it also allows the development of a denser and more diverse understorey layer, providing food and cover both for juveniles and for adults.

FEEDING ECOLOGY

Land turtles are primarily opportunistic feeders, their diets often reflecting seasonal and regional availabilities of food (e.g. Klimstra and Newsome, 1960; Hansen et al., 1976; Luckenbach, 1982). However, turtles are capable of discriminating colours and of feeding selectively in their environment (e.g. Quaranta and Evans, 1949; Swingland and Frazier, 1979; Macdonald and Mushinsky, 1988). In a study in Florida, *Gopherus polyphemus* was considered intermediate between foraging specialists and generalists (Macdonald and Mushinsky, 1988).

On Maracá, the two species of *Geochelone* primarily ate fruits throughout the year. They heavily supplemented their diet with flowers in the dry season, and with animal matter, live and dead vegetative plant material (leaves, vines, roots and leaf-litter), fungi, soil, sand and pebbles year-round. Species and sex differences were not detected, except for a higher overall consumption of fruit by male than by female *G. carbonaria* (significant in FO only). Seasonal differences, analysable only for *G. carbonaria*, were pronounced in fruit and flower consumption, and reflected overall availability in the forest. The pronounced differences between foraging observations and scat examinations suggest that the diet information presented here is preliminary. Given both the complexity of the rainforest and its ample food abundance, and the broad diet of the *Geochelone*, additional food items should be identified with further study.

Despite the seasonal changes observed, the diet of the *Geochelone* did not solely reflect abundance in the forest. The tortoises frequently ate at least two fruits that were rare in the study area. They rejected the flowers of *Allamanda nobilis*, which were available simultaneously with, and closely resembled, the favoured flowers of *Cochlospermum orinocense*. Furthermore, they consumed flowers of *Mauritia flexuosa* only in the late dry season, although the flowers were available through much of the year (Table 13.6). Both species of *Geochelone* also showed keen discriminating capabilities in the enclosure, often coming directly to a favoured food item.

In the enclosure, where soil was harder clay and sand was not available, tortoises did not swallow their foods whole as they did in the forest. As suggested by many (e.g. Sokol, 1971; Luckenbach, 1982; Marlow and Tollestrup, 1982), sand may be important as an abrasive agent enhancing digestion, and its absence in the enclosure may have contributed to the observed shift in behaviour.

Given this tendency of the *Geochelone* to swallow foods whole, and to frequent treefalls regularly (ideal environments for the growth and establishment of seedlings), they are probably effective seed dispersers. This also has been suggested for *Gopherus polyphemus* (Macdonald and Mushinsky, 1988) and *Testudo graeca* (Cobo and Andreu, 1988).

A large portion of the diet of the tortoises was high in cell-wall concentration (especially cutin), was relatively low in fermentability and cell contents (starch, soluble sugars and protein) and was low in total mineral concentration. However, these foods were relatively abundant in the forest and were high in Ca concentrations. In contrast, the few preferred items were low in cell-wall contents, high in cell contents and fermentability, and high in overall minerals, P, and total N concentrations. Items rejected were very similar to the preferred foods in nutrient composition, although their total mineral, P and N concentrations were not as high (Figure 13.2). This similarity may have been due, in part, to the high proportion of flowers represented in both the non-food and the preferred samples. The observed tendency for tortoises to eat several foods simultaneously in captivity, their seemingly haphazard behaviour in the field, and the diversity of items recovered from single scat samples may indicate that tortoises reach an adequate balance of nutrients by ingesting a wide variety of foods.

Based on the nutrient profiles obtained in this study, concentrations of the nutrients analysed cannot be used to predict diet choice. Yet three points should be considered. First, toxic and secondary compounds were not examined in this study, and they might be important in determining the diet of the tortoises (this is probably the case in the rejection of *Allamanda nobilis* flowers). It is likely, however, that fruiting trees have not evolved specific noxious compounds to deter the tortoises, which are potential seed dispersers. Second, fermentability, used here as a relative measure of microbial degradability of the foods, may not reflect digestibility in the tortoises' guts. However, both species of *Geochelone* rely on a hindgut microbial fermentation, similar to rumen fermentation, to degrade the cell-wall fraction of their diet (Bjorndal, 1989). Third, the relative concentrations of a nutrient in a series of food items do not necessarily reflect the relative availabilities of that nutrient in the food items. Minerals, especially, vary greatly in availabilty, and their absorption in the gut is a complex interaction of the pH, the presence or absence of other minerals, and the chemical form of the mineral (Van Soest, 1982; Robbins, 1983). Appropriate data on relative availabilities, together with nutrient analyses of a broader sample of items regularly eaten, preferred and rejected, should help clarify the nutritional basis for food preferences.

## ACKNOWLEDGEMENTS

I thank Dr P.E. Vanzolini for invaluable support at every stage, Dr P. Nogueira Neto for use of SEMA facilities, and G.M. de Oliveira, L. Pestana, J. and E. Lima da Silva, and Valquimar Felix de Souza ('Amazonas') for help in the field. Dr K. Bjorndal (University of Florida) conducted the nutrient analyses and J. Moore generously provided use of his laboratory. For specimen identifications, thanks are due to Dr M.F. da Silva, INPA (plants), Dr R.B. Foster, Field Museum of Natural History (seeds), and Dr J.W. Fitzpatrick, FMNH (bird bones). Grants from the Chicago Zoological Society, the Hinds Fund of the University of Chicago, and the National Academy of Science helped fund the project. I am deeply grateful to R. Moskovits for her immeasurable assistance in the field, and W. Milliken for adapting the information for this publication.

## NOTE

1. This paper is intended as an overview of the tortoise population and ecology of Maracá, and includes data which have previously been published on more specific aspects of the subject (Moskovits and Kiester, 1987; Moskovits, 1988; Moskovits and Bjorndal, 1990).

## REFERENCES

Auffenberg, W. (1965). Sex and species discrimination in two South American tortoises. *Copeia*, 1965, 335–342.

Auffenberg, W. (1971). A new fossil tortoise, with remarks on the origin of South American Testudinines. *Copeia*, 1971, 335–342.

Bjorndal, K.A. (1989). Flexibility of digestive responses in two generalist herbivores, the tortoises *Geochelone carbonaria* and *Geochelone denticulata*. *Oecologia*, 78, 317–321.

Brattstrom, B.H. (1965). Body temperature of reptiles. *American Midland Naturalist*, 73, 376–422.

Burger, J. (1977). Determinants of hatchling success in the diamond back terrapin, *Malaclemys terrapin*. *American Midland Naturalist*, 97, 444–446.

Castaño-Mora, O.V. (1985). Notas adicionales sobre la reproducción y el crecimiento de los morrocoyes (*Geochelone carbonaria* y *G. denticulata*, Testudines, Testudinidae). *Lozania (Acta Zoologica Colombiana)*, 52, 1–5.

Castaño-Mora, O.V. and Lugo-Rugeles, M. (1981). Estudio comparativo del comportamiento de dos especies de morrocoy: *Geochelone carbonaria* y *Geochelone denticulata* y aspectos comparables de su morfologia externa. *Cespedesia*, 10, 55–122.

Caughley, G. (1977). *Analysis of vertebrate populations*. Wiley, Chichester.

Cobo, M. and Andreu, A.C. (1988). Seed consumption and dispersal by the spur-thighed tortoise *Testudo graeca*. *Oikos*, 51, 267–273.

Defler, T.R. (1983). A remote park in Colombia. *Oryx*, 17, 15–17.

Fretey, J. (1977). *Les Cheloniens de Guyane Française. 1. Etude Préliminaire*. Thesis, University of Paris, Paris.

Hansen, R.M., Johnson, M.K. and van Devender, T.R. (1976). Foods of the desert tortoise, *Gopherus agasizii*, in Arizona and Utah. *Herpetologica*, 32, 247–251.

Klimstra, W.D. and Newsome, F. (1960). Some observations on the food coactions of the common box turtle (*Terrapene c. carolina*). *Ecology*, 41, 637–647.

Legler, J.M. (1963). Tortoises (*Geochelone carbonaria*) in Panama: distribution and variation. *American Midland Naturalist*, 70, 490–503.

Luckenbach, R.A. (1982). Ecology and management of the desert tortoises (*Gopherus agassizii*) in California. In: *North American tortoises: Conservation and ecology*, ed. R.B. Bury, pp. 1–37. US Fish and Wildlife Service, Wildlife Research Report 12, Washington, DC.

MacDonald, L.A. and Mushinsky, H.R. (1988). Foraging ecology of the gopher tortoise, *Gopherus polyphemus*, in a sandhill habitat. *Herpetologica*, 44, 345–353.

Marlow, R.W. and Tollestrup, K. (1982). Mining and exploitation of natural mineral deposits by the desert tortoise, *Gopherus agassizii*. *Animal Behaviour*, 30, 475–478.

Medem, F. (1962). La distribución geográfica y ecología de los Crocodylia y Testudinata en el Departamento del Chocó. *Revista Academia de Colombia, Ciencias Exactas, Fisicas y Naturales*, 11, 279–303.

Medem, F., Castano, O.V. and Lugo-Rugeles, M. (1979). Contribución al conocimiento sobre la reproducción y el crecimiento de los 'Morrocoyes' (*Geochelone carbonaria* y *G. denticulata*; Testudines, Testudinidae). *Caldasia*, 12, 497–511.

Moll, D. and Tucker, J.K. (1976). Growth and sexual maturity of the red-footed tortoise, *Geochelone carbonaria*. *Bulletin of the Maryland Herpetological Society*, 12, 96–98.

Morgan, D.D.V. and Bourn, D.M. (1981). A comparison of two methods of estimating the size of a population of giant tortoises in Aldabra. *Journal of Applied Ecology*, 18, 37–40.

Moskovits, D.K. (1985). *The behavior and ecology of the two Amazonian tortoises,* Geochelone carbonaria *and* Geochelone denticulata, *in Northwestern Brazil*. PhD Dissertation, University of Chicago.

Moskovits, D.K. (1988). Sexual dimorphism and population estimates of the two Amazonian tortoises (*Geochelone carbonaria* and *G. denticulata*) in northwestern Brazil. *Herpetologica*, 44, 209–217.

Moskovits, D.K. and Bjorndal, K.A. (1990). Diet and food preferences of the tortoises *Geochelone carbonaria* and *G. denticulata* in northwestern Brazil. *Herpetologica*, 46, 207–218.

Moskovits, D.K. and Kiester, A.R. (1987). Activity levels and ranging behaviour of the two Amazonian tortoises, *Geochelone carbonaria* and *Geochelone denticulata*, in northwestern Brazil. *Functional Ecology*, 1, 203–214.

Poole, R.W. (1974). *An introduction to quantitative ecology*. McGraw-Hill, New York.

Pritchard, P.C.H. and Trebbau, P. (1984). The turtles of Venezuela. *Society for the Study of Amphibians and Reptiles, Contributions to Herpetology*, No. 2, 1–403.

Quaranta, J.V. and Evans, L.T. (1949). The visual learning of *Testudo vicina*. *Anatomical Record*, 105, 580.

RADAMBRASIL (1975). *Departamento Nacional de Produção mineral, Projeto Radambrasil. Folha NA.20 Boa Vista e parte das folhas NA.21 Tumucumaque, NB.20 Roraima e NB.21*. Rio de Janeiro.

Robbins, C.T. (1983). *Wildlife feeding and nutrition*. Academic Press, New York.

Schwartz, C.W. and Schwartz, E.R. (1974). The three-toed box turtle in central Missouri: its population, home range, and movements. *Missouri Department of Conservation, Terrestrial Series*, 5, 1–28.

Seber, G.A.F. (1973). *The estimation of animal abundance*. Griffin, London.

Snedigar, R. and Rokosky, E.J. (1950). Courtship and egg laying of captive *Testudo denticulata*. *Copeia*, 1950, 46–48.

Sokol, O. (1971). Lithophagy and geophagy in reptiles. *Journal of Herpetology*, 5, 69–70.

Southwood, T.R.E. (1978). *Ecological methods with particular reference to the study of insect populations*, 2nd edition. W.H. Freeman, San Francisco.

Swingland, I.R. and Coe, M. (1978). The natural regulation of giant tortoise populations on Aldabra atoll: reproduction. *Journal of Zoology, London*, 186, 285–309.

Swingland, I.R. and Frazier, J.G. (1979). The conflict between feeding and overheating in the Aldabran giant tortoises. In: *A handbook in biotelemetry and radio tracking*, eds C.J.W. Amlaner and D.W.W. Macdonald, pp. 611–615. Pergamon Press, London.

Terborgh, J. (1983). *Five New World primates. A study in comparative ecology*. Princeton University Press.

Thompson, M.B. (1983). Populations of the Murray river tortoise, *Emydura* (Chelodina). The effect of egg predation by the red fox, *Vulpes vulpes*. *Australian Wildlife Research*, 10, 363–371.

Van Soest, P.J. (1982). *Nutritional ecology of the ruminant*. O. & B. Books, Corvallis, Oregon.

Williams, E.E. (1960). Two species of tortoises in northern South America. *Breviora*, 120, 1–13.

# 14 The Frogs of the Ilha de Maracá

MÁRCIO MARTINS
*Universidade de Amazonas, Manaus, Brazil*

## SUMMARY

Faunal surveys in northern Roraima, Brazil, revealed the presence of 38 species of frogs, 28 of them known to occur on the riverine island of Maracá. The region surveyed is part of the zoogeographical region called Guianan, and its anuran fauna has a complex history and includes many endemic species as well as widespread South American or Amazonian species.

Four bufonids, one dendrobatid, ten hylids, ten leptodactylids, one microhylid and two pseudids were recorded for Maracá. Ten additional species were recorded for other localities less than 130 km from Maracá, and most of them may also occur on the island. The anuran fauna of northern Roraima is poor when compared with other Amazonian regions, probably because of its long dry season.

Most species found in northern Roraima are widespread in the *llanos* and savanna habitats of the region and of southern and central Venezuela. Considering 30 relatively well-identified species from northern Roraima, 48% of them are widespread in South America, 27% widespread in Amazonia and 23% are endemic to the Guianan region. Patterns of habitat utilization in Maracá are similar to those found in other neotropical regions, except for the paucity of hylids inside the forests.

Maintenance of the Ilha de Maracá as a reserve will ensure the preservation of many species of frogs from northern Roraima, some of them endemic to the Guianan region and absent in other Brazilian regions.

## RESUMO

Levantamentos de fauna no norte de Roraima, Brasil, revelaram a presença de 38 espécies de anuros, 28 delas ocorrendo na ilha fluvial de Maracá. A região estudada faz parte da região zoogeográfica chamada de Guiana e sua fauna de anuros possui uma historia complexa e inclui várias espécies endêmicas, além de diversas que ocorrem em grande parte da América do sul ou da Amazônia.

Quatro bufonídeos, um dendrobatídeo, 10 hilídeos, 10 leptodactilídeos, um microhilídeo e dois pseudídeos foram encontrados em Maracá. Dez espécies adicionais foram encontradas em outras quatro localidades a menos de 130 km de Maracá e a maioria delas deve ocorrer na ilha. A fauna de anuros do norte de Roraima é pobre se comparada a outras regiões da Amazônia, provavelmente devido à estação seca longa daquela região.

A maioria das espécies encontradas no norte de Roraima são amplamente distribuídas nos lhanos e savanas desta região e do centro e sul da Venezuela. Considerando 30 espécies identificadas com segurança razoável para o norte de Roraima, 48% delas ocorrem em grande parte da América do Sul, 27% ocorrem na maior parte de Amazônia e 23% são endémicas à região das Guianas. Os padrões de utilização de ambiente em

Maracá são semelhantes àqueles encontrados em outras localidades neotropicais, exceto pela escassez de hilídeos nas florestas.

A Ilha de Maracá como reserva assegurar a preservação de várias espécies de anuros do norte de Roraima, algumas delas endêmicas à região das Guianas e ausentes em outras regiões do Brasil.

# INTRODUCTION

The State of Roraima of northern Brazil is situated in the Guianan region (Hoogmoed, 1979), and its frog fauna includes both endemic members of this region and species that occur throughout Amazonia. Hoogmoed (1979) summarized and characterized the herpetofauna of the Guianan region, stressing the presence of several endemic species (52% of the frogs) and 'our scant and fragmentary knowledge' of this group. The frog fauna of the Guianan region has a complex history, which can be summarized as follows:

- Some species are endemic and probably originate from old stocks of the area.
- Some species show relationships to Andean species, indicating common lowland ancestors.
- Some species invaded the region from the southern lowlands, via forests and savannas (Hoogmoed, 1979; see also Lescure, 1975, 1977).

The frog fauna of the Guianan region is relatively well known when compared with other regions in Amazonia. Among the studies which have characterized this fauna are those of Heatwole *et al.* (1965), Rivero (e.g. 1965, 1971), Lescure (e.g. 1975, 1976, 1977), Staton and Dixon (1977), and Hoogmoed (e.g. 1979; Hoogmoed and Gorzula, 1979).

The Ilha de Maracá's 100 000 hectares are covered almost entirely by forests. O'Shea (1989) provided a detailed description of the island and its various habitats, summarizing the herpetofauna and its associations with the different habitats occurring on the island.

Here I present a checklist of the frog fauna of the Ilha de Maracá, based on my own observations made before the Maracá Rainforest Project began, and those of Mark O'Shea who worked on the project (O'Shea, 1988, 1989 and pers. comm.). Species that occur in localities around Maracá and have yet to be recorded for the island itself, but which probably also occur there, are treated separately. Additional information on the species cited here can be found in the literature dealing with the frogs of the Guianan region (see above).

# METHODS

My collections and observations of frogs on the Ilha de Maracá were undertaken between October 1985 and August 1986. During this period my visits to the island

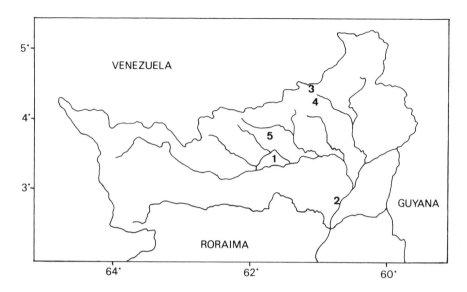

**Figure 14.1.** Map of northern Roraima State, northern Brazil, showing the location of the Ilha de Maracá (1) and four additional localities where observations were made: Boa Vista (2), Serra Pacaraima (3), Surumu (4), and Serra Tepequém (5)

were made every one or two months. Although my work was restricted to the easternmost corner of the island in the vicinity of the Ecological Station (Figure Pr.1, p. xvii), several different habitats were surveyed: the gallery forests of the Furos de Maracá and Santa Rosa; the swampy area below these forests; and the *terra firme* forest inside the island (including creeks and lakes). Additional collections were made at four other localities in Roraima (Figure 14.1) during the same period. The four localities were Boa Vista (130 km from Maracá), Serra Pacaraima (122 km), Surumu (107 km) and Serra Tepequém (55 km).

Frogs were collected at various times of the day, mostly by hand, and processed by the usual methods (Pisani and Villa, 1974). Those collected in Roraima are deposited in the herpetological collections at the Núcleo de Pesquisas de Roraima (INPA, Boa Vista), the Museu de Zoologia da Universidade de São Paulo and the Museu de História Natural da Universidade Estadual de Campinas (São Paulo).

I recorded habitat, microhabitat and data on reproduction for most of the individual frogs collected. For each species listed below, I present the number of specimens collected at each locality (with the number of specimens collected by O'Shea given in parentheses when appropriate), its apparent abundance on Maracá, a short description giving characteristic features to facilitate identification, habitat occupied, reproduction, distribution and remarks on taxonomy, indicating some papers which refer to the species. Specific names follow Frost (1985) and recent revisions (indicated in each account). I use the genus *Hyla* conservatively instead of *Ololygon* for those species related to *Hyla rubra*, following suggestions by Almeida and Cardoso (1985) and A. Cardoso (pers. comm.). Frog sizes are presented as snout–vent length (SVL).

# DISCUSSION

## SPECIES RICHNESS

The complex history and high diversity of topography and climate of South America have produced an extraordinarily rich and diverse herpetofauna (see review in Duellman, 1979). South America has almost 1000 anuran species (belonging to approximately 100 genera and 11 families), nearly 95% of them endemic to the continent (Duellman, 1979) and almost half of them in tropical and subtropical forests (Lynch, 1979). Local anuran faunas within Amazonia are very rich, with up to 50–80 species (see for example Appendix 2 in Duellman, 1989). However, local faunas of the savannas and northern forests of the Guianan region (cf. Hoogmoed, 1979) are relatively species-poor when compared with Amazonia, although several of their species may be present in considerable abundance. Lescure (1976), for instance, found 70 anuran species throughout French Guiana, and Heatwole *et al.* (1965) and Hoogmoed and Gorzula (1979) found 22 and 26 species respectively at different localities in southeastern Bolivar State, Venezuela. Staton and Dixon (1977) found 16 species at two localities in the central *llanos* of Venezuela, and their paper lists 38 species for several localities within northeastern Roraima.

What are the factors responsible for this relatively low anuran richness in the Guianan region? Vegetational heterogeneity and climatic stability are evidently determinant factors for local anuran richness (Duellman, 1989). The vegetational structure on Maracá is, at least at first sight, similar to most Amazonian localities I know. It can therefore be supposed that climate is the crucial factor for the low anuran diversity in Roraima. Annual amounts of rainfall *per se* do not explain these differences: 1751 mm were recorded for Boa Vista versus 2075 mm for the Manaus region, Amazonas (RADAMBRASIL, 1975, 1978), and in the latter area I found approximately 70 anuran species. However, at Boa Vista the rains are concentrated in three months (May to July) whereas around Manaus they tend to fall over a period of eight months (October to May). The environmental stability of a relatively uniform climate allows specialization amongst animal groups (see Duellman, 1989), and it is expected that the contrary is also true. In conclusion, the low anuran richness of Maracá and Roraima may be a consequence of the strong seasonality of the region, which constrains specialization.

## FAUNAL COMPARISONS AND DISTRIBUTION PATTERNS

The frog fauna of the savannas of Roraima and the *llanos* and Gran Sabana of Venezuela is very constant throughout the region. Of the 38 species listed here for northeastern Roraima, Heatwole *et al.* (1965) and Hoogmoed and Gorzula (1979) found 20 (out of a total of 26) and 17 (out of 22) respectively, for two localities in Estado Bolivar, and Staton and Dixon (1977) found 14 (out of 16) at two localities in the central *llanos* of Venezuela (see also Table 1 in Hoogmoed and Gorzula, 1979).

Considering the 38 species identified below, 48% are widespread in South America, 27% are widespread in Amazonia, 23% are endemic to the Guianan region (cf.

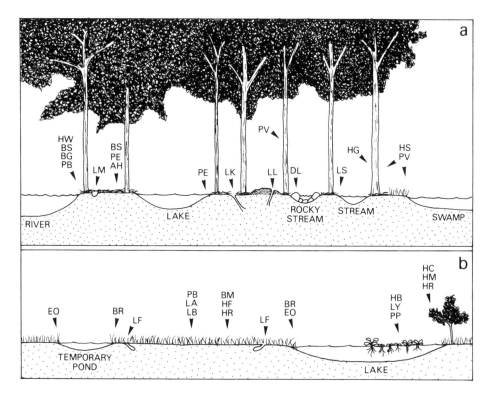

**Figure 14.2.** Schematic profiles of forest (a) and an open area (b) on Maracá, showing water bodies and associated anuran species. See species accounts for details of each species. AH = *Adenomera hylaedactyla*; BG = *Bufo guttatus*; BM = *B. marinus*; BR = *B. granulosus*; BS = *Bufo* sp.; DL = *Dendrobates leucomelas*; EO = *Elachistocleis ovalis*; HB = *Hyla boesemani*; HC = *H. crepitans*; HF = *H. fuscomarginata*; HG = *H. geographica*; HM = *H. microcephala*; HR = *H. rubra*; HS = *Hyla* sp.; HW = *H. wavrini*; LA = *Leptodactylus macrosternum*; LB = *L. bolivianus*; LF = *L. fuscus*; LK = *L. knudseni*; LL = *Lithodytes lineatus*; LM = *Leptodactylus mystaceus*; LS = *Leptodactylus* sp.; LY = *Lysapsus limellus*; PB = *Pseudopaludicola boliviana*; PE = *Physalaemus ephippifer*; PP = *Pseudis paradoxa*; PV = *Phrynohyas venulosa*

Hoogmoed, 1979), and 10% are widespread in the northern extremes of South America. As observed by Hoogmoed and Gorzula (1979) for El Manteco, the anuran fauna of northeastern Roraima is a composite of species typical of the *llanos* and savannas and those widespread in Amazonia and South America. Thus, the recurrence of a combination of widespread and Guianan endemic species is responsible for the found similarities discussed above.

## COMMUNITY STRUCTURE

The structure of anuran communities on Maracá must be analysed separately by habitats (Figure 14.2). In forests and forest borders, nine species (seven leptodactylids and two bufonids) were found within or below leaf-litter (each of them associated with lakes, rivers and/or streams), four hylids were found on the vegetation of swamps and streams, and one dendrobatid was associated with rocky habitats (Figure 14.2a). In open savanna areas, one leptodactylid, one bufonid, and one microhylid used temporary ponds, whereas five hylids, four leptodactylids, two bufonids, two pseudids, and one microhylid were found in permanent lakes (Figure 14.2b). Three hylids perched on plants at margins, two pseudids and one hylid were found on or within floating vegetation, and all others were found on or below the ground.

The patterns above are similar to those found in other neotropical localities (see for example Hoogmoed and Gorzula, 1979; Cardoso *et al.*, 1989; Duellman, 1989) except for the extremely low proportion of hylids inside the forests. Hylids are generally abundant in forests because of their ability to explore the vertical space (see Cardoso *et al.*, 1989). Historical factors may be responsible for the paucity of hylids in forests on Maracá: the ancestral stocks that gave rise to the local fauna may have been poor in forest-dwelling hylids.

## CONSERVATION

Considering its situation and size, the Ilha de Maracá as a reserve is very important for the preservation of the anuran fauna of northern Roraima. This fauna, although poor when compared to central Amazonia, includes some species endemic to the Guianan region, most of which are absent from other Brazilian regions. However, the Ilha de Maracá is almost entirely covered by forests, and consequently its maintenance as a reserve will primarily preserve those species associated with forests. Thus, to preserve the entire or most of the anuran fauna of northern Roraima, additional reserves including extensive areas of savannas and *tepuis* must be created.

## SPECIES ACCOUNTS

(numbers in parentheses refer to the records of Mark O'Shea)

### FAMILY BUFONIDAE

The family Bufonidae is cosmopolitan in tropical and temperate parts of the world, except for Oceanic, Australo-Papuan, and Madagascan regions (Frost, 1985). The family includes 25 genera, three of which occur in northern South America and are represented by several species (Frost, 1985). Only the genus *Bufo* was observed on Maracá.

### *Bufo granulosus* (Spix 1824)

*Material* – Maracá, 12 (6); Boa Vista, 3; Surumu, 15.
*Abundance* – Common.

*Diagnosis* – A medium-sized *Bufo* (♂♂ 43–51 mm, ♀ 80 mm) of the *granulosus* group; dorsum greyish-tan with scattered darker marks; skin of the dorsum, flanks, and limbs with pointed tubercles; venter light brown with darker blotches in the gular region in males; skin of the dorsum granulated.
*Habitat* – This species was observed at night in small temporary lakes and ponds in the savanna regions of Maracá, Boa Vista and Surumu.
*Reproduction* – Males of *B. granulosus* were heard calling on rainy nights from late February through mid-August; they call from shallows and from the margins of lakes and ponds.
*Distribution* – Northern South America to Argentina and southeastern Brazil (Frost, 1985).
*Remarks* – *Bufo granulosus* as recognized today may be a composite of many subspecies and/or species (see Gallardo, 1965; Frost, 1985). The population from Maracá is certainly the same described as *B. g. humboldti* by Gallardo (1965), based on specimens from São Marcos, Rio Uraricoera (see also Bokermann, 1967).

### *Bufo guttatus* (Schneider 1799)

*Material* – Maracá, (1); Surumu, 1.
*Abundance* – Rare.
*Diagnosis* – A large *Bufo* (♂ c. 80–100 mm) of the *guttatus* group; dorsum light brown; flanks and sides of head black; limbs dark brown above; venter dark brown with small yellow spots irregularly distributed.
*Habitat* – This species was observed at night in gallery and *terra firme* forests at Surumu, and was collected in *terra firme* forest at the western end of the Ilha de Maracá by O'Shea.
*Reproduction* – Young of various sizes were collected in December. In Venezuela, *B. guttatus* breeds at the beginning of the rainy season (Hoogmoed and Gorzula, 1979).
*Distribution* – Widespread in Amazonia (Frost, 1985).
*Remarks* – Rivero (1965) and Cochran and Goin (1970) analysed the distribution of *Bufo guttatus* in Venezuela and Colombia respectively.

### *Bufo marinus* (Linnaeus 1758)

*Material* – Maracá, 6; Boa Vista, 1; Pacaraima, 1; Surumu, 6.
*Abundance* – Extremely common.
*Diagnosis* – A large *Bufo* (c. 100–150 mm) of the *marinus* group; dorsum brown to brownish-tan sometimes with scattered darker spots; skin at the dorsum with several scattered tubercles; large triangular parotid glands; venter cream with brown marbling; skin of the dorsum granulated.
*Habitat* – This species was observed at night in lakes in the savannas and forests of Maracá, Boa Vista, Pacaraima and Surumu.
*Reproduction* – At Boa Vista, *B. marinus* began to breed with the first rains in February and March, and resumed in early June, with a peak in early May. Recently

metamorphosed juveniles were found from mid-March through mid-June. The clutch consists of a string of thousands of eggs. Tadpoles are black and form shoals that feed in shallows.

*Distribution* – Southern Texas to southern Amazonia in Brazil (Frost, 1985).

*Remarks* – Cei (1972) reviewed the South American *Bufo*, and Zug and Zug (1979) summarized the natural history of *B. marinus*.

### *Bufo* sp.

*Material* – Maracá, (4).
*Abundance* – Unknown.
*Diagnosis* – O'Shea found only young of this *Bufo* of the *typhonius* group on Maracá. In preservative the specimens appear as follows: dorsum brown with lighter and/or darker irregular spots; a row of pointed tubercles from the eyes through the groin, separating the dorsum from the flanks; venter cream with dark brown marbling on the belly and sometimes also in the gular region.
*Habitat* – Young of this species were collected by O'Shea in damp leaf-litter in forest close to the Cachoeira da Onça, and on a sandy beach at the Corredeira de Fumaça on the Ilha de Maracá. On both occasions they were encountered in the morning.
*Reproduction* – Young were collected in August.
*Distribution* – Unknown.
*Remarks* – This species was listed by O'Shea (1988) as *Dendrophryniscus* sp. While awaiting a revision of the *Bufo typhonius* group by M.S. Hoogmoed, I prefer to give no name to this species.

## FAMILY DENDROBATIDAE

The family Dendrobatidae is essentially neotropical, occurring from Nicaragua to southeastern Brazil (Frost, 1985). The five genera included in this family occur in northern South America (Myers, 1987; Martins and Sazima, 1989). Only one genus was found on Maracá.

### *Dendrobates leucomelas* (Steindachner 1864)

*Material* – Maracá, (1); Tepequém, 5.
*Abundance* – Rare.
*Diagnosis* – A medium-sized *Dendrobates* (*c*. 35 mm) of the *histrionicus* group; dorsum bright yellow with two deep black transversal bars and small deep black spots that also occur on the limbs; venter deep black with variable-sized yellow spots.
*Habitat* – This species was found by day in rocky habitats at the Cachoeiras de Fumaça and Purumame on the Ilha de Maracá (O'Shea, 1989), and on Serra Tepequém.

*Reproduction* – At Serra Tepequém I found males of *D. leucomelas* amongst rocks in July 1986, and at creek margins in May 1987. Local people say that this species congregates at creek margins at the beginning of the rainy season. O'Shea (1989) heard this species calling in the afternoon from horizontal fissures in large rock slabs.

*Distribution* – *Dendrobates leucomelas* is endemic to the Guianan region, where it occurs from eastern Venezuela to the Guianas (Frost, 1985).

*Remarks* – Hoogmoed and Gorzula (1979), Daly *et al.* (1987) and Zimmermann and Zimmermann (1988) provide valuable information on this species.

## FAMILY HYLIDAE

The family Hylidae occurs in the Americas, Australia, Tasmania, New Guinea and the Solomon Islands (Frost, 1985). Of nearly 40 genera included in this family, some 15 of them occur in northern South America, and are represented by many species. Three genera of hylids were found on Maracá.

### *Hyla boesemani* (Goin 1966)

*Material* – Maracá, 2.
*Abundance* – Rare.
*Diagnosis* – A medium-sized *Hyla* (*c*. 30–40 mm) of the *x-signata* group; dorsum greenish to yellowish brown, sometimes with small lighter spots; canthus rostralis with a dark brown stripe that extends from the snout to the tympanum; venter yellowish cream; snout round in dorsal and lateral views.
*Habitat* – *Hyla boesemani* was found at night in swamps and ponds in savannas on Maracá.
*Reproduction* – I heard males of *H. boesemani* calling from water hyacinths (*Eichhornia* sp.) in a swamp in July.
*Distribution* – Guyana, Surinam, southern Venezuela and Amazonas, Pará (Frost, 1985) and Roraima, Brazil.
*Remarks* – Information on this species is found in Hödl (1977).

### *Hyla crepitans* (Wied 1824)

*Material* – Maracá, 2 (3); Pacaraima, 13.
*Abundance* – Common.
*Diagnosis* – A medium-sized *Hyla* (*c*. 50–60 mm) related to the *circundata* and *boans* groups; dorsum generally brown by night and green by day; venter cream; iris golden with a blue stripe externally; prepollical spines in males.
*Habitat* – This species was found at night in small to large temporary or permanent ponds in the savannas and forest borders of Maracá and Serra Pacaraima.

*Reproduction* – Males of *H. crepitans* were heard calling from May to August. They called from the ground, or perched at low levels on plants at pond margins. The clutch consists of a roughly round monolayer film of eggs deposited on the water surface.
*Distribution* – Central America through northeastern Brazil (Frost, 1985).
*Remarks* – Kluge (1979) reviewed the 'gladiator frogs' of Middle America and Colombia, including *H. crepitans*.


*Hyla fuscomarginata* (A. Lutz 1925)

*Material* – Maracá, 1 (a male whose calls were recorded); Boa Vista, 1.
*Abundance* – Common.
*Diagnosis* – A small *Hyla* (♂♂ 19–23 mm) of the *staufferi* group; snout acuminate; dorsum yellowish tan with longitudinal brown stripes; venter whitish cream.
*Habitat* – This species was found at night in temporary and permanent ponds in the savannas of Maracá and Boa Vista.
*Reproduction* – Males of *H. fuscomarginata* were heard from May through August. They called from low heights on plants at pond margins.
*Distribution* – Southern Venezuela through southeastern Brazil (Frost, 1985; pers. obs.).
*Remarks* – Individuals of this species from Venezuela were described as *Ololygon trilineata* by Hoogmoed and Gorzula (1979).


*Hyla geographica* (Spix 1824)

*Material* – Maracá, (10).
*Abundance* – Rare.
*Diagnosis* – A medium to large *Hyla* (♂♂ 44–48 mm, ♀♀ 54–56 mm) of the *geographica* group; dorsum reddish brown sometimes with darker irregular shaped blotches, and/or a mid-dorsal longitudinal stripe, and/or small white spots; flanks black speckled with white; transversal dark brown stripes on the posterior limbs; venter yellowish cream; bones green.
*Habitat* – This species was found at night in forest streams on Maracá (O'Shea, 1988).
*Reproduction* – No data available.
*Distribution* – Tropical South America east of the Andes (Frost, 1985).
*Remarks* – Duellman (1973) reviewed the *Hyla geographica* group.


*Hyla microcephala* (Cope 1886)

*Material* – Maracá, 29 (10); Pacaraima, 4; Tepequém, 9.
*Abundance* – Very common.

*Diagnosis* – A small *Hyla* (♂♂ 19 mm, ♀ 25 mm) of the *microcephala* group; dorsum reddish brown to red; limbs orange–yellow; venter cream; vocal sac light green.
*Habitat* – This species was found at night in temporary and permanent water bodies with emergent sedges and grasses on Maracá, and at Pacaraima and Serra Tepequém.
*Reproduction* – Males of *Hyla microcephala* called throughout the rainy season (May to September). Calling males perched low (up to 1 m) on sedges and grasses.
*Distribution* – Southern Mexico throughout Central America and South America east of the Andes to southeastern Brazil (Frost, 1985).
*Remarks* – Duellman (1974) briefly reviewed *Hyla microcephala*.

## *Hyla rubra* (Laurenti 1768)

*Material* – Maracá, 14 (12); Boa Vista, 11; Surumu, 1.
*Abundance* – Very common.
*Diagnosis* – A small to medium-sized *Hyla* (♂♂ 35–38 mm, ♀ 37–40 mm) of the *rubra* group; dorsum highly variable, generally brownish tan with irregular darker blotches; venter white with gular region light red in males; limbs reddish cream below; posterior surfaces of thighs dark brown to black with yellow spots.
*Habitat* – *Hyla rubra* was found at night in temporary and permanent ponds and lakes in the savannas of Maracá, Boa Vista and Surumu.
*Reproduction* – Males of *H. rubra* were heard calling throughout the rainy season (May to September). Males called from the ground or, rarely, from low heights on plants at pond and lake margins. Eggs were deposited as a monolayer film on the water surface in shallows. The first tadpoles were observed in late May and recently metamorphosed juveniles in July and August.
*Distribution* – Trinidad, St Lucia and Panama throughout Amazonia to eastern Brazil (Frost, 1985).
*Remarks* – *Hyla rubra* as known today is possibly a composite of many species (Frost, 1985).

## *Hyla wavrini* (Parker 1936)

*Material* – Maracá, (2).
*Abundance* – Apparently rare.
*Diagnosis* – A large *Hyla* of the *boans* group (♂♂ 89–113 mm, ♀♀ 75–81 mm, Hoogmoed, 1990); dorsum greyish to greenish tan, sometimes bearing a reddish brown vertebral line and/or irregular darker spots or stripes; flanks cream with transverse dark brown bars; venter cream, gular region greyish brown with dark lines or spots.
*Habitat* – This species was found at night by O'Shea in riverine forest around Maracá.
*Reproduction* – No reproductive activity was observed. However, *H. wavrini* seems to reproduce throughout the year in several Amazonian localities (Martins and Moreira, in press).

*Distribution* – Amazonian Brazil and Venezuela (Hoogmoed, 1990; Martins and Moreira, in press).
*Remarks* – *H. wavrini* has been considered a synonym of *H. boans* until recently.

## *Hyla* sp.

*Material* – Maracá, (3).
*Abundance* – Apparently rare.
*Diagnosis* – A medium-sized *Hyla* (♂♂ 39–41 mm) related to *H. marmorata*; in preservative the dorsum is brown with irregularly distributed dark brown spots and bars; upper surface of thighs with three dark brown bars and spots; chin white, bordered by brown reticulations; belly and chest cream with scattered brown reticulations.
*Habitat* – O'Shea found this species at night in a *baixada* (damp depression in the forest) close to the Cachoeira da Onça on the Ilha de Maracá, where it was sitting on leaves up to 2 m above the ground.
*Reproduction* – No reproductive activity was observed.
*Distribution* – Unknown (see remarks).
*Remarks* – This species is similar to *H. marginata* in dorsal pattern and size. However, it differs from the latter and related species in several other characteristics, indicating that it may be an undescribed species.

## *Phrynohyas venulosa* (Laurenti 1786)

*Material* – Maracá, 6 (3).
*Abundance* – Common (see Reproduction).
*Diagnosis* – A large *Phrynohyas* (♂♂ 65–74 mm); paired lateral vocal sacs; dorsum brownish tan with irregular darker marks; flanks and anterior surfaces of limbs cream reticulated with brown; venter cream.
*Habitat* – This species was found at night on Maracá in swamps, ponds, lakes and streams in forests, or savannas close to forests.
*Reproduction* – *Phrynohyas venulosa* is an explosive breeder; males were heard only on 9–11 May 1986. Hundreds of males congregated in a swamp during these three days. Males called from the ground or from the vegetation up to 3 m.
*Distribution* – Mexico to southern Brazil including Trinidad and Tobago (Frost, 1985).
*Remarks* – Duellman (1971) reviewed the South American species of the genus *Phrynohyas*.

## *Phyllomedusa hypocondrialis* (Daudin 1802)

*Material* – Maracá, (1, photographed by W. Milliken); Pacaraima, 9; Tepequém, 5.
*Abundance* – Apparently rare.

*Diagnosis* – A medium-sized *Phyllomedusa* (♂ 48 mm) of the *hypocondrialis* group; dorsum green; anterior and posterior surfaces of thighs with dark transversal bars; venter cream.

*Habitat* – On Pacaraima and Tepequém, *P. hypocondrialis* was found in permanent and temporary pools on mountain slopes. The habitat in which it was found on Maracá was not recorded.

*Reproduction* – Males were heard calling in July on both Pacaraima and Tepequém. They called from low (1–2 m) in the marginal vegetation. Pyburn and Glidewell (1971) provided information on the reproductive biology of this species in Colombia.

*Distribution* – South America east of the Andes, from the Guianas to Paraguay, Argentina and southeastern Brazil (Frost, 1985).

## FAMILY LEPTODACTYLIDAE

The family Leptodactylidae is neotropical, occurring from southern North America to southern South America, including the West Indies (Frost, 1985). Of the 51 genera of leptodactylids, some 20 occur in northern South America with hundreds of species (Frost, 1985). Five leptodactylid genera were found on Maracá.

### *Adenomera* sp.

*Material* – Maracá, 1.
*Abundance* – Moderately common.
*Diagnosis* – A small *Adenomera* (♂ 23 mm); dorsum greyish brown with darker small spots; transversal dark brown stripes on the limbs; venter cream.
*Habitat* – *Adenomera* sp. was found by day within the leaf-litter in forests on Maracá.
*Reproduction* – No data available.
*Distribution* – Unknown (see Remarks).
*Remarks* – This species could not be assigned to any known species of *Adenomera*, reviewed by Heyer (1973).

### *Leptodactylus bolivianus* (Boulenger 1898)

*Material* – Maracá, 16 (12); Boa Vista, 2; Pacaraima, 2.
*Abundance* – Common.
*Diagnosis* – A large *Leptodactylus* (♂♂ 71–73 mm, ♀♀ 74–81 mm) related to the *ocellatus* group; two longitudinal glandular folds on the dorsum; dorsum brown with dark brown folds; venter cream with brown reticulation in the gular region; dark brown transversal bars on thighs and tibia; males with robust forelimbs and two pairs of black spines in each hand.
*Habitat* – This species was found at night in lakes in savannas close to forests on Maracá, and at Boa Vista and Serra Pacaraima. O'Shea (1988) also observed this species in *terra firme* forest and riverine habitats.

*Reproduction* – *Leptodactylus bolivianus* probably breeds from July to September, when males congregate at lake margins.
*Distribution* – South America south of Colombia, including the Guianas and Trinidad (Hoogmoed and Gorzula, 1979; Frost, 1985).
*Remarks* – Gallardo (1964) reviewed *L. bolivianus*.


**Leptodactylus fuscus (Schneider 1799)**

*Material* – Maracá, 9 (3); Boa Vista, 15; Surumu, 10.
*Abundance* – Very common.
*Diagnosis* – A small *Leptodactylus* (♂♂ 36–40 mm, ♀♀ 37–43 mm) of the *fuscus* group; six longitudinal glandular folds on the dorsum; dorsum greyish brown with darker blotches, sometimes with a lighter mid-dorsal longitudinal stripe (polymorphic); venter white; gular region with dark reticulations at each side in males.
*Habitat* – This species was found at night in ponds and lakes in the savannas of Maracá, Boa Vista and Surumu.
*Reproduction* – Males of *L. fuscus* were heard from the first rains in February and March through to mid-June, but clutches were found only from early May to mid-June. Males called in groups of three to five individuals on the ground at pond and lake margins. Clutches consist of eggs embedded in white foam and are deposited in subterranean nests dug by males at pond margins. Heavy rains or pond water flood these nests, allowing tadpoles to reach the pond. Tadpole development lasts nearly three weeks. A detailed description of the reproductive biology of *L. fuscus* is found in Martins (1988).
*Distribution* – Panama throughout South America east of the Andes to southern Brazil (Frost, 1985).
*Remarks* – Heyer (1978) reviewed the *fuscus* group of the genus *Leptodactylus*.


**Leptodactylus knudseni (Heyer 1972)**

*Material* – Maracá, 2 (1).
*Abundance* – Rare.
*Diagnosis* – A large *Leptodactylus* (♂ 127 mm) of the *pentadactylus* group; dorsum reddish brown sometimes with small darker blotches; triangular dark brown bars on the lips; males with a spine in each hand; venter cream with irregular brown reticulations.
*Habitat* – This species was found at night in the leaf-litter in forests on Maracá.
*Reproduction* – A male *L. knudseni* was heard in May; it called from a hole within roots in the forest floor.
*Distribution* – Amazon basin, Colombia, Venezuela and the Guianas (Frost, 1985).
*Remarks* – Heyer (1979) reviewed the *L. pentadactylus* group, including *L. knudseni*, and Hero and Gallati (1990) provided the characteristics that distinguish the latter species from *L. pentadactylus*.

### *Leptodactylus macrosternum* (Miranda Ribeiro 1926)

*Material* – Maracá, 22 (2); Boa Vista, 30; Surumu, 7.
*Abundance* – Common.
*Diagnosis* – A large *Leptodactylus* (♂♂ 74–80 mm, ♀♀ 76–83 mm) of the *ocellatus* group; ten longitudinal glandular folds on the dorsum, the second pair sometimes indistinct; dorsum greyish brown with darker blotches; venter cream with brown reticulations below the thighs and ventrolateral and gular regions; males with robust forelimbs and a pair of spines in each hand.
*Habitat* – This species was found at night in temporary and permanent lakes in the savannas of Maracá, Boa Vista and Surumu.
*Reproduction* – Males of *L. macrosternum* were heard in June and July. Males called from shallows, and eggs were embedded in white foam and deposited over the shallow waters. Tadpoles formed shoals that were guarded by their mothers during the initial stages of development.
*Distribution* – Amazonia to southern Brazil (Frost, 1985).
*Remarks* – Gallardo (1964) reviewed the species related to *L. ocellatus*, including *L. macrosternum*.


### *Leptodactylus mystaceus* (Spix 1824)

*Material* – Maracá, 8 (6).
*Abundance* – Moderately common.
*Diagnosis* – A medium-sized *Leptodactylus* (♂♂ 43–47 mm) of the *fuscus* group; dorsum reddish brown sometimes with darker blotches; transversal darker bars on the thighs, tibia, and feet; white supralabial stripe; canthus rostralis with a black stripe extending from the tip of the snout to the tympanum; venter cream with dark brown reticulation in the gular region of males.
*Habitat* – This species was found at night in stream valleys in *terra firme* forests and gallery forests on Maracá.
*Reproduction* – Males of *L. mystaceus* were heard from May to July. Males called from small constructed nests below leaf-iitter.
*Distribution* – Amazonia through northern Paraguay (Frost, 1985).
*Remarks* – O'Shea (1989) cited *L. mystaceus* and *L. amazonicus* for Maracá, but these are synonymous (see Heyer, 1983). Heyer (1978) reviewed the *L. fuscus* group, including *L. mystaceus*.


### *Leptodactylus* sp.

*Material* – Maracá, 4 (10).
*Abundance* – Common.
*Diagnosis* – A small to medium-sized *Leptodactylus* (♂♂ 32–34 mm, ♀♀ 37–39 mm) of the *melanonotus* group; interrupted glandular folds on the dorsum;

dorsum dark brown with a triangular darker spot between and behind the eyes; narrow cream bars on the lips; venter cream with brown reticulation; males with a pair of spines in each hand.

*Habitat* – *Leptodactylus* sp. was found at night in streams and swampy areas in forests on Maracá.

*Reproduction* – Males of this *Leptodactylus* sp. were heard in October 1985 and August and September 1986.

*Distribution* – Unknown (see Remarks).

*Remarks* – O'Shea (1989) assigned this species to the *Leptodactylus wagneri* complex, and also cited *L. podicipinus* (the name I have used for this species in previous lists). However, while awaiting a revision of the *L. wagneri* complex by W.R. Heyer, I prefer to give no name to this species. Heyer (O'Shea, pers. comm.) suspects that there are (at least) two species of the *L. wagneri* complex in the material which he examined from Maracá. However, when examining some of the same specimens I could not distinguish more than one species.

### *Lithodytes lineatus* (Schneider 1799)

*Material* – Maracá, 1 (1).

*Abundance* – Extremely rare.

*Diagnosis* – A medium-sized frog (♂ c. 40 mm) of the monotypic genus *Lithodytes*; dorsum and flanks black with a cream dorsolateral stripe extending from the tip of the snout to the groin; iris reddish brown; upper surface of limbs light brown with irregular black spots; bright red flash-marks in the inguinal region; venter greyish brown with cream spots laterally.

*Habitat* – This species was found by day in a nest of *Atta* ants in *terra firme* forest on Maracá, and O'Shea recorded it by night in a swamp in the forest.

*Reproduction* – A male *L. lineatus* was heard calling from a nest of *Atta* ants in the morning of 6 July 1986.

*Distribution* – Amazonia, Venezuela and the Guianas (Frost, 1985).

*Remarks* – Schlüter and Regös (1981) and Regös and Schlüter (1984) provided valuable information on *L. lineatus*.

### *Physalaemus ephippifer* (Steindachner 1864)

*Material* – Maracá, 32 (45).

*Abundance* – Very common.

*Diagnosis* – A small *Physalaemus* (♂♂ 26–28 mm) of the *cuvieri* group; dorsum brown with darker spots; flanks dark brown from loreal to inguinal regions; transversal dark brown bars on the upper surface of limbs; venter cream with brown reticulations in the pectoral and gular regions.

*Habitat* – This species was found by day in the leaf-litter, and by night in lakes in the forests of Maracá and Pacaraima.

*Reproduction* – Males of *P. ephippifer* were heard calling from the shallows of a lake in May.

*Distribution* – Lower Amazonian region and the Guianas (Frost, 1985).
*Remarks* – O'Shea (1989) called this species '*Physalaemus* sp.'. Hoogmoed and Gorzula (1979) listed some specimens probably of this species as *P. enesefae* Heatwole, Solano & Heatwole, 1965, but, based on the original descriptions, *P. enesefae* is probably a synonym of *P. ephippifer* (see also Frost, 1985).

### *Pseudopaludicola boliviana* (Parker 1927)

*Material* – Maracá, 39 (23); Surumu, 25.
*Abundance* – Very common.
*Diagnosis* – A diminutive species of *Pseudopaludicola* (♂♂ 13–15 mm); dorsum dark brown with either small darker spots or a mid-dorsal longitudinal cream to reddish brown stripe (polymorphic); transversal dark bars on the lips and upper surfaces of thighs and feet; venter cream with pale reticulations at the gular region and below the thighs; large conical tubercle on the heel.
*Habitat* – This species was observed at night in ponds, lakes and swamps in the savannas of Maracá, Boa Vista and Surumu, and by day in the gallery forests.
*Reproduction* – At Boa Vista, males of *P. boliviana* were heard calling in June and July. They called from the ground within the marginal vegetation of water bodies.
*Distribution* – Guyana, Surinam, Brazil (Roraima), Venezuela, Colombia, Bolivia and Paraguay (Frost, 1985; pers. obs.).
*Remarks* – O'Shea (1989) called this species *Pseudopaludicola pusillus* (following an identification made by W.R. Heyer), and cited another species of this genus (*Pseudopaludicola* sp.). However, considering the extensive polymorphism observed in this species, I suspect that there is only one species of *Pseudopaludicola* on Maracá. Lynch (1989) reviewed the species of *Pseudopaludicola* that occur in northern South America.

## FAMILY MICROHYLIDAE

The family Microhylidae is nearly cosmopolitan in tropical and temperate regions, except for the Palaearctic region and the West Indies (Frost, 1985). Of the 61 known genera eight occur in northern South America, represented by several species. Only one genus was found on Maracá.

### *Elachistocleis ovalis* (Schneider 1799)

*Material* – Maracá, 2 (10); Boa Vista, 2.
*Abundance* – Common.
*Diagnosis* – A small species of the genus *Elachistocleis* (♂♂ 22–24 mm, ♀♀ 25–28 mm); dorsum dark grey with small greyish punctuations scattered throughout, sometimes with a narrow whitish longitudinal stripe; snout acuminate in dorsal and lateral view; orange flash-marks at the inguinal region; venter grey with small yellow spots.

*Habitat* – *Elachistocleis ovalis* was observed at night in lakes in the savannas of Maracá and Boa Vista, and also in the forest on Maracá.
*Reproduction* – At Boa Vista, males of *E. ovalis* were heard from May to July. They called from shallows or from the margins. In May a male of this species was found calling from a small pond in a dry stream inside the forest.
*Distribution* – Panama to southern Argentina (Frost, 1985, but see Remarks).
*Remarks* – O'Shea (1989) identified this animal to the genus level only (*Elachistocleis* sp.). Following Hoogmoed and Gorzula (1979), who found this species at El Manteco, Venezuela, I tentatively identified it as *Elachistocleis ovalis* pending a revision of the genus.


## FAMILY PSEUDIDAE

The family Pseudidae includes four species that occur exclusively in South America (Frost, 1985). Its two genera both occur in northern South America, and both were found on Maracá.


### *Lysapsus limellus* (Cope 1862)

*Material* – Maracá, 1 (7); Boa Vista, 60; Surumu, 5.
*Abundance* – Extremely common.
*Diagnosis* – A small pseudid of the genus *Lysapsus* (♂♂ 17–21 mm, ♀♀ 23–27 mm); dorsum green with brown longitudinal stripes; no webbing in the hands; feet fully webbed; eyes facing upward; transversal brown bars on the limbs; venter yellowish green.
*Habitat* – This species was found both at night and during the day, in lakes and swamps in the savannas of Maracá, Boa Vista and Surumu.
*Reproduction* – Males of *L. limellus* were heard throughout the year. They called from floating plants and from the margins. Tadpoles were also observed throughout the year.
*Distribution* – The Guianas south to Paraguay (Frost, 1985).
*Remarks* – Gallardo (1961) reviewed the species of *Lysapsus*.


### *Pseudis paradoxa* (Linnaeus 1758)

*Material* – Maracá, 1 (2); Boa Vista, 7; Surumu, 1.
*Abundance* – Very common.
*Diagnosis* – A medium to large species (♂♂ 42–52 mm, ♀♀ 53–57 mm) of the monotypic genus *Pseudis*; dorsum green with irregular dark brown marks; venter and posterior surfaces of thighs cream with brown reticulations; eyes facing upward; hands without webbing; feet fully webbed.
*Habitat* – *Pseudis paradoxa* was found both by night and by day in lakes in the savannas of Maracá, Boa Vista and Surumu.

*Reproduction* – Males of *P. paradoxa* were heard from March to July, with a peak in June and July. They called from everywhere in the water. Tadpoles were found from May to July.
*Distribution* – Northern South America south to Paraguay (Frost, 1985).
*Remarks* – Gallardo (1961) reviewed the genus *Pseudis*.

## ADDITIONAL SPECIES FROM RORAIMA

Besides the 25 species listed above, I observed 10 additional species at Boa Vista, Serra Pacaraima, Serra Tepequém, and Surumu. Most of these species may occur on Maracá, but only additional field surveys would reveal them. Data on these 10 species are summarized in the following paragraphs.

A small unidentified *Colostethus* (♂ 20 mm) was found on the rocks of a small stream on the slopes of Serra Tepequém. Also at Serra Tepequém I found the following hylids: *Hyla minuta* Peters, 1872, a small species (♂♂ 23–25 mm SVL) widespread in South America (Frost, 1985); and *Phyllomedusa bicolor* (Boddaert, 1772), a large species (♂ 110 mm SVL) widespread in Amazonia (Frost, 1985). A medium-sized, unidentified *Eleutherodactylus* (♂ 28 mm SVL) of the *fitzingeri* group was also found at Serra Tepequém.

At Boa Vista, I found two additional hylids in habitats very similar to those found on Maracá: *Hyla nebulosa* Spix, 1824, a medium-sized treefrog (♂ 33 mm SVL) of the *rostrata* group that occurs in the Guianas, eastern Amazonia, and eastern Brazil (Frost, 1985); and *Hyla exigua* (Duellman, 1986), a small species (♂ 19 mm SVL) described from the Gran Sabana of Venezuela (see Duellman, 1986). The leptodactylid *Pleurodema brachyops* (Cope, 1869) was also found in large numbers at temporary and permanent lakes and ponds in the savannas of Boa Vista and Surumu; Martins (1989) described the defensive behaviour of this species from the vicinity of Boa Vista. The large *Rana palmipes* Spix, 1824 (♂ 86 mm, ♀ 103 mm SVL), the only South American ranid, was also found in that area. This species occurs from Mexico to northern South America (Frost, 1985).

Finally, at Serra Pacaraima, I found the hylid *Hyla raniceps* (Cope, 1862), a medium-sized species (♂ 47 mm SVL) of the *albopunctata* group that is widespread in South America, and the leptodactylid *Leptodactylus longirostris* Boulenger, 1882, a medium-sized species (♂ 40 mm SVL) of the *fuscus* group that occurs in Venezuela, the Guianas and Brazilian Amazonia.

## ACKNOWLEDGEMENTS

I am especially grateful to: Celso Morato de Carvalho for encouragement and facilities to work in Roraima, and for help in field and laboratory work; Guttemberg M. de Oliveira (IBAMA, Boa Vista) for his friendship, permission to work on Maracá, and companionship during my 1987 trip to Tepequém; and Mark O'Shea for kindly providing data on Maracá frogs.

Mark O'Shea, Marcelo Gordo and my wife Silvia Egler carefully read and commented on the manuscript; their helpful suggestions are gratefully acknowledged; however, the identifications presented herein are my own responsibility. Museum comparisons were facilitated by Drs Paulo E. Vanzolini (Museu de Zoologia, Universidade de São Paulo), Ulisses Caramaschi (Museu Nacional, Rio de Janeiro), and Adão J. Cardoso (Museu de História Natural, Universidade Estadual de Campinas). My work in Roraima was partially funded by Convênio CNPq/INPA/Governo de Roraima.

## REFERENCES

Almeida, C.G. and Cardoso, A.J. (1985). Variabilidade em medidas dos espermatozóides de *Hyla fuscovaria* (Amphibia, Anura) e seu significado taxonômico. *Revista Brasileira de Biologia*, 45, 387–391.

Bokermann, W.C.A. (1967). Notas sobre a distribuição de '*Bufo granulosus*' Spix, 1824 na Amazônia e descrição de uma subespécie nova (Amphibia, Bufonidae). *Actas do Simpósio sobre a Biota Amazônica (Zoologia)*, 5, 103–109.

Cardoso, A.J., Andrade, G.V. and Haddad, C.F.B. (1989). Distribuição espacial em comunidades de anfíbios (Anura) no sudeste do Brasil. *Revista Brasileira de Biologia*, 49, 241–249.

Cei, J.M. (1972). *Bufo* of South America. In: *Evolution in the genus Bufo*, ed. W.F. Blair, pp. 82–90. University of Texas Press, Austin.

Cochran, D.M. and Goin, C.J. (1970). Frogs of Colombia. *Bulletin of the US National Museum*, 288, 1–655.

Daly, J.W., Myers, C.W. and Whittaker, N. (1987). Further classification of skin alkaloids from Neotropical poison frogs (Dendrobatidae), with a general survey of toxic/noxious substances in the Amphibia. *Toxicon*, 25, 1023–1095.

Duellman, W.E. (1971). A taxonomic review of the South American frogs, genus *Phrynohyas*. *Occasional Papers. Museum of Natural History. University of Kansas*, 4, 1–21.

Duellman, W.E. (1973). Frogs of the *Hyla geographica* group. *Copeia*, 1973, 515–533.

Duellman, W.E. (1974). A reassessment of the taxonomic status of some Neotropical hylid frogs. *Occasional Papers. Museum of Natural History. University of Kansas*, 27, 1–27.

Duellman, W.E. (1979). The South American herpetofauna: a panoramic view. In: *The South American herpetofauna: its origin, evolution, and dispersal*, ed. W.E. Duellman, pp. 1–28. Museum of Natural History. University of Kansas, Monograph No. 7.

Duellman, W.E. (1986). Two new species of *Ololygon* (Anura: Hylidae) from the Venezuelan Guyana. *Copeia*, 1986, 864–870.

Duellman, W.E. (1989). Tropical herpetofaunal communities: patterns of community structure in Neotropical rainforests. In: *Vertebrates in complex tropical systems, ecological studies, Vol. 69*, eds M.L. Harmelin-Vivien and F. Bourlière, pp. 61–88. Springer-Verlag, New York.

Frost, D. (ed.) (1985). *Amphibian species of the world*. Association of Systematic Collections, Kansas.

Gallardo, J.M. (1961). On the species of Pseudidae (Amphibia, Anura). *Bulletin of the Museum of Comparative Zoology*, 125, 111–134.

Gallardo, J.M. (1964). Consideraciones sobre *Leptodactylus ocellatus* (L.) (Amphibia, Anura) y especies aliadas. *Physis*, 24, 373–384.

Gallardo, J.M. (1965). The species *Bufo granulosus* Spix (Salientia: Bufonidae) and its geographic variation. *Bulletin of the Museum of Comparative Zoology*, 134, 107–138.

Heatwole, H., Solano, H. and Heatwole, A. (1965). Notes on the amphibians from the Venezuelan Guayanas with description of two new forms. *Acta Biologica Venezuelica*, 4 (12), 349–364.

Hero, J-M. and Gallati, U. (1990). Characteristics distinguishing *Leptodactylus pentadactylus* and *L. knudseni* in the central Amazon rainforest. *Journal of Herpetology*, 24, 226–228.

Heyer, W.R. (1973). Systematics of the *marmoratus* group of the frog genus *Leptodactylus* (Amphibia, Leptodactylidae). *Natural History Museum of Los Angeles County Contributions in Science*, 251, 1–50.

Heyer, W.R. (1978). Systematics of the fuscus group of the frog genus *Leptodactylus* (Amphibia, Leptodactylidae). *Natural History Museum of Los Angeles County Contributions in Science*, Bulletin 29, 1–85.

Heyer, W.R. (1979). Systematics of the *pentadactylus* species group of the frog genus *Leptodactylus* (Amphibia: Leptodactylidae). *Smithsonian Contributions to Zoology*, 301, 1–43.

Heyer, W.R. (1983). Clarification of the names *Rana mystacea* Spix, 1824, *Leptodactylus amazonicus* Heyer, 1978 and a description of a new species, *Leptodactylus spixi* (Amphibia: Leptodactylidae). *Proceedings of the Biological Society of Washington*, 96, 270–272.

Hödl, W. (1977). Call differences and calling site segregation in anuran species from central Amazonian floating meadows. *Oecologia*, 28, 351–363.

Hoogmoed, M.S. (1979). The herpetofauna of the Guianan region. In: *The South American herpetofauna: its origin, evolution, and dispersal*, ed. W.E. Duellman, pp. 241–279. Museum of Natural History. University of Kansas, Monograph No. 7.

Hoogmoed, M.S. (1990). Resurrection of *Hyla wavrini* Parker (Amphibia: Anura: Hylidae), a gladiator frog from northern South America. *Zoologische Mededelingen*, 64, 71–93.

Hoogmoed, M.S. and Gorzula, S.J. (1979). Checklist of the savanna inhabiting frogs of the El Manteco region with notes on their ecology and the description of a new species of treefrog (Hylidae, Anura). *Zoologische Mededelingen*, 54 (13), 183–216.

Kluge, A.G. (1979). The gladiator frogs of Middle America and Colombia – a reevaluation of their systematics (Anura: Hylidae). *Occasional Papers. Museum of Zoology, University of Michigan*, 688, 1–24.

Lescure, J. (1975). Biogéographie et écologie des Amphibiens de Guyane Française. *Comptes Rendus des Séances, Societé de Biogeographie*, 440, 68–82.

Lescure, J. (1976). Contribution à l'étude des Amphibiens de Guyane française VI. Liste préliminaire des Anoures. *Bulletin, Museum National d'Histoire Naturelle, (Paris) 3e série*, 377, 68–82.

Lescure, J. (1977). Diversité des origines biogéographiques chez les Amphibiens de la region guyanaise. *Publications des Zoologiques Laboratoires de l' école Normale Supérieure*, 9, 53–65.

Lynch, J.D. (1979). The amphibians of the lowland tropical forests. In: *The South American herpetofauna: its origin, evolution, and dispersal*, ed. W.E. Duellman, pp. 189–215. Museum of Natural History. University of Kansas, Monograph No. 7.

Lynch, J.D. (1989). A review of the Leptodactylid frogs of the genus *Pseudopaludicola* in northern South America. *Copeia*, 1989, 577–588.

Martins, M. (1988). Biologia reprodutiva de *Leptodactylus fuscus* em Boa Vista, Roraima (Amphibia: Anura). *Revista Brasileira de Biologia*, 48, 969–977.

Martins, M. (1989). Deimatic behaviour in *Pleurodema brachyops*. *Journal of Herpetology*, 23, 305–307.

Martins, M. and Moreira, G. (in press). The nest and tadpole of *Hyla wavrini* Parker (Amphibia: Anura). *Memorias, Instituto Butantan*.

Martins, M. and Sazima, I. (1989). Dendrobatídeos: cores e venenos. *Ciência Hoje*, 9 (53), 34–38.

Myers, C.W. (1987). New generic names for some Neotropical poison frogs (Dendrobatidae). *Papeis Avulsos de Zoologia*, 36, 301–306.

O'Shea, M. (1988). *The herpetofauna of Ilha de Maracá, Territory of Roraima, Brazil. An annotated checklist of known species.* Mimeographed report.

O'Shea, M. (1989). The herpetofauna of Ilha de Maracá, State of Roraima, northern Brazil. In: *Reptiles: proceedings of the UK Herpetological Society's symposium on captive breeding*, pp. 51–71. British Herpetological Society.

Pisani, G.R. and Villa, J. (1974). Guia de tecnicas de preservacion de anfibios y reptiles. Society for the Study of Amphibians and Reptiles. *Circular Herpetologica*, 2, 1–24.

Pyburn, W.F. and Glidewell, J.R. (1971). Nests and breeding behaviour of *Phyllomedusa hypocondrialis* in Colombia. *Journal of Herpetology*, 5, 49–52.

RADAMBRASIL (1975). *Levantamento de recursos naturais*. Vol. 8, Folha NA.20 Boa Vista. DNPM, Rio de Janeiro.

RADAMBRASIL (1978). *Levantamento de recursos naturais*. Vol. 18, Folha SA.20 Manaus. DNPM, Rio de Janeiro.

Regös, J. and Schlüter, A. (1984). Erste Ergebnisse zur Fortplanzungsbiologie von *Lithodytes lineatus* (Schneider, 1799) (Amphibia: Leptodactylidae). *Salamandra*, 20, 252–261.

Rivero, J.A. (1965). The distribution of Venezuelan frogs. V. The Venezuelan Guayana. *Caribbean Journal of Science*, 4, 411–420.

Rivero, J.A. (1971). Notas sobre los anfibios de Venezuela. I. Sobre los hilidos de la guayana venezolana. *Caribbean Journal of Science*, 11, 181–193.

Schlüter, A. and Regös, J. (1981). *Lithodytes lineatus* (Schneider, 1799) (Amphibia: Leptodactylidae) as a dweller in nests of the leaf cutting ant *Atta cephalotes* (Linnaeus, 1758) (Hymenoptera: Attini). *Amphibia–Reptilia*, 2, 117–121.

Staton, M.A. and Dixon, J.R. (1977). The herpetofauna of the central llanos of Venezuela: noteworthy records, a tentative checklist and ecological notes. *Journal of Herpetology*, 11, 17–24.

Zimmermann, H. and Zimmermann, E. (1988). Etho-taxonomie und zoogeographische artengruppenbildung bei Pfeilgiftfroschen (Anura: Dendrobatidae). *Salamandra*, 24, 125–160.

Zug, G.R. and Zug, P.B. (1979). The marine toad, *Bufo marinus*: a natural history resumé of native populations. *Smithsonian Contributions to Zoology*, 284, 1–58.

# 15 Social Wasps (Hymenoptera, Vespidae) of the Ilha de Maracá

ANTHONY RAW
*Universidade de Brasília, Brazil*

## SUMMARY

In four habitats studied on the Ilha de Maracá, 36 species of social wasps (Hymenoptera, Vespidae) were identified, of which 22 were recorded for the first time for Roraima, bringing the total for the State to 42 species. Individual wasp species were largely restricted to particular habitats (forest floor, small forest clearings, forest edge and savannas). Several species nested in the forest but hunted in clearings and at the forest edge. Some species of the forest floor used smaller clearings (<200 m$^2$), while the wasps which hunted at the forest edge were more common in larger ones (>300 m$^2$). The density of wasps hunting in clearings increased with clearing size. Two new species are recorded, and three species and one subspecies are recorded for the first time in Brazil. The nests of 11 species are also described, and geographical distributions are discussed.

## RESUMO

Foram identificadas 36 espécies de vespas sociais (Hymenoptera, Vespidae) em quatro habitats na Ilha de Maracá, das quais 22 foram registradas pela primeira vez em Roraima, aumentando o total do estado para 42 espécies. A maioria de espécies de vespas foram restritas a habitats particulares (chão da floresta, clareiras pequenas, beira da floresta e campos). Várias espécies construiram seus ninhos na floresta mas caçaram na beira ou nas clareiras. Algumas espécies da floresta usavam clareiras menores (<200 m$^2$), enquanto espécies que caçam na beira da mata foram mais comuns em clareiras maiores (>300 m$^2$). A densidade de vespas caçando nas clareiras aumentou de acôrdo com o tamanho da clareira. Duas espécies novas são registradas, e três espécies e uma subespécie são registradas pela primeira vez no Brasil. Os ninhos de 11 espécies são também descritos, e distribuições geográficas são discutidas.

## INTRODUCTION

Lying at the northern edge of the Amazon basin and bordered on the east, north and west by the Guiana–Venezuela Highlands, northern Roraima is an interesting biogeographical area. The Ilha de Maracá is typical of the forested parts of the region, with a seasonal climate in which there is an annual drought of four or five months (November to March).

---

*Maracá: The Biodiversity and Environment of an Amazonian Rainforest.*
Edited by William Milliken and James A. Ratter. © 1998 John Wiley & Sons Ltd.

Most animal species show preferences for particular habitat types, as for example the social wasp species of central Brazil, many of which are restricted either to forest, *cerrados* or open grasslands (Raw, 1992). (I use the word 'habitat' in the sense which includes variables such as shade, humidity and availability of types of prey and nest sites.) In addition to carrying out an inventory of the social wasps of the Ilha de Maracá, I spent two weeks between January and February 1988 investigating possible preferences for habitat and clearing size. In areas of neotropical forest cleared for shifting cultivation, social wasps have been found to hunt for insects among the crop plants (Raw, 1988). On Maracá I examined the presence of 21 species of wasps in forest clearings with natural plant regeneration, in particular investigating the influence of clearing size on the species that used them.

## THE STUDY AREA

Social wasps were collected mostly within 3 km of the Ecological Station at the eastern end of the island in the semi-evergreen forest, at the forest edge, and on the savannas. Wasps were also collected in small areas where forest had been cleared to study forest regeneration (see Figure 23.2, p. 436).

The following habitats were studied:

1. Forest floor – well shaded by the canopy. Its plant cover is mostly open in nature, with small palms, bushes, Marantaceae, etc.
2. Small forest clearings. These might have been the result of tree falls. Wider parts of the forest paths are probably ecological equivalents of clearings, and were treated as such during this study.
3. The forest edge. There is an abrupt change between tall forest and open savanna over the space of 2–5 m. Several species of plants occur only at this boundary.
4. Savannas. These are mostly areas subject to seasonal flooding. They are primarily grasslands, with clumps of shrubs and trees on slightly higher ground above the flood level.

Unfortunately it was not possible to collect wasps in the canopy.

## METHODS

### GENERAL COLLECTING AND STUDY OF HABITAT PREFERENCES

General searching and collecting were pursued with the aim of discovering the largest number of species in each habitat, and the locations of their nests. In addition to hand-netting wasps, the nests and occupants of 10 species were collected. Collecting was not carried out quantitatively by time or by area, so the information presented in Table 15.1 shows only the presence or absence of wasps and their nests in the habitats studied.

## STUDY OF PREFERENCES FOR CLEARING SIZE

Wasps were also collected in areas of forest which had been cleared and marked to study forest regeneration. The clearings were located in a 500 × 400 m relatively uniform piece of forest, with a canopy height of 30–35 m. The clearings were of four size groups: (I) 36, 38 and 44 m$^2$; (II) 123, 133 and 176 m$^2$; (III) 312, 332 and 380 m$^2$; and (IV) three clearings of 2500 m$^2$ each. In order to obtain comparative data, the three smallest clearings (size I) were sampled a second time four days later, giving a total for these clearings of 236 m$^2$. Only half the area of each clearing in size III was sampled (area sampled 512 m$^2$), and a quarter of each in size IV (1875 m$^2$). All the sample areas were searched at a rate of 10 m$^2$ per minute, excluding time spent collecting specimens and marking the location of nests. For comparison, the data are presented in Table 15.2 as the number of wasps per 100 m$^2$.

# RESULTS

Thirty-six species of wasps were collected in the four habitats. Details of these species are presented at the end of this chapter.

## GEOGRAPHICAL DISTRIBUTIONS

Four distribution patterns are discernible among the species and subspecies present on Maracá:

1. *Widespread taxa* – More than half (21) of the species range over most of tropical South America, south to Bolivia and central Brazil and beyond. Of these *Polistes brevifissus, Mischocyttarus labiatus, Pseudopolybia vespiceps, Angiopolybia pallens, Stelopolybia angulata angulata, S. fulvofasciata, S. testacea, Apoica flavissima, A. pallida, Polybia liliacea, P. dimidiata, P. ignobilis, P. rejecta, Chatergus chartarius* and *Synoeca surinama* are more or less common species, while *Mischocyttarus carbonarius carbonarius* is more rare. *Stelopolybia cajennensis, Apoica thoracica* and *Protopolybia exigua* also occur in Central America, while *Stelopolybia multipicta* and *Polybia sericea* range south to southern Brazil, Paraguay and Argentina.

2. *Taxa ranging over most of northern South America* – *Brachygastra smithii* is largely a northern species, ranging from Mexico through Central and northern South America, and south to Amazonian Ecuador, Peru, Bolivia and the neighbouring state of Acre in Brazil. The records of this species for the states of Mato Grosso and São Paulo mentioned by Richards (1978) need to be verified. *Polybia occidentalis venezuelana* might also be included here, as it occurs in Venezuela, Colombia and eastern Peru.

3. *Taxa of the Guiana Highlands* – Two species and two subspecies found on Maracá are confined to the Guiana Highlands and their outlying hills. *Polybia dimorpha* occurs in Surinam and French Guiana, but *Mischocyttarus alboniger*

was apparently known previously only from the Nassau Mountains, Surinam. *Polistes versicolor kaiteurensis* and *Mischocyttarus injucundus bimarginatus* are restricted to the extreme north of Roraima and the neighbouring part of Guyana. All but *Polistes versicolor kaiteurensis* are new records for Brazil.

4. *Species occurring around the rim of the Amazon basin* – *Stelopolybia ornata*, *Metapolybia unilineata* and *Synoeca virginea* occur along the northern and western edges of the Amazon, but do not enter the central portion of the basin. *Mischocyttarus surinamensis surinamensis* and *M. prominulus* generally follow this pattern, apparently being more common around the edge, though they have been collected occasionally in the basin as well. *M. s. surinamensis* is found in French Guiana, Surinam, Guyana, Brazil (Pará, Roraima (this study), Amazonas, Acre, Rondônia (pers. obs.)), eastern Peru and Bolivia. *M. prominulus* occurs in Guyana, Brazil (Pará and Amazonas) and eastern Ecuador. The records except my own are from Richards (1978). Professor Richards also collected *M. s. surinamensis* in Mato Grosso.

There are insufficient records of *Mischocyttarus granadaensis*, *M. maracaensis* and *Polybia roraimae* to comment on their distributions, while the *Leipomeles* species occurring on Maracá has not been identified.

## HABITAT PREFERENCES

The habitats in which wasp species were encountered are shown in Table 15.1. The question of *Polybia occidentalis* is considered below. Of the 36 species, 22 were found only in one type of habitat and 12 were found in two types, while *Polybia dimidiata* and *Stelopolybia fulvofasciata* were found in three different types. Three, six, nine and four species were seen only in forest floor, clearing, forest edge and savanna habitats respectively. Five were found on the forest floor as well as in clearings and at the forest edge. None of the three species inhabiting the savanna was seen near the forest.

The presence of three species was determined by the occurrence of their abandoned nests. A nest of a *Leipomeles* species and one of *Protopolybia exigua* (?) were discovered on the undersides of leaves of small trees at the forest edge. A nest of *Chartergus chartarius* was also found at the forest edge.

Three nests of *Polybia roraimae* were discovered on the undersides of leaves of *Astrocaryum gynacanthum* Mart. (Palmae), *Ischnosiphon arouma* (Aubl.) Koern. (Marantaceae), and an unidentified shrub in the forest and in clearings. One nest of *Polybia occidentalis venezuelana* was located on a vine, *Uncaria guianensis* (Aubl.) Gmel. (Rubiaceae), in the savanna. This was the only example discovered of a species inhabiting both the forest and savanna (see Discussion).

## PREFERENCES FOR CLEARING SIZE

Studies of the densities of the 19 species of social wasps hunting in four sizes of forest clearings (Table 15.2) showed that their density increased with clearing size

**Table 15.1.** The social wasps encountered in four habitats on the Ilha de Maracá

| Species | Forest floor | Forest clearing | Forest edge | Savanna |
|---|---|---|---|---|
| *Angiopolybia pallens* (Lepeletier) | H | H | | |
| *Apoica flavissima* Van der Vecht | | | H | |
| *A. pallida* (Olivier) | | | N | |
| *A. thoracica* du Buysson | | | H | |
| *Brachygastra smithii* (de Saussure) | | | HN | |
| *Chartergus chartarius* (Olivier) | | | n | |
| *Leipomeles* sp. | | | n | |
| *Metapolybia unilineata* (von Ihering) | | | HN | |
| *Mischocyttarus alboniger* Richards | N | | | |
| *M. carbonarius carbonarius* (de Saussure) | H | | | |
| *M. granadaensis* Zikan | | N | | |
| *M. injucundus bimarginatus* Cameron | | | HN | |
| *M. labiatus* (F.) | H | H | | |
| *M. maracaensis* Raw | | H | HN | |
| *M. prominulus* Richards | | * | | |
| *M. surinamensis surinamensis* (de Sausssure) | | | H | |
| *Polistes brevifissus* Richards | | | | HN |
| *P. versicolor kaieteurensis* Bequaert | | HN | HN | |
| *Polybia dimidiata* (Olivier) | HN | H | H | |
| *P. dimorpha* Richards | | | HN | |
| *P. ignobilis* (Haliday) | H | H | | |
| *P. liliacea* (F.) | | H | H | |
| *P. occidentalis venezuelana* Soika | | | | HN |
| *P. rejecta* (F.) | H | N | | |
| *P. roraimae* Raw | N | HN | | |
| *P. sericea* (Olivier) | | | | H |
| *Protopolybia exigua* (?) (de Saussure) | | | n | |
| *Pseudopolybia vespiceps* (de Saussure) | | | | H |
| *Stelopolybia angulata* (F.) | H | | | |
| *S. cajennensis* (F.) | | H | HN | |
| *S. fulvofasciata* (Degeer) | H | H | H | |
| *S. multipicta* (Haliday) | | H | H | |
| *S. ornata* (Ducke) | | H | | |
| *S. testacea* (F.) | | H | | |
| *Synoeca surinama* (L.) | | H | H | |
| *S. virginea* (F.) | | H | H | |

H = wasps in flight and presumably hunting; N = presence of occupied nest; n = presence of abandoned nest; * = one male collected.

($P<0.01$ using Yates' exact test). The highest density was 9.3 wasps per 100 m² in clearing size IV, while densities in sizes III, II and I were 96, 65 and 31% of this number respectively.

The species of wasps were separated into three groups, depending on the size of the clearings in which they were seen. Those of the forest floor (Table 15.2: Group A) used only the forest clearings of <200 m² (sizes I and II). Three species (Group B) were collected in three or four sizes of clearings, while those which hunt at the

Table 15.2. The densities of 19 social wasps hunting in four sizes of forest clearings on the Ilha de Maracá. For comparison, the data are given as the number of wasps per 100 m$^2$. To obtain comparative data the three smallest clearings (size I) were sampled a second time, four days later, and only half the area of each clearing in size III and a quarter of each in size IV were sampled. All the areas studied were examined at a rate of 10 m$^2$ per minute

| Clearing size | I | II | III | IV |
|---|---|---|---|---|
| Range in size (m$^2$) | 36–44 | 123–176 | 312–380 | 2500 |
| Area samples (m$^2$) | 236 | 432 | 512 | 1875 |
| Group A | | | | |
| Angiopolybia pallens | 0.8 | 0.9 | | |
| Mischocyttarus granadaensis | | 0.2n | | |
| M. labiatus | 0.4 | 0.2 | | |
| M. maracaensis | | 0.2 | | |
| Group B | | | | |
| Polistes versicolor kaieteurensis | | 1.9n | 1.4 | 1.0n |
| Polybia dimidiata | 1.7 | 1.2 | 0.6 | 0.6 |
| Stelopolybia fulvofasciata | | 1.4 | 1.2 | 0.9 |
| Group C | | | | |
| Brachygastra smithii | | | | 0.6n |
| Mischocyttarus injucundus bimarginatus | | | | 0.5n |
| M. surinamensis surinamensis | | | 0.6 | 0.1n |
| Polybia dimorpha | | | | 0.6 |
| P. ignobilis | | | | 0.2 |
| P. liliacea | | | 0.4 | 0.5 |
| P. rejecta | | | 0.2 | 0.1 |
| P. roraimae | | | 12.5 | 1.2n |
| Stelopolybia cajennensis | | | | 0.7 |
| S. multipicta | | | 0.4 | |
| Synoeca surinama | | | 1.2 | 1.7 |
| S. virginea | | | 0.4 | 0.6 |
| Total density | 2.9 | 6.0 | 8.9 | 9.3 |

n = Species whose occupied nests were located in the sampled area.

forest edge (Group C) used only the clearings of >300 m$^2$ (sizes III and IV) ($P<0.001$ using Yates' exact test).

The nest of *Polybia dimidiata* was located in the forest 3 m from one of the largest clearings, while two nests of *Polybia liliacea* were located in the canopy 550 m and 700 m to the west and northwest of the clearings. In addition to the various wasp species seen hunting, an occupied nest of the nocturnal species *Apoica pallida* (Olivier) was discovered in one of the largest clearings.

## DISCUSSION

Twenty-two of the 36 species collected on the Ilha de Maracá are new records for Roraima. In his major publication on tropical social wasps, Richards (1978) listed

20 species for the State, so the total number of species now recorded is 42. The following six species mentioned by Richards were not seen on Maracá: *Polistes canadensis* (L.), *Polistes billardieri* F., *Apoica pallens* (F.), *Brachygastra lecheguana* (Latreille), *Stelopolybia pallipes* (Olivier) and *Polybia belemensis* Richards. The first four are species which frequent savannas, of which there are few on the island (see Milliken and Ratter, Chapter 5). Collecting on other parts of the island and during the rainy season should reveal additional species.

Collecting throughout the territory, particularly in different habitats, will undoubtedly increase the number of recorded species substantially, as is suggested by the comparative numbers of 120 and 137 species recorded in the neighbouring states of Amazonas and Pará (data from Richards, 1978). Three species and one subspecies were previously known only from northern South America and are new records for Brazil. Two species, *M. maracaensis* and *P. roraimae* were first collected during the present survey and have been described elsewhere (Raw, in press a). The nests described below are mostly typical of the genera. However, the small projections on the apex of the cocoon of *Mischocyttarus alboniger* are unusual. Nests encountered of *M. injucundus bimarginatus*, *M. granadaensis*, *M. maracaensis* and *M. alboniger* were founded by two or more females.

The species diversity of social wasps on Maracá is not considered to be high when compared to other Amazon forests (pers. obs.) and it is suspected that the activities of the numerous raiding columns of the army ant, *Eciton burchelli*, limit the number of the wasps on the island. Of the 29 species investigated on Maracá, 12 used the forest, 24 used the forest edge and five used the savanna. Furthermore, four were restricted to forest, 13 to the forest edge and one to savanna. Much the same pattern seen on the island has been recorded elsewhere in Brazil (Raw, 1988, 1992). The situation is illustrated by the presence of the two closely related species of *Polybia* (*Myrapetra*). Apparently they are confined to different habitats; *P. o. venezuelana* occurring in savanna and *P. roraimae* occurring in forest. There is no evidence to date that sympatric species have substantially different diets (Raw, 1988). However, one study conducted near Brasília (Raw, in press b) suggests that at least one group of neotropical social wasps comprises a 'founder-controlled community' (*sensu* Begon *et al.*, 1996), where none of the species assumes a dominant role and all are subject to a 'competitive lottery'. The species occupy spatial gaps as they become available and any species may occupy any gap regardless of which species had vacated it.

Both colour phases of *Pseudopolybia vespiceps* occur on Maracá. Richards (1978, p. 225) considered them to be morphs of a single species 'which possibly are really two species but do not differ in any way structurally'. The distribution of the morph *vespiceps* is more southerly and Richards cited records in Brazil in the States of Bahia, Espirito Santo and Rio de Janeiro. It is also found in Mato Grosso and the Federal District (pers. obs.). The morph *testacea* Ducke is more northerly, occurring in Guyana, Surinam, French Guiana and Ecuador, and in Brazil in Amazonas, Maranhão, Pará and Pernambuco. Both morphs were recorded by Richards (1978) from Peru and from Brazil (States of Goiás, Minas Gerais and São Paulo). The record of the morph *vespiceps* from the Ilha de Maracá is 1500 km further north than any previous one. These two colour phases warrant a detailed ecological study where they are sympatric.

On Maracá, several social wasp species nested in the forest near the ground. Advantages of this habitat may include the availability of suitable nesting substrates, lower temperatures and higher humidity. Nests of six species were discovered attached to the undersides of leaves of *Astrocaryum* spp. and *Maximiliana maripa* (Correa) Drude (Palmae), *Ischnosiphon arouma* and other understorey plants within 2 m of the ground. Nests of the following species were not encountered during the present study: *Stelopolybia angulata* and *S. ornata*, which generally nest in hollow trees; many *Metapolybia* species, *Synoeca surinama* and *S. virginea* which nest on trunks; and *Apoica thoracica* which suspends its nests from the branches of understorey trees (Richards, 1978 and pers. obs.).

The colonies of species which nested in larger clearings and at the forest edge were subject to higher temperatures. In the largest clearings shade temperatures often exceeded 46°C, while in the undisturbed forest they rarely reached 36°C (J. Thompson and D. Scott, pers. comm.).

The only species known to nest in the canopy which hunted near the ground was *Polybia liliacea*. This species constructs large nests, sometimes 2 m long, near the tops of trees. The several thousands of wasps of such a colony undoubtedly hunt over a large area, including the larger clearings. The species was never seen near the ground in the forest nor in smaller clearings, which might be explained by the difficulty a wasp could experience returning to the nest, in having to fly vertically upwards between the trees carrying its prey. There may be other canopy species not encountered during the present study because they are able to restrict their hunting to the stratum where their nests are located. It is suspected that *Apoica flavissima* is a canopy species and only one female was caught in a clearing where it was attracted to light.

The four savanna species, *Polistes brevifissus*, *Pseudopolybia vespiceps*, *Polybia occidentalis* and *P. sericea*, are commonly encountered in various types of open vegetation in Brazil (pers. obs.).

The increasing density of hunting wasps with increased size of clearing was presumably a result of the greater numbers of prey available in the larger clearings. This in turn might have been a result of greater primary production in areas which received longer periods of sunlight. The presence of a large colony of the nocturnal species *Apoica pallida* in one of the largest clearings suggested the presence of substantial numbers of nocturnal prey there.

The sizes of forest clearings influenced the 19 species of social wasps that used them. The wasps typical of the forest floor were found only in the smaller clearings, while the wasps which generally hunt at the forest edge were found in the larger clearings. Only three species are known to have used a wide range of clearing size.

## ANNOTATED CHECKLIST

Thirty-six species have been determined.

### *Polistes (Anaphilopterus) brevifissus* (Richards)

About 25 occupied and 40 abandoned nests were attached to the underside of a wooden bridge 100 m from the river on the main track to the Ecological Station. The nest of this species has not been described previously. Most nests contained up to $c$. 100 cells. Each was approximately diamond-shaped, suspended by the peduncle from one end and somewhat rounded below. Larger nests were similar above and below, but the central part was more parallel-sided. The largest nest seen (abandoned) was 40 cm long and estimated to comprise $c$. 1000 cells. The cells were 6.6 to 7.2 mm internal diameter, and 22 or 23 mm long. The cap of the cocoon was hemispherical and extended 3 or 4 mm beyond the rim of the cell. It was made of a tough, pale fawn silk.

Numerous adult females were collected, but no males were seen. The females varied in size, with the fore wing 10 to 14.4 mm long. They hunted among shrubs and small trees along the track within 400 m of their nests, and over the open *campos* (savannas) close by. This species was described from Mato Grosso. Its distribution is patchy, occurring in Brazil in Amapá, Pará, Mato Grosso, Mato Grosso do Sul, Bahia, Espirito Santo, Rio de Janeiro, Minas Gerais and Rio Grande do Sul, and in Argentina in Buenos Aires and Tucuman.

*Polistes (Anaphilopterus) versicolor kaieteurensis* (Bequaert)

Eight occupied and 24 abandoned nests of this subspecies were discovered in forest clearings of >150 m$^2$. The abandoned nests were all of eight cells or less. The largest nest seen was 9 cm long with at least 13 adult females and 8 adult males present. It was attached to a leaf of an *Astrocaryum gynacanthum* palm. The nest was similar in shape to the smallest ones of *P. brevifissus* described above. Cells are 4.7 to 4.9 mm wide and 15 to 17 mm long. The cap of the cocoon was hemispherical and extended 1 to 3 mm beyond the rim of the cell. It was made of a fairly tough, whitish silk. Two adults of a species of *Seminota* (Hymenoptera, Trigonalidae) emerged from the nest.

There are three subspecies. *P. v. versicolor* (Olivier) is widespread, ranging from Costa Rica and Grenada, West Indies, south to Paraguay and Argentina. The two other subspecies have restricted distributions. *P. versicolor flavoguttatus* Bequart occurs in valleys on the eastern slopes of the Andes in Ecuador, Peru and Bolivia. The subspecies *kaieteurensis* is restricted to the extreme north of Roraima and the neighbouring part of Guyana.

*Mischocyttarus (Mischocyttarus) labiatus* (F.)

Two females were collected at the edge of the forest clearings. Richards (1945, 1978) described the nest as being typical of the subgenus, but often having the peduncle >30 mm long. As he mentioned, it is a forest species. In Brazil the species has been recorded previously from Amazonas, Mato Grosso and Pará. It also occurs in Goiás and the Federal District of Brasília (pers. obs.).

*Mischocyttarus (Clypeopolybia) carbonarius carbonarius* (de Saussure)

One female of this species was discovered hunting in the forest. It closely resembles *Stelopolybia angulata* (see below).

*Mischocyttarus (Haplometrobius) alboniger* (Richards)

Three nests of this species were discovered in the forest. Two were on leaves of *Astrocaryum aculeatum* G.F.W. Mey. and one was on *A. gynacanthum*; they were situated 120 to 150 cm above the ground. The nests of this species have not been previously described. They were mottled pale and mid-grey and mid-brown, and very fragile. They contained three, seven and ten cells respectively. In all three the roughly circular comb was suspended from a lateral cell to the peduncle, which was 1.5 to 2.0 mm long and 0.5 to 0.6 mm wide. The three nests were similar, as all contained at least one shallow cell with an egg and at least one sealed cocoon.

An individual female was collected with two of the nests and there were two females associated with the third.

To illustrate nest development, the nest of 10 cells comprised the following: one empty cell 0.5 mm deep; three cells 2.4 to 3.3 mm deep containing an egg; four cells 4.4 to 6.7 mm deep containing a larva; and two sealed cells 10.8 and 11.00 mm long. One of the last two cells, suspended from the peduncle, bore a thin, fragile white cap of the cocoon. The cap of the other had been covered with a thin layer of material similar to that used to construct the nest, and on this surface there were five flanges 0.5 to 1.2 mm high and 0.5 to 1.1 mm broad. The flanges were as thick as the cell walls. The additional coating of material on the apex of the cocoon, and the flanges it bore, were presumably constructed by the adult wasp.

Apparently the species was previously known only from the type female collected in the Nassau Mountains, Surinam, and deposited in the Rijksmuseum van Natuurlijke Historie, Leiden, The Netherlands. The adult female is very pale brownish pink with red–brown and dark brown markings on the dorsal side. This is an unusual combination of colours for a neotropical social wasp, hence the specific name.

### *Mischocyttarus (Kappa) injucundus bimarginatus* (Cameron)

Eight nests were discovered at the edges of the larger forest clearings. All were located 60 to 130 cm from the ground, and were attached to the undersides of leaves of two cashew trees (*Anacardium occidentale* L.), on an introduced grape vine and on unidentified forest trees. One of the nests was located in a tree in which there was also an active colony of *Mischocyttarus maracaensis*. The four largest nests were collected. The carton of all four was dark brown. The combs of three of the nests were roughly circular in outline, while the largest nest was four cells wide and followed the outline of the lanceolate leaf to which it was attached. Each comb was suspended from its centre by a peduncle 5 to 8 mm long. There were 12, 36, 30 and 71 cells in the nests of which 0, 3, 1 and 38 cells respectively were sealed. The cap of the cocoon was dark brown, slightly convex, and constructed on the rim of the cell. The numbers of adult females and males associated with the four nests were respectively 19♀ and 7♂; 11♀ and 2♂; 2♀ and 1♂, and 4♀ and 0♂.

This was the most commonly encountered species of *Mischocyttarus* in the study area. The species closely resembles *Polybia rejecta* (see below). There are three subspecies with apparently isolated distributions. *M.i. injucundus* (de Saussure) occurs in French Guiana, and in Brazil in Amapá. It is also said to occur in Pará, Maranhão and possibly Bahia. The subspecies *bimarginatus* occurs in Guyana and Trinidad, West Indies, while *tingomariae* Richards is restricted to eastern valleys of the Andes in southern Colombia, Ecuador and northern Peru. This is the first record of *M. i. bimarginatus* in Brazil.

### *Mischocyttarus (Kappa) granadaensis* (Zikan)

One nest of this species was discovered, and is described for the first time. It was in a small clearing, attached to the underside of a leaf of *Maximiliana maripa* (Palmae) 2.2 m from the ground. The peduncle was 1.8 mm long and 1.5 mm wide, and located at one end of the oval comb. The nest was rusty-brown and fragile. It comprised 10 shallow cells 4 to 6 mm long and 3 mm wide. Four contained young larvae and six contained eggs. Two adult females were present on the nest. This species was previously known only from two females collected in Colombia. It closely resembles *Polybia dimorpha* (see below).

### *Mischocyttarus (Haplometrobius) prominulus* (Richards)

Several wasps were seen hunting at the edges of large clearings in places where *M. injucundus bimarginatus* and *M. maracaensis* hunted. However, no nest was discovered.

### Mischocyttarus (Monocyttarus) maracaensis (Raw)

Several females of *M. maracaensis* were seen hunting at the edge of forest clearings. One nest was discovered attached to the underside of a leaf of a sapling at the edge of a clearing. The nest was 1.7 m from the ground, and comprised 19 cells which were 3 mm wide. Three cells were 8 to 15 mm long and contained larvae and 16 were 2 to 8 mm long and contained eggs. Two adult females were present on the nest. An active colony of *M. injucundus bimarginatus* was found in the same tree.

### Mischocyttarus (Haplometrobius) surinamensis surinamensis (de Saussure)

Several wasps were seen hunting at the edges of large clearings in places where *M. injucundus bimarginatus* and *Mischocyttarus maracaensis* hunted. However, no nest was discovered. It is known from Surinam, Guyana, Peru and Amazonas, Pará, Acre, Rondônia and Mato Grosso in Brazil.

### Pseudopolybia vespiceps (de Saussure)

Both colour phases of this species were collected in Malaise traps by Dr Forbes Benton. It is the most northern record of the morph *vespiceps* (see Discussion). This species occurs in Guyana, Ecuador, Peru and throughout Brazil.

### Leipomeles sp.

An abandoned nest of this genus was discovered on the leaf of an understorey tree next to a path. There are two species in the genus, and both occur in Guyana, Surinam, French Guiana, Brazil (in Amazonas and Amapá) and in eastern Peru, so the nest might have been made by either *L. dorsata* (F) or *L. nana* (de Saussure). Neither species has been previously recorded from Roraima.

### Angiopolybia pallens (Lepeletier)

Several adult females of this species were collected as they were hunting in the forest, and in small clearings where the wasps often construct nests on the undersides of leaves. It is a widespread species in moist and semi-deciduous forests in Panama, Trinidad and tropical South America.

### Stelopolybia angulata angulata (F.)

Three females were collected as they hunted over the forest floor. This is a widely distributed subspecies found in moist tropical forests of Colombia, Venezuela, the Guianas and throughout the Amazon basin, and in parts of the Atlantic coastal forest.

### Stelopolybia cajennensis (F.)

Two nests of this species were discovered. An abandoned nest was attached to the rafters of a house at the northern end of the main forest trail, on the Santa Rosa branch of the Rio Uraricoera. The second nest was built in an unused plastic water pipe (45 mm diameter) at the Ecological Station. The nest comprised five separate combs occupying 370 mm of the length of the pipe. The longitudinal axes of the cells were horizontal, and the combs were arranged with two on one side and three on the other, with the cell openings facing each other. The combs were brown, roughly oval in outline, 64, 94, 112, 169 and 205 mm long and 18 to 28 mm wide, and attached to the substrate by 5, 5, 18 and 23 peduncles 3 to 6 mm long. Mature cells were 10 to 12 mm long and 2.9 to 3.0 mm wide. The cap of the cocoon was hemispheral and extended beyond the rim of the cell. It was pale brown and opaque.

Almost all the adult wasps were collected with the nest. There were 439 females and 23 males (5% males). One hundred females were dissected, of which 12% had enlarged ovaries and are considered to be queens. This species is found in Mexico, Central America, most of Amazonia and southward to Peru and Bolivia and to Bahia and São Paulo in Brazil.

### *Stelopolybia fulvofasciata* (Deger)

Several adult females were collected as they hunted in forest clearings. No nest was discovered. This species occurs in much of northern South America and is common throughout the Amazon basin.

### *Stelopolybia multipicta* (Haliday)

Six adult females were collected as they hunted at the edges of forest clearings near the Ecological Station, and near the savanna at the northern end of the main forest trails to the Santa Rosa branch of the Rio Uraricoera. No nest was discovered. This is a widespread species occurring from Mexico throughout much of Brazil to Argentina and Uruguay.

### *Stelopolybia ornata* (Ducke)

Two females of this species were collected in a clearing. The species occurs in Guyana, Surinam, Brazil (Amazonas) and Amazonian Ecuador, Peru and Bolivia.

### *Stelopolybia testacea* (F.)

One female was collected on Maracá in a Malaise trap by Dr Forbes Benton. This species is a widespread inhabitant of moist tropical forests in northern South America and is common in most of the Amazon basin.

### *Apoica (Apoica) flavissima* (Van der Vecht)

Two females were collected at lights out at the Ecological Station. This species occurs in northern South America and most of Amazonia, and in the Atlantic coastal forests from Paraiba to São Paulo.

### *Apoica (Apoica) pallida* (Olivier)

A nest of this species was discovered by Dr Jill Thompson. It was built 60 cm above the ground on the branch of a sapling, and shaded by leaves. The nest was typical of the genus, being a simple exposed comb. It was an almost perfect disc 90 mm in diameter, of a mottled grey, rust and pale brown. Its upper surface was a slightly eccentric cone whose apex was 35 mm above the bases of the outer cells. Mature cells were 23 to 26 mm long and 4.6 mm wide. The caps of the cocoons were white, very slightly convex and 1 or 2 mm inside the rim of the cell.

The nest was collected with 148 adult females and one adult male. However, it was estimated that there were less than 450 wasps associated with the nest. After being disturbed the wasps not captured all left the immediate vicinity. This is a widespread species occurring in Central America, Trinidad, northern South America and much of Amazonia.

### *Apoica (Apoica) thoracica* (du Buysson)

One female was collected at a light at the Ecological Station. This is a deep forest species found in Central America, northern South America and the Amazon, through Goiás and the Federal District of Brasília to Espirito Santo and Paraná.

*Protopolybia exigua* (de Saussure) (?)

An abandoned nest of this genus was collected, without occupants, attached to the leaf of a small tree at the edge of a clearing. It is presumed to be of *P. exigua*, as Richards (1978) mentioned its occurrence in Roraima. *P. exigua* occurs throughout tropical Central and South America and is the commonest member of the genus in most of Brazil.

*Polybia (Polybia) liliacea* (F.)

Two nests were located c. 20 m high in forest trees. Both were oval in cross-section, c. 1.5 m long and c. 0.5 m wide. A species of *Azteca* ant had constructed a nest around the trunk of one of the trees 15 m from the ground, which presumably protected the wasp nest from raids of the army ant *Eciton burchelli* (Westwood), which was common in the forest. It is presumed that the wasps hunt in the forest canopy. However, they were often seen hunting and visiting ripe fruits in the larger forest clearings. This species has been recorded from Central America, much of tropical South America and most of Brazil except the northeast and Atlantic coast.

*Polybia (Cylindroeca) dimidiata* (Olivier)

One nest was discovered by Dr Jill Thompson. It was typical of the species, being built around the trunk of a sapling in forest shade. The nest was 50 cm long and its base was 20 cm above the ground. It was located 3 m from the edge of a forest clearing. The wasps were frequently encountered as they hunted in the clearings. This species occurs in the Guianas and throughout Amazonia and central Brazil southward to Rio de Janeiro and São Paulo.

*Polybia (Formicicola) rejecta* (F.)

A few wasps of this widespread species were collected in the forest and at the edges of clearings near the Ecological Station and near the savanna, and at the northern end of the main forest trail to the Santa Rosa branch of the Rio Uraricoera. No nest was discovered. This species occurs in Central America and Trinidad southward to Peru and Bolivia. It is found in most of tropical Brazil except the northeast.

*Polybia (Myrapetra) dimorpha* (Richards)

A nest of this species was discovered by Dr Jill Thompson. It was built 1 m above the ground on the branch of a sapling, and shaded by leaves. The phragmocyttarous nest was roughly a truncated cone 70 mm long, 27 mm wide at the apex and 68 mm wide at the base. The envelope was greyish brown with a matt surface. The nest entrance was a circular aperture on the underside of the nest, close to its edge. There were three combs; the two lower ones each bore one circular aperture equivalent to eight cells near the side, to allow access to the upper one. The flat or slightly convex cap of the cocoon was thin and white and constructed at the cell entrance or up to 2.6 mm inside the rim. The nest was collected with most of the wasps. There were 148 females, but no male was collected.

This species is slightly smaller than *P. (Apopolybia) jurinei* de Saussure, but closely resembles it in shape and colour. However, the envelope of the latter's nest is flaky, whereas that of *P. dimorpha* is somewhat uneven and slightly roughened (like that of *P. occidentalis* and related species), but is not flaky. This species is found in the Guianas, Ecuador and Peru. This is the first record of the species from Brazil.

*Polybia (Myrapetra) occidentalis venezuelana* (Soika)

One nest of this species was collected, attached to a vine of *Uncaria guianensis*. (Rubiaceae) growing in a clump of trees in the savanna, 80 m from the forest edge. It was collected with

145 adult females. The wasps were commonly seen hunting in the savanna. *P. o. venezuelana* has been recorded from Venezuela, Colombia and Peru. This is the first record of the subspecies from Brazil.

### *Polybia (Myrapetra) roraimae* (Raw)

A previously undescribed species was discovered and the adult and the nest are described elsewhere (Raw, in press a). Three nests of *P. roraimae* were found; two in forest and one in a clearing. One was suspended from the branch of a sapling 60 cm from the gound and completely shaded by leaves, the second was attached to the underside of a leaf of an *Ischnosiphon* species (Marantaceae) 50 cm from the ground, and the third was attached to the underside of a leaf of an *Astrocaryum gynacanthum* palm 80 cm from the ground. The nests contained 150 to 172 adult females. The nests were located only 1.5 km from the nest of *P. occidentalis venezuelana* mentioned above. The nest is similar in size and structure to that of *P. occidentalis* (Olivier).

### *Polybia (Trichothorax) sericea* (Olivier)

Three specimens were collected by Dr Forbes Benton and one was collected by me, in open grassland near the Ecological Station. This is one of the most widespread species of social wasps in South America and in some savannas it is the most abundant social wasp.

### *Polybia (Trichothorax) ignobilis* (Haliday)

A few wasps were collected in the forest and clearings. This is an extremely widespread species in Central America and much of South America and is especially abundant in semi-deciduous forests and disturbed habitats.

### *Brachygastra smithii* (de Saussure)

Five occupied nests of this species were discovered. Two were built in bushes in the area of savanna near the northern end of the main forest trail to the Santa Rosa branch of the Rio Uraricoera. One nest was built among secondary growth in a large forest clearing. Two nests were attached to grape vines growing on pergolas in front of the buildings of the Ecological Station. The nest was phragmocyttarous. The envelope was a mottled grey–brown. Over the upper part of the nest the envelope was up to 5 mm thick, though below it was like that of nests of *B. augusti* (de Saussure) and *Polybia occidentalis*. All the nests were similar in shape, being roughly cylindrical with a rounded top and flattened base. They were 70 to 105 mm long and 50 to 65 mm wide. The entrance was an oval hole 18 to 25 mm long and 6 to 11 mm wide, located at the side near the base of the nest. Two nests 75 and 90 mm long were collected. Each contained four combs. The mature cells were 7 to 9 mm long and 3.3 mm wide. The cap of the cocoon was white and hemispherical, and extended up to 4 mm beyond the rim of the cell. One nest was collected at 0700 hrs when the wasps were still inactive. All 453 wasps collected were females. This species is found from Mexico to Bolivia, Peru and Brazil (Acre, Amazonas, Pará and Mato Grosso). The record from São Paulo should be verified.

### *Chartergus chartarius* (Olivier)

One nest was found by Dr Forbes Benton. It had been abandoned and only one dead adult female was discovered inside it. It had been built on a tree at the edge of a small forest clearing. A second nest of this genus, and presumed to be of this species, was attached to a tree at a height of *c*. 25 m close to the Santa Rosa branch of the Rio Uraricoera. Previously *C. chartarius* had been recorded in Brazil from southern Amazonia and central Brazil. However, the species is also known from Colombia, the Guianas, Ecuador, Peru and Bolivia.

### *Metapolybia unilineata* (von Ihering)

Two nests of this species were attached to the underside of the water tank at the Ecological Station. They were circular in outline and 15 and 20 cm diameter. The nests resembled the typical ones of the widespread species *M. cingulata* (F.). The envelope is mottled dark grey–brown with numerous transparent windows, and bears eaves up to 1 cm wide. The species was previously known from northern Brazil (Amazonas) and the Guianas. I have also collected this species in Acre and Rondônia.

### *Synoeca surinama* (L.)

This widespread species was one of the most frequently encountered in the larger forest clearings where the wasps hunted and visited ripe fruits. It is known from Colombia and Trinidad southward to Peru and Bolivia, and is widespread and common in moist tropical Brazilian forests away from the coast.

### *Synoeca virginea* (F.)

This species was often seen with *S. surinama*, hunting and visiting ripe fruits in the larger forest clearings. It is found in Colombia, the Guianas, Ecuador and Peru. In Brazil it is largely restricted to the Amazon, occurring in Amazonas, Acre, Rondônia, Pará and Mato Grosso. The record from Piauí should be verified.

## ACKNOWLEDGEMENTS

I thank the Royal Geographical Society of London which financed my visit. Several people kindly pointed out the locations of nests of wasps. Dr Forbes Benton allowed me to examine the specimens of Vespidae he had collected on various parts of the island. Dr James A. Ratter identified the plants. Dr John Proctor, Dr Jill Thompson and Mr Duncan Scott allowed me to collect in the clearings where they were conducting their studies. Dr Thompson was particularly helpful in conducting me around the forest clearings.

## REFERENCES

Begon, M., Harper, J.L. and Townsend, C.R. (1996). *Ecology, individuals, populations and communities*, 3rd edition. Blackwell Scientific Publications, Oxford.

Raw, A. (1988). Social wasps (Hymenoptera: Vespidae) and insect pests of crops of the Suruí and Cinta Larga indians in Rondônia, Brazil. *The Entomologist*, 107, 104–109.

Raw, A. (1992). The forest–savanna boundary and habitat selection by Brazilian social wasps. In: *Nature and dynamics of forest–savanna boundaries*, eds P.A. Furley, J. Proctor and J.A. Ratter, pp. 499–511. Chapman & Hall, London.

Raw, A. (in press a). Two new species of social wasps (Hymenoptera: Vespidae) from Roraima, northern Brazil. *Revista Brasileira de Zoologia*.

Raw, A. (in press b). An investigation into 'the third trophic level': Neotropical social wasps' use of spatial memory and odours of freshly damaged leaves when hunting (Hymenoptera, Vespidae). *Revista Brasileira de Zoologia*.

Richards, O.W. (1945). A revision of the genus *Mischocyttarus* de Saussure (Hymen., Vespidae). *Transactions of the Royal Entomological Society of London*, 95, 295–462.

Richards, O.W. (1978). *The social wasps of the Americas*. British Museum (Natural History), London.

# 16 Insects of the Ilha de Maracá Further Contributions I: General Entomology

## INTRODUCTION

A substantial number of entomological studies were carried out on the Ilha de Maracá, covering a broad range of insect orders. Thousands of species were recorded and numerous new species described, and the lengthy process of identifying and describing the specimens is still going on. The results of many of these studies have been published elsewhere, and are enough to fill a book in themselves. This chapter presents and discusses the results of six of these studies, which, between them, give an idea of the type of work which was carried out and the results which have been obtained.

## INTRODUÇÃO

Um grande número de estudos entomológicos foram realizados na Ilha de Maracá, abrangendo uma ampla gama de grupos taxonômicos. Milhares de espécies foram registradas, e numerosas espécies novas foram descritas, e o processo lento de identificação e descrição dos exemplares ainda continua. Os resultados de vários destes estudos já foram publicados, sendo suficientes para encher um livro por si mesmos. Este capítulo apresenta os resultados de seis destes estudos, os quais, entre si, fornecem uma idéia dos tipos de trabalho que foram realizados, e dos resultados obtidos.

## Biological data on the Passalidae (Coleoptera) of the Ilha de Maracá

### PAULO F. BÜHRNHEIM AND NAIR OTAVIANO AGUIAR
*Universidade do Amazonas, Manaus, Brazil*

### SUMMARY

Two visits were made to the Ilha de Maracá, the first during the rainy season (3–13 May 1987) and the second in the dry season (23–30 November 1987). On the first visit four species of Passalidae were collected: *Passalus glaberrimus* Eschsch., 1829, *Passalus interruptus* (Lin., 1758), *Passalus interstitialis* Eschsch., 1829, and *Verres furcilabris*

(Eschsch., 1829). On the second visit we added one more record, *Paxillus leachi* (MacLeay, 1819). Data on the biology of these species, observed during the visits, are furnished and discussed.

## RESUMO

Foram realizadas duas excursões na Ilha de Maracá, Estação Ecológica da Secretaria Especial do Meio Ambiente, em Roraima, Brasil, uma durante a estação chuvosa, no período de 03 a 13 de maio 1987 e a outra na estação seca de 23 a 30 de novembro 1987. Na primeira foram colecionadas quatro espécies de Passalídeos (Coleoptera): *Passalus glaberrimus* Eschsch., 1829, *Passalus interruptus* (Lin., 1758), *Passalus interstitialis* Eschsch., 1829, e *Verres furcilabris* (Eschsch., 1829). Na segunda oportunidade, além das espécies encontradas na primeira, foi colecionada mais uma, *Paxillus leachi* (MacLeay, 1819). Dados sôbre a bionomia destas espécies, obtidos durante os dois colecionamentos, são fornecidos e discutidos.

## INTRODUCTION

There are no previous records of Passalidae for Roraima, so the five species belonging to three different genera reported in this paper represent the starting point of our knowledge of this group in the State.

## MATERIALS AND METHODS

Two visits were made to the Ilha de Maracá, the first (3–13 May 1987) during the rainy season and the second (23–30 November 1987) in the dry season.

Passalidae were collected from rotten tree trunks in both the forest and savanna areas between the Ecological Station and the Casa de Santa Rosa (see Figure Pr.1, p. xvii). All trunks found on the track between these points and on the trail system in the south of the island were examined. The localization of galleries in the trunk tissues (bark, sapwood and heartwood) was noted, as was the number of individuals of passalids in each phase of development. All specimens collected, including eggs, were fixed in Pampel solution (42% distilled $H_2O$: 44% ethanol [96%]: 8% formalin: 5–6% glacial acetic acid) and stored in 70% ethanol. Night collections were also made using a black lightbox mounted on a white sheet, both near the bank of the Rio Uraricoera and in the forest alongside the track to Santa Rosa.

The collections are lodged in the entomological collections of the Instituto Nacional de Pesquisas da Amazônia (INPA) and of the zoology laboratory of the University of Amazonas (ZUA).

## RESULTS

The following five species were collected in the investigation: *Verres furcilabris* (Eschsch., 1829), *Passalus glaberrimus* Eschsch., 1829, *P. interruptus* (Lin., 1758), *P. interstitialis* Eschsch., 1829, and *Paxillus leachi* (MacLeay, 1819).

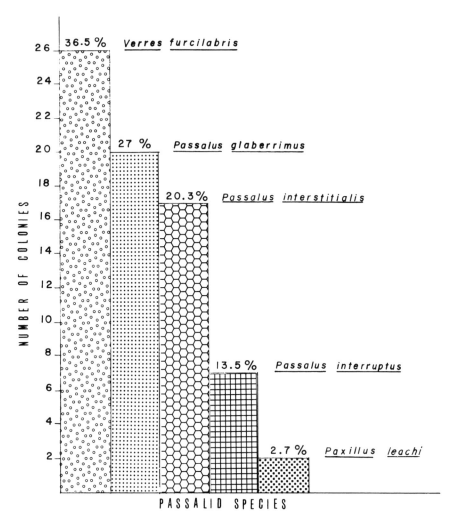

**Figure 16.1.** Relative abundance of colonies of passalid species found on the Ilha de Maracá, Roraima, during the periods 3–13 May and 25–30 November 1987

*Verres furcilabris* was the most abundant species in terms of colonies, comprising 36.5% of the total number found on the two visits (Figure 16.1). Thirty adults were found in 15 colonies examined in May: two of these colonies had a total of 16 larvae, one had four pupae and another contained many eggs. In November the nine colonies examined contained a total of 16 adults and two of them contained five larvae each (Table 16.1). This species was not found in the bark and was most frequent in the sapwood (Table 16.2).

*Passalus glaberrimus* comprised 27% of all the colonies examined (Figure 16.1). In the 11 colonies found in May there were only 33 adults while there were 30 adults in the six colonies encountered in November (Table 16.1). The great majority

Table 16.1. Colonies of passalid species with adults and immature stages found in fallen tree trunks on the Ilha de Maracá during the periods 3–13 May and 23–30 November 1987

| Passalid species | 3–13 May 1987 | | | | 23–30 November 1987 | | | |
|---|---|---|---|---|---|---|---|---|
| | No. colonies (No. specimens) | | | | No. colonies (No. specimens) | | | |
| | Adults | Larvae | Pupae | Eggs | Adults | Larvae | Pupae | Eggs |
| *Verres furcilabris* | 15 (30) | 2 (16) | 1 (4) | 1 (many) | 9 (16) | 2 (5) | | |
| *Paxillus leachi* | 11 (33) | | | | 2 (22) | | | |
| *Passalus glaberrimus* | 7 (23) | 2 (22) | 1 (2) | 1 (many) | 6 (30) | | 1 (2) | |
| *P. interstitialis* | 4 (9) | | | | 5 (34) | 3 (many) | 1 (3) | |
| *P. interruptus* | | | | | 3 (11) | | | |
| Colonies with 2 species found in the same gallery | | | | | | | | |
| *Verres furcilabris* with *Passalus glaberrimus* | 2 (8) (9) | ? 2 (15) ? | (2) 1 | | (4) 1 (1) | ? 1 (2) ? | | |
| *Passalus interruptus* with *P. interstitialis* | 3 (9) (6) | ? 1 (11) ? | | | | | | |

? = larvae could not be distinguished so they may have been of either species.

**Table 16.2.** Localization of passalid colonies in the trunk tissues of trees on the Ilha de Maracá

| Passalid species | Localization of the colonies in the trunks (%) | | | | |
|---|---|---|---|---|---|
| | Under bark | Under bark/ sapwood | Sapwood | Sapwood/ heartwood | Heartwood |
| *Paxillus leachi* | 100.0 | | | | |
| *Passalus interstitialis* | 71.4 | 21.4 | 7.1 | | |
| *P. interruptus* | 22.2 | 11.1 | 44.4 | 22.2 | |
| *P. glaberrimus* | 17.6 | | 70.6 | 5.9 | 5.9 |
| *Verres furcilabris* | | | 40.9 | 27.3 | 31.8 |
| *V. furcilabris/P. glaberrimus* | | 33.3 | 33.3 | 33.3 | |
| *P. interruptus/P. interstitialis* | | | | 100.0 | |

of the colonies were in the sapwood, with a few under the bark and very rare occurrence in the heartwood (Table 16.2).

*Passalus interstitialis* comprised 20.3% of the total number of colonies found (Figure 16.1). In seven colonies examined in May there were a total of 23 adults; two contained 22 larvae, while in one there were two pupae and some eggs in only a single colony. In the November collection five colonies with 34 adults were found: of these three contained some larvae and one two pupae (Table 16.1). Most of the colonies were found under the bark; they were rare in sapwood and absent from heartwood (Table 16.2).

*Passalus interruptus* constituted 13.5% of the colonies examined (Figure 16.1). Only four colonies with a total of nine adults were found in May, while in November three colonies contained 11 adults and in one of them there were three pupae (Table 16.1). Colonies of this species were not found in the heartwood but occurred in the other layers of the trunk, showing the greatest frequency in the sapwood (Table 16.2).

*Paxillus leachi* was only found in November, and the two colonies, containing 22 adults without any immature forms (Table 16.1), represented only 2.7% of total colonies examined (Figure 16.1). Both colonies were situated under bark (Table 16.2).

*Verres furcilabris* and *Passalus glaberrimus* were found sharing colonies on three occasions. Two colonies were examined in May which contained eight adults of the former and nine of the latter, together with 15 unidentified larvae and, in one colony, two pupae of *V. furcilabris*. In November only one colony was found with these two species in the same gallery, containing four adults of *V. furcilabris* and one of *P. glaberrimus* together with two unidentified larvae (Table 16.1). These shared colonies occurred between the bark and the sapwood, in the sapwood, and between sap- and heartwood (Table 16.2).

*Passalus interruptus* and *P. interstitialis* were also found together in three colonies, all in May. These colonies contained nine adults of the former species and six of the latter, as well as 11 unidentified larvae in one of the colonies (Table 16.1). The colonies were all situated between the sap- and heartwood of the trunk (Table 16.2).

Table 16.3. Numbers of passalid colonies per trunk examined on the Ilha de Maracá

| Colonies of passalids per trunk | No. trunks | % |
|---|---|---|
| Trunks containing one colony | | 73.7 |
| *Verres furcilabris* | 18 | |
| *Passalus interstitialis* | 8 | |
| *P. interruptus* | 3 | |
| *P. glaberrimus* | 13 | |
| Trunks containing two colonies | | 22.8 |
| *P. interruptus* and *P. interruptus* | 1 | |
| *P. interruptus* and *P. interstitialis* | 1 | |
| *P. interstitialis* and *V. furcilabris* | 1 | |
| *V. furcilabris* and *P. glaberrimus* | 4 | |
| *Paxillus leachi* and *P. interstitialis* | 1 | |
| Species together in same gallery | | |
| *P. interstitialis* and *P. interruptus* | 3 | |
| *V. furcilabris* and *P. glaberrimus* | 2 | |
| Trunks containing three colonies | | 3.5 |
| *V. furcilabris* and *P. interstitialis* with (in same gallery) *V. furcilabris* and *P. glaberrimus* | 1 | |
| *P. interruptus*, *P. interstitialis* and *Paxillus leachi* | 1 | |
| Total | 57 | 100.0 |

A total of 57 trunks containing 68 colonies were examined during our collection programme, comprising 36 with 42 colonies in May and 21 with 26 colonies in November. The number of colonies per trunk varied from one to three, with 73.7% containing only a single colony; two colonies of the same or different species were found in 22.8% of trunks, in five of which two species occurred in the same gallery; while 3.5% of trunks contained three colonies of different species, including in one case a colony with two species in the same gallery (Table 16.3).

The only capture made using the black lightbox was a male of *Verres furcilabris* collected in the forest at 2100 hrs on 4 May 1987. No specimens were collected using the light-traps at the Ecological Station or on the river bank.

## DISCUSSION AND CONCLUSIONS

The diversity of passalid beetles found on the Ilha de Maracá and recorded by Bührnheim and Aguiar (1991) was low, and practically the same in the rainy and dry seasons. Few workers in the neotropics have carried out continuous studies in a single area, but Morón (1979) collected nine species of five genera in tropical woodlands at Los Tuxtlas, Mexico, while Morón *et al.* (1985) recorded 13 species of seven genera during four (three-monthly) visits to an area of evergreen tropical woodland at Boca del Chajul, Chiapas, Mexico. In addition, Castillo (1987) found 14 species of eight genera in 18 collections over two years in Veracruz, Mexico, and Fonseca (1988) recorded 11 species of three genera during 13 consecutive months of collecting in tropical forest near Manaus, Amazonas, Brazil. The diversity encountered by these workers is higher than that of the present study, and may reflect a richer fauna and/or a more intensive sampling programme.

Immature specimens were more frequent during the rainy season, both in numbers of individuals and stages (eggs were only found for two species). Fonseca (1988), working with the three larval stages of *Passalus convexus* Dalman, 1817, and *P. latifrons* Percheron, 1841, near Manaus, found the greatest total number of larvae of the former species from May to August (the change from the rainy to the dry season) and of the latter species in July.

*Passalus interstitialis* and *Paxillus leachi*, which have been recorded in two localities in Mexico and two in Brazil (Morón, 1979; Morón *et al.*, 1985; Castillo, 1987; Bührnheim and Aguiar, 1991; Aguiar and Bührnheim, 1992), were always found in colonies under the bark at those localities, although the former species was rarely found in sapwood at the Ilha de Maracá.

Although only one species of Proculini, *Verres furcilabris*, was found, it was notable that its colonies occurred more frequently in the heartwood but sometimes also in the sapwood (Table 16.2). This conforms to the observations of Castillo (1987) that seven species of this tribe in Mexican tropical woodlands showed a strong preference for the interior of trunks.

Morón *et al.* (1985) found up to five passalid species colonizing the same trunk, while Castillo (1987) and Fonseca (1988), in 161 and 35 trunks respectively, found up to four. On Maracá the maximum number of species per trunk out of 57 trunks examined was three. However, the average frequency per trunk coincided (Table 16.3), the majority having only one species and the proportion declining according to number of species sheltered.

## ACKNOWLEDGEMENTS

The authors wish to thank Dr José Albertino Rafael, INPA, for support, and the technicians of the same institute, Elias Bindá Brasil, João Vidal and Luiz de Sales Aquino, for valuable help in the field.

## REFERENCES

Aguiar, N.O. and Bührnheim, P.F. (1992). Pseudoscorpiones (Arachnida) em associação forética com Passalidae (Insecta, Coleoptera) no Amazonas, Brasil. *Amazoniana*, 12 (2), 187–205.

Bührnheim, P.F. and Aguiar, N.O. (1991). Passalídeos (Coleoptera) da Ilha de Maracá, Roraima. *Acta Amazonica*, 21, 25–33.

Castillo, M.L. (1987). *Descripción de la comunidad de Coleoptera Passalidae en el bosque tropical perenifólio de la región de 'Los Tuxtlas', Veracruz*. Thesis, Universidad Autónoma de Mexico, Mexico, DF, 89 pp.

Fonseca, C.R.V. (1988). Contribuição ao conhecimento da bionomia de *Passalus convexus*

Dalman, 1817, e *Passalus latifrons* Percheron, 1841 (Coleoptera: Passalidae). *Acta Amazonica*, 18 (1–2), 197–222.

Morón, M.A. (1979). Fauna de Coleópteros Lamelicornios de la Estación de Biologia Tropical, 'Los Tuxtlas', Veracruz, UNAM, Mexico. *An. Inst. Biol. UNAM, Mexico, ser. zoología,* 50 (1), 375–454 (*apud* Castillo, 1987).

Morón, M.A., Villalobos, F.J. and Deloya, C. (1985). Fauna de Coleópteros Lamelicornios de Boca del Chajul, Chiapas, Mexico. *Folia Entomol. Mex.*, 66, 57–118.

# Pollinators, pollen robbers, nectar thieves and ant guards of *Passiflora longiracemosa*

FORBES P. BENTON
*CEPLAC/CEPEC/SECEN, Itabuna, Brazil*

## SUMMARY

The pollination biology and plant–insect interactions of the forest vine *Passiflora longiracemosa* (Passifloraceae), a common species at the eastern end of the Ilha de Maracá, were studied by observation and experiment. The flowers were found to be pollinated by two species of hermit hummingbirds (*Phaethornis superciliosus* and *P. squalidus*). *Tetragona* and *Trigona* bees which visit the flowers were shown experimentally to be pollen robbers and nectar thieves respectively, and not to play a role in pollination. The nectaries on the sepals of the flowers were observed to attract at least five species of ants, of which *Ectatomma tuberculatum* was the most regular visitor. This species was seen to repel *Acromyrmex subterraneus*, a leaf-cutting ant, from the flowers.

## RESUMO

Interações com insetos e a polinização do cipó *Passiflora longiracemosa* (Passifloraceae), uma espécie comum nas florestas da parte leste da Ilha de Maracá, foram estudadas por meio de observação e experiência. Ficou evidente que esta espécie é polinizada por duas espécies de beija-flor (*Phaethornis superciliosus* e *P. squalidus*), e foi mostrado por experiência que as abelhas que visitam as flores, dos gêneros *Tetragona* e *Trigona*, agem como ladrões de pólen e néctar, respectivamente, sem desempenhar um papel na polinização. Cinco espécies de formigas visitaram os nectários nos cálices da flor, dos quais *Ectatomma tuberculatum* foi a mais freqüente. Esta espécie foi observada repelindo um ataque de *Acromyrmex subterraneus* (uma espécie de saúva) das flores do cipó.

## INTRODUCTION

The flowers of *Passiflora longiracemosa* Ducke (Passifloraceae) are conspicuous in the forest of the Ilha de Maracá. They are brightly coloured and borne in clusters directly on the vine stem, from ground level to lower canopy level. The leaves, however, are found only at canopy level. The author has observed the vine at the extreme western end of the island (Anta Camp trail) and along the Preguiça trail (Figure Pr.1, p. xvii), as well as in the forest near the Ecological Station. What follows refers to vines within a few minutes' walking distance of the Station.

**Figure 16.2.** *Tetragona* bee collecting pollen from the anthers of *Passiflora longiracemosa*

## POLLINATORS AND POLLEN ROBBERS

The reddish pink coloration of the calyx tube (hypanthium) of *P. longiracemosa* is suggestive of hummingbird pollination (Faegri and van der Pijl, 1979). The hermit hummingbirds *Phaethornis superciliosus* and *Phaethornis squalidus* were observed hovering and feeding at the open flowers. Several times a bird was seen to hover for a few moments in turn at each available open flower: either those borne on the same cluster or those of nearby clusters. Nearby clusters were usually on different vines, and the vines are quite widely dispersed. The behaviour pattern of darting from one flower cluster to another was observed both in the morning and in the afternoon.

More commonly observed visitors to the open flowers, however, were two species of *Tetragona* bees, *T. dorsalis* (Smith) and *T. handlirschii* (Friese). Typically a bee alighted on the open anthers and scraped off the bright yellow pollen with its forelegs. It hovered for a few moments and realighted on the same anther or another anther of the same flower to resume pollen-scraping, this sequence being repeated many times. Gradually the size of the pollen balls in the pollen baskets on the hind legs increased (Figure 16.2). It was while the bee was hovering that the pollen was transferred to the pollen baskets, its legs being free to effect transfer at these times.

In view of the abundance of the bees and the ease with which they removed pollen, some simple experiments were conducted to determine their role in pollination. Firstly, the timing of flower opening and closing was established. Individual flowers were marked with a short length of wool, loosely knotted around the flower-stalk, and observed daily. As shown by Table 16.4, there is no evidence that

**Table 16.4.** Duration of flower opening in *Passiflora longiracemosa*

| Cluster | Flower | November | | | | December | | | | | | | |
|---|---|---|---|---|---|---|---|---|---|---|---|---|---|
| | | 27 | 28 | 29 | 30 | 1 | 2 | 3 | 4 | 5 | 6 | 7 | 8 |
| 1 | 1 | ON | C | OE | OW | OW | | | | | | | |
| | 2 | | OO | S | | | | | | | | | |
| | 3 | | | OO | C | E | | | | | | | |
| | 4 | | | OO | C | C | OE | OE | | | | | |
| | 5 | | | OO | C | C | S | | | | | | |
| | 6 | | | | OO | C | C | S | | | | | |
| | 7 | | | | | OO | C | F | | | | | |
| | 8 | | | | | OO | C | S | | | | | |
| 2 | 9 | | | | | OO | C | F | | | | | |
| | 10 | | | | | | OO | C | F | | | | |
| | 11 | | | | | | OO | C | F | | | | |
| | 12 | | | | | | OO | C | | | | | |
| | 13 | | | | | | | | | OO | F | | |
| | 14 | | | | | | | | | OO | F | | |
| | 15 | | | | ON | C | C | F | | | | | |
| | 16 | | | | ON | C | C | S | | | | | |
| 3 | 17 | | | | ON | C | C | S | | | | | |
| | 18 | | | | | OO | C | F | | | | | |
| | 19 | | | | | OO | C | S | | | | | |
| | 20 | | | | | | OO | C | F | | | | |
| 4 | 21 | | | | ON | C | C | OE | F | | | | |
| | 22 | | | | ON | C | C | E | | | | | |

ON  Flower open (opening not observed).
OO  Flower open (opening on day shown).
C  Flower closed.
OE  Ovary enlarged.
OW  Ovary well developed (young fruit).
S  Flower shed (ovary not enlarged).
F  Fruit.
E  Calyx tube eaten.

individual flowers remain open for longer than one day. Once closed they remain closed. However, the opening times of flowers on any one cluster are normally staggered over several days. A mass of bright yellow pollen is present as soon as the anthers are fully open (0830–0930 hrs). In two flowers examined there was plenty of pollen at 0930 hrs but only a minimal amount at 1400 hrs. Many other flowers examined had only traces of pollen on the anthers by late afternoon.

A *Tetragona* bee was occasionally observed to alight on the stigma between pollen-scraping bouts. To test for the possibility that self-pollination by the bee might lead to fertilization, bagging experiments were performed to assess the compatibility status of flower clusters on five different vines. Plastic bags were placed over clusters and closed with string. A window, which had previously been cut out of each bag, had been closed by gluing a piece of fine cotton cloth over it. It was expected that this would reduce excessive humidity inside the bag. Flowers opening after the bags were in position were either self-pollinated or cross-pollinated by brushing the pollen-covered anthers, held in forceps by their filaments,

Table 16.5. Hand pollinations of *Passiflora longiracemosa* flowers

| | Crossed | | | Selfed | | |
|---|---|---|---|---|---|---|
| Cluster | Flowers pollinated | Flowers shed | Fruits produced | Flowers pollinated | Flowers shed | Fruits produced |
| 1 | 3 | 2* | 1 | 4 | 4 | 0 |
| 2 | 3 | 1 | 2 | 13 | 13 | 0 |
| 3 | 0 | 0 | 0 | 7 | 7 | 0 |
| 4 | 0 | 0 | 0 | 2 | 2 | 0 |
| 5 | 0 | 0 | 0 | 5 | 5 | 0 |
| Total | 6 | 3 | 3 | 31 | 31 | 0 |

* The ovary of one flower had doubled in size.

onto the stigmas. In the case of cross-pollinations, only anthers from bagged clusters on other vines were used. Flowers were marked as before, after pollination, and re-bagged for a further 24 hours. All selfed flowers were shed within four days (Table 16.5), and their ovaries were still the same size as at flower opening time. If fruit development is going to occur, the ovary is very obviously enlarged within two to three days of flower opening. The evidence from these five vines is that *P. longiracemosa* is self-incompatible and that any pollen a bee might deposit on the stigma would not result in fertilization. This is in agreement with previous observations on *Passiflora* (e.g. Masters, 1870).

It is possible that the *Tetragona* bee might effect fertilization of flowers if it transferred pollen to the stigma of a flower on another vine plant. To test for this possibility, chicken-wire cages were placed around flower clusters to permit bee visits but exclude those by hummingbirds. Chicken-wire, with a mesh size of approximately 2.5 × 2.0 cm, was shaped into a cage around each cluster and secured with string. Each day the cage was opened so as to mark, as before, flowers that had opened. The *Tetragona* bees were often observed inside the cages removing pollen from the anthers. Visual confirmation that they were entering the cages was not necessary, since pollen on the anthers of caged flowers diminished as rapidly as it did in uncaged control clusters. All of the pollen depletion in the cages can be attributed to the *Tetragona* bees since the pollen mass on the anthers of bagged flowers in the previous experiment remained throughout the day, i.e. there is no natural shedding of pollen during the course of the day. The caged clusters and the uncaged clusters (controls) were well spaced out in the forest. Hermit hummingbirds were observed visiting open flowers of three of the uncaged clusters. Table 16.6 summarizes the results of the caging experiments. It can be concluded that pollination can only occur by hummingbird visits. The *Tetragona* bees appear to be playing an exclusively negative role in pollination.

At least one other species of *Passiflora* vine (*P. vitifolia* Kunth), also pollinated by hummingbirds, shares common traits with *P. longiracemosa*. These include: the small numbers of bright scarlet flowers produced regularly at daily intervals; flowers that last less than 24 hours; and vines that are widely scattered (Janzen, 1968). These traits appear to be characteristic of many flowers pollinated by 'trapliner'

Table 16.6. *Passiflora longiracemosa* flowers with and without hummingbird visits

|  | No. | Flowers marked | Flowers shed with no increase in ovary size | Flowers shed with increase in ovary size | Fruits produced |
|---|---|---|---|---|---|
| Caged clusters | 15 | 61 | 61 | 0 | 0 |
| Uncaged clusters | 17 | 52 | 41 | 1 | 10 |

hummingbird species which visit flowers sequentially. The *Tetragona* bees, on the other hand, probably return to their nests after loading up with pollen at a flower or flower cluster; a behaviour pattern which would not promote cross-pollination between flowers of different vines.

## NECTAR THIEVES

Several times a bee, *Trigona williana* Friese, larger in size than the *Tetragona* pollen robbers, was observed feeding at the floral nectaries of *P. longiracemosa*. The bee gained access to the nectar by cutting a small hole at the base of the calyx tube (hypanthium).

## ANT GUARDS AGAINST ANTS

Nectaries in the form of small swellings exuding microscopic droplets of nectar occur on the sepals (Figure 16.3). Most of the flower clusters examined were attended by ants, and the following species were observed feeding at the nectaries: *Wasmannia auropunctata*, *Solenopsis* sp., *Daceton armigerum*, *Ectatomma edentatum* and *Ectatomma tuberculatum*. The last species was the most frequently observed. Individual ants moved around each flower-head visiting each nectary in turn to feed on the exudate. Ant attendance typically persisted over the period of anthesis, coincident with nectar production. One flower examined was still secreting droplets the day after it opened.

It has long been suspected that extra-floral nectaries confer protection against herbivory through the agency of aggressive and predacious ants attracted to them (Delpino, 1874, *fide* Bentley, 1977), and recent studies have provided evidence to support this (see Hölldobler and Wilson, 1990). However, it is not usually known which are the particular plant enemies being attacked or repelled.

One cluster of *P. longiracemosa* was observed being cut up by a leaf-cutting ant (*Acromyrmex subterraneus*). Very few ants were on the cluster, or on the sparsely populated trail leading to it, but, nevertheless, at the end of a three-day period little was left of the cluster. Both open and younger unopened flowers had been cut up, and another nearby cluster was similarly destroyed. No species of ant was in attendance at the nectaries during the destruction of these clusters. On a vine a few metres distant from the destroyed clusters, a solitary worker of *A. subterraneus* was

**Figure 16.3.** Ant collecting nectar from the calyx lobes of *Passiflora longiracemosa*

observed moving onto a flower-cluster with *Ectatomma tuberculatum* in attendance. A worker of the latter moved down from the flower-head to meet the leaf-cutting ant, and after some grappling the leaf-cutting ant fell to the ground. *E. tuberculatum* was also seen predating *Eciton burchelli* workers on the island; solitary *Ectatomma* ants snatched small *Eciton* workers from the base column (trail connecting the swarm raid with the bivouac). It is possible therefore that *E. tuberculatum*, while on the flower-clusters, predates *Acromyrmex* ants as well as repelling them, although it is very likely that the roles are reversed in the middle of a swarm raid, the predator becoming the prey.

The observations thus suggest that the particular selection pressure that may have driven the evolution of sepal nectaries in *P. longiracemosa* was protection from leaf-cutter ants (and probably other herbivores). The author has observed flower herbivory in other plant species by lepidopterous larvae, and also predation of caterpillars by *Ectatomma* ants elsewhere in Brazil. Holes chewed through the calyx tube wall or hypanthium (around the middle and not at the base as in the case of the nectar-thieving bee), with damage also to the inner floral parts, were observed in several flowers of *P. longiracemosa* but the animal responsible was not discovered.

## ACKNOWLEDGEMENTS

The author would like to thank William Milliken, Pe. Jesús Moure and Jacques Delabie for identification of the vine, bees and ants respectively.

# REFERENCES

Bentley, B.L. (1977). Extrafloral nectaries and protection by pugnacious bodyguards. *Ann. Rev. Ecol. Syst.*, 8, 407–427.

Faegri, K. and van der Pijl, L. (1979). *The principles of pollination ecology.* 3rd revised edition. Pergamon Press, Oxford.

Hölldobler, B. and Wilson, E.O. (1990). *The ants.* Bellknap Press of Harvard University Press, Cambridge, Massachusetts.

Janzen, D.H. (1968). Reproductive behavior in the Passifloraceae and some of its pollinators in Central America. *Behaviour*, 32, 33–48.

Masters, M.J. (1870). Contributions to the natural history of the Passifloraceae. *Trans. Linn. Soc.*, 27, 593–645.

# Field observations on Phoridae (Diptera) associated with ants on the Ilha de Maracá

FORBES P. BENTON
*CEPLAC/CEPEC/SECEN, Itabuna, Brazil*

## SUMMARY

Observations were made on the relationships between parasitic phorid flies (Diptera: Phoridae) and ants on the Ilha de Maracá. Detailed descriptions of these relationships are presented, including accounts of the parasitic attacks by phorids of the genera *Apocephalus*, *Cremersia*, *Diocophora*, *Neodohrniphora* and *Pseudacteon* on the leaf-cutter ant *Atta cephalotes*, the army ant *Eciton burchelli*, and other ants of the genera *Camponotus*, *Labidus*, *Pheidole* and *Azteca*. In addition, associations between non-parasitic phorid species of the genera *Apterophora*, *Dohrniphora*, *Ecitophora* and *Ecituncula* with *Atta cephalotes*, *Eciton burchelli* and *E. hamatum* are described.

## RESUMO

Foram feitas observações nas interações entre forídeos (Diptera: Phoridae) e formigas na Ilha de Maracá. São apresentadas descrições detalhadas destes relacionamentos, inclusive registros de ataques parasíticos por forídeos dos gêneros *Apocephalus*, *Cremersia*, *Diocophora*, *Neodohrniphora* e *Pseudacteon* na saúva *Atta cephalotes*, na correição *Eciton burchelli*, e em outras formigas dos gêneros *Camponotus*, *Labidus*, *Pheidole* e *Azteca*. Adicionalmente, são descritas associações entre forídeos não-parasíticos dos gêneros *Apterophora*, *Dohrniphora*, *Ecitophora* e *Ecituncula* com *Atta cephalotes*, *Eciton burchelli* e *E. hamatum*.

## INTRODUCTION

Associations of phorid flies (family Phoridae) with ants are particularly numerous and varied in their nature in tropical habitats, this being a reflection of the large biomass of ants present. It is within those major ant genera which form the principal contingent in terms of numbers and biomass that most cases of myrmecophily are known. Typically the association is one of true endoparasitism of the ant by the phorid. Where this endoparasitism has not been verified by rearing of the phorid larva from the adult ant, it may be inferred from the behaviour of the adult phorid female or from the morphology of her ovipositor. In the parasitic species the ovipositor is a more or less strongly chitinized rigid structure with a hollow needle-like tip for injecting eggs into the host (Borgmeier, 1931).

**Table 16.7.** Provisional list of the species of Phoridae collected on the Ilha de Maracá. Full identification of some specimens awaits taxonomic revisions

| | |
|---|---|
| *Acontistoptera* sp. nr. *mexicana* Malloch | *D. longirostrata* (Enderlein) |
| *Apocephalus hispidus* Borgmeier | *D. luteicincta* Borgmeier |
| *A.* sp. nr. *luteihalteratus* Borgmeier | *D.* sp. nr. *luteicincta* Borgmeier |
| *A.* spp. | *D. paraguayana* (Brues) |
| *Apodicrania termitophila* (Borgmeier) | *D.* sp. nr. *shannoni* Borgmeier |
| *A.* spp. | *D.* sp. nr. *simplex* (Borgmeier & Prado) |
| *Apterophora borgmeieri* Prado | *D. subsulcata* (Borgmeier & Prado) |
| *A.* sp. | *D. sulcata* Borgmeier |
| *Calamiscus* sp. | *D. ventralis* (Borgmeier & Prado) |
| *Chaetopleurophora scutellata* (Brues) | *D.* spp. |
| *Cremersia* sp. nr. *setitarsus* Borgmeier | *Ecitophora collegiana* Borgmeier |
| *Cremersia spinicosta* (Malloch) | *Ecituncula tarsalis* Borgmeier |
| *Diocophora appretiata* (Schmitz) | *Holopterina longipalpis sedula* Borgmeier |
| *D. disparifrons* Borgmeier | *Melaloncha* sp. |
| *D.* sp. | *Megaselia (Aphiochaeta)* sp. nr. *carlynensis* (Malloch) |
| *Dohrniphora anterosetalis* (Borgmeier & Prado) | *M. (A.) luteicauda* (Borgmeier) |
| *D.* sp. nr. *bisetalis* Borgmeier | *M. (Megaselia) imitatrix* Borgmeier |
| *D.* sp. nr. *brunnea* Borgmeier | *M. (M.) picta* (Lehmann) |
| *D. buscki* Malloch | *M. (A.) setigera* (Brues) |
| *D. dispar* (Enderlein) | *M.* spp. |
| *D. divaricata* (Aldrich) | *Neodohrniphora* spp. |
| *D. ecitophila* (Borgmeier) | *Pseudacteon* sp. |

# PARASITIC PHORID ATTACKS ON ANTS

Parasitic phorid species were found in abundance on the Ilha de Maracá (Table 16.7), and field observations were made of their attacks on the species of ants listed below.

## *Atta cephalotes*

On the edge of the forest there was a large and active nest of this leaf-cutting ant at the start of the main trail leading from the Ecological Station to Santa Rosa (see Figure Pr.1, p. xvii). During the observation period (January 1988) the ants were nocturnal. At the end of the day as the first ants were leaving the nest, and at first light as the last ants returned with their loads, large numbers of phorids were in evidence for several consecutive days. They were flying close to the ants as they moved along their trail. Each fly followed an ant at close quarters for a short while, accompanying any deviation taken in its path. The fly would then either make a single jab at the ant and abandon it, or it would make repeated jabs before moving on. Sometimes the ant would halt in its path and the fly would make what seemed to be a jab of longer duration. The jabbing consisted of a rapid touching of the ant by the fly's abdomen. In cases where the ant halted, the more prolonged jabs were directed at the head of the ant. Flies were aspirated and they proved to be females of five species. One was an undescribed species of *Apocephalus* Coquillett, morphologically close to *luteihalteratus* Borgmeier. The other four belonged to the *curvinervis* (Malloch) species group of the genus *Neodohrniphora* Malloch. These two genera are known to contain species which attack *Atta* (Borgmeier, 1931), and the jabbing can be assumed to be actual or attempted ovipositions. The flies were looked for at night with the aid of a gas lamp, when ant activity on the trail was intense, but none were seen.

On 9 November 1987 another trail of *A. cephalotes* with a moderate level of activity was observed, roughly between 1000 hrs and midday. The trail emanated from a nest approximately half-way between the Ecological Station and the largest of the seasonal lakes to the west. However, there was little evidence of parasitic phorids – only four were seen and two were collected. One was the undescribed *Apocephalus* sp., which was sitting on a piece of leaf that was being carried by a leaf-cutting ant. The other was the same as one of the *Neodohrniphora* species, and was flying along the ant trail. A further specimen of this *Neodohrniphora* species was collected while it was making cursory observations of an *A. cephalotes* nest entrance along the Preguiça trail. It was collected around the middle of the day and was the only phorid seen. Since *A. cephalotes* was in some circumstances diurnal on the island, it is tempting to speculate that the exclusively nocturnal appearance at the first nest mentioned may have been provoked by the high density of phorids present locally at the time (January 1988).

Wetterer (1990) and Lewis *et al.* (cited by Wetterer) also observed, in other geographical regions, that some colonies of *A. cephalotes* forage diurnally whilst others forage nocturnally. Lewis *et al.* found that the pattern of diel foraging activity was consistent along a particular trail for weeks or months at a time, but that diel changes could occur suddenly. However, they could find no physical environmental factor which could account for the changes. On the other hand, Wetterer implicated local abundance of diurnal phorid parasites as one possible factor accounting for these changes from diurnal to nocturnal foraging.

## *Eciton burchelli*

This 'swarm raiding' army ant was encountered on almost every walk in the forest during the study period. Sometimes two or three different colonies were seen on the same day. The related *Eciton hamatum*, a 'column raider', was similarly common. At the swarm front of one *E. burchelli* colony on the Preguiça trail, phorids were observed attacking clusters of worker ants which were attempting to extract an unidentified prey item from a dead branch. Microscopic examination of these flies revealed that they were females of an undescribed species which may warrant the erection of a new genus to accommodate it. Females of the same species were found in the Borgmeier collection of the Museu de Zoologia, Universidade de São Paulo, lodged under the genus *Cremersia* Schmitz, but bearing the label *Apocephalus* n.sp. They were collected on Barro Colorado Island, Panama. The morphology of the female ovipositor is to some extent intermediate between that of *Cremersia* and *Apocephalus*.

## Camponotus sp.

A swarm raid of *Eciton burchelli* was observed that had ascended a tall dead tree in the forest near the Ecological Station and struck an ant colony. Workers of a species of *Camponotus* rained down on the forest floor, where they were induced to flee from those *Eciton* workers on the ground. The fleeing *Camponotus* attracted large numbers of parasitic phorids, which attacked the *Camponotus* ants by jabbing at them. They were females of *Apocephalus hispidus* Borgmeier and an undescribed *Apocephalus* species closely related to *hispidus*. The genus *Apocephalus* is known to parasitize *Camponotus* ants (Pergande, 1901; quoted by Borgmeier, 1931).

At the periphery of another swarm raid of *E. burchelli*, two females of the phorid *Diocophora disparifrons* Borgmeier were collected as they pursued *Camponotus* ants that were fleeing from the *Eciton* workers. This phorid species is known to attack *Camponotus* ants that have been 'ferreted out' by other *Eciton* species (Borgmeier, 1959). Two other species of *Diocophora* were observed on the Ilha de Maracá, one being *D. appretiata* (Schmitz) and the other an undescribed species. Whenever looked for, these two species were always found lining the base column connecting the *Eciton* swarm raid with its bivouac. Both males and females of the two species flew up and down the base column, and were often to be seen resting in close proximity to the ants. Frequently they sat on dead twigs and leaves on the ground overhanging the column. The present author has never observed the flies strike at the *Eciton* ants. The same two species of *Diocophora* have also been observed by the author in the Atlantic coastal forests of Bahia, where they were exhibiting a similar behaviour pattern in association with *E. burchelli*. Borgmeier (1959) mentions the association of *D. appretiata* with *Eciton* ants, and the possibility that it may attack *Camponotus* ants, as has been observed in the case of *D. disparifrons*. Although the present author has not found *D. disparifrons* lining the base column and is unaware of it being so reported in the literature, the facts suggest that the females of other *Diocophora* species may be at the ready, alongside the base column, to pursue and paratisize their appropriate hosts (which may be *Camponotus* ants) once they have been flushed out of their nests by the *E. burchelli* workers.

*Camponotus* is regarded as one of the major ant genera in the neotropics, and observations on the Ilha de Maracá corroborate this statement. Three bivouacs of an *E. burchelli* colony were located while it was in the nomadic phase. The bivouacs were in the forest near the Ecological Station. As is characteristic for the species in the nomadic phase, the colony spent only 24 hours at each bivouac site. At each of these three abandoned bivouac sites the prey vestiges were predominantly those of *Camponotus* ants, identifiable by their characteristic head capsules. At the bivouac site of another colony of *E. burchelli* it was the same prey item that prevailed. *Camponotus* ants were one of the commonest ants to be seen foraging at night, but they were not seen foraging during the day. The parasitic species of Phoridae have well-developed eyes, and the characteristic behaviour of flying in close proximity to their ant host, accompanying any path deviation taken by the ant prior to oviposition, is suggestive of the importance of vision in pinpointing the host. It is probable therefore that they would not be able to do this at night.

It may be no coincidence that at least three species of parasitic phorids in two genera appear to have evolved a dependence on *E. burchelli* to flush out *Camponotus*, a host which would otherwise be hidden and inaccessible during the day.

## *Labidus coecus*

On two occasions colonies of this army ant were observed under attack by phorids. One colony was being attacked by two species of *Cremersia*, one of which was *C. spinicosta* (Malloch) or an extremely closely related species. The second species was an undescribed *Cremersia* sp., quite close to *C. setitarsus* Borgmeier. The other colony was being attacked only by the same undescribed *Cremersia* sp. Species of *Cremersia* are known to attack army ants of the genera *Labidus*, *Nomamyrmex* and *Neivamyrmex* (Borgmeier, 1963). Trails of *L. coecus* colonies are typically found partly on and partly below the ground. Sections of the trails on the surface are shielded by galleries whose walls and roofs are constructed from loosely assembled soil particles. Such was the case for the two colonies observed on the Ilha de Maracá. On the exposed parts of the trail and where the gallery had been breached, the *Cremersia* spp. were hovering close to the ants before attacking them. On either side of the moving column of ants, where it was exposed, there were stationary ants with their heads raised, mandibles open and antennae waving. Such 'guard' ants probably serve to deter phorid parasites and other arthropods.

## *Pheidole* sp.

On 15 November 1987, along the Filhote trail, two female phorids of an as yet unidentified species of the genus *Apocephalus* were flying close to an ant trail of *Pheidole* sp. Species of *Apocephalus* are known to attack *Pheidole* (Feener, 1981).

## *Azteca* sp.

During the afternoon of 23 January 1988 a swarm raid of *Eciton burchelli* was observed in the forest close to the Ecological Station. The swarm raid ascended a tree with a large carton nest of *Azteca* sp. affixed to the trunk just below the level of the canopy. Above the ant nest in the crown of the tree were four wasps' nests, one of which was considerably larger than the others. After viewing with binoculars and later consultation with Dr Anthony Raw, the wasps were tentatively identified as *Polybia liliacea*. The *Eciton* swarm raid did not plunder either the *Azteca* or *Polybia* nests. The greater part of the swarm raid was confined to lianas suspended from some of the branches, and relatively few *Eciton* ants were swarming over the carton nest of the *Azteca*, which apparently prevented access to the wasps' nests in the crown. As the swarm raid descended the tree, *Azteca* ants streamed down the trunk and lianas after it. Individual *Eciton* ants at the trailing fringe of the swarm were pinned down by small groups of *Azteca* ants, which themselves attracted phorid flies. The flies jabbed at the *Azteca* ants in the manner characteristic of parasitic Phoridae. These phorids were observed at eye level, and were females of a *Pseudacteon* Coquillett sp. After a while the *Azteca* released their hold on the *Eciton* and retreated back up the trunk, and by this time all that was left of the swarm raid were a few stragglers.

A different *Pseudacteon* species was observed attacking an *Azteca* sp. that was nesting in a *Cordia nodosa* (Boraginaceae) shrub, along the Anta Camp trail at the western end of the island. The swollen nodes of one of these shrubs were cut open, and females of the *Pseudacteon* species were attracted to the *Azteca* ants released. Species of *Pseudacteon* are known to parasitize ants of the genera *Solenopsis*, *Lasius*, *Dorymyrmex* and *Crematogaster* (Borgmeier, 1963), but these are the first observations of attacks on *Azteca*.

## NON-PARASITIC PHORIDAE ASSOCIATED WITH ANTS

The ovipositors of the Phoridae mentioned below are membraneous and hence not adapted for injecting eggs. This is one of the reasons for their assignment to the category 'non-parasitic'.

### *Eciton burchelli*

At the end of the afternoon of 20 January 1988, a colony of this species in the nomadic phase was observed as it began to enter a new bivouac (the same colony is referred to above under the heading *Camponotus*). Workers were pouring into a hollow tree trunk, at the entrance hole of which two species of Phoridae were aspirated: *Ecituncula tarsalis* Borgmeier and *Ecitophora collegiana* Borgmeier. A total of 101 males and eight females were obtained. Both species have wingless females and two couples were still *in copula* after immersion in alcohol at the time of collection. Probably all eight females were being airlifted by their male partners into the new bivouac. It is conceivable, however, that some proportion of the uncoupled males may have been flying in the opposite direction, away from the bivouac, in order to airlift females still at the previous night's bivouac, or positioned along the emigration trail, to the new bivouac. The implication of this is that a male may airlift more than one female during a colony emigration, and possibly obtain a copulation for each female airlifted. This is at least one reasonable explanation for the unequal sex ratio mentioned.

### *Eciton hamatum*

There was a bivouac of this ant on the Preguiça trail on 31 October 1987. Around the edge of the bivouac the following phorids were aspirated: *Ecitophora collegiana* females, *Apterophora* Brues sp. males, males of two unidentified species and a single female of *Dohrniphora ecitophila* Borgmeier.

### *Atta cephalotes*

Near to the *Atta cephalotes* trail observed on 9 November 1987 (referred to above), a few females of an *Apterophora* sp. were collected. Females of this genus are apterous. When offered freshly killed small insects in a humid chamber in the laboratory, the *Apterophora* sp. females fed avidly. They imbibed fluid, probably haemolymph, through their long thin proboscises.

## REFERENCES

Borgmeier, T. (1931). Sobre alguns Phorídeos que parasitam a saúva e outras formigas cortadeiras. *Arch. Inst. Biol. S. Paulo*, 4, 209–228.

Borgmeier, T. (1959). Neue und wenig bekannte Phoriden aus der neotropischen Region, nebst einigen Arten aus dem Belgischen Congo-Gebeit (Diptera, Phoridae). *Studia Ent.*, 2, 129–208.

Borgmeier, T. (1963). Revision of the North American Phorid flies. Part 1. The Phorinae, Aenigmatiinae and Metopininae, except Megaselia. *Studia Ent.*, 6, 1–256.

Feener, D.H. (1981). Competition between ant species: outcome controlled by parasitic flies. *Science*, 214, 815–817.

Wetterer, J.K. (1990). Diel changes in forager size, activity, and load selectivity in a tropical leaf-cutting ant, *Atta cephalotes*. *Ecological Entomology*, 15, 97–104.

# An Entomological Curiosity

FORBES P. BENTON
*CEPLAC/CEPEC/SECEN, Itabuna, Brazil*

## SUMMARY

Results are presented of observations on the interactions between a forest tree (*Couratari oblongifolia*, Lecythidaceae), a beetle (Attelabinae) and a leaf-cutter ant (*Atta cephalotes*) on the Ilha de Maracá. During a period of 24 hours the beetles stripped almost all of the young leaves from the tree, rolling them into small packets which fell to the ground. Each packet contained one egg of the beetle. Most of these packets were actively collected by the ants, which carried them to their nest. Observation showed that there was sufficient food in the leaf packet for complete development of the beetle, and it is not clear whether being taken to the *Atta* nest conferred an advantage or a disadvantage on the eggs.

## RESUMO

São apresentados os resultados de observações feitas nas interações entre uma árvore da floresta (*Couratari oblongifolia*, Lecythidaceae), um pequeno besouro (Attelabinae) e uma saúva (*Atta cephalotes*) na Ilha de Maracá. Durante um período de 24 horas, os besouros cortaram quase todas as folhas novas da árvore, produzindo pacotes enrolados que caíram no chão. Cada pacote contintia um ovo. A maioria destes pacotes foram coletados ativamente pelas formigas, e levados para o ninho. Observações mostraram que o pacote deu recursos suficientes para o desenvolvimento completo do besouro e não ficou claro se o deslocamento para o ninho de saúvas conferiu vantagem ou desvantagem.

## INTRODUCTION

Near the Ecological Station on the Ilha de Maracá, along the main trail to Santa Rosa (Figure Pr.1, p. xvii), small packages of rolled leaves were noticed dropping onto the forest floor one afternoon (January 1988). They were falling from the upper canopy of a tall tree of *Couratari oblongifolia* Ducke & Knuth (Lecythidaceae) that was in leaf flush. The packages were made from young leaves that had been rolled into small closed cigar-like cylinders averaging approximately 7.5 × 5.0 mm. Through binoculars, small beetles were observed in the canopy swarming and sitting on leaves. Very large numbers of the leaf cigars had accumulated on the forest floor as dusk fell. Each of several leaf cigars examined had one small white egg (or sometimes two) stuck to an inner roll of the leaf. The following morning few leaf cigars remained on the ground, but by the afternoon leaf cigar production was in full activity again.

The canopy of the *Couratari* tree was directly over an *Atta cephalotes* (leaf-cutter ant) trail, and that night it was possible to confirm that the ants were picking up the leaf cigars and carrying them underground into their nest. The main trail of the ants

Figure 16.4. *Atta cephalotes* ants carrying 'leaf cigars' made from *Couratari oblongifolia* by Attelabinae beetles

led to a tree where intense leaf-cutting and collection was taking place, about 100 m beyond the *Couratari* trunk. There was a junction in the main ant trail near the *Couratari* trunk, from which a secondary ant trail led off, branching abruptly to form an indefinite trail network, coinciding with the projection of the overhead *Couratari* canopy. In this area ants were picking up the fallen leaf cigars, and laden ants returning to the nest from the secondary trail were carrying leaf cigars exclusively (Figure 16.4). Some of the beetles were netted in the canopy and proved to be a species of weevil (Attelabinae).

The question of interest is whether there is a selective advantage to weevil eggs in leaf cigars that are taken below ground. That such large numbers of the leaf cigars were removed, and that they were removed by way of a secondary trail, suggests that the phenomenon may not have been mere coincidence. It is known that in some Brazilian Attelabinae there is more than sufficient food in the rolled leaf cigar for the larva to complete its development (Bondar, 1937). This was confirmed in the species from the Ilha de Maracá when adults were reared from leaf cigars brought to the laboratory. Most of the interior of the cigar was consumed by the larva, leaving the outer wall intact. By the time the adults were ready to emerge, the leaf cigars had dried up completely. A small exit hole was cut by the adult in the side wall. It appears therefore that there are no nutritional benefits enjoyed by the weevils taken into the ant nest.

In some Brazilian Attelabinae, the leaf cigars remain attached to the host plant until the weevil has completed its development. The weevil larvae of some of these species are attacked by hymenopterous parasites (Lima, 1956). Attelabinae species that allow the leaf cigars to fall to the ground as soon as cigar construction is completed may benefit by escape in evolutionary time from parasitism. It might further be speculated that removal of the cigars below ground by leaf-cutting ants also reduces the probability of parasitism.

It has to be said that leaf-cutting ants will pick up and carry to their nest vegetable material of varied origins. It is possible therefore that there is no adaptive significance to the weevil in being removed below ground by leaf-cutting ants, and that all such weevils may be doomed.

## ACKNOWLEDGEMENTS

The author would like to thank Cosme Damião A. de Mota and José Ramos of INPA Botany Department for climbing a tree to net the beetles.

## REFERENCES

Bondar, G. (1937). Observações sobre Curculionideos enroladores de folhas (Col.). *Rev. Ent.*, 7 (2–3), 141–144.

Lima, A. da C. (1956). Insetos do Brasil. 10. Coleópteros (4). *Esc. Nac. Agron. Série didática*, no. 12, 1–373.

# Litter-consuming termites on the Ilha de Maracá

ADELMAR G. BANDEIRA
*Instituto Nacional de Pesquisas da Amazônia, Manaus, Brazil*

## SUMMARY

A study of leaf-litter-consuming termites on the Ilha de Maracá revealed the presence of seven species: *Ruptitermes reconditus, R. silvestrii, R.* sp. nov., *Syntermes chaquimayensis, S. molestus, S. parallelus* and *Nasutitermes guayanae*. The average diameter of the circular cuts made in leaves by these species was 3.0 mm for *S. molestus*, 3.5 mm for *S. parallelus*, 5.5 mm for *S. chaquimayensis* and approximately 1 mm (when circular) for the *Ruptitermes* species and *N. guayanae*. Higher levels of detritivorous activity were observed in primary forest than in artificial clearings, principally by *Ruptitermes*. Litter-consuming termites appear to play an important role in removal of fine litter and its incorporation in the soil on the Ilha de Maracá, although this was not quantified experimentally.

## RESUMO

Um estudo dos térmites consumidores de liteira na Ilha de Maracá, Roraima, revelou sete espécies: *Ruptitermes reconditus, R. silvestrii, R.* sp. nov., *Syntermes chaquimayensis, S. molestus, S. parallelus* e *Nasutitermes guayanae*. O diâmetro médio dos cortes circulares nas folhas foi de 3.0 mm para *S. molestus*, 3.5 mm para *S. parallelus*, 5.5 mm para *S. chaquimayensis* e de aproximadamente 1 mm (quando circulares) para as espécies de *Ruptitermes* e *N. guayanae*. Encontrou-se maior atividade detritívora, principalmente de *Ruptitermes*, em mata primária, comparada com clareiras artificiais. Os térmites consumidores de liteira em Maracá aparentaram ter papel fundamental na remoção e incorporação da liteira fina ao solo, embora não tenha sido calculada experimentalmente a taxa desta atividade.

## INTRODUCTION

Termites are the most important group of invertebrates in nutrient recycling in tropical regions, and are responsible for decomposition of over 50% of organic matter in some tropical ecosystems (Abe, 1980).

Four families of termites are found in the Brazilian Amazon: Kalotermitidae and Rhinotermitidae, which are strictly xylophagous, the monospecific Serritermitidae which live (as 'tenants') with *Cornitermes* spp. (Termitidae), feeding on detritus collected by their 'hosts' (Araújo, 1977), and Termitidae, the largest group, which includes over 90% of the Amazonian species. The last group shows the greatest diversity of feeding preferences, although the majority are xylophagous (Bandeira, 1989).

It is already known that in Amazonian *terra firme* forests the *Syntermes* species feed on leaf-litter (Bandeira, 1979; Mill, 1982; Luizão and Schubart, 1986). Bandeira and Torres (1985) also observed a species of *Nasutitermes* cutting fallen leaves close to Belém, and *Ruptitermes* have been observed foraging in litter in Mato Grosso by Mathews (1977) and at three Amazonian sites (Iquê-Jurena (Mato Grosso), Anavilhanas (Amazonas) and the Ilha de Maracá (Roraima)) by Mill (1982).

In this chapter the species of litter-feeding termites on the Ilha de Maracá are reported, and qualitative and quantitative aspects of their feeding are analysed. The abundance of foraging groups in two forest sites and two artificial clearings is also compared.

## METHODS

Collections and observations were made in primary forest plots (Nos 3 and 6) and artificial clearings (Nos 1 and 2) established for the studies of forest regeneration and litter decomposition reported by Scott *et al.* (1992) and Thompson *et al.* (1992). The clearings (50 × 50 m) were made in February 1987 (Figure 23.2, p. 436). Each plot was divided into 10 × 10 m quadrats. The first termite collection was made in February 1988, when species activity in each quadrat was recorded for all of the plots. Termites were collected among the litter, below fallen trunks, and in epigeal nests. Collections were preserved in 75% ethanol. The methodology of sampling from contiguous 10 × 10 m quadrats was also employed by La Fage *et al.* (1973) to observe subterranean termite foraging activity in rolls of lavatory paper in Arizona.

In each quadrat, 10 random leaf collections were made for analysis of termite activity, using a thick wire spike to impale the litter. After preliminary observations and analysis of activity in the field, the leaves were classified into four categories: (i) attacked by *Syntermes*, when showing circular cuts 3.0–5.5 mm in diameter, (ii) attacked by *Ruptitermes* and/or *Nasutitermes*, when showing circular cuts approximately 1.0 mm in diameter, (iii) attacked by *Syntermes* and *Ruptitermes/ Nasutitermes*, and (iv) not showing signs of termite attack.

The $\chi$-squared statistical test (2 × 2 contingency table) (Silveira Neto *et al.*, 1976) was used to test for significant differences between the forest and clearing plots for: (i) frequency of quadrats showing termite attack; (ii) proportion of attacked leaves (all termite species); and (iii) leaf attack by *Syntermes* compared with *Ruptitermes/ Nasutitermes*.

## RESULTS

### FEEDING PREFERENCES

The only termites found eating leaves exclusively were *Syntermes* spp., including *S. chaquimayensis*, *S. molestus* and *S. parallelus*. The *Ruptitermes* species, which belong to the soldier-less termite group (Apicotermitinae), were represented by *R. reconditus*, *R. silvestrii* and *R.* sp. nov., all of which were found to be eating both

Table 16.8. Abundance (per hectare) of groups of *Ruptitermes* and *Syntermes* foraging in primary forest and in artificial clearings (2500 m$^2$) on the Ilha de Maracá (percentages in parentheses)

| Species | Primary forest | Artificial clearings |
|---|---|---|
| *Ruptitermes reconditus* | 26 (31.0) | – |
| *R. silvestrii* | 18 (21.4) | 16 (44.5) |
| *R.* sp. nov. | 8 (9.5) | 4 (11.1) |
| Subtotal | 52 (61.9) | 20 (55.6) |
| *Syntermes chaquimayensis* | 22 (26.2) | 8 (22.2) |
| *S. molestus* | – | 4 (11.1) |
| *S. parallelus* | 10 (11.9) | 4 (11.1) |
| Subtotal | 32 (38.1) | 16 (44.4) |
| Total | 84 (100) | 36 (100) |

leaves and the interiors of pieces of wood in an advanced state of decomposition (and always in contact with the ground). *Nasutitermes guayanae* was occasionally observed eating dry leaves, but was more commonly seen to eat wood (like the other 14 species of this genus encountered in the study area). *Syntermes* and *Ruptitermes* forage in the litter and carry pieces of leaves into underground nests.

One can identify the species of *Syntermes* by the diameters of their circular cuts in the leaves, being 3.0 mm (mean) for *S. molestus*, 3.5 mm for *S. parallelus*, and 5.5 mm for *S. chaquimayensis*. *Ruptitermes* spp. and *N. guayanae* do not always make circular cuts, and when they do these are approximately 1.0 mm in diameter. *Syntermes* was found to attack old, recently fallen or even green leaves in the litter, whereas *Ruptitermes* has a preference for old leaves infected by fungi. *Nasutitermes guayanae*, like *Syntermes*, eats both old and young leaves.

## FOREST VERSUS CLEARING

The densities of foraging termite groups for *Ruptitermes* and *Syntermes* were estimated as 52 and 32 per hectare respectively for the forest plots, and 20 and 16 for the clearings (Table 16.8). The most common species were *R. reconditus* and *S. chaquimayensis* in primary forest, and *R. silvestrii* and *S. chaquimayensis* in clearings. The number of termites per foraging group was generally between 200 and 300 for the *Ruptitermes* spp., 10 for *S. chaquimayensis*, 40 for *S. parallelus* and over 100 for *S. molestus*.

The frequencies of 10 × 10 m quadrats containing leaves attacked by *Ruptitermes/Nasutitermes* were 100% in forest plots and 92% in the clearings, and those for *Syntermes* attack were 70 and 82% respectively (Table 16.9). There was no significant difference between the data from forest and clearings. The average proportion of leaves attacked by *Syntermes*, *Ruptitermes/Nasutitermes*, and *Syntermes/Ruptitermes/Nasutitermes* was 10, 20 and 4% in the primary forest sites and 15, 18 and 5% in the clearings respectively. Summing these figures without distinguishing between

**Table 16.9.** Frequency of quadrats (10 × 10 m) containing leaves attacked by *Syntermes* and other termites (*Ruptitermes* and *Nasutitermes*) in primary forest and artificial clearings (2500 m²) on the Ilha de Maracá (expected values in parentheses)

| Environment | *Syntermes* | *Ruptitermes* and *Nasutitermes* | Total |
|---|---|---|---|
| Primary forest | 70 (75.12) | 100 (94.88) | 170 |
| Clearings | 82 (76.88) | 92 (97.12) | 174 |
| Total | 152 | 192 | 344 |

$\chi^2 = 1.2$, n.s.

**Table 16.10.** Percentages of leaves attacked or not attacked by termites in primary forest and artificial clearings (2500 m²) on the Ilha de Maracá

| Environment | Leaves attacked | Leaves not attacked | Total |
|---|---|---|---|
| Primary forest | 33 | 64 | 100 |
| Clearings | 38 | 62 | 100 |
| Total | 71 | 129 | 200 |

**Table 16.11.** Percentages of collected leaves attacked by *Syntermes* and by other termites (*Ruptitermes* and *Nasutitermes*) in primary forest and artificial clearings (2500 m²) on the Ilha de Maracá

| Environment | *Syntermes* | *Ruptitermes* and *Nasutitermes* | *Syntermes*/*Ruptitermes*, *Nasutitermes* |
|---|---|---|---|
| Primary forest | 10 | 20 | 4 |
| Clearings | 15 | 18 | 5 |

consumer categories, one finds 34% of attacked leaves in forest and 38% in clearings, which does not represent a significant difference (Table 16.10). The consumption of leaves by *Syntermes* was significantly lower than that by other species (Table 16.11).

The collection frequency of *Nasutitermes guayanae* was 88% in the forest plots and 84% in the clearings. The only 'cartonado' nest of this species was encountered in one of the clearings. All the other collections were made beneath the bark of tree trunks, in covered tunnels on the trunks and in galleries in dead wood.

The predation of *Syntermes* by the ant *Pachycondyla commutata* was frequently observed, a phenomenon which is already well known (Wheeler, 1936; Mill, 1984a). On the other hand, no such predation was observed on *Ruptitermes* spp., although these were foraging openly among the leaves without soldiers for protection. The workers of these species contain defensive substances which are liberated (when under threat) by the rupture of reservoirs in their labial glands, which extend back to the anterior part of the abdomen (see also Mathews, 1977; Mill, 1984b).

## DISCUSSION AND CONCLUSION

The species of *Syntermes* encountered on Maracá by Mill (1984a), who also worked in the eastern part of the island, were *S. calvus*, *S. molestus* and *S. solidus*. During the present study, in spite of the fact that collections were made outside the study plots (including at night, up to 2200 hrs), *S. calvus* and *S. solidus* were not encountered. The species collected during these forays were the same as those in the plots.

No experiment was carried out to evaluate quantitatively the consumption of leaves by termites. To obtain an indication of foraging activity by these insects, the percentage of partially eaten leaves was estimated (34% in the forest and 38% in clearings), as discussed previously. This measurement was made in February, during the dry season, when there was an accentuated leaf fall due to the abundance of deciduous tree species in the forests of that part of Maracá (see Milliken and Ratter, Chapter 5). The comparison of percentages of attacked leaves does not give a precise idea of the relative activity of termites in forest and clearing sites, since the litter fall was considerably higher in the forest than in the clearings (which only received leaves from the trees around their edges). The abundance of termites, and thus the overall activity, was actually higher in the forest where the environmental conditions are apparently more favourable for them.

The leaf-eating termites do not generally consume everything, but tend to leave the thicker nerves and sometimes part of the lamina. The ability to identify the feeding species depended on the quantity of leaf remaining for examination. The method of collecting leaves with a spike did not pick up small pieces, and thus the results are more significant if interpreted as indicating the initial (rather than overall) termite activity.

The average litter production in Brazilian tropical forests is around 8 tonnes/ha/yr (Dantas and Phillipson, 1989). On Maracá, Scott *et al.* (1992) estimated a fall of 10.27 t/ha/yr of fine litter in primary forest, including two of the sites used in this study, during the same period (1987–88). According to Proctor (1983), approximately 70% of the detritus produced in equatorial forests consists of leaves, but this proportion could be higher if large palm fronds were included.

Luizão and Schubart (1986) estimated that termites of the genus *Syntermes* were responsible for the removal of over 40% of the leaves in an experimental site close to Manaus. These authors did not consider the eventual participation of other termites in the removal process, but Abe (1980) showed that termites in general are responsible for over 50% of detritivorous activity in Malaysian forests. Thus, considering the production and decomposition data, one could expect that termites may be consuming over 5 t/ha/yr of fine litter on Maracá. This is of considerable importance, since it may be the principal route by which organic material is incorporated in the soil.

Although it is well known that *Syntermes* and *Ruptitermes* are litter consumers, there are no previous reports in the literature on the destination of the leaves carried into their nests, nor of the depth of these nests. This is certainly an important subject for further study, since it is known that Amazonian soils are generally poor in nutrients, these being concentrated in the relatively thin organic horizon.

One cannot discount the possibility that other species of *Nasutitermes*, as well as *N. guayanae*, also consume dry leaves on Maracá. Normally the termites of this genus forage with the protection of tunnels or other structures, and it is therefore difficult to observe their foraging activity. It should also be mentioned that the gross litter (trunks, branches, etc.) is also largely consumed by termites. Various genera participate in this process, of which the most important on Maracá is *Nasutitermes*.

To conclude, it is believed that litter-consuming termites play a fundamental part in the decomposition of this material on Maracá, and thus play a crucial role in ecology. However, the exact magnitude of their contribution to this process has yet to be quantified.

## ACKNOWLEDGEMENTS

Thanks are due to Cláudio Sena, who helped with the fieldwork, to Herbert Schubart, João Ferraz, Lucille Anthony and Flavio Luizão for critical review of the manuscript, and to Eliana Tamar Ribeiro initial typing of the text (all from INPA). William Milliken translated the manuscript into English.

## REFERENCES

Abe, T. (1980). Studies on the distribution and ecological role of termites in a lowland rain forest of West Malaysia. 4. The role of termites in the process of wood decomposition in Pasoh Forest Reserve. *Rev. Ecol. Biol. Soc.*, 17 (1), 23–40.

Araújo, R.L. (1977). Further notes on the bionomics of *Serritermes* (Isoptera). *Revta Bras. Ent.*, 21 (2), 31–32.

Bandeira, A.G. (1979). Notas sobre a fauna de cupins (Insecta: Isoptera) do Parque Nacional da Amazônia (Tapajós), Brasil. *Bol. Mus. Para. Emílio Goeldi, nov. sér. Zool.*, 96, 1–12.

Bandeira, A.G. (1989). Análise de termitofauna (Insecta: Isoptera) de uma floresta primária e de uma pastagem na Amazônia Oriental, Brasil. *Bol. Mus. Para. Emílio Goeldi, sér. Zool.*, 5 (2), 225–241.

Bandeira, A.G. and Torres, M.F.P. (1985). Abundância e distribuição de invertebrados do solo em ecossistemas da Amazônia Oriental. *Bol. Mus. Para. Emílio Goeldi, sér. Zool.*, 2 (1), 13–38.

Dantas, M. and Phillipson, J. (1989). Litterfall and litter nutrient content in primary and secondary Amazonian 'terra firme' rain forest. *Journal of Tropical Ecology*, 5, 27–36.

La Fage, J.P., Nutting, W.L. and Haverty, M.I. (1973). Desert subterranean termites: a method for studying foraging behaviour. *Environmental Entomology*, 2 (5), 954–956.

Luizão, F.J. and Schubart, H.O.R. (1986). Produção e decomposição de liteira em floresta de terra firme da Amazônia Central. *Acta Limnol. Bras.*, 1, 575–600.

Mathews, A.G.A. (1977). *Studies on termites from the Mato Grosso State, Brazil*. Academia Brasileira de Ciências, Rio de Janeiro.

Mill, A.E. (1982). Faunal studies on termites (Isoptera) and observations on their ant predators (Hymenoptera, Formicidae) in the Amazon basin. *Revta Bras. Ent.*, 26 (3/4), 253–260.

Mill, A.E. (1984a). Predation by the ponerine ant *Pachycondyla commutata* (Hymenoptera: Formicidae) on termites of the genus *Syntermes* in the Amazon Basin. *Journal of Natural History*, 18 (3), 405–410.

Mill, A.E. (1984b). Exploding termites – an unusual defensive behaviour. *Entomologist's Monthly Magazine*, 120, 179–183.

Proctor, J. (1983). Tropical forest litterfall. I. Problems of data comparison. In: *Tropical rain forest: ecology and management*, eds S.L. Sutton, T.C. Whitmore and A.C. Chadwick. pp. 267–273. Blackwell, Oxford.

Scott, D.A., Proctor, J. and Thompson, J. (1992). Ecological studies on a lowland evergreen rainforest on Maracá Island, Roraima, Brazil. II. Litter and nutrient cycling. *Journal of Ecology*, 80, 705–717.

Silveira Neto, S., Nakano, O., Barbin, D. and Villa Nova, N.A. (1976). *Manual de ecologia dos insetos*. Editora Agronômica Ceres, Piracicaba.

Thompson, J., Proctor, J. and Scott, D.A. (1992). A semi-evergreen forest on Maracá Island I. Physical environment, forest structure and floristics. In: *The rainforest edge*, ed. J. Hemming, pp. 19–29. Manchester University Press.

Wheeler, W.M. (1936). Ecological relations of ponerine and other ants to termites. *Proc. Amer. Acad. Arts. Sci.*, 71 (3), 159–243.

# Butterflies of the Ilha de Maracá

OLAF H.H. MIELKE AND MIRNA M. CASAGRANDE
*Universidade Federal de Paraná, Curitiba, Brazil*

## SUMMARY

A survey of the butterflies of the Ilha de Maracá was conducted during the wet and dry seasons, in forest and in savanna. A total of 453 species were collected, mostly from the forest understorey, including one species new to science (*Pythonides maraca*) and five new records for Brazil. An annotated species list is provided in an appendix.

The composition of the butterfly fauna appears to be relatively typical of Amazonia, and although the species collected probably only represent 50% of those present, it is not exceptionally diverse. Only 14 of the species collected are typical of savanna vegetation, and there was no evidence of endemism. Butterfly populations were found to be considerably lower during the dry season than during the rainy season.

## RESUMO

Levantamentos das borboletas da Ilha de Maracá foram realizados nas estações seca e chuvosa, em savana e floresta. Um total de 453 espécies foi coletado, principalmente do sub-bosque, inclusive uma espécie nova para a ciência (*Pythonides maraca*) e cinco registros novos para o Brasil. É fornecida uma lista anotada num apêndice.

A composição da Lepidopterofauna da Ilha parece ser mais ou menos típica da Amazônia, e, embora a coleção provavelmente represente apenas 50% da fauna existente, não é excepcionalmente diversa. Apenas 14 do total são espécies típicas da vegetação da savana ou do cerrado, e não há evidência de endemismo na fauna. A população de borboletas foi mais baixa na estação seca do que na estação chuvosa.

## INTRODUCTION

A list of the species of butterflies (Lepidoptera) collected during the authors' four visits to the Ilha de Maracá during the Maracá Rainforest Project, together with additional species collected by the entomological team of Dr José Albertino Rafael (INPA), is presented in Appendix 6, p. 467. All of the specimens were collected on the principal trail leading from the Furo de Maracá (through the Ecological Station) to the Furo de Santa Rosa, and in the adjacent network of tracks (Figure 23.2, p. 436). The three principal vegetational units encountered along this route (described in detail by Milliken and Ratter, Chapter 5) include:

1. Secondary vegetation alongside the causeway traversing the seasonally flooded savanna below the Ecological Station.
2. Riverside forest, and *terra firme* forest between the Ecological Station and Santa Rosa.
4. Typical *Curatella americana/Byrsonima crassifolia* savanna at Santa Rosa.

The total number of collections was 2724, and the most productive collecting expedition was in August 1987 (the first visit). The least productive was in May 1988 (the last visit). It may, however, be that the optimal time would have been in July, since by the end of August the majority of the butterflies caught had already been severely damaged.

The specimens are lodged in the collection of the Departamento de Zoologia of the Universidade Federal do Paraná and the Instituto Nacional de Pesquisas da Amazônia.

## MATERIALS AND METHODS

Collections were made using butterfly nets and traps, the latter consisting of cylindrical wire cages approximately 50 cm high baited with fermented bananas and sugar-cane juice.

## RESULTS

A list of the species of Lepidoptera collected on the Ilha de Maracá is presented in Appendix 6, together with data on capture dates and habitats. The 2724 specimens collected represent 453 species, including the new species *Pythonides maraca* (Figure 16.5) which was described for the first time by Mielke and Casagrande (1991). These species represent the following families and subfamilies:

- Papilionidae (8 Papilioninae)
- Pieridae (4 Pierinae, 8 Coliadinae, 1 Dismorphiinae)
- Nymphalidae (9 Ithomiinae, 2 Danainae, 37 Satyrinae, 12 Brassolinae, 2 Morphinae, 55 Nymphalinae, 2 Apaturinae, 18 Charaxinae)
- Libytheidae (1)
- Lycaenidae (65 Riodininae, 46 Lycaeninae)
- Hesperiidae (4 Pyrrhopyginae, 85 Pyrginae, 94 Hesperiinae)

## DISCUSSION

Bearing in mind that the number of collections made and the number of collecting days spent on the Ilha de Maracá (27) were relatively low, it is not currently possible to make a realistic estimate of the overall diversity of butterflies on the island. As a very rough estimate, however, the 453 species captured to date can probably be taken to represent approximately 50% of the total. Studies of the diversity of Lepidoptera are still in their infancy, but three such projects are currently under way in the headwaters of the Rio Madeira. The three sites, two of which are in Peru (Tambopata, Rio Tambopata and Manu, Rio Madre de Dios), and one in Brazil (Rondônia), have so far yielded 1203, 905 and 1500 species

# BUTTERFLIES

**Figure 16.5.** *Pythonides maraca* – a new species of butterfly for the Ilha de Maracá. 1–2 = male holotype (dorsal and ventral); 3–4 = female holotype (dorsal and ventral)

respectively (Lamas, 1981; Brown, 1984; Emmel and Austin, 1990), the last being the highest diversity recorded anywhere in the world. Areas whose vegetation has been altered by man often provide more habitats (successional stages of secondary vegetation, etc.) and thus generally yield higher numbers of species than those where the vegetation is undisturbed. Although some of the vegetation at the eastern end of the Ilha de Maracá appears to have suffered such alteration, the majority of its area is undisturbed and it is unlikely to support as diverse a fauna as that of the Madeira basin.

The diversity of the vegetation on the Ilha de Maracá appears to contribute little to the diversity of its Lepidoptera; most of the butterfly species were collected from forest habitats and only 14 of the 453 species are typical of savanna. The areas of savanna on the island are too small to maintain large numbers of butterfly species, but the species which were found there do not differ significantly from those one would expect to find in other areas of Brazilian savanna or *cerrado* vegetation. There are no families or sub-families characteristic of these habitats, but certain genera are typical of them such as *Eurema* (Pieridae), *Danaus*, *Anosia*, *Junonia*, *Colaenis* and certain species of '*Euptychia*' (Nymphalidae), *Audre* and certain species of '*Thecla*' (Lycaenidae), and *Cogia*, *Chiomara*, *Pyrgus*, *Heliopetes*, *Lerodea*, *Pompeius*, *Wallengrenia*, *Hylephila*, *Polites*, *Copaeodes*, *Vehilius*, *Nastra*, *Vidius*, *Mnaseas* and *Corticea* (Hesperiidae).

The forest butterfly fauna of Maracá is fairly typical for Amazonia, but the groups which are particularly useful for comparing these faunas (e.g. *Parides* (Papilionidae) and Nymphalidae (Ithomiinae)), although normally abundant and diverse, were scarcely captured on the island. The only species of *Parides* taken there (*P. sesostris*) is common throughout Amazonia, and within the Ithomiinae only *Hypothyris ninonia colophonia* is typical of Roraima and Venezuela. The western end of the island was not visited, but as the diversity of forest tree species is greater (Milliken and Ratter, Chapter 5), the faunal diversity is also likely to be higher.

The pronounced dry season experienced on the Ilha de Maracá inevitably has a significant effect on the butterfly population; the most successful collecting period during the present study was in August (two or three months after the beginning of the rains), and subsequent visits during drier periods yielded relatively few specimens. There does not, however, appear to be a typical dry season butterfly fauna. During the last visit (in May) nine species were added to the list (*Phoebis sennae*, *Cepheuptychia cephus*, *Manataria maculata*, *Eunica caresa*, '*Thecla*' sp., *Urbanus esma*, *Mylon pelopidas*, *Timochares trifasciata trifasciata* and *Camptopleura auxo*), but of these all but two (*Manataria maculata* (10 specimens) and *E. caresa* (4 specimens)) were collected only once and their capture could probably be attributed to chance. *E. caresa* is possibly a migratory species, so *M. maculata* may be the only species truly typical of the end of the dry season. The specimens collected were in very poor condition, and it seems likely that this species flew in March since they were not encountered during an earlier visit in February.

There is evidently a significant difference between the butterfly fauna of the forest canopy and of the understorey, but since the majority of the collections made during this study were from the understorey (in many cases it was not possible to identify the species seen flying in the canopy) we cannot evaluate these differences in detail. Amongst the Papilionidae one could observe the species of *Eurytides* and *Papilio* flying high in the forest, but sometimes they came down to visit flowers or damp sand. Certain genera of Nymphalidae also include canopy-flying species, including *Prepona*, some *Callicore* spp. and *Megeuptychia antonoe*, the last of which is particularly interesting since all other Satyrinae fly in the understorey. There did not, however, appear to be any typical canopy species among the Pieridae.

Although most of the species collected represent new records for the State of Roraima, only five (one in the Nymphalidae and four in the Hesperiidae) were previously unrecorded for Brazil. The discovery of new Amazonian butterfly species is a relatively rare occurrence in Brazil, since these insects have been fairly comprehensively collected in the past by amateurs and professionals alike. The only species from the Ilha de Maracá which was new to science, *Pythonides maraca*, is not endemic to the island but has also been collected in the State of Amazonas and in Venezuela. Such lack of endemic species is unremarkable bearing in mind that the vegetational and topographic units of the reserve are essentially continuous with those of adjacent regions. The only species which has been collected in no Brazilian locality other than Maracá is *Polites vibicoides* (Hesperiidae), a species which was described from Surinam in 1983.

## ACKNOWLEDGEMENTS

We are sincerely grateful to Drs R.J. Vane-Wright and P.R. Ackery (BMNH) for enabling us to compare our material with the collections of that institution.

## REFERENCES

Brown, K.S. Jr (1984). Species diversity and abundance in Jaru, Rondônia (Brazil). *News Lepid. Soc., Lawrence*, 1984 (3), 45–47.
Emmel, T.C. and Austin, G.T. (1990). The tropical rainforest butterfly fauna of Rondônia, Brazil: species diversity and conservation. *Trop. Lep., Gainesville*, 1 (1), 1–12.
Lamas, G. (1981). La fauna de mariposas de la reserva de Tambopata, Madre de Dios, Peru (Lepidoptera, Papilionoidea e Hesperioidea). *Revta Soc. Mex. Lepid., Mexico*, 6 (2), 23–39.
Mielke, O.H.H. and Casagrande, M.M. (1991). Lepidoptera: Papilionoidea e Hesperioidea na Ilha de Maracá, Alto Alegre, Roraima, parte do Projeto Maracá, com uma lista complementar de Hesperiidae de Roraima. *Acta Amazonica*, 21, 175–210.

# 17 Insects of the Ilha de Maracá Further Contributions II: Medical Entomology

## Sandflies (Diptera: Psychodidae) of the Ilha de Maracá

ELOY G. CASTELLÓN[1], NELSON A. ARAÚJO FILHO, NELSON F. FÉ[2] AND JOSETTE M.C. ALVES[3]

[1]*Instituto Nacional de Pesquisas da Amazônia, Manaus, Brazil;* [2]*Instituto de Medicina Tropical de Manaus, Manaus, Brazil;* [3]*Secretaria Estadual de Educação e Cultura, Manaus, Brazil*

### SUMMARY

The results of the first survey of the sandflies (Diptera, Psychodidae) of the Ilha de Maracá are presented. A total of 55 species, of two genera (*Brumptomyia* and *Lutzomyia*) were recorded, of which the most common were *L. davisi* and *L. squamiventris squamiventris* (accounting for 49 and 29% of the captures respectively). These 55 species represent 80% of the sandflies known to exist in the State of Roraima. Trapping was conducted solely with CDC traps, and further studies using other techniques would inevitably raise the species count substantially.

Two of the species collected, *Lutzomyia flaviscutellata* and *L. umbratilis*, are known to carry cutaneous leishmaniasis, and a further seven are suspected of doing so. One of the species, *L. longipalpis*, is known to carry visceral leishmaniasis elsewhere in Roraima, but this disease has not been recorded in the Maracá area.

### RESUMO

Apresentam-se os resultados do primeiro levantamento de flebotomíneos (Diptera, Psychodidae) da Ilha de Maracá. Foi registrado um total de 55 espécies de dois gêneros (*Brumptomyia* e *Lutzomyia*), das quais as mais comuns foram *L. davisi* e *L. squamiventris squamiventris* (49 e 29% das capturas, respectivamente). Estas espécies representam 80% dos flebotomíneos conhecidos no Estado de Roraima. Os insetos foram capturados apenas com armadilhas CDC, e outros estudos usando outras técnicas certamente dariam um aumento na contagem de espécies.

Duas das espécies coletadas (*Lutzomyia flaviscutellata* e *L. umbratilis*) são conhecidas como portadores do agente de leishmaniose tegumentar, e mais sete são suspeitas do mesmo. Além disso, uma das espécies (*L. longipalpis*) é conhecida como transmissor de leishmaniose visceral em Roraima, mas não foi registrada incidência desta doença grave nas proximidades de Maracá.

## INTRODUCTION

Existing data on the sandfly fauna of the State of Roraima are limited to the publications of Martins *et al.* (1963), who carried out an inventory in the southern and central regions of the State, Fraiha *et al.* (1974), who reported on their collections from the road between Boa Vista and Caracaraí (13 km below the Rio Mucajaí), and Castellón *et al.* (1989, 1991) who published details of the sandfly fauna of the southern, central and northern regions.

Here data are presented on the sandflies collected from the Ilha de Maracá in 1987 and 1988. These were the first such data collected from the area.

## MATERIALS AND METHODS

Specimens were collected between 1800 and 0600 hrs using CDC (light) traps suspended at heights of 1, 5 and 10 m in the forest. The processes used in the laboratory mounting of the specimens are described in Ryan (1986). All of the material is lodged in the collection of the Instituto Nacional de Pesquisas da Amazônia (INPA), Manaus.

## RESULTS

A total of 16 289 specimens of sandflies were collected on the Ilha de Maracá. A list of the 55 species of two genera recorded on the island is presented in Table 17.1, together with details of the known distributions of the species in the State of Roraima.

## DISCUSSION

The 55 species and subspecies of sandflies so far collected on the Ilha de Maracá represent a high proportion (80%) of the 69 known for the State of Roraima. Of the 14 species that were not recorded on the island, four were collected from the southern region of the State (*Lutzomyia monstruosa*, *L. rorotaensis*, *L. ruii* and *L. scaffi*), three from the central region (*L. baityi*, *L. sherlocki* and *L. spathotrichia*), five from the north (*L. driesbachi*, *L. evandroi*, *L. longipalpis*, *L. longispina* and *L. olmeca bicolor*), and two species had previously been collected from both the north and south/central regions (*L. carrerai carrerai* and *L. dasypodogeton*). Of these, *L. olmeca bicolor* was collected only in Malaise traps (which were not used on the Ilha de Maracá).

The most abundant species on Maracá were *Lutzomyia davisi* (also common in the north of the State), with 49.2% of the total capture, and *L. squamiventris squamiventris* (common in the central region) with 28.9%. Seven other species constituted a further 14.2% of the captures (*L. aragaoi*, *L. claustrei*, *L. chagasi*, *L. choti*, *L. panamensis*, *L. spinosa* and *L. squamiventris maripaensis*), and the remaining 7.8%

Table 17.1. Species of sandflies collected on the Ilha de Maracá

| | Records for other regions of Roraima | | |
|---|---|---|---|
| | Central/South (Castellón et al., 1989) | North (Castellón et al., 1991) | Central/South (Martins et al., 1963) |
| *Brumptomyia avellari* (Costa Lima) | | | |
| *B. pintoi* (Costa Lima) | | | |
| *B. spinosipes* (Floch & Abonnenc) | | | |
| *B. travassosi* (Mangabeira) | | | |
| *Lutzomyia abonnenci* (Floch & Chassignet) | | | * |
| *L. amazonensis* (Root) | * | * | * |
| *L. anduzei* (Rozeboom) | * | * | * |
| *L. antunesi* (Coutinho) | * | * | * |
| *L. aragaoi* (Costa Lima) | | * | |
| *L. ayrozai* (Barreto & Coutinho) | * | * | |
| *L. barrettoi* (Mangabeira) | | | |
| *L. begonae* (Ortiz & Rojas) | * | * | |
| *L. brachypyga* (Mangabeira) | | | |
| *L. campbelli* (Damasceno, Causey & Arouck) | | | * |
| *L. c. cayennensis* (Floch & Abonnenc) | * | | |
| *L. chagasi* (Costa Lima) | * | * | |
| *L. choti* (Floch & Abonnenc) | * | | |
| *L. claustrei* (Abonnenc, Leger & Fauran) | * | * | |
| *L. davisi* (Root) | * | * | * |
| *L. dendrophila* (Mangabeira) | * | * | * |
| *L. dubitans* (Sherlock) | | | * |
| *L. eurypyga* (Martins, Falcão & Silva) | * | * | * |
| *L. flaviscutellata* (Mangabeira) | * | | * |
| *L. furcata* (Mangabeira) | * | | * |
| *L. gomezi* (Nitzelescu) | * | | * |
| *L. h. hirsuta* (Mangabeira) | * | | |
| *L. infraspinosa* (Mangabeira) | | | |
| *L. inpai* (Young & Arias) | * | * | |
| *L. lichyi* (Floch & Abonnenc) | | | * |
| *L. lutziana* (Costa Lima) | | | * |
| *L. mangabeirana* (Martins, Falcão & Silva) | * | | * |
| *L. micropyga* (Llanos, Martins & Silva) | * | | * |
| *L. nordestina* (Mangabeira) | * | | |
| *L. pacae* (Floch & Abonnenc) | | | * |
| *L. panamensis* (Shannon) | | | * |
| *L. paraensis* (Costa Lima) | * | * | |
| *L. peresi* (Mangabeira) | * | | * |
| *L. punctigeniculata* (Floch & Abonnenc) | | | * |
| *L. runoides* (Fairchild & Hertig) | | | |
| *L. saulensis* (Floch & Abonnenc) | * | * | * |
| *L. sericea* (Floch & Abonnenc) | * | * | |
| *L. serrana* (Damesceno & Arouck) | | | |
| *L. shannoni* (Dyar) | * | * | * |
| *L. spinosa* (Floch & Abonnenc) | * | | * |
| *L. squamiventris maripaensis* (Floch & Abonnenc) | * | * | |
| *L. s. squamiventris* (Lutz & Neiva) | | * | * |

*continues overleaf*

Table 17.1. (continued)

| | Records for other regions of Roraima | | |
|---|---|---|---|
| | Central/South (Castellón et al., 1989) | North (Castellón et al., 1991) | Central/South (Martins et al., 1963) |
| L. triacantha (Mangabeira) | | | |
| L. trichopyga (Floch & Abonnenc) | * | * | |
| L. trinidadensis (Newstead) | * | | * |
| L. trispinosa (Mangabeira) | * | | |
| L. tuberculata (Mangabeira) | * | * | * |
| L. ubiquitalis (Mangabeira) | * | * | * |
| L. umbratilis (Ward & Frahia) | * | * | |
| L. walkeri (Newstead) | * | * | |
| L. williamsi (Damasceno, Causey & Arouck) | | | |

was made up of 46 species. Six species were represented by one specimen only, and can be considered as rare species for the localities and/or periods of collection.

Comparatively few collections have been made in the region, and these are insufficient to provide a basis for the comparison of the diversity of its sandfly fauna with those of other parts of Amazonia. It is clear that further collecting efforts using different trapping techniques would add significantly to the known sandfly fauna of the region.

Sandflies are of considerable importance epidemiologically, as vectors of leishmaniasis. Of the species collected on the Ilha de Maracá, two are known vectors of the disease (*Lutzomyia flaviscutellata* and *L. umbratilis*), and seven others are suspected of being so (*L. amazonensis*, *L. anduzei*, *L. antunesi*, *L. ayrozai*, *L. davisi*, *L. hirsuta* and *L. paraensis*). Given that *L. davisi* was the most commonly recorded species of sandfly on the Ilha de Maracá, there is clearly the potential for infection in the area. Nevertheless studies of the distribution of leishmaniasis in the State of Roraima have shown the Municipio de Alto Alegre, in which the Ilha de Maracá lies, to have the lowest prevalence of the disease in the State (Araújo Filho and Castellón, in press). It has been demonstrated that the areas of highest incidence of leishmaniasis correspond to those in which deforestation and agriculture are prevalent (Araújo Filho, 1981), but also that the influx of *garimpeiros* (gold prospectors) into the State has brought about a significantly increased incidence of the disease (Araújo Filho and Castellón, in press). The current spread of agriculture in the vicinity of the Ilha de Maracá and the use of the Rio Uraricoera by large numbers of *garimpeiros* may therefore effect an increase in leishmaniasis in its environs.

The most serious form of the disease, visceral leishmaniasis or 'calazar', has also been recorded in the State (Guerra et al., 1989). The vector responsible appears to be *Lutzomyia longipalpis*, whose presence in Roraima was first cited by Páes et al. (1989). However, this species is generally found in association with Indian (Macuxi) villages in the dissected terrain of the lower reaches of the Guiana Complex, where dogs act as reservoirs for the disease (Castellón and Domingos, 1991). *Lutzomyia longipalpis* has not been recorded for the Ilha de Maracá.

## ACKNOWLEDGEMENTS

The authors wish to acknowledge João Ferreira Vidal and Luis Sales de Aquino for their field collecting. The work was financed by INPA-SCT and the INPA/RGS/SEMA Maracá Rainforest Project.

## REFERENCES

Araújo Filho, N.A. (1981). Leishmaniose Tegumentar Americana e o desmatamento da Amazônia. *Acta Amazonica*, 11 (1), 187–189.

Araújo Filho, N.A. and Castellón, E.G. (in press). Leishmaniose Tegumentar Americana (LTA) no Estado de Roraima, Brasil. Dados epidemiológicos. *Acta Amazonica*.

Castellón, E.G. and Domingos, E.D. (1991). Calazar em Roraima: aspectos epidemiológicos e ecológicos. *Mem. Inst. Oswaldo Cruz*, 86 (3), 375.

Castellón, E.G., Araújo Filho, N.A. de, Fé, N.F. and Alves, J.M.C. (1989). Flebotomíneos (Diptera: Psychodidae) no Estado de Roraima, Brasil. I. Espécies coletadas nas regiões Sul e Central. *Mem. Inst. Oswaldo Cruz*, 84, Supl. IV, 95–99.

Castellón, E.G., Araújo Filho, N.A. de, Fé, N.F. and Alves, J.M.C. (1991). Flebotomíneos (Diptera: Psychodidae) no Estado de Roraima, Brasil. II. Espécies coletadas na região Norte. *Acta Amazonica*, 21, 45–50.

Fraiha, H., Ward, R.D., Loureiro, C.A. and Soares, G.M. (1974). Flebotomídeos Brasileiros – IV. Nota sobre *Psychodopygus chagasi* (Costa Lima, 1941) (Diptera: Phlebotomidae). *Rev. Brasil. Biol.*, 34, 89–91.

Guerra, M.V.F., Araújo Filho, N.A., Páes, M.G., Barros, M.L.B., Sá, R.C. and Ramos, E.D. (1989). Aspectos clínicos do calazar no Estado de Roraima, Brasil. *XI Congr. Bras. Parasitol.*, Rio de Janeiro, p. 17.

Martins, A.V., Falcão, A.L. and Silva, J.E. da (1963). Notas sobre os flebótomos do Território de Roraima, com a descrição de três novas espécies. *Rev. Brasil. Biol.*, 23, 333–348.

Paes, M.J., Araújo Filho, N.A. de, Guerra, M.V.F., Barros, M.L. and Fé, N.F. (1989). Ocorrência da *Lutzomyia longipalpis* (Lutz and Neiva, 1912) no Estado de Roraima. *IX Congr. Bras. Parasitol.*, Rio de Janeiro, p. 165.

Ryan, L. (1986). *Flebótomos do Estado do Pará–Brasil*. (Diptera: Psychodidae: Phlebotominae). Doc. Téc. No. 1. Instituto Exandro Chagas – Fundação SESP – Ministério da Saúde.

# Triatomine bugs on the Ilha de Maracá

TOBY V. BARRETT
*Instituto Nacional de Pesquisas da Amazônia, Manaus, Brazil*

## SUMMARY

The results of a rapid survey of the triatomine bugs and associated fauna in the buildings of the Ecological Station on the Ilha de Maracá are presented. *Triatoma maculata* was found to infest the roof spaces of these buildings, which it shared with bats (*Molossus molossus* and *Myotis nigricans*) and birds (*Thraupis palmarum* and *Progne chalybea*). Although this species of triatomine bug is known to carry trypanosomiasis (*Trypanosoma cruzi*), none of the specimens collected on Maracá were found to be carrying the disease.

## RESUMO

São apresentados os resultados de um levantamento de triatomíneos nos alojamentos da Estação Ecológica da Ilha de Maracá. Uma infestação de *Triatoma maculata* foi encontrada nos forros destes prédios, partilhando o espaço com morcegos (*Molossus molossus* e *Myotis nigricans*), e aves (*Thraupis palmarum* e *Progne chalybea*). Enquanto esta espécie de inseto é reconhecida como portador de tripanosomiase (*Trypanosoma cruzi*), nenhum dos exemplares coletados na Ilha de Maracá carregava o agente da doença.

## INTRODUCTION

The Ecological Station on the Ilha de Maracá, which was inaugurated in April 1978, comprises five principal buildings each of which is roofed with corrugated asbestos sheeting. The windows are all sealed with mosquito netting. The buildings lie just a few metres from the edge of the *terra firme* forest, in which the *inajá* palm *Maximiliana maripa* (Correa) Drude is abundant.

During the Maracá Rainforest Project, some of the scientists resident in the Ecological Station reported the presence of triatomine bugs in the buildings. They were concerned about the possibility of exposure to *Trypanosoma cruzi*, the parasite that can cause Chagas' disease and which is transmitted by these insects. The results that follow are taken from the report on a site-visit made to assess the situation on 11 April 1987.

## RESULTS

Investigations within the roof spaces of the buildings revealed an abundance of bats and birds' nests, particularly in the corners. The bats were principally free-tailed or mastiff bats of the family Molossidae, namely *Molossus molossus*, but there were also a few 'little brown bats', *Myotis nigricans*, of the family Vespertilionidae. These bat species, of which specimens were taken (INPA 1109 and INPA 1116) and identified by Rogério Gribel of INPA, Manaus, are discussed by Robinson (Chapter 9). Both species were breeding in the roof space, *Molossus molossus* using the cavities where the corrugated roofing panels met the walls.

The birds' nests were relatively large – from 5 to 20 litres in volume – and were chiefly composed of twigs and straw-like material. They belonged to two species: the palm tanager *Thraupis palmarum* and the grey-breasted martin *Progne chalybea* (kindly identified by Duncan Scott). In some cases the gaps between these nests and the walls were being used as roosting sites by the *Molossus* bats.

Other signs of vertebrate activity encountered in the roof space included a collection of fruits, a small mummified rodent, and the eggs of a lizard which were found beneath a bird's nest. No mammalian nest was found, possibly because of the high temperatures which these roof spaces experience during the day.

Triatomine bugs of the species *Triatoma maculata* were found in the following localities (Roman numerals refer to nymphal instars):

(a) Roof space; loose bricks harbouring bats; one I, two III.
(b) Roof space; bird's nest; one I, two II, one III.
(c) External switchbox; nest of a small bird; seven III.
(d) Roof space; bird's nest close to *Molossus* bats; two I, four II, ten III, two IV, four V.
(e) Roof space; bird's nest; one I, four III, one IV, one V, one adult male, one adult female.

In addition, numerous hatched eggs, exuviae and dead nymphs and adults were found in the birds' nests. Ten of the live specimens (all from site a, and two V and five III from site d) were dissected and examined for the presence of *Trypanosoma* in the digestive tract. The blood of one of the bats encountered in their presence was also examined. All tests proved negative. The remaining live bugs were used to found a laboratory colony at INPA.

## DISCUSSION

The focus of infestation of *Triatoma maculata* in the Ecological Station was found to be the birds' nests in the roof spaces of the buildings. This is an interesting case of the incorporation of a natural triatomine habitat into a building; a similar situation occurs in North America, where nests of *Neotoma* woodrats have been reported in the attics and walls of houses (Ryckman, 1981).

The primary source of this focus of *T. maculata* was not investigated, but elsewhere this species occurs in the wild under tree bark, in birds' nests, occasionally in bromeliads, and in palms (especially *Scheelea*, *Attalea* or *Maximiliana* spp.[1]), where it is often associated with *Rhodnius* spp., the latter invariably exhibiting higher trypanosome infection rates (Tonn *et al.*, 1978; Barrett, 1991). In the semi-arid lowlands of Venezuela *T. maculata* is very abundant and is associated particularly with domestic chickens, although there is some overflow into houses.

The presence of *T. maculata* in the forested area of the island is of interest, because both this species and the closely related *T. pseudomaculata* are more characteristic of the drier regions to the north and south, respectively, of the Amazon rainforest (Lent and Wygodzinsky, 1979). Another triatomine collected at the Ecological Station, *Eratyrus mucronatus*, is an Amazonian species inhabiting large hollow trees throughout the region. Adults of this species were collected at lights on external walls in the evening.

*Trypanosoma cruzi* is widespread in the western hemisphere, in many mammals of different orders and in most species of Triatominae. It is prudent to regard all these bugs as potential vectors of Chagas' disease and to take precautions against the contamination of people, food and drink with the infective metacyclic trypanosomes in the bugs' faeces. On the other hand, less than 50 cases of autochthonous human trypanosomiasis, acquired by direct contact with silvatic cycles of transmission, are known from the Brazilian Amazon region. This is a very low number in comparison with the 15–18 million people in Latin America who are thought to be infected as a result of living in houses in which breeding colonies of triatomines transmit the parasite among people and domestic animals. The principal threat to the Amazon region in this respect is still the potential spread of domestic populations of *Rhodnius prolixus* from the north, or *Triatoma infestans* from the south (Barrett, 1991).

## NOTE

1. These ill-defined palm genera are probably best regarded as synonyms amongst references in the entomological literature.

## REFERENCES

Barrett, T.V. (1991). Advances in the Triatominae bug ecology in relation to Chagas' disease. *Advances in Disease Vector Research*, 8, 143–176.

Lent, H. and Wygodzinsky, P. (1979). Revision of the Triatominae (Hemiptera, Reduviidae) and their significance as vectors of Chagas' disease. *Bulletin of the American Museum of Natural History*, 163, 123–520.

Ryckman, R.E. (1981). The kissing bug problem in Western North America. *Bulletin of the Society of Vector Entomologists*, 6, 167–169.

Tonn, R.J., Otero, M.A., More, R., Espinola, H. and Carcavallo, R.U. (1978). Aspectos biológicos, ecológicos y distribución geográfica de *Triatoma maculata* (Erichson, 1848) (Hemiptera, Reduviidae) en Venezuela. *Boletin de la Dirección de Malariologia y Saneamiento Ambiental*, 18, 16–24.

# *Anopheles* species of the Ilha de Maracá: Incidence and distribution, ecological aspects and the transmission of malaria

## ILÉA BRANDÃO RODRIGUES[1] AND WANDERLI P. TADEI[2]

[1]*Instituto Nacional de Pesquisas da Amazônia, Manaus, Brazil;* [2]*UNESP, São José do Rio Preto, Brazil*

## SUMMARY

Data are presented for the incidence and distribution of *Anopheles* species on and in the vicinity of the Ilha de Maracá. Seventeen species were recorded, belonging to the subgenera *Nyssorhynchus* (9 species), *Arribalzagia* (5), *Anopheles* (2) and *Lophopodomya* (1). *Anopheles albitarsis* was the most frequent species on the island, representing 55% of the captures. *A. darlingi* constituted only 2% but was captured at various points on the island, though predominantly in the vicinity of the Ecological Station. Immature forms were captured amongst the roots of *Eichhornia*, *Scirpus*, *Nymphaea*, *Mayaca*, *Utricularia* and *Elaeocharis* spp. No larvae were captured among the roots of *Chara* spp. Malarial infection data for the species, measured by the ELISA test, were positive only for *Anopheles darlingi* (*Plasmodium falciparum*), giving a percentage infection of 4% (of 24 individuals captured). Results were negative for *A. albitarsis*, *A. argyritarsis*, *A. oswaldoi*, *A. intermedius* and *A. apicimacula*. The possibility of an outbreak of malaria on the island exists, depending on (1) population densities, (2) intensity of human/vector contact, and (3) the presence of *A. darlingi*.

## RESUMO

São relatados dados sobre a incidência e distribuição das espécies de *Anopheles* registradas ao longo da Reserva Ecológica da Ilha de Maracá. Foram registradas 17 espécies, pertencentes aos subgêneros *Nyssorhynchus* (9 espécies), *Arribalzagia* (5), *Anopheles* (2) e *Lophopodomiya* (1). *Anopheles albitarsis* foi a mais freqüente na ilha, representando 55% do total colecionado e *A. darlingi*, apenas 2% porém, esta última de occorência reduzida, foi capturada em diferentes pontos da ilha, especialmente nas redondezas da Estação Ecológica. As formas imaturas foram capturadas junto às raízes de *Eichhornia*, *Scirpus*, *Nymphaea*, *Mayaca*, *Utricularia* e *Elaeocharis* spp. Entre as raízes de *Chara* spp. não foram capturadas larvas. Os dados da infecção natural das espécies, avaliada pelo teste da ELISA, deram positivo apenas para *A. darlingi* (*Plasmodium falciparum*), perfazendo um percentual de 4% de infecção (24 espécimes no total). Os resultados foram negativos para *A. albitarsis*, *A. argyritarsis*, *A. oswaldoi*, *A. intermedius* e *A. apicimacula*. Foi admitido que um surto de malária poderia occorer na dependência de (1) densidade populacional das espécies, (2) intensidade do contato homem/vetor, e (3) presença de *A. darlingi*.

## INTRODUCTION

Studies of the distribution and density of *Anopheles* species in conservation areas are of considerable relevance to our understanding of the transmission of malaria in areas undergoing anthropogenic activity. In this paper, ecological aspects and the transmission of malaria are investigated in the context of the Amazonian *Anopheles* vectors of the disease.

## METHODS

A survey of *Anopheles* species was carried out at 97 points on the Ilha de Maracá, at 34 and 63 of which adult (winged) and immature specimens were taken respectively. Collections were made on both banks of the Furo de Santa Rosa and the Furo de Maracá. Further collections were made in the interior of the island, at sites dictated by the availability of access trails and the occurrence of tributary streams. Collection sites included a variety of habitats, incorporating dense forest, open savanna and stream margins, and whenever possible both adult and immature specimens were taken at the same locality. In order to obtain data on the seasonality of the species, collections were made of adult individuals at various points on the access track to the Ecological Station and around the Casa de Santa Rosa (Figure Pr.1, p. xvii). Two lagoons (known as Chavascal I and Chavascal II) alongside the access track to the Ecological Station were chosen for the collection of immature specimens for analysis of their distribution, seasonality and ecological characteristics, considering (i) physical/chemical parameters, (ii) aquatic vegetation (macrophytes) and (iii) phycological data (algae). Chavascal I was a borrow-pit where material had been removed for construction of the causeway, and Chavascal II was an area of natural standing water.

## RESULTS

The collections were made between May 1987 and June 1988, and of a total of 3495 specimens, of which 1161 were larval and 2334 were adult, the following 17 species pertaining to the subgenera *Nyssorhynchus* Blanchard, 1902, *Arribalzagia* Theobald, 1903, *Anopheles* Meigen, 1818, and *Lophopodomiya* Antunes, 1937, were identified:

*Anopheles* (*Nyssorhynchus*) *albitarsis* Lynch Arribalzaga, 1878
*A.* (*N.*) *argyritarsis* Robineau-Desvoidy, 1827
*A.* (*N.*) *braziliensis* (Chagas, 1907)
*A.* (*N.*) *darlingi* Root, 1926
*A.* (*N.*) *nuneztovari* Gabaldon, 1940
*A.* (*N.*) *oswaldoi* (Peryassú, 1922)
*A.* (*N.*) *rangeli* Root, 1926; Gabaldon, Cora-Garcia & Lôpez, 1940
*A.* (*N.*) *triannulatus* (Neiva & Pinto, 1922)
*A.* (*Arribalzagia*) *apicimacula* Dyar & Knab, 1906
*A.* (*N.*) *evansae* (Brèthes, 1926)
*A.* (*A.*) *intermedius* (Peryassú, 1908)
*A.* (*A.*) *mediopunctatus* (Theobald, 1903)
*A.* (*A.*) *punctimacula* Dyar & Knab, 1906

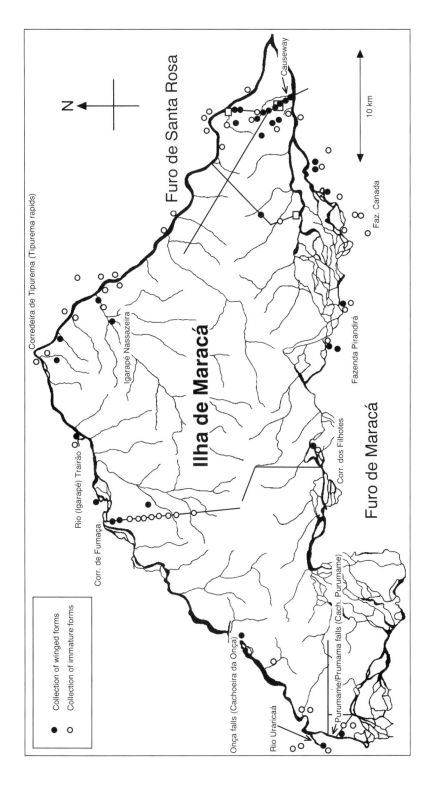

**Figure 17.1.** Points of collection of *Anopheles* spp. on the Ilha de Maracá

*A. (A.) shannoni* Davis, 1931
*A. (Anopheles) mattogrossensis* Lutz & Neiva, 1911
*A. (A.) peryassui* Dyar & Knab, 1908
*A. (Lophopodomiya) squamifemur* Antunes, 1937

Amongst the species detected on the island, *Anopheles albitarsis* represented 55% of the total collection and the other two most common species were *A. argyritarsis* and *A. triannulatus* (17 and 14% respectively). Most of the other species, with the exception of *A. oswaldoi* (4%), represented less than 2% of the collection. *A. darlingi*, a species of particular interest on account of its transmission of malaria, was an infrequent species (2%), although it was collected both at the Casa de Santa Rosa and on the Furo de Maracá. There was a relatively high incidence of this species in the buildings of the Ecological Station, including on the inside of the window netting of the sleeping quarters. Its immature forms were collected in four localities on the Furo de Santa Rosa and only one in the interior of the island (the Igarapé Nassazeira, see Figure 17.1). This species represented 19% of the larvae collected on the Furo de Santa Rosa, but only 4% in the interior.

Considering the points along the Furo de Santa Rosa and the interior of the island, there was a high level of negative records for both larval and adult stages of mosquitoes, with the exception of the vicinity of the Ecological Station (which is not far from the Furo de Maracá). On the Furo de Maracá itself there was not only a higher occurrence of mosquitoes recorded, but also a generally higher density. The collection results at the relatively disturbed sites, i.e. the vicinity of the Ecological Station and the Casa de Santa Rosa, were higher than at the others, and the species most commonly caught were the same as at sites on the Furo de Maracá, clearly related to the *fazendas* (ranches) along its banks.

The biting activity of the species showed a crepuscular peak, with greatest intensity around 1700 hrs, but this varied between the dry and the rainy seasons. Studies of seasonality were made for both larval and adult stages. Two points were selected for studies of adults on the access track to the Ecological Station and two at the Casa de Santa Rosa. Of the latter, one was in the forest and one was in the grassy savanna. *Anopheles albitarsis* and *A. argyritarsis* were recorded throughout the year on the access track. *A. argyritarsis* was less abundant in the dry season, but *A. albitarsis* showed little seasonal change. *A. darlingi* was only recorded during the rainy season.

Species diversity and frequency were low at the collection points in forest along the Furo de Santa Rosa (only *Anopheles albitarsis*, *A. triannulatus*, *A. braziliensis* and *A. mediopunctatus* were captured), and data are inadequate to reach conclusions on seasonal frequencies. The diversity recorded in the grassy savanna, however, was higher, comprising 10 species of which the most common were *A. albitarsis*, *A. argyritarsis* and *A. braziliensis*. Although the first of these was present in almost all of the samples, *A. argyritarsis* was not observed during the dry season whereas *A. braziliensis* was predominantly encountered in that season. The same seasonal trends were observed in these species in the vicinity of the Ecological Station.

In spite of the differences in their origins, the water bodies used for the seasonality studies of immature forms (Chavascal I and Chavascal II) were similar in the majority of the physico-chemical parameters measured, with contrasting values obtained only for the levels of humic substances, silicates and ammonia. In terms of the aquatic vegetation, Chavascal II showed a higher diversity and abundance of species throughout the year. The following genera were recorded: *Nymphoides* (Chav. II); *Eleocharis* (Chav. I/II); *Utricularia* (Chav. I/II); *Mayaca* (Chav. I/II); *Eichhornia* (Chav. I/II); *Echinodorus* (Chav. II); *Scirpus* (Chav. I/II); *Nymphaea* (Chav. II) and *Scleria* (Chav. II).

Mosquito frequency data showed greater seasonal stability in Chavascal II than in Chavascal I, which underwent greater environmental fluctuation between the dry and rainy seasons. During the dry season Chavascal I virtually disappeared, leaving only small breeding sites within its area, whereas in Chavascal II the water level merely dropped with a consequent reduction in area. In spite of this, the incidence of anophelines remained stable throughout the dry season, a period when the occurrence of adult forms is reduced. However, although there was stability in the overall count, changes were observed in larval densities at the species level in each habitat. Drastic reductions in larval densities were observed for the two most common species in the breeding sites, *Anopheles albitarsis* and *A. argyritarsis*.

Another aspect of the observations which requires examination was the fact that although *A. argyritarsis* was a very frequent species in the samples of adult forms, so few were encountered in larval stages that it was not possible to draw any conclusions about their behaviour. Collections of adult forms were made with the objective of recording the pattern of biting activity of the species, and the captured specimens were dissected for subsequent analysis of infection levels with human *Plasmodium*, by immunoenzymatic tests with monoclonal antibodies (ELISA). Three hundred and thirteen specimens were examined from Maracá, obtained by collection with human bait. These represented the following species: *Anopheles darlingi* (24 individuals); *A. albitarsis* (102); *A. argyritarsis* (124); *A. oswaldoi* (20); *A. triannulatus* (18); *A. nuneztovari* (2); *A. mediopunctatus* (2); *A. intermedius* (19) and *A. apicimacula* (2). All of the results were negative for *Plasmodium vivax* and only one specimen of *A. darlingi*, which was collected inside one of the Ecological Station buildings in February 1988, was positive for *P. falciparum*.

## DISCUSSION

The data obtained on Maracá indicate a generally very low density of anophelines, with the exception of the vicinity of the Ecological Station and the Casa de Santa Rosa, both of which have undergone some historical environmental disturbance (minor in the first case and major in the second). These results are similar to those obtained in the vicinity of the Balbina hydroelectric project in the State of Amazonas, where low anopheline incidence was recorded at the majority of collection sites, with the exception of residential areas where environmental modification had taken place.

Considering the species diversity at different points on the Ilha de Maracá and the incidence of the same, we can deduce that the part of the island between the access track (causeway) to the Ecological Station and the Casa de Santa Rosa constitutes an area whose anopheline fauna has been altered in its diversity, probably as a result of external interference. This hypothesis is supported by the similarity in anopheline diversity in this area to that recorded along the Furo de Maracá, where ranches abut the island. This similarity is particularly noticeable for the data from the Casa de Santa Rosa.

The considerable seasonal fluctuations in anopheline population densities measured at the Ecological Station and at the Casa de Santa Rosa, with population reductions in the rainy season, are in accord with those in the literature, which also cite reductions in anopheline population densities during the rainy season resulting from modifications in their reproduction sites (Deane *et al.*, 1948; Galindo *et al.*, 1956; Standfast and Barrow, 1968; Moran, 1981; Tadei *et al.*, 1988). The results of the studies of mosquitoes' activity cycles also corresponded to those related by other authors in other parts of Amazonia (Galvão *et al.*, 1942; Deane, 1947; Tadei and Correia, 1982; Tadei *et al.*, 1983; Deane *et al.*, 1948; Roberts *et al.*, 1981; Charlwood and Hayes, 1978; Tadei *et al.*, 1988).

The changes in population densities recorded for the adult forms do not correspond to changes in the immature forms. On the contrary, total larval densities remained relatively stable. The small changes which were observed, primarily in Chavascal I, were probably the result of environmental changes. In the Municipality of Ariquemes (Rondônia), Tadei (1988) similarly demonstrated that changes in larval densities resulting from seasonal changes were not directly related to the populations of the adult form of *Anopheles darlingi*.

The data on aquatic plants at the collection sites enabled us to verify that mosquito larvae may be found in association with the roots of *Eichhornia* spp., as reported in the literature by Forattini (1962). At one collection point in Chavascal I, in which *Chara* spp. were found, no larvae were encountered and this corresponds with the results of Hinman (1928), Hoehne (1948) and Forattini (1962), who reported the absence of larvae as a result of the toxic activity of these plants – possibly demonstrating a larvicidal function.

The presence of *Plasmodium* in *Anopheles darlingi* and the absence in the other species collected corresponds with the results of other studies in the literature, in which *A. darlingi* is cited as the principal vector of human malaria transmission in Amazonia, responsible for transmission in the interior and in four-fifths of the whole country (Coutinho, 1946; Deane, 1947; Deane *et al.*, 1948; Rachou, 1958; Forattini, 1962; Ferreira, 1964).

Therefore, considering the data in the literature on infected species of *Anopheles*, and comparing them with the species recorded on the Ilha de Maracá, we can surmise that after *A. darlingi* (their principal vector), the next most likely species to play a role as malaria vector is *A. albitarsis*, by virtue of its high population density in the vicinity of the Ecological Station and the fact that it has already been shown by monoclonal antibody tests to carry malaria (Arruda *et al.*, 1986). *A. argyritarsis* should also be considered as a possible vector, by virtue of its high density, although it has never been shown to be infected with *Plasmodium*. On the other hand, most of the other species on the island, such as *A. triannulatus*, *A. oswaldoi* and *A. nunez-tovari* (among others), could potentially become vectors, depending upon the following parameters: (i) population densities; (ii) intensity of contact with human malaria carriers; and (iii) the presence of a residual population of *A. darlingi* – as discussed by Tadei *et al.* (1988) for mosquito populations in the State of Amazonas.

However, the occurrence of an outbreak of malaria on the island is relatively unlikely, given that there is little throughflow of people carrying transmissable forms. All the same, one cannot exclude the possibility of cases of the disease, since there is sporadic transit of potential carriers up the Furo de Santa Rosa to *garimpos* (gold mines) upstream. The fact that one case of an individual of *Anopheles darlingi* infected with *Plasmodium falciparum* was recorded during the period of study reinforces this point.

## REFERENCES

Arruda, M.M., Carvalho, M., Nussenzweig, R.S., Maracic, F., Ferreira, A.W. and Cochrane, A. (1986). Potential vectors of malaria and their different susceptibility to *Plasmodium falciparum* and *Plasmodium vivax* in Northern Brazil identified by immunoassay. *Am. J. Trop. Med. Hyg.*, 35, 873–881.

Charlwood, J.D. and Hayes, J. (1978). Variações geográficas no ciclo de picada do *Anopheles darlingi* Root no Brasil. *Acta Amazonica*, 8 (4), 601–603.

Coutinho, J.O. (1946). Contribuição para o conhecimento das espécies do subgênero *Kerteszia* (*Diptera–Culicidae*) – sua importância na transmissão da malária. Doctoral thesis, Universidade de São Paulo.

Deane, L.M. (1947). Observações sobre a malária na Amazônia brasileira. *Rev. Esp. Saúde Pub.*, 1, 3–60.

Deane, L.M., Causery, O.R. and Deane, M.P. (1948). Notas sobre a distribuição e a biologia dos anofelinos das regiões nordestinas e Amazônicas do Brasil. *Rev. Bras. Serv. Espc. Saúde Publ.*, 1, 827–965.

Ferreira, E. (1964). Distribuição geográfica dos anofelinos no Brasil e sua relação com o Estudo Atual da Eradicação da Malária. *Rev. Bras. Malariol. D. Trop.*, 16 (3), 329–348.

Forattini, O.P. (1962). *Entomologia médica*. Fac. Hig. Saúde Publ. da Univ. de São Paulo, São Paulo.

Galindo, P., Trapido, H., Carpenter, S.J. and Blanton, F.S. (1956). The abundance cycles of arboreal mosquitoes during six years at a sylvan yellow fever locality in Panama. *Annals of the Entomological Society of America*, 49, 543–547.

Galvão, A.L.A., Damasceno, R.G. and Marques, A.P. (1942). Algumas observações sobre a biologia dos anofelinos de importância epidemiológica de Belém, Pará. *Arquivos de Higiene*, 12, 51–111.

Hinman, E.H. (1928). *Chara fragilis* and mosquito development. *American Journal of Hygiene*, 7 (2), 279–296.

Hoehne, F.C. (1948). *Plantas aquáticas*. Instituto de Botânica de São Paulo.

Moran, E.F. (1981). *Developing the Amazon*. Indiana University Free Press, Bloomington.

Rachou, R.G. (1958). Anofelinos de Brasil: comportamento das espécies vetoras de malária. *Rev. Bras. Malariol. D. Tropical*, 10 (2), 145–181.

Roberts, D.R., Hoch, A.L., Peterson, M.E. and Pinheiro, F.P. (1981). Programa multidisciplinária de vigilância de las infermidades infecciosas en zonas colindantes con la carretera transamazonica en Brasil. IV – Estudio entomologica. In: *A handbook of Amazonian species of* Anopheles (Nyssorhynchus), eds M.E. Faran and K.J. Linthikum. *Mosquito Systematics*, 13 (1), 1–91.

Standfast, H.A. and Barrow, G.J. (1968). Studies on the epidemiology of arthropod-borne virus infections at Mitchell River Mission, Cape York Peninsula, North Queensland. I. Mosquito collections, 1963–1966. *Transactions of the Royal Society for Tropical Medicine and Hygiene*, 62, 418–429.

Tadei, W.P. (1988). Biologia de anofelinos Amazônicos. XI. Estudos em populações de *Anopheles* e controle da malária em Ariquemes (Rondônia). *Acta Amazonica*, 18 (Supl. Polonoroeste).

Tadei, W.P. and Correia, J.N. (1982). Biologia de anofelinos Amazônicos. IV. Observações sobre a atividade de picar de *Anopheles nunez-tovari* Gabaldon (Diptera, Culicidae). *Acta Amazonica*, 12 (1), 71–74.

Tadei, W.P., Mascarenas, B.M. and Podestá, M.G. (1983). Biologia de anofelinos Amazônicos. VIII. Conhecimentos sobre a distribuição de espécies de *Anopheles* na região de Tucuruí – Marabá (Pará). *Acta Amazonica*, 13 (1), 103–140.

Tadei, W.P., Carvalho, M.B., Santos, J.B.M.F., Ferreira, A.V. and Nussenzweig, K.S. (1988). Dinâmica da transmissão da malária em Rondônia e susceptibilidade das espécies de *Anopheles* aos *Plasmodium vivax* e *Plasmodium falciparum*, identificada por testes imunoenzimática. *24º Congresso da Sociedade Brasileira de Medicina Tropical*, 26.

# 18 Arachnids of the Ilha de Maracá

## Notes on the spiders of the Ilha de Maracá

ARNO ANTÔNIO LISE
*Pontifícia Universidade Católica do Rio Grande do Sul, Porto Alegre, Brazil*

### SUMMARY

Surveys of the spider populations on the Ilha de Maracá were carried out during the wet and the dry seasons. Approximately 270 species were collected, including several new to science (e.g. *Echinotheridion urarum*). The most important families, in terms of numbers of species, were the Salticidae, Araneidae, Theridiidae, Thomisidae, Pisauridae, Ctenidae and Heteropodidae. Spider populations were lowest during the dry season, when they presumably retreat to humid microhabitats. Densities of weaving species were highest in relatively open forest habitats or along man-made trails. Results indicate a relatively rich arachnofauna for the island, and future studies would inevitably yield further species.

### RESUMO

Levantamentos de populações de aranhas na Ilha de Maracá foram realizados na estação chuvosa e na seca. Foram coletadas aproximadamente 270 espécies, inclusive várias espécies novas (e.g. *Echinotheridion urarum*). As famílias mais importantes, em termos de número de espécies encontradas, foram Salticidae, Araneidae, Theridiidae, Thomisidae, Pisauridae, Ctenidae e Heteropodidae. As populações de aranhas foram menores na estação seca, quando provavelmente se retiram para micro-habitats úmidos. A densidade de espécies tecedores foi maior em floresta aberta, ou ao longo de trilhas. Resultados indicam a existência de uma aracnofauna relativamente diversa na ilha, e estudos futuros inevitavelmente renderão espécies adicionais.

### INTRODUCTION

Studies of the spiders of the Ilha de Maracá were conducted during both the dry season and the rainy season in 1987. Collections were made in the vicinity of the Ecological Station at the eastern end of the island (Figure Pr.1, p. xvii). Although the time available for sampling and behavioural analyses of the spiders was limited, and insufficient to reach detailed and all-embracing conclusions, it is at least possible to put forward some preliminary comments on these matters.

Three principal methods were used for sampling the spiders: for sampling arboreal species in the forest an entomological 'beating umbrella' was used; in the

*Maracá: The Biodiversity and Environment of an Amazonian Rainforest.*
Edited by William Milliken and James A. Ratter. © 1998 John Wiley & Sons Ltd.

savannas a 'sweep net' was used; and for the collection of ground-dwelling species 'pitfall traps' were employed. Whenever possible during the collection, notes on the behaviour and biology of the animals were made, although as a result of the limited time available these were not made to a rigorous format.

The first collections were made between 18 and 30 March, and for the first few days the forest was extremely dry and yielded very few spiders. A subsequent fall of rain brought about a substantial increase in humidity in the forest for approximately three days, with a consequent increase in the abundance of spiders, but the rainy season had yet to begin and the forest reverted to its previous dry state.

The second collecting period, in June 1987 (17th to 24th), was in the middle of the wet season. It rained almost every day during this period, as a result of which collecting with beating umbrellas and sweep nets was rendered impractical, and the pitfall traps filled with water. Again very few spiders were encountered during this period – even those species which had been found in abundance in March were absent or very poorly represented. The third and final collecting effort was made in December of the same year (4th to 11th), more or less in the middle of the dry season. The forest was completely dry and the number of collections made was even lower than in June.

# RESULTS

It was evident that the quantity and regularity of rainfall strongly influence the densities of spider populations in the habitats studied. It is not known what happens to the spiders during dry periods, and this is a question which deserves further investigation. That they enter a latent phase both in the dry season and in the wet season seems highly unlikely (no substantial quantities of egg-masses were observed during these periods), and it is probable that they retreat to microhabitats with higher humidities than the ambient. Likely microhabitats include the soil – though the somewhat sparse covering of leaf-litter in these forests probably offers minimal protection from drying – and leaf-sheaths of the abundant forest palms. There was, however, no conclusive evidence to support these hypotheses.

One might also suggest that the spiders migrate to the forest canopy during the rainy season (in the dry season the humidity in this environment would be too low), but if this were the case one might expect to have found webs, and these were not observed.

Maracá appears to support a rich arachnofauna in the context of the little that is so far known of the spiders of Amazonia. Of the 115 families listed by Platnick (1989) for the world, around 70 must be present in the neotropics. Thirty-two of these families, comprising a probable 120 genera with approximately 270 species, are represented in the collections made on the Ilha de Maracá (Appendix 7, p. 479). Further collecting on the island, particularly if it were to include canopy fogging techniques, would undoubtedly add very significantly to this tally. It is known that certain spider species exhibit strong heliophilic tendencies, and small-scale canopy fogging carried out during entomological studies on Maracá by Dr José A. Rafael (INPA) did yield spiders which were not encountered during our studies.

From what is currently known of Maracá's arachnofauna, it appears that the most important families in terms of number of species are the Salticidae, Araneidae, Theridiidae, Thomisidae, Pisauridae, Ctenidae and Heteropodidae. The degree to which it has so far been possible to identify the collections has been severely limited by the current lack of literature and taxonomic specialists in neotropical arachnology. It will be many years before the species list can be completed. However, the representatives of certain families have been studied intensively, and several new species have been identified from Maracá. The first of these to be described was *Echinotheridion urarum* Buckup and Marques, 1989. Others are awaiting publication.

A significant proportion of the spiders collected on Maracá were arboreal species. These included web-builders, weavers, and epiphytic species which hunt directly from the surface of the vegetation. The weavers included the Araneidae, Theridiidae, Tetragnathidae and Pisauridae. The number of web-building Pisauridae collected is worth mentioning, since most of the members of this family are either terrestrial or associated with running water.

The greatest density of weaving species was found in areas where the forest was comparatively open (i.e. close to the forest edge) or along man-made trails used as flight paths by their insect prey. The changes of abundance of spiders in the forest canopy were impossible to ascertain from the ground. Generally speaking the population densities of spiders decrease as the vegetation density rises and the availability of light and of open spaces suitable for web-building diminishes. This decrease also holds true for the epiphytic and terrestrial species which do not weave webs. Changes in species diversity between edge and forest habitats were not observed.

Of the web-building spiders, the most commonly encountered were the Araneidae, Tetragnathidae and Theridiidae:

*Araneidae* Ten species of *Micrathena* were collected which construct their webs close to the ground (1–1.5 m), preferentially at the forest margin and along well-lit forest trails. Of these, *M. schreibersi* was particularly abundant. Only three species of *Cyclosa* were collected, and these prefer shaded locations in the forest interior. A few specimens of *Verrucosa* were collected, which are being described as two new species. These were encountered at the margins of treefall gaps in the forest, approximately 3 m above the ground. They prefer broad, well-lit spaces, which is where *Parawixia* species are also found. There appear to be three species of *Hypognatha* on the island, whose webs are abundant in the forest except where it is particularly dense.

*Tetragnathidae* Among this family the most commonly encountered are members of the genus *Leucage*. They are easily identified by their horizontal webs, in which they position themselves upside down in the centre. When disturbed they rapidly flee to the vegetation at the edges of their webs, in which they effectively conceal themselves.

*Theridiidae* The most prominent of these are representatives of the genus *Achaearanea*, which generally build an irregular or mixed web in which a leaf is folded in the form of a funnel or cone, serving as a refuge. *A. trapezoidalis* builds a web which gives the effect (visually) of plastic. Members of this family are found throughout the forest, but their habitat or plant species preferences were not observed.

In addition to the above, web-building spiders found much more rarely on Maracá included the family Uloboridae (*Miagrammopes*, *Philoponella* and *Uloborus*). Of these, *Philoponella* and *Uloborus* occur relatively frequently in all habitats studied on the island. Their webs are inconspicuous and not infrequently use the webs of larger spiders for support.

The most commonly seen epiphytic spiders (which are visual hunters) were the Salticidae. The Anyphaenidae and certain Clubionidae also figure in this category, but although in many areas these are highly abundant, there are few on Maracá. There were also very few representatives of the groups which are typically soil-dwellers, i.e. Mygalomorphae, Ctenidae, Lycosidae, Zodariidae and Pisauridae. The Mygalomorphae, which include the largest spiders of Amazonia, were chiefly represented by the Theraphosidae. These are very rare on the island both in terms of number of species and abundance, and were found beneath fallen logs, in or below the rolled-up leaf-sheaths of palms (where certain Ctenidae also occurred), or inside the leaf-sheaths of terrestrial bromeliads.

*Leprolocus spinifrons* was the only member of the Zodariidae encountered on the island. It lives in the leaf-litter, and is caught with difficulty because of its small size and the speed with which it moves across the surface of the soil. It was particularly abundant in March, but disappeared completely in July and December.

Bearing in mind all that has been written regarding the Ilha de Maracá, both as an ecosystem and as a source of diverse scientific information, it is hoped that the attention of the national and international authorities, as well as the scientific community, will be focused upon the importance of its preservation. It is hoped also that scientific investigations will continue on the island, as there is still a vast amount that remains to be discovered.

## REFERENCES

Platnick, N.I. (1989). *Advances in spider taxonomy 1981–1987.* A supplement to Brignoli's *A catalogue of the Aranae described between 1940 and 1981.* Manchester University Press.

# Pseudoscorpions (Arachnida) of the Ilha de Maracá

NAIR OTAVIANO AGUIAR AND PAULO F. BÜHRNHEIM
*Universidade do Amazonas, Manaus, Brazil*

## SUMMARY

Seventeen species of pseudoscorpions belonging to 12 genera and seven families were collected on the Ilha de Maracá during visits in 1987 and 1988. The pseudoscorpions were collected in areas of forest and savanna in habitats such as soil, fallen tree trunks and standing tree crowns using various types of traps and collection techniques. Five of the species were collected in phoretic association with insects.

## RESUMO

Foram coletadas na Ilha de Maracá, Roraima, 17 espécies de pseudoscorpiões, pertencentes a 12 gêneros de 7 famílias, durante excursões realizadas em 1987 e 1988. Os pseudoscorpiões foram colecionados em áreas de mata e savana, tanto no solo, quanto em troncos caídos e copas de árvores, empregando vários tipos de armadilhas e técnicas de coleta. Cinco daquelas espécies foram coletadas em associação forética com insetos.

## INTRODUCTION

The first pseudoscorpions known from Amazonia were *Cordylochernes scorpioides* (Lin., 1758), in phoretic association with the cerambycid beetle *Acrocinus longimanus* (Lin., 1758), and *Lustrochernes intermedius* (Lin., 1758), recorded from the State of Pará by Ellingsen (1905). However, research aimed at understanding the diversity and ecology of these arachnids in Brazilian Amazonia started little more than 10 years ago with studies by Mahnert (1979, 1985a, b), Adis and Mahnert (1985, 1990) and Mahnert and Adis (1985) of material collected in areas of inundated forest, principally in Amazonas but also in Pará. In addition, Mahnert *et al.* (1986) provided a key for the families which occur in the Amazonian region.

None of these studies mentioned pseudoscorpions from Roraima, with the exception of *Brazilatemnus browni* Muchmore 1975 which was recorded on the Rio Branco by Mahnert (1979).

Many species of pseudoscorpions exhibit phoretic behaviour, associating themselves with different flying insects. Some of these relationships have been described in Amazonia in the states of Pará, Amazonas and Acre by a number of workers (Ellingsen, 1905; Mahnert and Aguiar, 1986; Mahnert, 1987; Aguiar and Bührnheim 1991, 1992a, b).

In this communication we present data on a total of 17 species of pseudoscorpions collected on the Ilha de Maracá. The collections were made during our two visits to the island in 1987 and by other workers in the same area in 1987 and 1988.

## METHODS

All the pseudoscorpions examined were collected on the Ilha de Maracá in forest and savanna habitats lying between the Ecological Station and the Casa de Santa Rosa (Figure Pr.1, p. xvii).

The majority of collections were made on two visits, the first during the rainy season, 3–15 May 1987, and the second in the dry season, 21–30 November 1987. Pseudoscorpions were collected in a number of ways: from the bark of trees; from colonies of passalid beetles in fallen tree trunks; by using the pyrethrum-fogging technique during the dry season; and on flying insects captured by light traps at night or caught directly during the day. They were also collected in Malaise, Shannon and Pennsylvania BLB traps.

Pseudoscorpions were also obtained as phoretic associates from the cerambycid beetles collected by C.S. Motta *et al.* between 16 and 22 March 1988. These collections had been pickled in 70% ethanol immediately following capture. Other collections were made from soil using Berlese funnels.

The material studied is preserved in 70% ethanol and lodged in the Entomological Collection of the Laboratory of Zoology of the University of Amazonas.

## SPECIES COLLECTED

### SUBORDER CHTHONIINEA, FAMILY CHTHONIIDAE

**Pseudochthonius sp.**

This genus is made up of species from the north of South America, occurring in St Vincent (Antilles), Venezuela and in Ecuador (Beier, 1932a), as well as three species recorded from Brazilian Amazonia (Muchmore, 1970; Mahnert, 1979). The Brazilian species came from Amazonas and Pará and all were collected in litter or soil (Mahnert and Adis, 1985). Our collection consisted of two females and a nymph extracted by a Berlese funnel from a sample of forest soil.

Material examined: in a forest soil sample, 2 females and 1 nymph, 22 July 1987, col. L. Anthony *et al.*

*Lechytia chthoniiformis* **(Balzan 1890)**

This species was originally described from the basin of the Paraná and Paraguay rivers. Beier (1959, 1964) demonstrated that it had a wider distribution, recording it in Peru and Chile. On the Ilha de Maracá it was found in phoretic association with the cerambycid beetle *Stenodontes spinibarbis* (Lin., 1758) (Aguiar and Bührnheim, 1991, 1992c). In addition, a single male was collected in a gallery of the passalid beetle *Verres furcilabris* (Eschsch., 1829) in the fallen trunk of a forest tree.

Material examined: in a passalid beetle gallery in fallen tree trunk, forest – 1 male, 25 November 1987, col. N.O. Aguiar; on 4 specimens of *Stenodontes spinibarbis* (Lin., 1758), 5 males and 21 females, 16–22 March 1988, col. C.S. Motta *et al.*

## SUBORDER CHTHONIINEA, FAMILY TRIDENCHTHONIIDAE

*Tridenchthonius mexicanus* **(Chamberlin and Chamberlin 1945)**

This species was described from material collected in a rotting coffee bush trunk in Mexico. It has also been recorded in Costa Rica (Beier, 1976), in the southeast of Trinidad (Hoff, 1946) and in Brazil in the states of Pará and Amazonas (Mahnert, 1985a, b). The phoretic association of *T. mexicanus* with passalid beetles was recorded in Mexico (Reyes-Castillo and Hendrichs, 1975) and in Brazil (Amazonas state) where no less than nine Passalidae were recorded as carriers (Aguiar and Bührnheim, 1992a). On the Ilha de Maracá we discovered one female *T. mexicanus* being transported by the passalid *Passalus interstitialis* (Eschsch., 1829), and males, females and two immatures in passalid galleries in a fallen tree trunk in the forest.

Material examined: in a passalid beetle gallery in fallen tree trunk, forest, 2 females 3 May 1987, 2 females 4 May 1987, 1 male and 1 tritonymph 26 November 1987, 5 males plus 1 female and 2 deutonymphs 28 November 1987, col. N.O. Aguiar; on a specimen of *Passalus interstitialis* (Eschsch., 1829) caught by light trap, 1 female 21–22 November 1987, col. N.O. Aguiar.

## SUBORDER NEOBISIINEA, FAMILY OLPIIDAE

*Apolpium* **sp.**

Species of *Apolpium* are recorded from Central America (Costa Rica; Barro Colorado Island, Panama) to the north of South America in Venezuela, Colombia and Ecuador (Beier, 1932a; Hoff, 1945) and Brazil where Mahnert (1985b) records two species, *A. ecuadorense* Hoff, 1945 and *A. minutum* Beier, 1931, from Amazonas. Only one specimen of an *Apolpium* species was recorded on the Ilha de Maracá. It was a female extracted by Berlese funnel from a soil sample taken from the transition area between forest and savanna.

Material examined: in soil from forest–savanna transition, 1 female, 6 May 1988, col. L. Anthony *et al.*

*Pachyolpium* **sp.**

The species of *Pachyolpium* are widely distributed from Central America to Paraguay (Beier, 1932a, 1954; Mahnert and Schuster, 1981). Only one species is recorded from Brazilian Amazonia, *P. irmgardae* Mahnert, 1979, which was described from there. One female specimen of *Pachyolpium* sp. was found in the buildings of Ecological Station on the Ilha de Maracá.

Material examined: in the buildings of the Ecological Station, 1 female, 10 May 1987, col. Duncan Scott.

## SUBORDER CHELFERINEA, FAMILY CHEIRIDIIDAE

*Neocheiridium corticum* **(Balzan 1890)**

*Neocheiridium corticum* is the type species of the genus and was described from the Rio Apa in Paraguay. Until now it was only known from the type locality and there was no information on phoretic behaviour. In fact, the only species of the genus previously recorded in Amazonia was *N. triangulare* Mahnert and Aguiar, 1986, which was found in phoretic association with a sphingid moth. However, we examined two specimens of the cerambycid *Stenodontes spinibarbis* (Lin., 1758) collected by mercury vapour lamp at the Ecological Station. One of these carried a nymph of *N. corticum* and the other both a female and a nymph, thus extending the known range and demonstrating the phoretic behaviour of this pseudoscorpion (Aguiar and Bührnheim, 1991).

Material examined: on 2 specimens of *Stenodontes spinibarbis* (Lin., 1758), 1 nymph 19–20 March 1988, 1 female and 1 nymph 21–22 March 1988, col. C.S. Motta *et al*.

## SUBORDER CHELFERINEA, FAMILY ATEMNIDAE

### *Paratemnoides minor* (Balzan 1891)

This is the only species of *Paratemnoides* recorded in Brazilian Amazonia, occurring in the states of Amazonas and Pará (Feio, 1945; Mahnert, 1979). It has a wide distribution in the neotropics from Paraná in the south of Brazil to Ecuador and Venezuela (Beier, 1932b; Feio, 1945).

We found the species on the Ilha de Maracá under the bark of a fallen rotting trunk in forest. The trunk was about 3 m long and 15 cm in diameter and contained a population of innumerable individuals extending over the whole surface of the wood, apart from the part in contact with the soil. We collected a sample of specimens consisting of 9 male and 21 female adults, 24 tritonymphs, 34 deutonymphs and 57 protonymphs. The protonymphs were extracted from silk-lined chambers in the galleries under the bark occupied by the 'colony'. Two female specimens were also collected, one in a black lightbox trap to which it had probably been carried by a flying insect (although phoretic behaviour has not yet been reported in this species), and another on a leaf in the forest.

Material examined: in a black lightbox trap, 1 female, 4–5 May 1987, col. N.O. Aguiar; on a leaf in forest, 1 female, 27 November 1987, col. N.O. Aguiar; under the bark of a fallen rotten trunk in forest, 9 males, 21 females, 24 tritonymphs, 34 deutonymphs and 57 protonymphs, 30 November 1987, col. N.O. Aguiar.

## SUBORDER CHELFERINEA, FAMILY CHERNETIDAE, SUBFAMILY LAMPROCHERNETINAE

### *Americhernes* aff. *longimanus* (Muchmore 1976)

The genus *Americhernes* was created by Muchmore (1976) to accommodate three species from the United States, one also found in Mexico, previously placed in *Lamprochernes* Tômôsvary, 1882, and three new US species, including *A. longimanus*, and another from Puerto Rico. The distribution of the genus was widened by Mahnert (1979), when he described two further species, *A. incertus* and *A. bethaniae*, both from Amazonas close to Manaus. Four males and four females of an *Americhernes* very close to *A. longimanus* were collected from the crowns of savanna trees by the pyrethrum-fogging technique. *A. longimanus* itself is only known from Florida and Mississippi, so clearly more study of the Maracá specimens is required.

Material examined: from crown of savanna trees by the pyrethrum-fogging technique, 4 males and 4 females, 25 November 1987, col. N.O. Aguiar *et al.*

### *Cordylochernes* (?) sp.

Seventeen female specimens of a species of *Cordylochernes* (?) were collected, 14 from a Malaise trap and three from a Shannon trap.

Material examined: from a Malaise trap, 1 female, February 1987, col. L.S. Aquino; ibid., 13 females, 2–13 May 1987, col. N.O. Aguiar *et al.*; from a Shannon trap, 3 females, 2–13 May 1987, col. N.O. Aguiar *et al.*

### *Lustrochernes intermedius* (Balzan 1891)

*L. intermedius* has been recorded throughout South America from Venezuela to Argentina. In Amazonia it has been found by Mahnert (1979, 1985a, b) in Amazonas and Pará under bark, and by Adis and Mahnert (1985) in tree crowns. It has also been found in Amazonas in phoretic association with two species of passalid beetles (Aguiar and Bührnheim, 1992a) and with the cerambycid *Acrocinus longimanus* (Lin., 1758) (Aguiar and Bührnheim, 1992c). On Maracá two males and six females of *L. intermedius* were found on three individuals of the cerambycid *Stenodontes spinibarbis* collected in a light trap. Two males, two females and two tritonymphs were also collected in galleries of passalid beetles inside a fallen trunk. In addition a female was also found in a black lightbox trap to which it had probably been brought by a flying insect.

Material examined: in passalid galleries, 1 male, 4 May 1987; 1 female, 25 November 1987; 1 male, 1 female and 2 tritonymphs, 28 November 1987, col. N.O. Aguiar; in black lightbox trap, 1 female, 18–19 March 1988, col. C.S. Motta *et al.*; on 3 specimens of *Stenodontes spinibarbis* (Lin., 1758), 1 male and 3 females, 17–18 March 1988; 1 male and 3 females 21–22 March 1988, col. C.S. Motta *et al.*

### *Lustrochernes* aff. *reimoseri* (Beier 1932)

Aguiar and Bührnheim (1992a) recorded this pseudoscorpion in Amazonas in phoretic association with 11 species of passalid beetles, nine of the genus *Passalus* and two of *Veturius*. It was found on the Ilha de Maracá in phoretic association with another passalid, *Verres furcilabris* (Eschsch., 1829). The typical *L. reimoseri* was described from Costa Rica and has only been recorded from that country. Apart from the phoretic specimen, we found two other specimens of *L.* cf. *reimoseri* in passalid galleries in a fallen tree trunk. Despite having comparable characters to Beier's (1932a) species, there are others which set it apart and suggest that it is a closely related species.

Material examined: on *Verres furcilabris* (Eschsch., 1829) taken in a black lightbox trap, 1 male and 6 females, 4 May 1987, col. N.O. Aguiar; in a passalid gallery, 1 female 3 May 1987, 1 female 26 November 1987, col. N.O. Aguiar.

*Lustrochernes* sp.

A female and a tritonymph of a *Lustrochernes* species were collected from the crown of a forest tree by the pyrethrum-fogging technique.

Material examined: Forest tree-crown, collected by pyrethrum-fogging technique, 1 female and 1 tritonymph, 23–24 November 1987, col. N.O. Aguiar.

## SUBORDER CHELFERINEA, FAMILY CHERNETIDAE, SUBFAMILY CHERNETINAE

*Parachernes adisi* (Mahnert 1979)

The genus *Parachernes* has a very wide geographic distribution with species occurring in various regions of the world such as the southeast of Australia, Malaysia, West Africa (Island of Sao Tomé), part of the nearctic region, the south of the United States and through nearly the whole neotropical region from Mexico to the south of South America. Eight species have been recorded for the Brazilian Amazon, five of them, including *P. adisi*, described by Mahnert (1979) from Amazonas. Three females of *P. adisi* were collected in a Malaise trap, one in a Shannon trap, and five males, seven females and two immatures (a tritonymph and a protonymph) from forest tree-crowns by pyrethrum-fogging. The type and paratypes of this species were also collected by the same technique from the crowns of flooded white-water forest in Amazonas (Mahnert, 1979).

Material examined: in a Malaise trap, 1 female, February 1987, col. L.S. Aquino; ibid., 2 females, 2–13 May 1987, col. N.O. Aguiar *et al.*; in a Shannon trap, 1 female, 2–13 May 1987, col. N.O. Aguiar *et al.*; in forest tree-crown by pyrethrum-fogging, 5 males, 7 females, 1 tritonymph and 1 protonymph, 23–24 December 1987, col. N.O. Aguiar *et al.*

*Parachernes albomaculatus* (Balzan 1891)

This pseudoscorpion was until now only recorded for Venezuela and Colombia (Beier, 1932b, 1954). However, we have now collected on the Ilha de Maracá a female gripping the rear leg of a Neuroptera Chrysopidae. We have not found any previous reference to phoretic behaviour in this species.

Material examined: attached to Neuroptera Chrysopidae, 1 female, 6 May 1987, col. N.O. Aguiar

*Parachernes plumosus* (With 1908)

*P. plumosus* was described from Venezuela and Colombia, and later in Ecuador (Beier, 1959) and in Brazil, where it was collected on tree trunks in black- and white-water flooded forests in Amazonas (Mahnert, 1979; Mahnert and Adis, 1985). We now extend the distribution records to the Ilha de Maracá where one male, two females and a tritonymph were collected in a Malaise trap, to which they could have been brought by flying insects.

Material examined: in a Malaise trap, 1 male, 2 females and a tritonymph, 2–13 May 1987, col. N.O. Aguiar *et al.*

*Parachernes inpai* (Mahnert 1979)

Another *Parachernes* species was found in a Malaise trap hung in the crown of a tree in the forest. Only one female was collected.

Material examined: in a Malaise trap in the crown of a forest tree, 1 female, 3–13 May 1987, col. N.O. Aguiar *et al.*

SUBORDER CHELFERINEA, FAMILY WITHIIDAE

*Parawithius (Victorwithius) gracilimanus* (Mahnert 1979)

This species was collected in Amazonas from tree trunks and epiphytes in white-water-flooded forest (Adis and Mahnert, 1985; Mahnert and Adis, 1985). Aguiar and Bührnheim (1992a) also recorded phoretic association with five species of passalid beetles, all belonging to the genus *Passalus*. A female *P. (V.) gracilimanus* was collected on Maracá from a tree crown by the pyrethrum-fogging technique.

Material examined: in a tree crown collected by pyrethrum-fogging, 1 female, 23–24 November 1987, col. N.O. Aguiar *et al.*

# DISCUSSION AND CONCLUSIONS

All the pseudoscorpions reported here are recorded for the first time for the Ilha de Maracá, Roraima. Of the 11 identified to species level, four belong to the Venezuelan fauna (Beier 1932a, b, 1954) and six to that of Brazilian Amazonia (Mahnert, 1979, 1985a, b). *Lechytia chthoniiformis* and *Neocheiridium corticum* were previously only recorded in their type localities in Paraguay.

Diversity of pseudoscorpions in Brazilian Amazonia has been much studied in the last 10 years and, according to Adis and Mahnert (1990), the known fauna of the region comprises 10 families, 26 genera and 60 species. The diversity recorded on the Ilha de Maracá from our two short visits and from some material received from other workers is 7 families, 12 genera and 17 species. No doubt further study would yield many more taxa on the island.

The majority of species were found in forest and only the species of *Americhernes* occurred in the savanna. The diverse collecting techniques allowed sampling of species which live in different habitats, such as under the bark of rotting logs, in soil, in tree crowns, as well as the capture of phoretic species associated with flying insects (Table 18.1). Probably the pseudoscorpions found in the various types of

**Table 18.1.** Species and habitats of pseudoscorpions found on the Ilha de Maracá in 1987 and 1988

| Species | Habitats | Total no. individuals |
|---|---|---|
| *Pseudochthonius* sp. | s (3) | 3 |
| *Lechytia chthoniiformis* | t (1), p (26) | 27 |
| *Tridenchthonius mexicanus* | t (14), p (1) | 15 |
| *Apolpium* sp. | s (1) | 1 |
| *Pachyolpium* sp. | d (1) | 1 |
| *Neocheiridium corticum* | p (3) | 3 |
| *Paratemnoides minor* | p (1), a (1), t (145) | 147 |
| *Americhernes* aff. *longimanus* | a (8) | 8 |
| *Cordylochernes* sp. | p (17) | 17 |
| *Lustrochernes intermedius* | p (9), t (6) | 15 |
| *Lustrochernes* aff. *reimoseri* | p (7), t (2) | 9 |
| *Lustrochernes* sp. | a (2) | 2 |
| *Parachernes adisi* | p (4), a (14) | 18 |
| *Parachernes albomaculatus* | p (1) | 1 |
| *Parachernes plumosus* | p (4) | 4 |
| *Parachernes inpai* | p (1) | 1 |
| *Parawithius (V.) gracilimanus* | a (1) | 1 |
| Total | | 273 |

a = shrub, d = in building, p = phoretic, s = soil, t = fallen tree trunk.
Figures in parentheses are no. of individuals in each habitat.

traps had also been brought to them by flying insects, suggesting that more detailed future investigations of phoretic relationships would be profitable.

*Tridenchthonius mexicanus*, *Paratemnoides minor*, *Lustrochernes intermedius*, *Parachernes adisi* and *Parawithius (V.) gracilimanus*, already recorded in Brazilian Amazonia, were also collected on Maracá in habitats similar to those observed by Adis and Mahnert (1985), Mahnert and Adis (1985) and Aguiar and Bührnheim (1992a).

# REFERENCES

Adis, J. and Mahnert, V. (1985). On the natural history and ecology of Pseudoscorpiones (Arachnida) from an Amazonian blackwater inundation forest. *Amazoniana*, 9, 297–314.

Adis, J. and Mahnert, V. (1990). On the species composition of Pseudoscorpiones (Arachnida) from Amazonian dryland and inundation forests in Brazil. *Revue Suisse Zool.*, 97, 49–53.

Aguiar, N.O. and Bührnheim, P.F. (1991). Pseudoscorpiões foréticos de *Stenodontes spinibarbis* (Lin., 1758) (Coleoptera, Cerambycidae) e redescrição de *Lechytia chthoniiformis* (Balzan, 1890) (Pseudoscorpiones, Chthoniidae) da Ilha de Maracá – Roraima. *Acta Amazonica*, 11 (1), 425–433.

Aguiar, N.O. and Bührnheim, P.F. (1992a). Pseudoscorpiões (Arachnida) em associação forética com passalidae (Insecta, Coleoptera) no Amazonas, Brasil. *Amazoniana*, 12 (2), 187–205.

Aguiar, N.O. and Bührnheim, P.F. (1992b). *Dolichowithius mediofasciatus* Mahnert, 1979

(Arachnida, Pseudoscorpiones, Withiidae) em forésia com Platipodidae (Insecta, Coleoptera), no Amazonas, Brasil. *Amazoniana*, 12 (2), 181–185.
Aguiar, N.O. and Bührnheim, P.F. (1992c). Pseudoscorpioes foréticos de Cerambycidae (Coleoptera) e ocorrência de *Parachelifer* Chamberlin, 1932 (Pseudoscorpiones, Cheliferidae) na Amazônia. *Boletim do Museu Paraense Emílio Goeldi*, sér. Zool., 8 (2), 343–348.
Beier, M. (1932a). Pseudoscorpionida I. Subord. Chthoniinea et Neobisiinea. *Tierreich*, 57, xx + 258 pp.
Beier, M. (1932b). Pseudoscorpionida II. Subord. Cheliferinea. *Tierreich*, 58, xxi + 294 pp.
Beier, M. (1954). Eine Pseudoscorpioniden – Ausbeute aus Venezuela. *Memorie Mus. Civ. Stor. Nat. Verona*, 4, 131–142.
Beier, M. (1959). Zur Kenntnis der Pseudoscorpioniden – Fauna des Andengebietes. *Beitr. Neotrop. Fauna*, 1, 185–228.
Beier, M. (1964). Die Pseudoscorpioniden-Fauna Chiles. *Ann. Naturhistor. Mus. Wien*, 67, 307–375.
Beier, M. (1976). Neue und bemepkenswerte zentralamerikanische Pseudoskorpione aus dem Zoologischen Museum in Hamburg. *Ent. Mitt. Zool. Inst. Zool. Mus. Hamb.*, 5 (91), 5 pp.
Ellingsen, E. (1905). Pseudoscorpions from South America collected by Dr A. Borelli, A. Bertoni de Winkelried, and Prof. Goeldi. *Boll. Mus. Zool. Anat. Comp. Univ. Torino*, 20 (500), 17 pp.
Feio, J.L. de Araujo. (1945). Novos pseudoscorpiões da região neotropical (com descrição de uma Subfamília, dois gêneros e sete espécies). *Bolm. Mus. Nac. Rio de Janeiro, n.s., Zool.*, 44, 47 pp.
Hoff, C.C. (1945). New neotropical Diplosphyronida (Chelonethida). *Am. Mus. Novit.*, 1288, 17 pp.
Hoff, C.C. (1946). Three new species of Heterosphyronid Pseudoscorpions from Trinidad. *Am. Mus. Novit.*, 1300, 7 pp.
Mahnert, V. (1979). Pseudoskorpione (Arachnida) aus dem Amazonas-Gebiet (Brasilien). *Revue Suisse Zool.*, 86, 719–810.
Mahnert, V. (1985a). Pseudoscorpions (Arachnida) from lower Amazon Region. *Revt. bras. Ent.*, 29, 75–80.
Mahnert, V. (1985b). Weitere Pseudoskorpione (Arachnida) aus dem zentralen Amazonasgebiet (Brasilien). *Amazoniana*, 9, 215–241.
Mahnert, V. (1987). Neue oder wenig bekannte, vorwiegend mit Insekten vergesellchaftete Pseudoskorpione (Arachnida) aus Südamerika. *Mittg. Schweiz. Entomol. Ges.*, 60, 403–416.
Mahnert, V. and Adis, J. (1985). On the occurrence and habitat of Pseudoscorpiones (Arachnida) from Amazonian forest of Brazil. *Stud. Neotrop. Fauna Environ.*, 20, 211–215.
Mahnert, V. and Aguiar, N.O. (1986). Wiederbeschreibung von *Neocheiridium corticum* (Balzan, 1890) und Beschreibung von zwei neuen Arten der Gattung aus Südamerika (Pseudoscorpiones, Cheiridiidae). *Mittg. Schweiz. Entomol. Ges.*, 59, 499–509.
Mahnert, V. and Schuster, R. (1981). *Pachyolpium atlanticum* n. sp., ein Pseudoskorpion aus der gezeitenzone der Bermudas – morphologie und okologie (Pseudoscorpiones: Olpiidae). *Revue Suisse Zool.*, 88, 265–273.
Mahnert, V., Adis, J. and Bührnheim, P.F. (1986). Key to the families of Amazonian Pseudoscorpiones (Arachnida) (In English, German and Portuguese). *Amazoniana*, 10, 21–40.
Muchmore, W. (1970). An unusual *Pseudochthonius* from Brazil (Arachnida, Pseudoscorpionida, Chthoniidae). *Ent. News.*, 81, 221–223.
Muchmore, W. (1976). Pseudoscorpions from Florida and the Caribbean area. 5 *Americhernes*, a new genus based upon *Chelifer oblongus* Say (Chernetidae). *Fla. Ent.*, 59, 151–163.
Reyes-Castillo, P. and Hendrichs, J. (1975). Pseudoscorpiones assóciados con Pasálidos. *Acta Politécnica Mexicana*, 16 (72), 129–133.

# 19 Earthworms of the Ilha de Maracá

**GILBERTO RIGHI**
*Universidade de São Paulo, São Paulo, Brazil*

## SUMMARY

A list of the species of earthworms collected on the Ilha de Maracá is presented, together with a key to their identification (based on external characters). Seven species are reported for the first time from the area, viz Glossoscolecidae: *Pontoscolex (Meroscolex) roraimensis, P. (P.) nogueirai, P. (P.) cuasi, P. (P.) corethrurus, Glossodrilus baiuca* and *G. oliveirae*, and Octochaetidae: *Dichogaster bolaui. Glossodrilus motu* sp. nov. is described and illustrated.

## RESUMO

É apresentada uma lista das espécies de minhocas da Ilha de Maracá (Roraima, Brazil), e uma chave de identificação baseada em carácteres externos. Novas ocorrências na área são indicadas nas famílias Glossoscolecidae: *Pontoscolex (Meroscolex) roraimensis, P. (P.) nogueirai, P. (P.) cuasi, P. (P.) corethrurus, Glossodrilus baiuca* e *G. oliveirae*, e Octochaetidae: *Dichogaster bolaui. Glossodrilus motu,* sp. nov. é descrita e figurada.

## INTRODUCTION

Oligochaeta were collected in a variety of habitats at the eastern end of the Ilha de Maracá in December 1979, with the support of the Secretaria Especial do Meio Ambiente, and subsequently in 1987 (Projeto Maracá) as part of the research programme of Dr João Ferraz (INPA). Specimens were collected by manual digging and sorting. From the results of this fieldwork it became apparent that the area studied supports relatively few species of earthworms, often in relatively large populations (>25 individuals in 20 × 20 × 30 cm of soil), and that these worms are actively sought alongside water-courses and *buritizais* (*Mauritia flexuosa* L.f. palm swamps) by peccaries (*Tayassu pecari*) and various species of birds.

The abundance of the earthworms, the diversity of habitat types and the facilities of the Ecological Station make Maracá an ideal site for studies of tropical oligochaete ecology, such as population dynamics, humus production, soil turnover/fertilization, and comparison with termite activity. The last aspect is of particular interest since termites, notably those species with endogenous and suspended nests, are common in the study area (see Chapter 16). Such studies would assist in examination of the hypothesis that termites are the tropical analogues of temperate earthworms (Drummond, 1887) – an hypothesis that has been accepted without further testing by certain soil-fauna researchers (Kubiena, 1955; Mathews, 1977).

The information presented here consists of a species list, and a key to their identification using external characters. It is intended that this will provide a basis for the type of ecological research suggested above. Details of the distributions and descriptions of the species are available from the bibliographic references, except for the new species *Glossodrilus motu*. The description of the latter follows the methodology of Righi (1984).

## SURVEY RESULTS

### EARTHWORM SPECIES ON THE ILHA DE MARACÁ

The following species were recorded (numbers refer to the collecting localities listed in the key below and asterisks indicate first reports from the area):

GLOSSOSCOLECIDAE
*Pontoscolex (Meroscolex) roraimensis* Righi, 1984*     (1,2)
*Pontoscolex (P.) maracaensis* Righi, 1984
*Pontoscolex (P.) nogueirae* Righi, 1984*     (2,3,4)
*Pontoscolex (P.) cuasi* Righi, 1984*     (3,5,6,7)
*Pontoscolex (P.) corethrurus* (Müller, 1857); Gates, 1973; Righi, 1984*     (1,2,3,5, 8,9,10)
*Glossodrilus baiuca* Hamoui and Donatelli, 1983*     (6,7)
*Glossodrilus arapaco* Righi, 1982
*Glossodrilus oliveirae* Righi, 1982*     (1,6,7,10)
*Glossodrilus motu*, sp. nov.
*Glossodrilus tico* Righi, 1982

OCTOCHAETIDAE
*Dichogaster bolaui* (Michaelsen, 1891); Righi and Guerra, 1985, p. 149* (1)
*Dichogaster modiglianii* (Rosa, 1896); Righi and Guerra, 1985, p. 150

**Key to collecting localities (see Figure Pr.1, p. xvii)**
1. Savanna at Santa Rosa.
2. Swampy forest on the Preguiça trail, 2 km north of the Furo de Maracá.
3. Seasonally flooded area of black soil with *Mauritia flexuosa* palms, between the Santa Rosa savanna and the Casa de Santa Rosa.
4. *Terra firme* forest between the Santa Rosa savanna and the Furo de Santa Rosa (Ferraz transect).
5. Black soil close to flooded area between the Ecological Station and the Santa Rosa savanna.
6. Damp depression with *buritizal* (*Mauritia flexuosa* palm swamp) between the Santa Rosa savanna and the Furo de Santa Rosa (Ferraz transect).

7. Margin of the Furo de Santa Rosa (Ferraz transect).
8. Black soil close to a stream between the Ecological Station and the Santa Rosa savanna.
9. *Terra firme* forest close to the Santa Rosa savanna.
10. *Terra firme* forest between the Santa Rosa savanna and the Casa de Santa Rosa.


**Key for identification of the species**

The morphological characters indicated in the key refer to animals fixed in 10% formalin without anaesthetic, observed in water with a stereoscopic microscope at up to ×50 magnification.

| | | |
|---|---|---|
| 1 | Clitellum begins at segment XIV, XV or XVI, and ends between XXII and XXIX. Slow-moving animals which do not struggle when removed from the soil | 2 |
| | Clitellum begins at segment XIII and ends at XX. Very active animals which struggle when removed from the soil, undulating the whole body | 11 |
| 2 | Chaetae positioned in four pairs of longitudinal rows,[1] regularly spaced along body | 3 |
| | Chaetae positioned in four pairs of longitudinal rows, regularly spaced anteriorly but not posteriorly | 10 |
| 3 | One pair of sucker-shaped glandular pubertal thickenings occupying the ventral surface of segments 1/2 XVII–1/2 XVIII *Glossodrilus motu* | |
| | One pair of longitudinally elongate glandular pubertal thickenings (pubertal ridges) occupying the ventral surface of at least two segments, beginning at XVIII or further back. May or may not exhibit other pubertal markings | 4 |
| 4 | Pubertal ridges situated at 1/3 XXIII–XXVII, 1/2 XXVIII *Pontoscolex (Meroscolex) roraimensis* | |
| | Pubertal ridges begin at XVIII, XIX or XX | 5 |
| 5 | Pubertal ridges at 1/2 XX–1/2 XXIII | 6 |
| | Pubertal ridges begin at XVIII or XIX | 7 |
| 6 | Clitellum at XV–XXIV. Glandular pubertal thickenings at VIII, X, XI and XII; may be asymmetric or lacking in one or two segments *Pontoscolex (P.) maracaensis* | |
| | Clitellum at XIV, XV–XXV. No pubertal differentiation in the segments anterior of the clitellum *Pontoscolex (P.) nogueirae* | |
| 7 | Total length less than 30 mm. Spacing between chaetae *aa* at mid-body 2–3 times greater than spacing between chaetae *ab* *Glossodrilus tico* | |
| | Total length greater than 50 mm. Mid-body *aa* chaetae spacing more than 15 times *ab* chaetae spacing | 8 |
| 8 | Total length less than 70 mm. Clitellum at XVI–XXII *Glossodrilus arapaco* | |
| | Total length greater than 70 mm. Clitellum at XV, XVI–XXIV | 9 |

9  Total length 70–100 mm. Pubertal ridges at 1/2 XVIII–XXI
   *Glossodrilus baiuca*
   Total length 180–270 mm. Pubertal ridges at XVIII–XX
   *Glossodrilus oliveirae*
10 Chaetae *a* and *c* regularly positioned towards posterior, *b* and *d* alternating between two levels  *Pontoscolex (P.) cuasi*
   Chaetae positioned quincuncially towards posterior
   *Pontoscolex (P.) corethrurus*
11 Body uniformly pale green. Total length 20–30 mm. One pair of seminal furrows at 1/2 XVII–1/2 XIX in the line of ventral chaetae, enclosed by a glandular area in the form of two juxtaposed dumb-bells
   *Dichogaster modiglianii*
   Body whitish apart from a pink clitellum. Total length 20–45 mm. One pair of seminal furrows at 1/2 XVII–1/2 XIX in the line of ventral chaetae, lacking distinct glandular differentiation  *Dichogaster bolaui*

## *Glossodrilus motu* sp. nov.

*Material* – Roraima; Ilha de Maracá; seasonally inundated forest at margin of swamp, with black soil (very damp), collected at depth of 5 cm; one (clitellate) specimen, holotype ZU-1173, deposited at Department of Zoology, University of São Paulo; G. Righi coll. 31/10/1987.

*Description* – Length 37 mm. Diameter of anterior region (VII) 1.2 mm; 1.1 mm at clitellum; 1.2 mm in mid-body region and 0.9 mm in posterior region. Pigmentation absent. There are 111 segments. The prostomium and the majority of segment I are invaginated. In the oral cavity, opened by a longitudinal incision, the prostomium appears as a small, dorsal, anterior appendix. The chaetae are regularly distributed in four pairs of longitudinal series, beginning at segment II. They are sigmoid and elongate, with a distal nodule and a unicuspidate apex without ornamentation. Their length varies in the mid-body region from 102 to 127 $\mu$m (M=116), and in the posterior region from 138 to 169 $\mu$m (M=150). Chaetae distribution ratios in the mid-body region (XXX–XL) are as follows: *aa*; *ab*; *bc*; *cd*; *dd* = 17.8: 1.0: 5.0: 0.8: 21.8 (*ab* = 64 $\mu$m), and in the posterior region (XC–C), 14.5: 1: 4: 1.1: 13.6 (*ab* = 64 $\mu$m).

The clitellum is situated at XV–XXIII (9 segments), is milky-white and more turgid dorsally, and has distinct segmental furrows and chaetae. It is thickest in the *bc* space of 1/2 XVII–1/2 XVIII, joining ventrally with a pair of pubertal marks which lie in the line of ventral chaetae, centred at 17/18 (Figure 19.1). These marks are sucker-shaped and almost triangular, their margins raised and white and their centres slightly depressed and black. The ventral surface of XVII–XVIII, between the pubertal marks, is depressed. There are no other pubertal marks. The genital pores are microscopic. The nephridiopores are intersegmental or immediately above the *b* line of chaetae.

The septa 6/7–10/11 are conical and interpenetrated; 7/8–9/10 are thickened and muscular; whereas 6/7 and 10/11 are thinner and the other septa are fragile and flat.

Muscular bands, oblique within the body cavity, join the ventral body wall (alongside the nerve cord) with each lateral wall at XVI–XIX. These are most numerous from XVII–XVIII. Behind the short and wide anterior oesophagus lies (at VI) a broadly cylindrical muscular gizzard with a longitudinal keel. A pair of calciferous glands is situated dorso-laterally to the oesophagus at XII, distending the posterior septa to occupy the space of segments XII–XV. Each of these glands is ovoid, and divided into a glandular part and a membranous part. The glandular part, with a composite tubular structure, occupies the posterior half and a slender ventral strip of the anterior half. The membranous part, which is inconspicuous and empty, is continued dorsally and anteriorly by a short duct which joins the opposite gland and connects them both to the oesophagus at XII. The oesophagal/intestinal transition is situated at 15/16. The typhlosole begins at XVI, appearing as a dorsal lamina approximately one-third of the intestinal diameter in height, and slightly longitudinally undulate. There are no intestinal caeca.

Two pairs of capacious hearts are situated at X and XI. In each segment there is a pair of holonephridia with simple funnels. In the post-clitellary nephridia (Figure 19.1), loop III (bladder) shows a median strangulation in the region of its junction with the nephridiopore, isolating a lateral portion and a ventral portion. The anterior and lateral surfaces of the lateral portion are enveloped by loop I, which is long and folded in its ental extremity. The ventral portion of loop III is surrounded by the broad 'glandular canal', which opens posteriorly. Loop II is small and ventrally positioned. A pair of testes and large silvery seminal funnels lie free in the cavity of XI, full of coagulated spermatozoa. The pair of seminal vesicles, originating from the posterior face of septum 11/12, runs beneath the calciferous glands and rises at their posterior ends, covering them (and also the intestine at XVI) with a pattern of bands folded irregularly at each side. The male ducts run along the line of ventral chaetae up to 17/18, where they open into a pair of male pores inside the pubertal marks.

One pair of ovaries and of female funnels are found at XIII. The female pores were not located. Two pairs of spermathecae are situated at IX and X, opening at 8/9 and 9/10 immediately above the *b* line of chaetae. The spermathecae of X are twice as large as those of IX, and all are empty. Each spermatheca (Figure 19.1) is pyriform, without external distinction between duct and ampulla. In microscopic preparation in water, one can see (as a result of transparency) that the ampulla is twice as long as the thin duct, and that the cavity of the ampulla is crossed by numerous circular transverse pleats.

*Considerations* – The new species, by exhibiting intimately twinned chaetae and spermathecae at 8/9 and 9/10 in the line of ventral chaetae, thus belongs to the group made up of *Glossodrilus itajo* (Righi, 1971); *G. arapaco* Righi, 1982; *G. oliveirae* Righi, 1982 and *G. ortonae* Righi, 1988. These species are distingushed from *G. motu* by the following characters:

- *G. itajo* Bicuspid chaetae. Pubertal ridges at 1/2 XVI–1/2 XVIII. Male pores at 16/17.
- *G. arapaco* Pubertal ridges at 1/2 XVII–1/2 XX. One pair of copulatory papillae at XXI. Male pores at 18/19.

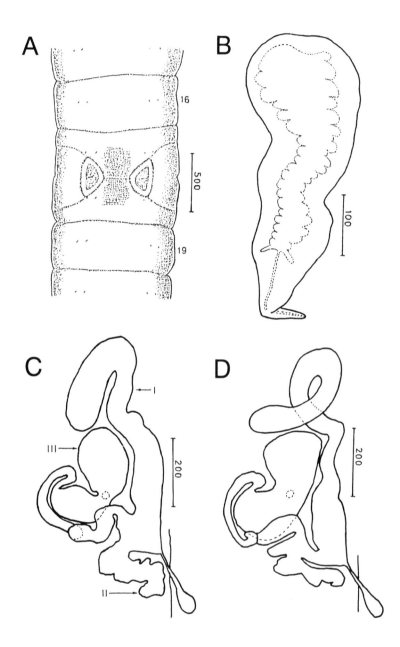

**Figure 19.1**. *Glossodrilus motu* sp. nov: (A) ventral surface of the segments XVI–XIX; (B) spermatheca of 9/10; (C–D) post-clitellary nephridia (I–III = nephridial loops). Scale in μm

- *G. oliveirae* Ornamented chaetae. Pubertal ridges at 2/3 XVIII–XX. Pubertal tumescences at *bc* of VIII–IX. Male pores at XIX.
- *G. ortonae* Pubertal ridges at 12 XVII–XX. Testis sacs present. Seminal vesicles bifid.

The name of the new species stems from the language of the Maiongong Indians, and means 'earthworm'.

## ACKNOWLEDGEMENTS

I am very grateful to Dr João Baptista Ferraz (INPA) for the opportunity to carry out this study, and to William Milliken for the English version of my paper.

## NOTE

1. The series either side are designated *a*, *b*, *c*, and *d* (beginning at the mid-ventral line).

## REFERENCES

Drummond, H. (1887). On the termite as the tropical analogue of the earthworm. *Proc. R. Soc. Edinb.*, 13, 137–146.

Gates, G.E. (1973). Contribution to a revision of the earthworm family Glossoscolecidae. I. *Pontoscolex corethrurus* (Müller, 1887). *Bull. Tall Timbers Res. Stat.*, 14, 1–12.

Hamoui, V. and Donatelli, R.J. (1983). Uma espécie nova de Oligochaeta Glossoscolecidae, *Glossogrilus (G.) baiuca*, sp. n., do Território de Roraima, Brasil. *Rev. Brasil. Biol.*, 43 (2), 143–146.

Kubiena, W.L. (1955). Animal activity in soils as a decisive factor in establishment of humus form. In: *Soil zoology*, ed. K. McE. Kevan, pp. 73–82. Butterworth Scientific Publ., London.

Mathews, A.G.A. (1977). *Studies on termites from the Mato Grosso State, Brazil*. Acad. Brasil. Ciênc., Rio de Janeiro.

Michaelsen, W. (1891). Oligochaeten des Naturhistorischen Museums in Hamburg, IV. *Jahrb. Hamburg. wiss. Anst.*, 8, 299–399.

Müller, F. (1857). *Lumbricus corethrurus*, Bürstenschwans. *Arch. Naturg.*, 23 (1), 113–116.

Righi, G. (1971). Sobre a Família Glossoscolecidae (Oligochaeta) no Brasil. *Arq. Zool., S. Paulo*, 20 (1), 1–96.

Righi, G. (1982). Adições ao gênero *Glossodrilus* (Oligochaeta, Glossoscolecidae). *Rev. bras. Zool. S. Paulo*, 1 (1), 55–64.

Righi, G. (1984). *Pontoscolex* (Oligochaeta, Glossoscolecidae) a new evaluation. *Stud. Neotrop. Fauna*, 19 (3), 159–177.

Righi, G. (1988). Uma coleção de Oligochaeta da Amazônia brasileira. *Papéis Avulsos Zool., S. Paulo*, 36 (30), 337–351.

Righi, G. and Guerra, R.A.T. (1985). Alguns Oligochaeta do norte e noroeste do Brasil. *Bolm. Zool., S. Paulo*, 9, 145–157.

Rosa, D. (1896). I lombrichi raccolti a Sumatra dal Dott. Elio Modigliani. *Ann. Mus. Stor Nat. Genova*, (2) 36, 502–532.

# 20 The Rotifera of shallow waters of the Ilha de Maracá

## WALTER KOSTE[1] AND BARBARA ROBERTSON[2]

[1]*Quakenbrück, Germany;* [2]*Instituto Nacional de Pesquisas da Amazônia, Manaus, Brazil*

## SUMMARY

Plankton samples collected in periodically dry, shallow-water systems on the Ilha de Maracá were investigated for rotifers. This survey was the first of its type in the State of Roraima. The rotifer community, of which 159 species were collected, was characteristic of small bodies of water with decomposing vegetation. Most of the species collected are cosmopolitan, but *Dicranophorus sebastus* (Harring & Myers, 1928) and *Lecane clara* (Bryce, 1892) are new for the neotropics, and *Lepadella christinei, L. tricostata* and *Testudinella robertsoni* are new to science.

## RESUMO

Amostras da fauna planctônica, coletadas em ecossistemas aquáticos, rasos e sazonais, na Ilha de Maracá, foram investigadas para a presença de rotíferos. Este levantamento foi o primeiro deste tipo realizado no Estado de Roraima. A comunidade de rotíferos, que incluiu 159 espécies, é típica de pequenas poças de água com vegetação em estado de decomposição. Em sua maioria, as espécies são cosmopolitas, mas *Dicranophorus sebastus* (Harring & Myers, 1928) e *Lecane clara* (Bryce, 1892) são registros novos para os neotrópicos, e *Lepadella christinei, L. tricostata* e *Testudinella robertsoni* são espécies novas.

## STUDY AREAS

Two principal savanna types have been described for the Ilha de Maracá (Milliken and Ratter, Chapter 5) of which one, the seasonally flooded savanna, was sampled during this survey (see Plate 4). This vegetation is essentially treeless and is dominated by Gramineae. During the peak of the rainy season (May, June and July), most of this savanna flooded to a maximum depth of approximately 30 cm. During the rest of the year it dried out, although wet patches did remain. When sampling was carried out in June the water was warm (30°C) and well oxygenated (5.45 mg l$^{-1}$). The pH was nearly neutral (c. 6), and the conductivity was very low (15 $\mu$S). There are also several seasonal ponds within the forest on the Ilha de Maracá, one of which was sampled during this study. Although essentially dry for much of the year, the pond was flooded to a depth of over a metre during the sampling period. Limnologically the water was very similar to that on the savanna; it was warm (29°C), well saturated (7.25 mg l$^{-1}$), slightly acidic (pH 5.1) and of low conductivity (10 $\mu$S).

## MATERIALS AND METHODS

All samples were collected between 18 and 22 June 1987 during the rainy season, with a 55 $\mu$m plankton net. The samples were qualitative, and all were immediately preserved in formalin (final concentration 7%). One sample (SP) was taken in the seasonal pond to the northwest of the Ecological Station, and three others (Sav 1–3) were taken in the seasonally flooded savanna area between the Ecological Station and the Furo de Maracá (Figure Pr.1, p. xvii).

For identification of the specimens, approximately 20 drops of every sample were mixed with glycerine (10% in distilled water) and spread on a slide. Specimens of interest were isolated and placed on another slide with a minute drop of 10% glycerine. Trophy mounts were prepared with 10% sodium hypochlorite, and permanent slides were prepared by the evaporation method described in Koste (1978).

## RESULTS AND DISCUSSION

Most of the rotifers were identified to species level, except for a few specimens of the genera *Cephalodella*, *Collotheca*, *Monommata*, *Notommata* and *Polyarthra* which suffered preservation artefacts. The 159 species of rotifers collected during this study are listed in Appendix 8 (p. 481), together with data on the environments in which they were found and on their general ecological preferences and distributions. The water bodies studied were generally full of plants or rotting vegetation, as a result of which they tended to support specialized microfauna characteristic of acidic but oxygen-poor waters. In general, clear water and white-water habitats are richest in terms of rotifer species, whereas black-water environments are relatively poor. In the 168 samples collected between 1941 and 1959 by Sioli (assisted between 1947 and 1948 by Dr R. Braun), 206 species of rotifers were recorded (Koste, 1972).

As there have been no previous studies of rotifers in the State of Roraima, it is difficult to compare and discuss our results. However, the species composition encountered on Maracá is similar to that found by the first author in other neotropical locations including Venezuela, Brazil, northern Ecuador, Peru, and Uruguay (excluding the higher altitude lakes). Also, given that the great majority (146) of the species are littoral forms (only 13 are planktonic), the assemblage resembles the composition typically encountered in small ponds, lagoons and temporary pools in Amazonia. Finally, the *Mytilina*, *Lepadella* and *Testudinella* complex, and the many Bdelloidea species present, represent a characteristic 'rotten mud' community associated with decomposing vegetation, such as that described by Koste and Robertson (1983) for a Central Amazonian lake.

Most of the species collected are cosmopolitan; only 27 of them are known solely from warm waters, i.e. are sub-tropical and tropical stenotherms. Ten species are endemic to South America: *Floscularia decora, Keratella americana, Lecane amazonica, L. astia, L. stichaea amazonica, L. wulferti, Lepadella donneri, L. tricostata, Platyias leloupi latiscapularis* and *Testudinella ohlei*.

## SPECIES OF PARTICULAR INTEREST

*Dicranophorus sebastus* was found in the seasonal pond, and prey rotifers such as *Lecane* spp. and *Lepadella rhomboides* were visible. Since Harring and Myers' original description (1928) there had been no further report of this species, which appears to have been known only from the acidic waters of the Nearctic.

*Lecane clara* is a cosmopolitan species which prefers *Sphagnum* but is also found in other psammal and littoral waters, always in the shallows. This is its first record for South America.

*Ptygura linguata* is a rare species which was found both in the seasonal pond and in the savannas. The tube was attached to plant fragments of *Utricularia* sp. and some unidentified algal clusters. *P. linguata* was first described from New Jersey, USA, but the first author rediscovered it in lagoons near Rio Nhamundá, a tributary of the Amazon River (Brandorff *et al.*, 1982). *Lepadella christinei* Koste, *L. tricostata* Koste and *Testudinella robertsoni* Koste are new species, described for the first time in Koste and Robertson (1990).

## ACKNOWLEDGEMENTS

The authors would like to thank Edinaldo Nelson dos Santos Silva for his indispensable help in collecting samples, and Pedro A. Suarez Mera for the limnological data. The first author is also grateful for the long-term loan of microscope facilities by the Deutsche Forschungsgemeinschaft, Bonn-Bad Godesburg, Germany.

## REFERENCES

Brandorff, G.O., Koste. W. and Smirnov, N.N. (1982). The composition and structure of Rotifera and crustacean communities of the Lower Rio Nhamundá, Amazonas, Brazil. *Stud. Neotrop. Fauna Environ.*, 17, 69–121.

Harring, H.K. and Myers, F.J. (1928). The Rotifer fauna of Wisconsin. IV. The Dicranophoridae. *Trans. Wisconsin Acad. Sci., Arts and Letters*, 23, 667–808.

Koste, W. (1972). Rotatorien aus Gewässern Amazoniens. *Amazoniana*, 3 (3/4), 258–505.

Koste, W. (1978). *Die Rädertiere Mitteleuropas (überordnung Monogononta)*. Begr. v. M. Voigt, 2 vols., Gebr. Borntraeger, Stuttgart.

Koste, W. and Robertson, B. (1983). Taxonomic studies of the Rotifera (Phylum Aschelminthes) from a Central Amazonian várzea lake, Lago Camaleão (Ilha de Marchantaria, Rio Solimões, Amazonas, Brazil). *Amazoniana*, 8 (2), 225–254.

Koste, W. and Robertson, B. (1990). Taxonomic studies of the Rotifera from shallow waters on the Island of Maracá, Roraima, Brazil. *Amazoniana*, 11 (2), 185–200.

# 21 Biological Indicators in the Aquatic Habitats of the Ilha de Maracá

CECILIA VOLKMER-RIBEIRO,[1] MARIA CRISTINA D. MANSUR,[1] PEDRO A.S. MERA[2] AND SHEILA M. ROSS[3]

[1]*Fundação Zoobotânica do Rio Grande do Sul, Porto Alegre, Brazil;* [2]*Instituto Nacional de Pesquisas da Amazônia, Manaus, Brazil;* [3]*University of Bristol, Bristol, UK*

## SUMMARY

Three distinct aquatic habitats were identified on the Ilha de Maracá: seasonal ponds, *igarapés* (streams) and man-made water bodies (borrow-pits). Samples of the algae, macrophytes, sponges and molluscs were taken in each of these environments during both the wet and the dry seasons, and samples of the water during the wet season. These were compared with samples taken in the same way from the Rio Uraricoera. The results of the physical and chemical analyses of the waters of the three habitats on the Ilha de Maracá fell within the characteristic values of certain Central Amazonian 'black-water' habitats, with low conductivity, acidic pH and extremely low levels of calcium, limiting or even impeding the occurrence of molluscs. The results obtained for the waters of the Rio Uraricoera fell within those characteristic of the 'white-water' class of Amazonian rivers. The sponge community was found to be particularly rich and abundant in the seasonal pond, where it was different from those found in temporary water bodies in Central Amazonia, and similar to those which formed spongillite deposits in savanna lakes in southwest Brazil.

## RESUMO

Três ambientes aquáticos distintos foram detectados na Ilha de Maracá: pequenos lagos temporários, igarapés e represamentos artificiais. Amostragens das algas, macrófitas, esponjas e moluscos foram levadas a efeito durante a estação úmida e a seca em uma de cada destas unidades, e da água durante a estação úmida. Os resultados foram comparados com os de amostragens do mesmo tipo tomadas no Rio Uraricoera. Os resultados das análises físicas e químicas das águas dos três ambientes na Ilha de Maracá ficaram dentro das características de alguns ambientes de 'águas negras' da Amazônia Central, com baixa condutividade, pH ácido e níveis de cálcio extremamente baixos, limitando ou mesmo impedindo a ocorrência de moluscos. Os resultados obtidos da amostragem tomada no rio Uraricoera conformam-se às caraterísticas dos rios amazônicos de 'águas brancas'. A comunidade de esponjas é particularmente rica e abundante no lago temporário amostrado, onde é, contudo, diferente daquelas dos ambientes aquáticos da Amazônia Central, mas similar as que formaram depósitos de espongilitos em lagoas nas savanas (cerrados) da região sudoeste do Brasil.

## INTRODUCTION

No extensive studies existed on the aquatic ecosystems of the Ilha de Maracá prior to the INPA/RGS/SEMA Maracá Rainforest Project. The authors' main aim in the present study was to investigate the aquatic habitats in order to provide baseline data for understanding the island's ecosystem, and its relation to the Amazonian environment.

## METHODS

Three principal types of aquatic environments were identified using an aerial photomosaic of the island. One representative example of each of these was chosen for the study of water, algae, macrophytes, sponges and molluscs. The three chosen sites were easily reached by land, using the trails leading from the Ecological Station into the forest. Samples were also taken from the Rio Uraricoera adjacent to the island. Biological sampling was conducted between 22 and 27 July 1987 (wet season) and between 3 and 12 December 1987 (dry season). Water samples were only collected during the wet season, since most of the island's aquatic habitats dried up at the height of the dry season. For each water sample, 10 chemical analyses were performed: pH, conductivity, dissolved oxygen, calcium, total hardness, silicon, iron, reactive phosphate (e.g. orthophosphate) and two types of inorganic nitrogen ($NO_3$-N and $NH_4$-N). Analyses were carried out using the DREL/5 Portable Chemical Laboratory made by the Hach Chemical Company, USA, unless otherwise stated. Zoological collecting procedures are reported in Volkmer-Ribeiro (1992) and Mansur and Valer (1992) respectively.

## DESCRIPTION OF THE SAMPLING SITES

The seasonal pond selected for sampling (the largest of the ponds shown in Figure 23.2, p. 436) is the source of a small *igarapé* (stream) which flows into the northern channel of the Rio Uraricoera, the Furo de Santa Rosa. The pond is approximately 200 m in diameter, and in the wet season the depth of water was no more than 2 m at its centre. It dried up at the height of the dry season, but from 3 to 13 December 1987 there was still a considerable quantity of water present. The pond is surrounded by dense forest, and its position is very clearly defined by a ring of *Mauritia flexuosa* L.f. (*buriti*) palm trees. In the pond itself stands of *Montrichardia linifera* Schott (Araceae), *Thalia* spp. (Marantaceae) and Cyperaceae form a dense fringe in the shallows. There are abundant grasses, and *Nymphaea* sp. (Nymphaeaceae) occurs in profusion on the surface of the water. Below the surface, *Utricularia* spp. (Lentibulariaceae) are the dominant submerged macrophytes. The pond bottom is covered by an accumulation of dead plant material, and its waters show a brownish colouration.

The *igarapé* which was sampled also has its source in a small seasonal pond, and flows out into seasonally inundated savanna and subsequently into the Uraricoera.

The samples were collected alongside the wooden 'buriti bridge' (site 25 in Figure 5.2, p. 76) on the track which links the Ecological Station with the Furo de Santa Rosa. The *igarapé* was no more than 30 cm deep between 22 and 27 July, and it was already dry by the sampling period in December.

The two man-made water bodies (borrow-pits) are the result of dredging the seasonally inundated savanna for material to build a causeway, on which the access track to the Ecological Station was laid (Plate 4a). Various *igarapés* flow into this marshy area, including the one chosen for sampling. In the wet season the borrow-pits had a variable depth of around one metre. In the dry season they were reduced to areas of very shallow water with a dense vegetation of *Eichhornia azurea* Kunth (Pontederiaceae), interspersed with Cyperaceae (principally *Eleocharis* spp.).

Samples from the Uraricoera were taken from its southern channel (the Furo de Maracá), close to the point where the causeway meets the river. During the wet season the level of the water rose several metres, and the current was very strong. Only during December was it possible to sample molluscs from the edge of the river, under a steep bank more than 2 m high.

## RESULTS AND DISCUSSION

The results of the analyses of the water samples are presented in Table 21.1; those for the algae and vegetation in Tables 21.2 and 21.3 respectively; and those for molluscs and sponges in Table 21.4.

### WATER ANALYSES

#### pH

The Rio Uraricoera, sampled next to the ferry landing point, showed the highest pH of all the waters analysed. This river carries significant quantities of eroded and suspended sediments from the upper reaches of its catchment. The sediments effectively elevate concentrations of weathered minerals such as Ca and Mg in the water, but at the same time there is no significant enrichment of humic acids. As a result, the pH is around neutral. The pH in the borrow-pit was slightly acidic. Both the seasonal pond and the *igarapé* had acidic values, illustrating the excessively leached nature of the soil and the accumulation of humic acids, resulting from the decomposition of litter from the surrounding forest and, in the case of the borrow-pits, of the abundant macrophytes.

### Conductivity

The Rio Uraricoera also showed the highest values for conductivity, again as a result of its suspended sediments. The levels in the borrow-pit were lower, and those of the pond and the *igarapé* were lowest with less than half the value of the river. In general all the values of conductivity were low. This reflects a system of high

**Table 21.1.** Analysis of water samples from the Rio Uraricoera and three aquatic habitats on the Ilha de Maracá, collected in July 1987 and made by S.M. Ross, using a portable DREL/5 Chemical Laboratory. Dissolved oxygen was measured by P.S. Mera in the field by the potentiometric method

| Samples | Temp. (°C) | pH | Conductivity ($\mu$S cm$^{-1}$) | Dissolved O$_2$ (mg l$^{-1}$) | NO$_3$-N (mg N l$^{-1}$) | NH$_4$-N (mg N l$^{-1}$) | P (mg l$^{-1}$) as PO$_4$ | Fe (mg l$^{-1}$) | Si (mg l$^{-1}$) as SiO$_2$ | Ca (mg l$^{-1}$) as CaCO$_3$ | Total hardness (mg l$^{-1}$) as CaO$_3$ |
|---|---|---|---|---|---|---|---|---|---|---|---|
| Rio Uraricoera | 29 | 6.88 | 28 | 8.24 | 0.12 | 0.16 | 0.075 | 0.56 | 12.75 | 4.5 | 10.0 |
| Shallow borrow-pit | 31 | 6.15 | 17 | 5.45 | 0.0 | 0.55 | 0.06 | 0.79 | 11.25 | 0 | 9.0 |
| Seasonal pond | | | | | | | | | | | |
| surface | 27 | 5.46 | 11 | | | | | | | | |
| bottom | 27 | 5.36 | 12 | 7.27 | 0.02 | 0.65 | 0.05 | 0.56 | 2.2 | 0 | 4.0 |
| *Igarapé* (stream) | 27 | 5.62 | 13 | 5.45 | 0.23 | 0.95 | 0.04 | 0.79 | 10.75 | 0 | 5.0 |

absorption efficiency by roots, combined with highly weathered soils whose low levels of organic matter and sesquioxides provide a relatively small capacity for cation and anion exchange.

**Dissolved oxygen (potentiometric)**

The highest levels of dissolved oxygen were found in the Rio Uraricoera and in the pond. This is not surprising, since the river has a fast current with much turbulence whilst the phytoplankton and the abundant submerged macrophytes produce oxygen in the pond water. The lowest values, found in the *igarapé* and the borrow-pit, are explained by the low volumes of water in both, plus the shadowing of the *igarapé* by the forest canopy, preventing photosynthetic oxygen production by plants. The high solar exposure of the borrow-pit water resulted in high water temperatures, with consequently low oxygen solubility.

**Calcium and total hardness (EDTA titration method)**

Only the Rio Uraricoera showed measurable Ca and Mg values (as total hardness). Both were quite low, and close to the limits of detection by this method. All the other sites showed Ca levels below the limits of detection. There was little measurable Mg in any of the samples, with the lowest values found in the seasonal pond and the *igarapé*.

**Dissolved silica (heteropoly blue method)**

All the Maracá soil samples revealed very high values of extractable silica (Ross, unpublished) and the soils in the region of the Maracá Ecological Station were highly weathered (Ross *et al.*, 1990), so it is not surprising that the Si levels were also high in the freshwater habitats. The river, the borrow-pit and the *igarapé* had quite high concentrations of silica, but that of the seasonal pond was only about one-fifth to one-sixth of that in the other areas sampled. This was probably due to the absorption of the available silica by plants, sponges, diatoms and other organisms which occur within and around the pond.

**Dissolved iron (phenanthroline method)**

The Fe content in the samples was high and similar in all areas, ranging from 0.56 mg $l^{-1}$ in the seasonal pond and the Rio Uraricoera to 0.79 mg $l^{-1}$ in the borrow-pit and the *igarapé*. These values were to be expected, as the soil in the region is very rich in iron oxides.

**Reactive phosphate (orthophosphate, ascorbic acid method)**

With so much iron in the surrounding soils, the scarcity of available phosphate in the surface waters is not surprising due to the high soil phosphate fixing capacity.

Table 21.2. Algae* identified from the seasonal pond and from the shallow borrow-pit, July 1987

|  | Seasonal pond (Total of 130 spp.) | | Shallow borrow-pit (Total of 71 spp.) | |
| --- | --- | --- | --- | --- |
|  | No. spp. | % | No. spp. | % |
| Cyanophyta | 6 | 4.62 | 5 | 6.94 |
| Euglenophyta | 9 | 6.92 | 5 | 6.94 |
| Chrysophyta |  |  |  |  |
| Chrysophyceae | 4 |  |  |  |
| Bacillarophyceae | 6 | 7.69 |  |  |
| Pyrrhophyta | 2 | 1.54 | 1 | 1.39 |
| Chlorophyta |  |  |  |  |
| Chlorophyceae |  |  |  |  |
| Volvocales | 1 | 0.77 | 1 | 1.39 |
| Ulotrichales | 1 | 0.77 |  |  |
| Oedogoniales | 2 | 1.54 | 2 | 2.78 |
| Chlorococcales | 7 | 5.38 | 10 | 13.89 |
| Tetrasporales | 1 | 0.77 | 1 | 1.39 |
| Conjugatophyceae |  |  |  |  |
| Desmidiales | 89 | 68.46 | 44 | 61.11 |
| Zygnematales | 2 | 1.54 | 2 | 2.78 |

* In July 1987 the dominant species were: *Hapalosiphon aureus*, *Peridinium gatunense*, *Dinobryon sertularia Cosmarium dimaziforme*, *Hyalotheca dissiliens*, *Pediastrum duplex*, *Coelastrum cambricum* and *Groenbladia undulata*. *Peridinium gatunense* (Pyrrhophyta) and *D. sertularia* (Chrysophyceae) were blooming in July 1987. Such species make up siliceous plates and as such place a high demand on the silica dissolved in the water. Desmidiales were the group which predominated in the seasonal pond.

The lowest value of phosphate was found in the *igarapé* (which also showed the highest values of Fe). The highest value for phosphate was found in the Rio Uraricoera, although this does not correspond with the lowest value of Fe. In a river as large as the Rio Uraricoera, additional sources of phosphate include tributaries and the input of animals and man.

**Inorganic nitrogen (nitrate ($NO_3$-N) and ammonium ($NH_4$-N) – $NO_3$-N by cadmium reduction method; $NH_4$-N by Nessler method)**

The highest concentrations of nitrate ($NO_3$-N) and the lowest of ammonium ($NH_4$-N) were found in the Rio Uraricoera. In other samples the ammonium total was significantly higher than the nitrate. Nevertheless all the inorganic N values were low (much lower than in temperate soils and in superficial waters). The zero value of nitrate recorded in the seasonal pond and the very low values recorded in the *igarapé* do not seem to be due to low concentration of oxygen, nor to limiting nitrification nor exchange denitrification. They are probably the result of an efficient mineralized nitrogen absorption method in the plant roots and micro-organisms in the soil surface (when it becomes available), coupled with the generally low levels of soil organic matter and the resultant relatively low potential for soil nitrogen mineralization.

Table 21.3. Analysis of the macrophytic vegetation from the seasonal pond, the shallow borrow-pit and the Rio Uraricoera (Furo de Maracá)

| Vegetation/macrophytes | Seasonal pond | Shallow borrow-pit | Furo de Maracá |
|---|---|---|---|
| *Mayaca* sp. 1 |  | XXX |  |
| *Mayaca* sp. 2 |  | XXX |  |
| Podostemaceae |  |  | XXX |
| *Utricularia foliosa* L. | X | XXX |  |
| *Utricularia* sp. 1 | X |  |  |
| *Utricularia* sp. 2 | X |  |  |
| *Echinodorus* sp. | X |  |  |
| *Eleocharis variegata* L. | XXX |  |  |
| *Ludwigia* sp. | X |  |  |
| *Mauritia flexuosa* L.f. | XX |  |  |
| *Montrichardia linifera* Schott | XXX |  |  |
| *Oryza* sp. | X |  |  |
| *Scirpus* sp. | X | XX |  |
| *Scleria* sp. | X |  |  |
| *Thalia geniculata* L. | XXX | XXX |  |
| *Nymphaea* sp. | XX | XXX |  |
| *Eichhornia diversifolia* (Vahl) Urban |  | X |  |
| *Eichhornia azurea* Kunth |  | XXX |  |
| *Eichhornia crassipes* (Mart.) Solms-Laut |  |  | XX |
| *Lemna minor* L., *Pistia stratiotes* L., *Salvinia auriculata* Aubl. |  |  | X |

X = frequent; XX = abundant; XXX = very abundant.

# FAUNA

## Sponge fauna

The sponge fauna on the island is quite varied, with only one species (*Trochospongilla variabilis*) common to all three environments. Another species, *Heteromeyenia stepanowi*, occurred only in the *igarapé* and the borrow-pit. The most notable aspect of the sponge fauna surveyed on the island (see Volkmer-Ribeiro, 1992) was the association of the five species identified in the seasonal pond (Table 21.4). There is thus a marked difference in the occurrence of sponges in the three studied habitats. As yet, none of these species has been collected in the Rio Uraricoera. *Radiospongilla amazonensis* was collected in the dry season from *igarapé* Matá-Matá, a small tributary of the Rio Uraricoera close to the Furo de Maracá. The only sponge specimen from the river itself was obtained from a pool among the rocks at the height of the dry season; it had no gemmules and probably belongs to the genus *Oncosclera* Volkmer-Ribeiro, 1970, which is characteristic of fast-flowing and turbulent waters.

## Malacofauna

The absence of molluscs in the pond and the *igarapé* (Table 21.4) can be explained by the acidic pH and absence of calcium. Fittkau (1981) relates the extreme poverty

**Table 21.4.** Sponges and molluscs from freshwater habitats on the Ilha de Maracá, and from the Rio Uraricoera

| | Seasonal pond | Stream (*igarapé*) | Shallow borrow-pit | River |
|---|---|---|---|---|
| Sponges | | | | |
| *Metania spinata* (Carter, 1881) | X | | | |
| *Corvomeyenia thumi* (Traxler, 1895) | X | | | |
| *Trochospongilla variabilis* Bonetto & Ezcurra de Drago, 1973 | X | X | X | |
| *Dosilia pydanielli* Volkmer-Ribeiro, 1992 | X | | | |
| *Radiospongilla amazonensis* Volkmer-Ribeiro and Maciel, 1983 | X | | | |
| *Heteromeyenia stepanowi* (Dybowsky, 1884) | | X | X | |
| *Oncosclera* sp. | | | | X |
| Molluscs | | | | |
| Gastropoda, Ampullaridae | | | X | |
| Gastropoda, Thiaridae | | | | X |
| Bivalvia, Hyriidae | | | | |
| *Diplodon suavidicus* (Lea, 1856) | | | | X |
| *Castalia ambigua ambigua* (Lamarck, 1819) | | | | X |
| *Triplodon corrugatus* (Lamarck, 1819) | | | | X |

of molluscs in Central Amazonia to the lack of freely accessible electrolytes, especially those of calcium. In the borrow-pit, where the pH is only slightly acidic, an empty Ampullaridae shell was found, as were various batches of mollusc spawn stuck to the leaves and stems of *Eichhornia azurea* above water level. This spawn, characteristic of *Pomacea* species, was salmon pink in colour, with the eggs covered by a calcareous layer. Despite the fact that calcium was not detected in the borrow-pit with the technique used, it is supposed that these molluscs find sufficient to form a shell and spawn during the rainy season, which coincides with their spawning period. According to Junk and Furch (1985), the rains of Amazonia contain low levels of calcium, and heavy falls can raise the calcium levels in the environment significantly. No mollusc shells or spawn were found in the borrow-pit in December. According to Castellanos and Fernandez (1976), it is common for the Ampullaridae to die after the spawning season. However, some survive by burying themselves under submerged roots, etc. Mansur and Valer (1992) present a survey of the mussels from the *igarapés* of the Rio Uraricoera close to the Ilha de Maracá, and also from the river itself and from the Rio Branco. Specimens of gastropods of the family Thiaridae were collected from the submerged roots and rocks in the soft, sandy, shallow substrate of the Rio Uraricoera next to the ferry landing place.

## GENERAL DISCUSSION AND CONCLUSIONS

On the basis of analytical results, the waters of the Rio Uraricoera and those of smaller water bodies on the Ilha de Maracá are rather different. The Rio Uraricoera runs through the ancient Guiana Shield, which according to Fittkau's (1971) geochemical division of the Amazon basin belongs (together with the Central Brazilian Shield) to the Peripheric Province. Both regions are similar in terms of the chemistry of their soils and waters, and of their mollusc fauna (Fittkau, 1981). Waters arising in these areas are characteristically poor in mineral elements and sediments, and low in pH values (Junk and Furch, 1985). Results of physical and chemical analyses of the water in this study indicate that the water quality of the Rio Uraricoera (pH 6.9; conductivity 28 $\mu$S cm$^{-1}$) lies somewhere between the low pH and low conductivity of black waters (pH 4–5.8; conductivity 7–10.5 $\mu$S cm$^{-1}$) and those of white waters (pH 7–7.5; conductivity 50–60 $\mu$S cm$^{-1}$) (Junk and Furch, 1985). In the Rio Uraricoera the pH is practically neutral and optimum for mollusc life. However, although calcium was present in detectable quantities, the content (4.5 mg l$^{-1}$) was low and less than the minimum concentration for molluscs, which according to Hynes (1978) should be about 20 mg l$^{-1}$. Nevertheless, Mansur *et al.* (1988) found reasonable numbers of mollusc species and individuals in lotic environments with calcium concentrations of only 5.8 and 6.8 mg l$^{-1}$ in south Brazil.

The chemical analyses of the waters of the Ilha de Maracá exhibit the characteristics of some 'black-water' rivers, *igapós* and *igarapés* of Central Amazonia, with low values of conductivity, acidic pH and extreme scarcity of calcium. Junk and Furch (1985) have already pointed to the occurrence of black waters in the northern peripheric region of the Amazon basin. The extreme example of this is represented on the island by the seasonal pond (pH 5.4; conductivity 11.5 $\mu$S cm$^{-1}$). It reveals a rich sponge fauna, with five species each from a different genus. Almost any object which could act as a substrate within the pond was to some extent encrusted with sponges, with *Metania spinata* predominant nearest the bottom and *Corvomeyenia thumi* nearer the surface. Species of *Utricularia* are the favourite substrate for the sponge *Corvomeyenia thumi*, and the fallen leaves of *buriti* palms are the preferred substrate for the other sponges listed in Table 21.4. Of the five sponges occurring in the pond only *Metania spinata* and *Corvomeyenia thumi* are green, owing to their association with zoochlorellas (Figures 21.1 and 21.2). As these two sponges predominate in the pond, they are presumably another source of the wealth of dissolved oxygen in the water, due to the photosynthetic activity of their associated green algae. The acidity of the water obviously prevents the establishment of a mollusc fauna, which is completely absent. The low content of silica found in the pond is explained by its utilization in the construction of the skeletons of sponges and the siliceous plates of some algae, which were also found there in abundance. On the other hand, the high levels of dissolved oxygen are explained by the abundance of green submerged plants and algae, the latter principally desmids.

The Amazon basin supports one of the richest, if not the richest, freshwater sponge fauna of the world (Volkmer-Ribeiro, 1981). Such a fauna had until now been recorded for the floodplains of Amazonian rivers, but the seasonal pond studied on Maracá is the first such habitat in Amazonia to demonstrate a similar richness in sponge diversity. The species composition of the sponge community in the seasonal pond was completely different from the ones found in the *igapós* of Central Amazonia where the drab yellow, brown, grey or black sponges encrust the trunks, branches and leaves of the flooded forest several metres above the forest floor. No

**Figure 21.1.** Living colony of the freshwater sponge *Metania spinata* (Carter, 1881), encrusting a piece of dead *Mauritia flexuosa* leaf from the bottom of the seasonal pond in the dry season. Photo: Dr Arno A. Lise

green sponges are found in such permanently shaded habitats (Volkmer-Ribeiro and de Rosa-Barbosa, 1972; Volkmer-Ribeiro, 1976; Volkmer-Ribeiro and Maciel, 1983; Volkmer-Ribeiro and Costa, 1992; Volkmer-Ribeiro and Tavares, 1993).

The sponge community in the seasonal pond was found to be remarkably similar to those which formed sub-fossil spongillite deposits in the savannas of southwest Minas Gerais, southern Goiás, northeastern Mato Grosso do Sul and northwestern São Paulo states (Volkmer-Ribeiro, 1992; Volkmer-Ribeiro and Motta, 1995). These deposits were formed inside ponds bordered by lines of *buriti* palms and capped with a peat layer and dense grassy vegetation. No mollusc shells were found inside such deposits, indicating an acidic depositional environment. The authors cited above attributed the absence of peat and spicular deposits in the Maracá pond to the seasonal drought and the minimal amount of silica detected in the pond water respectively. The periodic exposure and drying of the organic matter produced in the pond would enhance its oxidation and prevent its accumulation on the bottom. A rapid survey of the savanna ponds around Boa Vista and beside the road between Boa Vista and the Ilha de Maracá showed a similar botanical/sponge context to that of the seasonal pond on Maracá, with surrounding rings of *Mauritia flexuosa* and with *Montrichardia linifera*, *Thalia* spp., *Nymphaea* and *Utricularia* within the ponds themselves. One or other of the two predominant sponge species, and sometimes both, are abundant in these Boa Vista savanna ponds.

The similarities between the aquatic environment and sponge fauna detected for the Maracá seasonal pond and the southwestern savanna ponds now point to the

Figure 21.2. Branches of the freshwater sponge *Corvomeyenia thumi* (Traxler, 1895) resting upon *Utricularia* close to the surface of a seasonal pond in the dry season. Photo: Dr Arno A. Lise

possibility that the Maracá ponds represent remnants of a savanna recently overtaken by forest, as opposed to forest peat-bog ponds as initially suggested by Volkmer-Ribeiro (1992).

## ACKNOWLEDGEMENTS

The authors are indebted to Dr Arno A. Lise of Pontifícia Universidade Católica for the photographs of the sponges in the seasonal pond. The first and second authors acknowledge the granting of fellowships by Conselho Nacional de Desenvolvimento Científico e Tecnológico (CNPq), Brazil.

## REFERENCES

Castellanos, S.J.A. and Fernandez, D. (1976). Mollusca Gastropoda Ampullariidae. In: *Fauna de agua dulce de la Republica Argentina*, ed. R.A. Riguelet. FECIC, Buenos Aires.
Fittkau, E.J. (1971). Ökologische Gliederung des Amazonas-Gebietes auf geochemischer Grundlage. *Münster. Forsch. Geol. Paléont.*, 20/21, 35–50.
Fittkau, E.J. (1981). Armut in der Vierfalt. Amazonien als Lebensraum für Weichtiere. *Mitteilungen der Zoologischen Gesellschaft Branau, Branau am Inn*, 3 (13/15), 329–343.
Hynes, H.B.N. (1978). *The biology of polluted waters*. Liverpool University Press.
Junk, W.J. and Furch, K. (1985). The physical and chemical properties of Amazonian waters

and their relationships with the biota. In: *Key environments: Amazonia*, eds G.T. Prance and T.E. Lovejoy, pp. 1–17. Pergamon Press, Oxford.

Mansur, M.C.D. and Valer, R.M. (1992). Moluscos bivalves do Rio Uraricoera e Rio Branco, Roraima, Brasil. *Amazoniana*, 12 (1), 85–100.

Mansur, M.C.D., Veitenheimer-Mendes, I.L. and Almeida-Caon, J.E.M. de (1988). Mollusca, Bivalvia de um trecho do curso inferior do Rio Jacuí, Rio Grande do Sul, Brasil. *Iheringia*, Ser. Zool., 67, 87–108.

Ross, S.M., Thornes, J.B. and Nortcliff, S. (1990). Soil hydrology, nutrient and erosional response to the clearance of *terra firme* forest, Maracá Island, Roraima, northern Brazil. *Geographical Journal*, 156 (3), 267–282.

Volkmer-Ribeiro, C. (1976). A new monotypic genus of freshwater sponges (Porifera – Spongillidae) and the evidence of a speciation 'via' hybridism. *Hidrobiologia*, 50 (3), 271–281.

Volkmer-Ribeiro, C. (1981). Porifera. In: *Aquatic biota of Tropical South America, Part 2: Anarthropoda*, eds S.H. Hurlbert, G. Rodrigues and N.D. Santos, pp. 86–95. San Diego State University, California.

Volkmer-Ribeiro, C. (1992). The freshwater sponges in some peat-bog ponds in Brazil. *Amazoniana*, 12 (2), 317–335.

Volkmer-Ribeiro, C. and Costa, P.R.C. (1992). On *Metania spinata* (Carter, 1881) and *Metania kiliani* n.sp.: Porifera, Metaniidae Volkmer-Ribeiro, 1986. *Amazoniana*, 12 (1), 7–16.

Volkmer-Ribeiro, C. and Maciel, B.S. (1983). New freshwater sponges from Amazonian waters. *Amazoniana*, 8 (2), 255–264.

Volkmer-Ribeiro, C. and Motta, J.F.M. (1995). Esponjas formadoras de espongilitos em lagoas no Triângulo Mineiro e adjacências, com indicação de preservação de habitat. *Biociências*, 3 (2), 145–169.

Volkmer-Ribeiro, C. and Rosa-Barbosa, R. de (1972). On *Acalle recurvata* (Bowerbank, 1863) and an associated fauna of other freshwater sponges. *Rev. Brasil. Biol.*, 32 (3), 303–317.

Volkmer-Ribeiro, C. and Tavares, M.C.M. (1993). Sponges from the flooded sandy beaches of two Amazonian clear water rivers (Porifera). *Iheringia*, Sér. Zool., 75, 187–188.

# 22 Soil Properties and Plant Communities Over the Eastern Sector of the Ilha de Maracá

PETER A. FURLEY
*University of Edinburgh, Edinburgh, UK*

## SUMMARY

The nature and variation of soil properties on the Ilha de Maracá were investigated by relating soil characteristics to the vegetation types identified during botanical surveys, and by preliminary analysis of LANDSAT TM satellite data. Clear relationships were evident between the soil properties and the plant communities, suggesting that soils have a major influence over the nature and distribution of plant cover. The dominant controls are firstly minor changes in topography leading to major differences in drainage, and secondly variations in the nature of parent materials. Most soils are coarse-textured, well drained, acidic and nutrient-poor, with the exception of alluvial soils of present and past stream-courses and soil profiles developed over mesotrophic parent materials. On the whole, recognition of the vegetation through satellite imagery can be used for identification of soil types in the non-forested areas, but is less successful in picking up differences in the areas of continuous forest canopy.

## RESUMO

Foram investigados o cárater e a variação dos solos da parte leste da Ilha de Maracá, através de relacionamento das características do solo aos tipos de vegetação identificados pelo levantamento botânico, e análise preliminar das imagens de LANDSAT TM. Interrelações nítidas entre os tipos de solos e as comunidades vegetais sugerem que os solos excercem forte controle sobre a natureza e a distribuição da cobertura vegetal. Os controles dominantes são pequenas mudanças na topografia, que resultam em diferenças maiores na drenagem, e variações na natureza dos materiais de origem. A maioria dos solos são de textura grossa, boa drenagem, ácidos e pobres em nutrientes; as exceções são os solos aluviais dos cursos de drenagem (presentes e antigos) e solos desenvolvidos de materia-mãe mesotrófica. Em geral, identificação de tipos de vegetação por meio de imagens de satélite forneceu uma boa indicação dos tipos de solo nas áreas não florestais, mas não foi tao boa como indicadora de diferenças pedológicas nas áreas de mata contínua.

## INTRODUCTION

It has long been questioned whether the soils underlying tropical forest bear any consistent relation to the character of the plant communities and whether such associations might be detectable from remote sensing techniques. This chapter

*Maracá: The Biodiversity and Environment of an Amazonian Rainforest.*
Edited by William Milliken and James A. Ratter. © 1998 John Wiley & Sons Ltd.

assesses both propositions in the light of field surveys and satellite image processing, from an initial reconnaissance in 1985 and two further field investigations in February–March 1987 and February 1988. Although the surveys were based mostly in the eastern sector of the island, they included comparative work in surrounding areas and are related to the soil research undertaken by Nortcliff and Robison (Chapter 4) and the botanical surveys of Milliken and Ratter (Chapter 5).

The principal objectives of this research were: to analyse the soil properties of the main eastern vegetation associations identified during the botanical survey; to provide a better understanding of the edaphic controls on vegetation at the forest–savanna boundary; and to evaluate the success of image processing in extrapolating the findings to other parts of the island. The forest–savanna boundary work has been published in Furley and Ratter (1990) and some of the image processing investigations in Furley et al. (1994) and Dargie and Furley (1994); consequently these aspects will be only briefly summarized in this chapter.


SOILS OF NORTHERN RORAIMA

The only comprehensive investigation of soils in the region was carried out as part of the Projeto RADAM surveys in the early 1970s (MME/DNPM, 1975). These were based on Side Looking Airborne Radar (SLAR) with base maps made up of black-and-white mosaics of images at a scale of 1:250 000. A summary for northern Roraima, based on Projeto RADAM and extracted from the Atlas of Roraima (IBGE, 1981), is given in Figure 22.1.

The main group of soils identified by the RADAM survey is the red–yellow podzol (SNLCS in Brazilian terminology, and equivalent to Ultisol in USDA Soil Taxonomy (Soil Survey Staff, 1990)). These soils occur along the length of the river basins, capping the gently to strongly undulating terrain and overlying steeply sloping escarpments. They are frequently argillic, but are associated with coarser-textured red–yellow latosols (Oxisols) and concretionary lateritic soils, most of which are well weathered and often severely eroded into gullies and ravines. Surface horizons are grey-coloured, arenaceous and, except for the most acid profiles, tend to have medium to high exchangeable base contents.

The latosols occur in a widespread group lying to the east of the state, notably around Boa Vista. They occur over level to gently undulating terrain and are extensively weathered and leached, display variable textures and colour, and are consistently deep and porous. They underlie both forest and savanna, and are made up of at least three sub-groups:

I. The yellow latosols, mostly below 100 m altitude and liable to wet season flooding. They tend to be more clay-rich than most of the group and are generally deep, well drained and porous, with ochric A horizons and a weak structure, frequently associated with dystrophic hydromorphic laterites.
II. The red–yellow latosols (with a strong red colour in the B horizon), occupying higher ground. These soils are often very deep, with varying textures, and are also often associated with hydromorphic laterites and quartz sands.

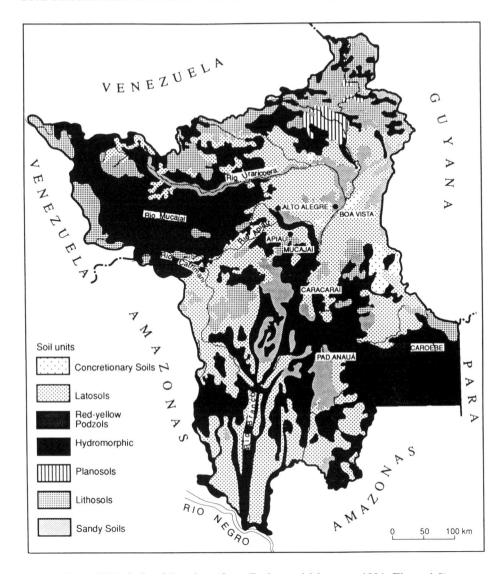

**Figure 22.1.** Soils of Roraima (from Furley and Mougeot, 1994, Figure 1.8)

III. The dark red latosols, found to the east of Boa Vista. These have more argillaceous textures associated with more eutrophic soils (structured terra roxa and eutrophic lithosols in Brazilian terminology). They are usually well drained, deep, friable and acidic, and possess elevated levels of iron oxides.

Sandy alluvial soils (Inceptisols and Entisols) are distributed along seasonally flooded stream channels. More than 95% of the soil is made up of quartz with associated minerals such as zircon, tourmaline and rutile. Such soils tend to have greyish surface horizons which exhibit only early stages of weathering, and textural variations in the profiles reflect this lower pedogenic activity and consequently have closer affinity with the initial parent material.

A number of smaller but important soil groups are also found in the north of the state. Around the northern peripheries there are numerous lithosols (mostly Entisols). These are shallow rocky soils, rarely possessing a B horizon and closely related to the underlying weathered parent material. Hydromorphic soils are common over the higher floodplains (*várzeas*) of the Rio Branco. These are characterized by an organic A horizon, accompanied by subsurface gleying and mottling typical of seasonal variation in the water-table. They are frequently associated with hydromorphic lateritic soils, resulting from both laterization and podzolization processes, and tend to be acid and poorly drained on low-lying level topography. There are also stretches of planosols (mostly Inceptisols) to the north of Boa Vista, formed over level surfaces around 100–600 m above sea level. Such soils develop clay accumulations in the B horizon, leading to impermeability.

The reconnaissance nature of the Projeto RADAM surveys is rapidly revealed by subsequent work at more detailed scales, as has been shown elsewhere in the Amazon (Furley, 1986). The RADAM soil map of the Ilha de Maracá, for example, indicates several soil groups based largely on radar image interpretation and with only one field site on the island. The final RADAM classification and map has therefore relied on a great deal of extrapolation. The results presented in this chapter and in Nortcliff and Robison (Chapter 4) support the view that there is considerable local soil variation, particularly associated with relief and drainage as well as reflecting underlying parent material differences.

## REMOTE SENSING

Interpretations were made of a mosaic of black-and-white aerial photographs at a scale of 1:70 000 and of LANDSAT MSS and TM satellite images, particularly bands 3, 4 and 5, to help characterize the principal vegetation and related soil types (Dargie and Furley, 1994).

Interpretation of the LANDSAT TM images provided a method of extrapolating the vegetation units (supervised classification based on the botanical ground surveys) to areas of the island which could not be surveyed in the time available. Based on the assumption that the pattern of relationships was similar to that identified in the soil–plant community work, some idea of the nature and distribution of soils in the eastern third of the island could be estimated.

Twelve land cover units were identified, plus cloud, cloud shadow and a limited number of unclassified units. The cover totals and details of procedures can be found in Furley *et al.* (1994). Open cover types, including water, swamp, wet and dry grassland (*campos*), bare ground and tree savanna, make up almost 60% of the LANDSAT 'scene' (180 × 180 km) which includes the eastern part of Maracá. On the Ilha de Maracá itself the non-forested vegetation, including water and swamp, totalled *c.* 15% with a markedly greater proportion of savanna in the eastern third and a strong seasonal flux between classes. Each of these non-forest land cover units could be clearly distinguished and the map provided the baseline for the location of some of the sample sites. However, over the areas covered by continuous forest, the problems of atmospheric water vapour combined with the pixel resolution of the imagery precluded the identification of more than fairly broad forest categories. For these areas, the satellite imagery provided no further information of use to soil investigation.

The principal plant communities which could be distinguished from image processing included:

Forest:
- Evergreen closed canopy
- Semi-deciduous closed canopy
- Intermediate forests
- Open canopy

Wet site vegetation:
- *Buritizal* (dominated by *buriti* palms *Mauritia flexuosa* L.f.)
- *Vazante* (shrubby/herbaceous communities along periodic stream lines) – see Plate 4a

Savanna (*campo*): dry or seasonally flooded – see Plate 1

Other land covers:
- Shallow water or water with emergent vegetation – see Plate 4b
- Deep water
- Trees overlapping water

Following a preliminary assessment of the distribution of these units, it was evident that the soil sampling framework would be constrained by the location and accessibility of representative vegetation tracts over the island and by the location of the detailed botanical surveys. For these reasons the study was focused upon areas within walking distance of the Ecological Station or within a day's travel by boat and trail.

A series of profiles was established to characterize small-scale variations in the flora, micro-topography and drainage. The remaining soil profiles were sited to represent distinct plant communities and therefore to characterize the units identified by the botanical work. This intensifies the level of the overview provided by Nortcliff and Robison in Chapter 4.

# SOIL ANALYSIS

Standard methods of soil analysis were employed following air-drying in the field and subsequent sieving to separate fractions greater and less than 2 mm eps (equivalent particle size diameter = fine earth). Where possible, the methods used by the Brazilian Soil Survey, Serviço Nacional de Levantamento e Conservação de Solos, were adopted (SNLCS, n.d.; SBCS, 1973).

Exchangeable acidity and pH
  pH — Standard electrode in 1:2 soil:water and 1:2 M/100 $CaCl_2$

  Total acidity (ex. H + ex. Al) — KCl as extractant using phenolphthalein as indicator

Organic matter
  Organic C% — Modified Walkley–Black wet digestion
  Loss on ignition (%)
  Total N% — Macro-Kjeldahl
  Readily available P (mg/100 g) — Spectrophotometer with ammonium molybdate and stannous chloride

Moisture and Base status
  Moisture equivalence (%)
  Base status — 1M ammonium acetate as extractant
  Base saturation (%)
  Exchangeable Ca, Mg, K — Atomic absorption
  Exchangeable Mg — Flame photometry
  Total bases
  Effective cation exchange capacity (ECEC) — – Ca + Mg + K + Na + ex. Al + ex. H

Iron oxides — Sodium dithionite–citrate procedure

Particle size — X-ray sedigraph and checks against standard hydrometer/pipette methods

The soil descriptions provided here add a further 37 sites to the soil survey undertaken by Nortcliff and Robison (Chapter 4). A comparison was made to reconcile the slight but important differences in the analytical techniques between the two soil surveys. Replicate randomly selected samples from sites in the present survey were tested by both methodologies. Overall it would appear that the results for cations and base status are sufficiently comparable for the two surveys to be considered as complementary (see Table 22.1).

## RESULTS

### SOIL PROPERTIES AND THEIR RELATIONSHIP TO VEGETATION

Based on the results of the botanical survey (Milliken and Ratter, 1989; Chapter 5), six broad vegetation communities were distinguished, within which there was considerable floristic and environmental diversity (Table 22.2). A number of representative sites over the eastern third of the island were examined (see Figure 22.2), whose physical and chemical characteristics are presented in detail in Appendix 9 (p. 485).

**A1. Predominantly dense forest at well-drained sites**

According to Milliken and Ratter (1989 and Chapter 5) a characteristic *terra firme* forest type is found on freely drained soils over the flattish plateau referred to as the

**Table 22.1.** Comparison of results of soil analysis methods used in this study (Edinburgh) and in Chapter 4 (N & R)*

| Sample no. | Method | pH (H$_2$O) | K (cmol/kg) | Na (cmol/kg) | Ca (cmol/kg) | Mg (cmol/kg) |
|---|---|---|---|---|---|---|
| Ecological Station (well site) | | | | | | |
| 5 | Edinburgh | 5.6 | 0.06 | 0.18 | 0.02 | 0.90 |
| 5 | N & R | 5.6 | 0.09 | 0.21 | 0.01 | 0.60 |
| Santa Rosa savanna site | | | | | | |
| 19 | Edinburgh | 5.3 | 0.04 | 0.16 | 0.01 | 0.15 |
| 19 | N & R | 4.9 | 0.06 | 0.17 | 0.00 | 0.15 |
| Santa Rosa *campo* site | | | | | | |
| 28 | Edinburgh | 5.6 | 0.01 | 0.14 | 0.02 | 0.04 |
| 28 | N & R | 6.1 | 0.01 | 0.21 | 0.01 | 0.08 |
| Santa Rosa shrub site | | | | | | |
| 35 | Edinburgh | 5.6 | 0.11 | 0.49 | 0.19 | 2.63 |
| 35 | N & R | 5.5 | 0.19 | 0.17 | 0.25 | 2.02 |
| Angico transect savanna site | | | | | | |
| 66 | Edinburgh | 6.6 | 0.22 | 0.29 | 0.34 | 9.70 |
| 66 | N & R | 6.9 | 0.25 | 0.12 | 0.26 | 6.31 |
| Angico transect forest–savanna site | | | | | | |
| 70 | Edinburgh | 7.0 | 0.15 | 0.75 | 0.18 | 7.39 |
| 70 | N & R | 6.4 | 0.20 | 0.59 | 0.09 | 4.21 |

* Although the results reveal differences according to the method of analysis, particularly in the mesotrophic Angico site, they show similar trends and are considered to lie within acceptable levels of variation for comparative purposes.

Eastern Shield by Nortcliff and Robison (Chapter 4). The Sapotaceae are the most important family in this area, and *Pradosia surinamensis* (Eyma) Penn., *Licania kunthiana* Hook. f. and *Tetragastris panamensis* (Engl.) Kuntze are common species in the canopy. Various palms such as *Maximiliana maripa* (Correa) Drude are typical of the understorey. Tree species diversity as a whole is low by Amazonian standards, with figures of 80 spp. $\geq$10 cm dbh in 1.5 ha near the Ecological Station.

The soils underlying these forests, characterized here by the profile close to the Ecological Station (Profile A1(c)), are generally mildly acidic (pH 6 in water as opposed to not more than 5.1 elsewhere), the surface $A_1$ horizon being ameliorated by an unusually thick organic litter. Exchangeable H and Al levels are low, increasing only slightly with depth. The organic properties reflect the richer conditions at the surface. Thus organic C% reaches 2–3% at the Ecological Station site but is usually closer to 1% elsewhere, dropping rapidly below the surface. Similarly, total N% is relatively high at the research station (0.3–0.4% as opposed to less than 0.2% elsewhere), and available P is consistently low, with around 0.2–0.3 mg/100 g in the $A_1$ horizons throughout the dense forest sites. Loss on ignition is more variable (3.5% at the Ecological Station to around 1% elsewhere), but shows an increase with depth (as does moisture equivalence), indicating more humid conditions lower in the profile. This is consistent with a decrease in the sand and an increase in the silt proportions of the fine earth fractions from the surface horizon downwards. The base status of these soils is generally low, although the Ecological Station site has 4.5 cmol/kg total bases in the surface horizon, around 70% of which

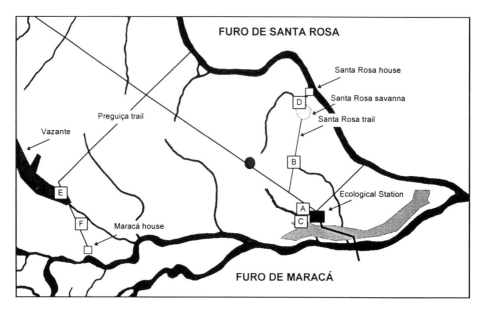

**Figure 22.2.** The eastern end of the Ilha de Maracá showing the main survey trails and the locations of soil profiles A to F

is made up of exchangeable calcium. Despite the low ECEC (effective cation exchange capacity), the base saturation of surface samples is high (97% in the case of the Ecological Station site), emphasizing the role of the organic nutrient turnover.

### A2. Predominantly dense forest, well drained, with mesotrophic soils

A distinctive deciduous variant of these forests is found at the Angico site (Profile A2), where there is a strong presence of *Anadenanthera peregrina* (L.) Speg. (Leguminosae). Such forests, whose occurrence is very limited on Maracá, are typified by mesotrophic soils, notably with higher calcium and magnesium levels (see Appendix 9). *Anadenanthera peregrina* is also to be found in savanna environments, always in association with a base-rich substrate (Ratter *et al.*, 1973, 1977, 1978; Furley and Ratter, 1988; Furley, 1996).

*Anadenanthera peregrina* has been noted elsewhere as an indicator of more mesotrophic conditions, and at the Angico transect total base levels of 6.9 cmol/kg were recorded, mostly consisting of exchangeable Mg and Ca. The *Peltogyne* forest (i.e. forest dominated by the leguminous tree *Peltogyne gracilipes* Ducke), common in the central and western parts of the island but rare in the east, was encountered on the Preguiça trail to the south of a strip of *vazante*. Here slightly acidic soils became markedly more alkaline with depth, these figures being paralleled by high base levels (4.2 cmol/kg at the surface to 16.6 cmol/kg at depth). The organic litter at this site is thick, with organic C% of 3 or more at the surface, and textures are sandy throughout.

**Table 22.2.** Profile sites chosen to characterize broad vegetation types on the Ilha de Maracá

---

A. Predominantly dense forest
    A1. *Terra firme* and well drained
        (a) Santa Rosa (2 profiles)
        (b) Preguiça trail (4 profiles)
        (c) Ecological Station area (3 profiles) (Appendix 9)
        (d) Cleared forest (1 profile)
    A2. *Terra firme* and well drained (mesotrophic site)
        (a) Angico transect (1 profile) (Appendix 9)
    A3. *Terra firme* margins and seasonally/semi-permanently poorly drained
        (a) Santa Rosa (2 profiles)
        (b) Ecological Station; *buriti* site (1 profile) (Appendix 9)
B. Predominantly low or open forest
    (a) Santa Rosa (1 profile)
    (b) Preguiça trail (1 profile)
    (c) Ecological Station area (2 profiles) (Appendix 9)
C. Forest edge
    (a) Santa Rosa (3 profiles)
    (b) Angico transect (1 profile) (Appendix 9)
D. Predominantly arboreal savanna
    (a) Santa Rosa (1 profile) (Appendix 9)
    (b) Angico transect (1 profile)
E. Predominantly thicket and shrub
    (a) Well drained; Santa Rosa (2 profiles)
    (b) Poorly drained: *vazante* (1 profile) (Appendix 9)
F. Predominantly herbaceous grassland (*campo*)
    (a) Santa Rosa (7 profiles)
    (b) Preguiça trail (2 profiles) (Appendix 9)

---

It is worth drawing attention to the fact that some parent materials have distinct effects on soil properties below structurally similar dense forest. In most cases, however, surface soils are mildly acidic, organic, and generally sandy, with high base levels presumably reflecting the rapid nutrient turnover from decomposing litter. Between profiles there are considerable differences depending upon the mineral contribution of the weathered subsurface rock. In the case of a cleared area (for example in the well dug close to the Ecological Station), samples were taken down to more than 3.5 m in readily identifiable, moderately weathered mica schist. This parent material generated more alkaline soils and, not surprisingly, high base levels (mostly Mg and to a lesser extent Ca) in the lower horizons.

### A3. Dense forest with seasonally wet soil conditions (moist forest)

These conditions are well illustrated by the vegetation to the northern end of the Santa Rosa transect and in the dense but low forest in depressions (*baixadas*) and shallow valleys dissecting the *terra firme* plateau. At the northern end of the Santa Rosa site, the sandy soils are acidic (surface horizons around pH 5 in water) with weak cation levels, marginally higher in the slightly organic surface horizon. Subsurfaces are gleyed from around 10 cm depth. Conditions are more alkaline at the *buriti* site, draining from the mica schists above, with higher base levels throughout the profile and a higher organic matter content at the surface. Once again, markedly gleyed horizons characterize the profile below 10 cm.

A typical example is given by Profile A3(b) which represents a footslope position close to a line of *buriti* palms (*Mauritia flexuosa*) lying within dense forest.

## B. Predominantly low or open forest (*campina*)

Several variants of low forest are evident at the eastern end of the island and have been termed *campina* forest. Such forests occur at the margins of grassy savannas (*campos*) such as at Santa Rosa, as low woodland and thicket (*carrasco*) such as between the two patches of *campo* on the Preguiça trail, at the margins of semi-permanently damp areas (such as *buritizal*), or in particularly sandy sites such as those near the Ecological Station.

The Santa Rosa *campinas* (Milliken and Ratter, 1989) are dense and some 8–12 m tall. Typical trees of the forest margin are mixed with species from dense forest and with ground bromeliads. The *Eugenia*-dominated thicket on the Preguiça trail consists of an extremely dense mixture of trees and shrubs 3–4 m tall, with a very clear transition to open *campo*.

The soils are well exemplified by those underlying the *manguezal*, an open low forest dominated by *Clusia renggerioides* Tr. & Pl. at the edge of a *buritizal* site on the trail to Santa Rosa (Profile B(c)). These seasonally wet situations were damp in the deeper horizons even at the end of a prolonged dry season. The plant community is described in detail by Milliken and Ratter (1989). The persistence of wet conditions appears to inhibit organic decomposition and surface soils have high organic C%, low total N% and predictably high loss on ignition and moisture equivalence levels. The organic matter is likely to contribute to the noticeably acid surface conditions (pH just over 4 in water), although the low exchangeable Al level means that the total exchangeable acidity is not extreme. Nutrients are in short supply (just over 1 cmol/kg) and available P is present only in trace amounts. Textures are extremely coarse and the vegetation would appear to be highly oligotrophic, dependent upon surface recycling of organic litter.

## C. Forest transitions to savanna and *vazante* formations

The eastern part of the Ilha de Maracá lies at the current boundary between forest and savanna, which is also conterminous with the contemporary limits of land occupation (Eden and McGregor, 1989; Eden et al., 1991). Soil variations were examined across three transects:

(a) Forest transition to seasonal and hyperseasonal savanna (terminology following Sarmiento, 1988).
(b) Forest–*vazante* transitions.
(c) Forest transition to savanna 'islands' over the interior of the island.

The soils have been considered in detail elsewhere (e.g. Furley and Ratter, 1990; Furley, 1992) and soil profiles strikingly reflect the transition from forest to non-forest communities.

Distinct spatial patterns of soil properties were evident from these transects and express the coincidence of environmental and pedological changes across the vegetation boundaries. The principal cause of soil change seems to be the drainage regime associated with the upper limits of seasonal flooding. This is further associated with the heights to which gleying reaches in the profiles. The change in soil physical and chemical properties is particularly evident where forest abuts onto seasonal and hyperseasonal savanna (a) and is usually marked by a small but critical topographic change with forest lying upslope on well-drained soils. At the *vazante* sites (b) which, from interpretation of the satellite data, are common throughout the island, the forest abuts sharply onto embryonic valley landforms. The dramatically different soils are presumed to have been derived from alluvial flood deposits which, in some cases, may represent ancient river courses with sediments derived from the whole catchment basin (i.e. allogenic to the island). Finally (c) the 'islands' of savanna, which are found throughout the eastern parts of the island, exhibit oligotrophic white sandy soils whilst retaining savanna communities in the face of invasion from surrounding arboreal vegetation.

The forest–savanna boundary at the Angico site (type (a) above) provides a good example of the sharp natural transition in vegetation resulting from a combination of soil parent material variation and soil water differences. The latter result from slight changes in topography and seasonal flooding. The boundary site (Profile C(b)) neatly illustrates the change from the upslope, well-drained, mesotrophic soils supporting semi-deciduous forest, to the downslope hyperseasonal savanna with higher exchange acidity and markedly lower base cation levels, lower organic matter, available P and iron concentrations.

## D. Predominantly arboreal savanna

Savannas are more frequent in the eastern sector of the island, although they make up a low proportion of the total vegetation. Detailed botanical description is found in Milliken and Ratter (1989 and Chapter 5). The soils of the savanna at Santa Rosa (Profile D(a)), where *Curatella americana* and *Byrsonima crassifolia* dominate the arboreal component of the vegetation, characterize the pedological conditions found at these sites (Appendix 9). The profile is slightly acid and gleyed below about 20 cm, because of the high groundwater levels prevalent during wet seasons. Exchangeable acidity is low. Organic levels are also depressed (around 1% organic C), perhaps reflecting the intense desiccation during the dry season. The low total N% gives C/N ratios of around 17 at the surface with bands of coarse-textured gravels in the subsurface. Very high exchangeable Mg levels are found in the lower profile (*c.* 120 cm depth) giving rise to high total base levels and base saturation which, with high exchangeable Al, lead to very high ECEC levels for this type of soil. This appears to be related to the presence of micaceous schist at depth in the profile, and contrasts with the typical soils of the *cerrado* (savanna) region in

### E. Predominantly herbaceous shrubs often forming dense thickets

The *vazante* site on the Preguiça trail (Profile E(b)) demonstrates a very different pattern of soils coincident with sharp changes in geomorphology and vegetation. The site has parallels in parts of the Santa Rosa basin and in some of the narrow water-filled or streamless drainage courses occupied by *buriti* palms.

The most striking feature of the *vazante* profile, taken as an illustration of this category, is the dark coloured organic and argillic nature of the profile (Appendix 9). Moisture loss, loss on ignition, organic C% and total N% figures are very high at the surface, sometimes extending down to the top of the B horizon. Although the soils are still sandy there is a marked clay + silt proportion giving a dense, blocky structure. The base cation and base saturation levels are high, apart from the leached $A_2$ horizon, and pH values rise from an acidic surface to alkaline figures at depths of a metre.

### F. Predominantly herbaceous grassland (*campo*)

Several sites at Santa Rosa and along the Preguiça trail were devoid of trees and possessed only a few small shrubs. The difference in soil terms between the grassland and the arboreal savanna is minimal. The dense grass cover found in the centre of the shallow drainage basin at Santa Rosa, presumably signalling the dampest site, still gives organic C% figures comparable to those of the nearby arboreal savanna and very little difference is observable in other soil properties. If anything, the Preguiça trail *campos* (Profile F(b)) are more extreme, with greater acidification and porosity in the extremely coarse-textured soils (white sands). Nutrient levels are again very low, indicating exceptionally oligotrophic conditions.

The quartz sand *campos* along the Preguiça trail admirably illustrate this strongly soil-influenced plant community (Appendix 9).

## CONCLUSIONS

### SOIL CHARACTERISTICS AND PLANT COMMUNITIES

The dense forest sites are either well drained or seasonally wet (A1 and A3). Surface horizons are acidic and usually organic but deficient in phosphorus. The base status of these soils is low but base saturation is frequently high. Soils are often alkaline at depth in the profile, associated with the occurrence of base-rich parent materials and finer textures. In well-drained mesotrophic forest (A2) the high base cation levels, notably calcium and magnesium, are related to micaceous parent materials and support deciduous trees. The open and low forest sites (B) are frequently damper (even at the end of the dry season). This inhibits organic matter decomposition giving rise to high carbon, nitrogen and loss on ignition figures. Surface soils are acidic although total exchangeable acidity is kept low and base nutrients and phosphorus are in short supply.

The forest transition sites illustrate the controlling effect of topography upon drainage and the impact of changes in parent material (C). In the case of the former, seasonal flooding constrains rooting, and trees tend to be located above the zone of regular wetting, indicated by gleying. On the other hand, where the parent material changes sharply as a result of alluvial or colluvial deposition, soil texture, permeability/porosity and nutrient availability strongly influence vegetation change (D). This is well illustrated by the oligotrophic soils of the 'island' savannas (F), where very high sand content, extreme porosity and severe nutrient deficiencies characterize the *campo* vegetation.

The alluvial sites provided totally different soil properties, and are typified by tall shrubby or thicket vegetation. The *vazante* site exemplifies this situation with some of the highest proportions of fine-textured soil throughout the profile and high organic matter and cation levels at the surface found on the island (E). The soils are typically acid close to the surface but become alkaline at depth.

## THE PATTERN OF SOIL PROPERTIES

The soils in the eastern part of the Ilha de Maracá are generally coarse-textured, frequently with a high proportion of particle sizes greater than the fine earth range, and some may represent ancient stream-courses. Coarse-textured soils underlie both dense and open forest, tree savanna and *campo*, and provide porous soils with good drainage.

A few locations possess significantly higher levels of silt and clay, although the overall texture remains sandy. These are notably found in the *vazante* floodplain soils and the *buriti* gallery soils, both of which have deposits of dark sandy clay parent material, assumed to be alluvial. In these situations, the usual high porosity is diminished and the rapid runoff observed during wet spells indicates considerable impermeability and leads to lowered rates of organic decomposition and therefore higher organic contents. The likelihood of periodic and seasonal flooding alternating with intense drying would prohibit tree growth in the case of the *vazante*, and permit a type of gallery formation in the smaller-scale case of the *buriti*-filled depressions.

The soils are consistently acidic, except for deeper subsurface horizons close to more alkaline mica schist parent materials. Mildly acidic surface horizons underlie virtually all the vegetation types in the east of the island. A number of the soils are dystrophic (poor in nutrients and allic or Al-rich).

With the exception of a limited number of alkaline sites, or sites affected by drainage from them, soils are uniformly oligotrophic (nutrient poor), with low levels of nutrient cations, phosphorus and nitrogen. Surface horizons are generally more organic, leading to varying degrees of amelioration of the mineral soil immediately below the litter layer.

Only one site, out of the 37 profiles examined, is an exception to the oligotrophic and occasionally dystrophic pattern of soils. The Angico transect revealed mesotrophic soils matched by a deciduous tree cover – a relationship which has been identified in a number of forest and *cerrado* communities within Brazil. On Maracá, such mesotrophic conditions are signalled by elevated levels of magnesium with lower but important increases in calcium.

Surface and soil water plays a major part in determining the distribution and character of the vegetation. The periodic or semi-permanent presence of subsurface water influences the location of the hyperseasonal savannas, the characteristic lines of *buriti* palms and the low *campina* nature of seasonally flooded forest.

## ACKNOWLEDGEMENTS

In the field, the help of Jim Ratter, William Milliken, Jill Thompson, Duncan Scott and John Proctor was particularly appreciated. David Agnew and Christopher Minty undertook the majority of the laboratory analyses.

## REFERENCES

Dargie, T.C.D. and Furley, P.A. (1994). Monitoring change in land use and the environment. In: *The forest frontier: settlement and change in Brazilian Roraima*, ed. P.A. Furley, pp. 68–85. Routledge, London.

Eden, M.J. and McGregor, D.F.M. (1989). *Ilha de Maracá and the Roraima region*, Maracá Rainforest Project, First Report – Geography. Royal Geographical Society, London.

Eden, M.J., Furley, P.A., McGregor, D.F.M., Milliken, W. and Ratter, J.A. (1991). Effect of forest clearance and burning on soil properties in northern Roraima, Brazil. *Forest Ecology and Management*, 38, 283–290.

Furley, P.A. (1986). Radar surveys for resource evaluation in Brazil: an illustration from Rondônia. In: *Remote sensing and tropical land management*, eds M.J. Eden and J.T. Parry, pp. 79–99. Wiley, Chichester.

Furley, P.A. (1992). Edaphic controls at the forest–savanna boundary. In: *Nature and dynamics of forest–savanna boundaries*, eds P.A. Furley, J. Proctor and J.A. Ratter, pp. 91–117. Chapman & Hall, London.

Furley, P.A. (1996). The influence of slope on the nature and distribution of soil and plant communities in the central Brazilian cerrado. In: *Advances in hillslope processes*, eds M.G. Anderson and S.M. Brooks, Vol.1, pp. 327–346. Wiley, Chichester.

Furley, P.A. and Mougeot, L.J.A. (1994). Perspectives. In: *The rainforest frontier: settlement and change in Brazilian Roraima*, ed. P.A. Furley, pp. 1–38. Routledge, London.

Furley, P.A. and Ratter, J.A. (1988). Soil resources and plant communities of the central Brazilian cerrado and their development. *Journal of Biogeography*, 15, 97–108.

Furley, P.A. and Ratter, J.A. (1990). Pedological and botanical variations across the forest savanna transition on Maracá Island. *Geographical Journal*, 156 (3), 251–266.

Furley, P.A., Dargie, T.C.D and Place, C.J. (1994). Remote sensing and the establishment of a geographic information system for resource management on and around Maracá Island. In: *The rainforest edge: plant and soil ecology of Maracá Island, Brazil*, ed. J. Hemming, pp. 115–133. Manchester University Press.

IBGE (1981). *Atlas de Roraima*. Instituto Brasileiro de Geografia e Estatística, Rio de Janeiro.

Milliken, W. and Ratter J.A. (1989). *The vegetation of the Ilha de Maracá*. First Report of the Vegetation survey of the Maracá Rainforest Project (RGS/INPA/SEMA). Royal Botanic Garden Edinburgh.

MME/DNPM (1975). *Projeto RADAM, levantamento de recursos naturais*, Vol. 8, Ministério das Minas e Energia, Departamento Nacional de Produção Mineral, Rio de Janeiro.

Ratter, J.A., Richards, P.W., Argent, G. and Gifford, D.R. (1973). Observations on the vegetation of northeastern Mato Grosso – 1. The woody vegetation types of the Xavantina, Cachimbo Expedition area. *Philosophical Transactions of the Royal Society*, Series B, 266, 449–492.

Ratter, J.A., Askew, G.P., Montgomery, R.F and Gifford, D.R. (1977). Sobre o cerradão de solos mesotróficos no Brasil central. In: *4º Simpósio sobre o cerrado*, ed. M. Ferri, pp. 303–316. Universidade de São Paulo.

Ratter, J.A., Askew, G.P., Montgomery, R.F and Gifford, D.R. (1978). Observations of forests of some mesotrophic soils in central Brazil. *Revista Brasileira de Botânica*, 1, 47–58.

Sarmiento, G. (1988) *The ecology of Neotropical savannas*. Harvard University Press, Cambridge, Massachusetts.

SBCS, Sociedade Brasileira de Ciência do Solo (1973). *Descrição dos solos no campo*. Campinas, São Paulo.

SNLCS (n.d.). *Sistema brasileira de classificação de solos, 2º aproximação*. Serviço Nacional de Levantamento e Conservação de Solos, EMBRAPA, Rio de Janeiro.

Soil Survey Staff (1990). *Keys to soil taxonomy,* 4e. Soil Management Support Services, Monograph No. 6, Blacksburg, Virginia.

# 23 Human Occupation on the Ilha de Maracá: Preliminary Notes

## JOHN PROCTOR[1] AND ROBERT P. MILLER[2]

[1]*University of Stirling, Stirling, UK;* [2]*University of Florida, Gainesville, USA*

## SUMMARY

Contemporary, historical, archaeological and ecological evidence for human occupation of the Ilha de Maracá is analysed, and its influence on the present vegetation is discussed. Parts of the island were originally inhabited by indigenous peoples (including the Sapará and Wayumará), as evidenced by early accounts and by the discovery of pot shards and stone axes. These were forced off the island or wiped out by introduced diseases, and replaced by colonizing smallholders who mostly lived on or close to the banks of the Rio Uraricoera. The last of these smallholders left the island in the 1970s, and the remains of some of their plantations are still visible.

Evidence of previous felling of at least part of the forest at the eastern end of the island is provided by patches of multi-trunked trees (coppice regeneration), concentrations of tree species known to be associated with forest regeneration (e.g. *Astrocaryum aculeatum*), charred wood in the soil and low forest basal areas and timber volumes. It is hard to separate some of these features from the effects of the marginal situation of the Maracá forest, but evidence suggests previous felling and, probably, savanna burning. This should be taken into account when considering the results of other studies conducted in the area.

## RESUMO

É analisada a evidência contemporânea, histórica, arqueológica e ecológica da ocupação humana na Ilha de Maracá, e é discutida sua influência na vegetação atual. Partes da ilha foram ocupadas por grupos indígenas no passado (p. ex. Sapará e Wayumará), como foi testemunhado por viajantes no século 19, e pela descoberta de pedaços de cerâmica e machados de pedra. Estes grupos fugiram de perseguição ou morreram de doenças introduzidas, sendo substituídos por colonos, de que a maioria morava na beira ou perto do Rio Uraricoera. Os últimos desses colonos deixaram a ilha na década de 70, e os restantes das suas roças ainda são visíveis.

Evidência de que pelo menos uma parte da floresta da parte leste de Maracá foi derrubada no passado é fornecida pela ocorrência de árvores com troncos múltiplos, concentrações de espécies de árvores reconhecidas pela associação com regeneração florestal (p. ex. *Astrocaryum aculeatum*), carvão no solo, e baixos área basal e volume de madeira. É difícil saber se estas características são resultados de distúrbio ou são normais numa mata bem na periferia da Hiléia, mas ainda assim a evidência sugere derrubada prévia e, provavelmente, também queima da savana. Estes fatos devem ser tomados em conta na interpretação de outros resultados de estudos científicos desta área.

# INTRODUCTION

The Ilha de Maracá is now uninhabited, apart from Ecological Station residents and visitors, and gold prospectors who camp sporadically (and illegally) near the river banks. The question of past dwellers on Maracá, however, and their influence on the forest, has remained unclear. Milliken and Ratter (1989) concluded their preliminary account of the vegetation of Maracá by stating that although the majority of the island has 'natural, almost completely undisturbed vegetation', the composition of the forest at the eastern end of the island suggested that it had been established relatively recently. Koch-Grünberg (1917 – transl. J. Hemming), referring to the end of the 19th century, pointed out that various tribes of Indians lived on the Ilha de Maracá and at Livramento nearby. Concerning the 20th century, we have the recent statement of Sr Gildo Magalhães (see later discussion) that the Ilha de Maracá was 'always much inhabited'.

Our aim in this paper is to draw together preliminary evidence from a range of sources concerning man's past presence and influence on the eastern part of the island. Most of the scientific work on Maracá, including the majority of that discussed in this volume, was carried out in the area to be discussed in this chapter.

# HISTORICAL EVIDENCE

According to Hemming (1990), there was a short-lived Spanish settlement (1773–75) 'at a creek called Caya-Caya off the north shore of the Santa Rosa channel, a few kilometres above the eastern tip of the island'. In an account of his travels through the area, Koch-Grünberg (1917) includes details of a visit made by Robert Schomburgk in 1837 to a Sapará village on the Santa Rosa channel, roughly opposite the northern tip of Maracá. Schomburgk travelled overland to the village, and there is no record of his having set foot on Maracá. He noted, however, that the Sapará and other tribes in the area had a degenerate appearance, and were suffering from many diseases and from fighting with a wild neighbouring tribe.

Koch-Grünberg visited the island himself in December 1911, and his account (1917) is an important source for the history of Maracá. At the time of his visit, only two males remained of the Wayumara tribe (which was known to Schomburgk), and they lived on the island. A small remnant of the Sapará were also thought to have remained on the island. However, when Koch-Grünberg's informant, Galvão, had arrived in the area (about 1880), there were (according to Koch-Grünberg, 1917) many Indians of different tribes on Maracá. The Indians were later driven off by a settler called Bessa, who claimed that the lands belonged to him. The Indians fled into the forest, but because they had no protection against the wet many of them (especially the children) died, leaving only a few remnants in the vicinity of the island. During Koch-Grünberg's visit Bessa was living downstream from the eastern tip of Maracá (Figure 23.1).

Koch-Grünberg acquired a Macuxi guide called Felipe, who was one of the few Indians still living on Maracá. They went up the Santa Rosa channel of the

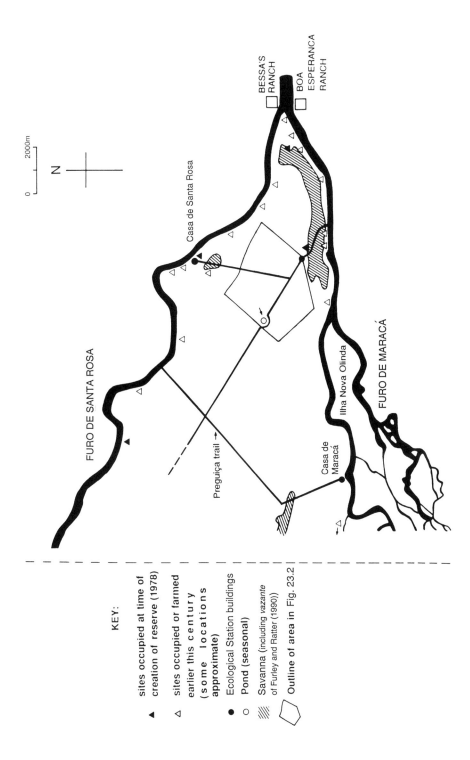

**Figure 23.1.** Recent human occupation on eastern Maracá. The trails were set up after the establishment of the Ecological Station in 1978

Uraricoera and after a journey of about four hours reached a port (probably the site of the present house at Santa Rosa). There was a 'wretched shack' and some new buildings there, which belonged to a man of the Taulipáng or Taurepang tribe called José Rocha and his brother-in-law. Felipe lived with one of the surviving Wayumara, Manduca, in huts in the interior of the island. A Taulipáng named Domingo of Arria awaited them there to trade manioc (presumably grown on the island) for powder and shot. Koch-Grünberg noted that the journey to Felipe's house was on a good path, through light forest, and over a small savanna (see Plate 1b). Felipe's house was a large hut in good condition.

The description of the site closely matches that of the forest–savanna boundary area studied by Thompson *et al.* (1992a), and Koch-Grünberg's 'light forest' is probably the semi-evergreen forest there, many of whose trees would have been leafless at the time of his visit. The remains of a manioc processing house, which was used in the 1970s by an Indian called Waldemiro, still exist and may be a relic of Felipe's time. Koch-Grünberg (1917) described how Manduca returned from his plantation to the house in the evening (he had his own home in the forest). He was accompanied by women who were loaded with sugar-cane and manioc. There was clearly a substantial disturbance to the forest and savanna by these people, but it must have been small compared with that of a few years previously when, as Koch-Grünberg informs us, there had been a large Indian village, Santa Rosa, under a headman named Ignácio. This large village was destroyed by Bessa, but in 1911 the extensive ruins were still visible on the banks. The Santa Rosa place name has survived for this part of the island although a search has yet to be made for remnants of the village.

The next explorer to visit the island was Hamilton-Rice in 1924. He mentioned no Indians living on Maracá, but an aerial photograph (Plate 45 in his papers of 1928) shows a farm about 1 km from the eastern tip of the island on the bank of the Santa Rosa channel. An orange grove still persists there. This farm was given by Bessa to his son-in-law (G. Magalhães, pers. comm.). Although Hamilton-Rice (1928) does not mention Bessa's crimes, he acknowledges that another of Bessa's sons-in-law, Cyro Dantas, responsible for recruiting Indians to crew the expedition's canoes, was distrusted and disliked 'principally on account of his father-in-law'.

## ARCHAEOLOGICAL EVIDENCE

No systematic archaeological survey of the island has been made, but several remains of artefacts have been found in the forest within 3 km of the Ecological Station (Figure 23.2). For example, W. Milliken and J.A. Ratter (pers. comm.) found pottery associated with low-stature, multi-stemmed forest on sandy soil. One of the authors (RPM) found the remains of an earthenware pot at the base of an *Enterolobium cyclocarpum* (Jacq.) Griseb. tree (a species known to be used by Indians for canoes and fruit, but normally riparian) at an inland site which is discussed later. Stone axe-heads were discovered in 1988 near the same site by Sebastião Andrade de Silva.

## FOREST ECOLOGICAL EVIDENCE

Extensive areas of riparian secondary forest, with species of *Cecropia*, *Jacaranda copaia* ssp. *spectabilis* (Mart. ex DC.) A. Gentry and *Maximiliana maripa* (Correa) Drude, all known to be favoured by disturbance (Balée and Campbell, 1990), line the banks for several kilometres on the eastern part of the island, both on the Maracá and Santa Rosa channels. They are clearly distinct from older riparian gallery forest, and from the taller forest inland. There are occasional mango trees. Notes on the vegetation of the old farms near Santa Rosa are given by Milliken and Ratter (1989). There is a large old clearing with the remains of a banana plantation about 700 m upstream from Santa Rosa, and J. Fragoso (pers. comm.) found many old implements including pots, pans, tools and a huge manioc frying pan about 200 m downstream from there. These riparian secondary forests are the remains of plantations, which were cultivated (at least in part) until 1980 and which are further discussed later.

A patch of about 1 ha of secondary forest occurs inland, near the forest regeneration plots discussed by Thompson *et al.* (1992b) (Figure 23.2). This was cleared by a farmer called Antonio Dantas (Zé Alves, pers. comm.) about 1976. It is still of small stature, species-poor, and with a low basal area (see Miller and Proctor, Chapter 6).

Much of the forest on the eastern part of Maracá, such as that described by Milliken and Ratter (1989), is believed to be much older secondary growth, pre-dating Bessa's assault on the Indians about 1880. The forest is very variable in stature in spite of a general similarity in the nutrient-poor, sandy soils (Ross, 1992), and its changing appearance as one walks through it suggests a mosaic of old regrowth of different ages, recovering from shifting cultivation. Multiple-stemmed trees, probably representing coppice regeneration, are frequently encountered in this area.

A patch of several hectares of forest about 1 km from the Ecological Station is particularly revealing. Here there are large individual trees of *Ceiba pentandra* (L.) Gaertn. and *Enterolobium cyclocarpum*. Both of these species are normally riparian on Maracá, and since both have a number of uses for Indians their unusual inland occurrence suggests that they may have been planted or grown from scattered or discarded seeds. Furthermore there is an abundance of the malodorous herb *Petiveria alliacea* L. (Phytolaccaceae) associated with this patch of forest, a species which is often employed in traditional medicine but which was not recorded elsewhere on the island. There is also an unusual abundance of large snails at this site, which suggests that the soil may be higher in calcium (hence favouring cultivation) than that of the surroundings. This is the site where (as was discussed earlier) axe-heads have been found (Figure 23.2), and where one of the *Enterolobium* individuals was growing by the remains of an earthenware pot.

Floristic evidence for past disturbance is common over much of the other forest at the eastern end of Maracá. There is an abundance of the palm *Astrocaryum aculeatum* G.F.W. Mey, which was described by Roosmalen (1985), from his experience in Guiana, as 'Always near human settlements, or if in undisturbed forest, a relict of former Indian settlements'. Other common species in these forests which are known to be favoured by human disturbance (Schultz, 1960; Balée, 1988;

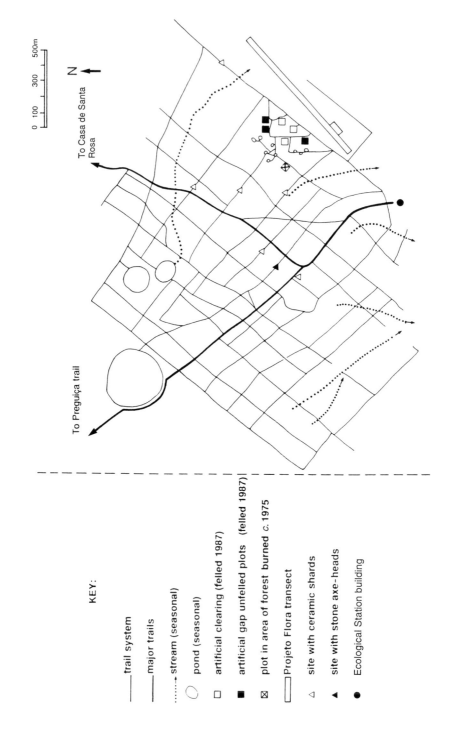

**Figure 23.2.** The locations of the Maracá trail system, permanent ecological plots including clearings and gaps, and sites of axe-heads and ceramic remains

Balée and Campbell, 1990) include *Apeiba schomburgkii* Szysz., *Genipa americana* L., *Guazuma ulmifolia* Lam., *Jacaranda copaia*, *Lindackeria paludosa* (Benth.) Gilg. and *Maximiliana maripa*.

Milliken and Ratter (1989 and Chapter 5), using the PCQ transect method for several forest areas at the eastern end of Maracá, found tree basal areas ranging from 14.5 to 54.5 $m^2ha^{-1}$. Thompson *et al.* (1992a, b) described six plots in evergreen forest with basal areas ranging from 21.7 to 26.7 $m^2ha^{-1}$, and one plot in semi-evergreen forest with a basal area of 21.1 $m^2ha^{-1}$. The low values are similar to those from two plots (21.1 $m^2ha^{-1}$ and 21.9 $m^2ha^{-1}$) surveyed in the Xingu basin area of Amazonia, which were believed to be anthropogenic (Balée and Campbell, 1990). By contrast, Dawkins (1958, 1959) regarded a basal area of 36 $m^2ha^{-1}$ as a pan-tropical average, and Pires and Prance (1985) have reported basal areas of more than 40 $m^2ha^{-1}$ from other 'exceptionally large' Amazonian forests. Although certain of the small-stature forests described by Milliken and Ratter are obviously influenced by soil and other ecological factors, this mosaic of forests of different statures found at the eastern end of Maracá probably reflects disturbances at various periods in its history.

The quantity of large dead wood (including standing dead trees) in the plots described by Thompson *et al.* (1992b, c) was low: evergreen forest contained a mean of 5.08 t ha$^{-1}$ (Scott *et al.*, 1992a); and semi-evergreen forest contained a mean of 8.68 t ha$^{-1}$ (Scott *et al.*, 1992b). These values are similar to that of 8.35 t ha$^{-1}$ recorded for a secondary forest in Ghana (John, 1973) but much lower than the values of 25.8 t ha$^{-1}$ recorded for primary forest near Manaus (Klinge, 1973), 23.8–57.8 t ha$^{-1}$ for primary forests in Sarawak (Proctor *et al.*, 1983) and 39 t ha$^{-1}$ for undisturbed Amazonian heath forest in Venezuela (Jordan and Murphy, 1982).

Charred logs have been found in a number of places on Maracá, including the permanent regeneration plots (about 1 km from the Ecological Station) studied by Thompson *et al.* (1992b). The logs are of very hard wood which can lie undecomposed after burning for over 100 years. Soil pits dug near the forest regeneration plots had abundant charcoal at a range of depths, and this is clearly a common occurrence in Amazonia (Sanford *et al.*, 1985).

It might be argued that most of the forest structural features described above for the plots of low basal area at the eastern end of Maracá, as well as the evidence from burning, simply reflect a stressed evergreen forest near its climatic boundary with savanna. However, the archaeological and floristic evidence (the plant indicators of human disturbance described above are not necessarily features of forest–savanna boundary vegetation) do imply an anthropogenic element on Maracá.

## EVIDENCE FROM INTERVIEWS

Our first interview was with Guttemberg Moreno de Oliveira (the present Administrator of the Island) on 18 April 1989. Guttemberg began his work just as the last four farmers were leaving the island after it had been made an ecological reserve in 1978. The last to leave (after accepting compensation from SEMA in 1980 and 1981) were: Severino (who lived about 2 km upstream from the port of Santa

Rosa); Waldemiro (at Santa Rosa); Zé Ceará (near the eastern tip of the island); and Antonio Dantas (about 400 m east of the causeway to the Ecological Station, on the northern edge of the wet savanna). Severino and Waldemiro both had plantations which remain in a derelict state today. Zé Ceará and Antonio Dantas had cattle herds of 300 head and 56 head respectively, when they left. These were grazed on the large seasonally flooded savanna area in the southeast of the island. Both men had farms near the banks of the Furo de Maracá, and these (along with those of Severino and Waldemiro) are visible today as some of the secondary forest referred to earlier. In addition, Dantas had cleared (perhaps experimentally) the hectare area further inland (mentioned earlier). Photographs relating to the farms are held in the office of IBAMA (which has replaced SEMA) in Boa Vista.

The second interview was held on 16 July 1991 with Severino, who was then aged 63 and lived at Aningal. He said that he lived further upstream on the Santa Rosa channel than any of the farmers known personally to him. He then gave a list of the men known to him who had farmed on the island. These were (in order downstream from Severino's farm to the tip of the island: Felipe, Chico Baixinho, Adalberto, Mocotó (who lived on the mainland), Waldemiro, Chico Bacaba (who lived on the mainland), Bernaldo (who lived on the mainland and tended the old orange grove visible in Plate 45 in Hamilton-Rice (1928)), and Peba, who lived at the tip of the island. Severino had grown bananas, cashews, lemons and manioc. We had learned from Guttemberg that Waldemiro had grown bananas, limes, mangoes and oranges; Waldemiro himself told J. Fragoso that he had grown coffee on the island; and we presume from the remains we observed that Waldemiro had grown manioc also.

On the southern side of the island, Severino recalled that the following kept farms (in order upstream): Antonio Dantas, Zé Ceará, Jaime, Belarmino (who lived near the present Maracá dock area from 1935 and was probably responsible for planting the large mango tree there), Arturzinho (who lived for about four years at the same site), and Manuel Picanço. Another person whose name Severino could not remember lived a little distance upstream from Picanço.

Severino then told of a journey he had made up the 'Igarapé Inajá'. The exact site of this tributary is not known, but Severino described its junction with the Santa Rosa channel as being near the first island in the river upstream from his old farm. About 'half a day's walk' into the interior of the island from the junction, there was much Brazilian cedar (*Cedrela odorata* L., Meliaceae). About six hundred logs of this were cut illegally in about 1980. Three hundred were floated out and sold, but during an attempt to float out the next fifty a man was caught and convicted of the theft. Two hundred and fifty large logs remain decomposing at the site, which was never found by the authorities. Interestingly, very few individuals of this species were discovered on the island during the Maracá Rainforest Project, but in Hamilton-Rice (1928) there is a map which marks a 'Caxoeira Cedro' (Cedar Rapids) on the Uraricoera, near this area. On his way to the Cedar area, Severino described seeing much secondary growth with relict yams (*Dioscorea* sp.) and broken clay pots.

On another excursion Severino had accompanied a gold prospector on a three-day journey across the island to the Maracá channel. After a half day's walk from his house, they came across *Peltogyne gracilipes* Ducke forest (the monodominant rainforest described botanically by Milliken and Ratter, Chapter 5). Within the *Peltogyne* forest they came later that day to an open lake (of great beauty) about 200 m across, surrounded by pure *Peltogyne* (this lake was seen from the air by J. Fragoso after hearing Severino's description). They then came to some low hills, and, well past the half-way stage of their journey, they found extensive secondary forests and the remains of a dwelling by a large stream.

The third interview took place on 20 July 1991 with Sr Gildo Magalhães in Boa Vista. Gildo was born in 1915. His father was Fausto (who is mentioned by Koch-Grünberg (1917)). He was born at the Boa Esperança ranch and lived there until 1975. This ranch is situated on the right bank of the river on the mainland opposite the house formerly occupied by Bessa (Bessa was Gildo's godfather and was described by the latter as 'having a smooth side like a rasp'!). The ranch was the base camp for Hamilton-Rice's 1924 expedition, which Gildo could remember well. Gildo recalled that the island of Maracá was 'always much inhabited'. He was able to shed light on the cultivated land around the orange grove (Hamilton-Rice, 1928, Plate 45). There was a farm there which was occupied by Bessa's son-in-law, Cícero de Almeida, from 1916 to 1924. Around the same time, just downstream from the orange grove, was a sugar-cane mill (with a plantation on the higher ground) belonging to Sr José da Costa Padilha. Another man called Manuel Galvão (perhaps related to the person of the same surname mentioned by Koch-Grünberg) later took many tiles to this area of the island with the intention of using them to roof a large house. The house was never built, but the tiles remain. Gildo agreed with Severino's list of more recent farmers on the east of Maracá, and added one more to it. His grandfather had settled a relative at a place on the island called Guariúba (the common name for the tree *Clarisia racemosa* Ruiz & Pav.), which is near the Paranã do Apuí (a small channel not located precisely but apparently several kilometres up the Furo de Maracá). The relative lived there for 15 years and had 'extensive' plantations (which in this case probably only refers to a few hectares).

These three interviews revealed important memories of man's activities on Maracá. We have no doubt that they are accurate, and the accounts immediately explain the large areas of secondary fringing forests we had observed around the east of the island. The observations on the remains of human occupation inland, and well to the west of previously known sites, are intriguing.

## CONCLUSIONS

It is likely that substantial areas of Maracá were affected by the presence of Indian groups prior to Bessa's assault on the Indian communities around 1880. The exact nature of their use of and impact on the vegetation will never be known, but it can probably be assumed that they practised slash-and-burn agriculture in the forest area, and exploited both the forest and savanna for hunting. As Moran (1990) points out, savannas are more advantageous for hunting (than forests), as they present better visibility and a variety of ecotones. It is highly likely that patches of savanna on Maracá were periodically burned, both to facilitate locomotion as well as to renew the herbaceous cover to attract game animals, principally deer.

The question of fire is in fact one of the central points of the controversy surrounding the origin and maintenance of savanna vegetation (Goodland and Ferri, 1979), and is a question very relevant to the vegetation history of Maracá. Furley and Ratter (1990), for example, support a hypothesis of recent and continuing forest advance on patches of savanna in the interior of the island. Initiation of this advance may have begun with the drastic decline in Indian populations on Maracá at the end of the last century. The persistence of savanna areas on the eastern tip of the island, in which there is little sign of forest advance, may be related to the fact that these areas are exactly those which were occupied when Maracá was transformed into an ecological reserve. The water-table obviously exerts a major influence on this vegetation type. Proctor (1992) has suggested that tree colonization (in the absence of fire and human disturbance) will lower the water-table through evapo-transpiration, thus favouring further forest expansion in areas where 'hyperseasonal' or 'seasonal' savanna (Sarmiento, 1992) has prevailed.

There has clearly been much disturbance in the 20th century to the riparian areas around the east of the island, with many plantations and, on the wet savanna on the island's southeast, cattle grazing.

More work is required on the history of man on Maracá, and future studies should include the less well-known western areas of the island. The eastern part of the island has been the location for much ecological work, and in interpreting their results the authors should consider that many of the ecosystems there may still be recovering from human disturbances.

Shifting cultivation is increasingly seen as being widespread in pre-Columbian Amazonia. For example, Fanshawe (1954) pointed out that shifting cultivation was once so widespread in what is now Guyana that it is hazardous to consider any forest there as being primary. Elsewhere, Schultz (1960) observed an association between sandy soils (more bleached but texturally similar to some of those at the eastern end of Maracá) and Indian remains and secondary forest in Surinam. Viewed in this light, the forests of Maracá are not unusual in the extent of their disturbance, which should be viewed as adding to their variety. The forest on Maracá supports a rich wildlife, and studies of the recovery processes in the areas that have apparently been disturbed in the past will yield guidelines for the management of disturbed but currently unprotected forests elsewhere.

## ACKNOWLEDGEMENTS

Dr D.M. Newbery is thanked for his translation of the German texts. We thank the Overseas Development Administration for providing funds for our work.

## REFERENCES

Balée, W. (1988). Indigenous adaptation to Amazonian palm forests. *Principes*, 32, 47–54.
Balée, W. and Campbell, D.G. (1990). Evidence for the successional status of liana forest (Xingu River Basin, Amazonian Brazil). *Biotropica*, 22, 36–47.

Dawkins, H.C. (1958). *The management of natural tropical high forest with special reference to Uganda.* Commonwealth Forestry Institute Paper 34.

Dawkins, H.C. (1959). The volume increment of natural tropical high forest and limitations on its improvement. *Empire Forestry Review*, 38, 175–180.

Fanshawe, D.B. (1954). Forest types of British Guiana. *Caribbean Forester*, 15, 73–111.

Furley, P.A. and Ratter, J.A. (1990). Pedological and botanical variations across the forest–savanna transition on Maracá Island. *Geographical Journal*, 156, 251–266.

Goodland, R.J.A. and Ferri, M.G. (1979). *Ecologia do Cerrado.* Itatiaia/EDUSP, Belo Horizonte.

Hamilton-Rice, A. (1928). The Rio Branco, Uraricuera, and Parima. *Geographical Journal*, 71, 113–143, 209–223, 345–357.

Hemming, J. (1990). How Brazil acquired Roraima. *Hispanic American Historical Review*, 70, 295–325.

John, D.M. (1973). Accumulation and decay of litter and net production of forest in tropical West Africa. *Oikos*, 24, 430–435.

Jordan, C.F. and Murphy. P.G. (1982). Nutrient dynamics of a tropical forest ecosystem and changes in the nutrient cycle due to cutting and burning. *Annual Report to US National Science Foundation, Institute of Ecology, University of Georgia*, pp. 122–166.

Klinge, H. (1973). Biomasa y materia orgánica del suelo en el ecosistema de la pluviselva centro-amazonica. *Acta Científica Venezolana*, 24, 174–181.

Koch-Grünberg, T. (1917). *Von Roraima zum Orinoco. Ergebnisse einer Reise in Nordbrasilien und Venezuela in den Jahren 1911–1913, Volume 2.* Ernst Vohsen Verlag, Berlin.

Milliken, W. and Ratter, J.A. (1989). *The vegetation of the Ilha de Maracá.* Royal Botanic Garden Edinburgh.

Moran, E.M. (1990). *A Ecologia Humana das Populações da Amazônia.* Vozes, Petrópolis.

Pires, J.M. and Prance, G.T. (1985). The vegetation types of the Brazilian Amazon. In: *Key environments: Amazonia*, eds G.T. Prance and T. Lovejoy, pp. 109–145. Pergamon Press, New York.

Proctor, J. (1992). The savannas of Maracá. In: *The rainforest edge*, ed. J. Hemming, pp. 8–18. Manchester University Press.

Proctor, J., Anderson, J.M., Fodgen, S.C.L. and Vallack, H.W. (1983). Ecological studies in four contrasting lowland rain forests in Gunung Mulu National Park, Sarawak II. Litterfall, litter standing crop and preliminary observations on herbivory. *Journal of Ecology*, 71, 261–283.

Roosmalen, M.G.M. van (1985). *Fruits of the Guianan flora.* Institute of Systematic Botany, University of Utrecht.

Ross, S.M. (1992). Soil nutrients and organic matter in forest and savanna habitats on Maracá Island, Northern Roraima. In: *The rainforest edge*, ed. J. Hemming, pp. 63–91. Manchester University Press.

Sanford, R.L., Saldarriaga, J., Clark, K., Uhl, C. and Herrera, R. (1985). Amazon rain-forest fires. *Science*, 227, 53–55.

Sarmiento, G. (1992). A conceptual model relating environmental factors and vegetation formations in the lowlands of tropical South America. In: *The nature and dynamics of forest–savanna boundaries*, eds P.A. Furley, J. Proctor and J.A. Ratter, pp. 583–601. Chapman & Hall, London.

Schultz, J.P. (1960). Ecological studies on rain forest in northern Suriname. *Mededelingen van het Botanisch laboratorium der Rijks – Universiteit te Utrecht*, 163, 1–267.

Scott, D.A., Proctor, J. and Thompson, J. (1992a). Ecological studies on a lowland evergreen rainforest on Maracá Island, Roraima, Brazil. II. Litter and nutrient cycling. *Journal of Ecology*, 80, 705–717.

Scott, D.A., Proctor, J. and Thompson, J. (1992b). A semi-evergreen forest on Maracá Island, Roraima, Brazil. II. Litter and nutrient cycling. In: *The rainforest edge*, ed. J. Hemming, pp. 30–44. Manchester University Press.

Thompson, J., Proctor, J., Viana, V., Ratter, J.A. and Scott, D.A. (1992a). The forest–

savanna boundary on Maracá Island, Roraima, Brazil. I. An investigation of two contrasting transects. In: *The nature and dynamics of forest–savanna boundaries*, eds P.A. Furley, J. Proctor and J.A. Ratter, pp. 367–392. Chapman & Hall, London.

Thompson, J., Proctor, J., Viana, V., Milliken, W., Ratter, J.A. and Scott, D.A. (1992b). Ecological studies on a lowland evergreen rainforest on Maracá Island, Roraima, Brazil I. Physical environment, forest structure and leaf chemistry. *Journal of Ecology*, 80, 689–703.

Thompson, J., Proctor, J. and Scott, D.A. (1992c). A semi-evergreen forest on Maracá Island I. Physical environment, forest structure and floristics. In: *The rainforest edge*, ed. J. Hemming, pp. 19–29. Manchester University Press.

# Appendices

## Appendix 1 (Chapter 5)
## Plant species mentioned in the account of the vegetation of the Ilha de Maracá

Where specimens have only been determined to generic or family level, collection numbers are often cited in parentheses (collector codes: M, Milliken *et al.*; R, Ratter *et al.*). A complete list of the species recorded on the Ilha de Maracá with habit and habitat notes is given in Milliken and Ratter (1989).

The following groups were identified by specialists (indicated wherever possible by the codes for their herbaria): Acanthaceae (R.K. Brummitt, K), Annonaceae (part) (P.J.M. Maas, U), Aquifoliaceae (S. Andrews, K), Araceae (S.J. Mayo, K), Asclepiadaceae (D. Goyder, K), Bignoniaceae (A. Gentry, MO), Boraginaceae (*Cordia*) (N. Taroda, UEC), Bromeliaceae (M. Arrais, TEPB, and J. Cowley, K), Burseraceae (part) (D. Daly, NY), Cactaceae (N. Taylor, K), Caryocaraceae (G.T. Prance, K), Chrysobalanaceae (G.T. Prance, K), Combretaceae (C. Stace, LTR), Compositae (N. Hind, K), Costaceae (P.J.M. Maas, U), Cyperaceae (D. Simpson, K), Elaeocarpaceae (M. Coode, K), Erythroxylaceae (T. Plowman, F), Euphorbiaceae (*Sapium*) (R. Kruij, U), Gesneriaceae (L.E. Skog, US), Gramineae (S.R. Renvoize, K), Guttiferae (B. Hammell, MO), Hippocrateaceae (A. Menneger, U), Lauraceae (H.H. Van der Werff, MO), Lecythidaceae (S.A. Mori, NY), Leguminosae (G.P. Lewis, K), Lentibulariaceae (P. Taylor, K), Loranthaceae (B. Stannard, K), Malpighiaceae (W.R. Anderson, MICH), Marantaceae (part) (L. Andersson, GB), Melastomataceae (J.J. Wurdack, US), Meliaceae (T.D. Pennington, K), Menispermaceae (R. Barneby, NY), Moraceae (C.C. Berg, BG), Myrtaceae (part) (L.R. Landrum, ASUF), Onagraceae (P. Raven, MO), Orchidaceae (P. Cribb, K), Palmae (J. Dransfield, K), Polygonaceae (J. Brandbyge, AAU), Rubiaceae (part) (B. Boom, NY), Rutaceae (*Zanthoxylum*) (J.R. Pirani, SPF), Sapindaceae (*Allophylus*) (T.D. Pennington, K), Sapotaceae (T.D. Pennington, K), Scrophulariaceae (D. Philcox, K), Sterculiaceae (C.L. Cristobal, CTES), Verbenaceae (part) (S. Atkins, K), Pteridophyta (P.J. Edwards, K), Fungi (B. Spooner, K, and D. Pegler, K). The other collections were identified by the authors.

*Abarema jupunba* (Willd.) Britton & Killip, Leg./Mim.
*Abuta grandifolia* (Mart.) Sandw., Meni.
*Abuta imene* (Mart.) Eichl., Meni.
*Acosmium tomentellum* (Mohl.) Yakovl., Leg./Pap.
*Adiantum pulverulentum* L., Adiant.
*Agave* sp., Agav.
*Albizia* sp. (*Pithecellobium* sp. aff. *elegans* Ducke), Leg./Mim.
*Alchornea schomburgkii* Kl., Euph.
*Alchorneopsis floribunda* (Benth.) M. Arg., Euph.
*Alexa canaracunensis* Pittier, Leg./Pap.
*Alseis longifolia* Ducke, Rub.
*Amaioua corymbosa* Kunth, Rub.

*Anacardium giganteum* Hancock ex Engl., Anac.
*A. occidentale* L., Anac.
*Anadenanthera peregrina* (L.) Speg., Leg./Mim.
*Anaxagorea acuminata* (Dun.) St Hil., Annon.
*Andira surinamensis* (Bondt.) Splitg. ex Pulle, Leg./Pap.
*Anetium citrifolium* (L.) Splitg., Vittar.
*Aniba taubertiana* Mez, Laur.
*Aniba*? sp. 1 (M625), Laur.
*Anthurium clavigerum* Poeppig, Arac.
*A. gracile* (Rudge) Schott, Arac.
*Apeiba schomburgkii* Szysz., Til.
*Apinagia tenuifolia* Van Royen, Podos.

*Aspidosperma* cf. *eteanum* Marcgr., Apoc.
*A. nitidum* Benth., Apoc.
*Asplenium serratum* L., Asplen.
*Astrocaryum aculeatum* G.F.W. Mey., Palm.
*A. gynacanthum* Mart., Palm.
*A. jauari* Mart., Palm.
*Astronium lecointei* Ducke, Anac.
*Bacopa reflexa* (Benth.) Edwall, Scroph.
*Bactris maraja* Mart., Palm.
*Bauhinia outimouta* Aubl., Leg./Caes.
*B. ungulata* L., Leg./Caes.
*Bixa orellana* L., Bix.
*Blechnum serrulatum* L.C. Rich., Blech.
*Bocageopsis multiflora* (Mart.) R.E. Fries, Annon.
*Bombacopsis quinata* (Jacq.) Dug., Bomb.
*Bowdichia virgilioides* Kunth, Leg./Pap.
*Brosimum lactescens* (Moore) C.C. Berg, Mor.
*B. rubescens* Taub., Mor.
*B. utile* (Kunth) Pittier ssp. *ovatifolium* (Ducke) C.C. Berg, Mor.
*Buchenavia tetraphylla* (Aubl.) Howard, Comb.
*Byrsonima coccolobifolia* (Spr.) Kunth, Malp.
*B. crassifolia* (L.) Kunth, Malp.
*B. schomburgkiana* Benth., Malp.
*B. spicata* (Cav.) DC., Malp.
*B. verbascifolia* L. Rich. ex A. Juss., Malp.
*Cabomba piauhiensis* Gardn., Cab.
*Calathea* spp., Marant.
*Calophyllum lucidum* Benth., Gutt.
*Campyloneurum latum* T. Moore, Polypod.
*Canna glauca* L., Cann.
*Caryocar villosum* (Aubl.) Pers., Caryo.
*Casearia sylvestris* Sw., Flac.
*Cassia moschata* Kunth, Leg./Caes.
*Cattleya violacea* (Kunth) Rolfe, Orch.
*Cecropia latiloba* Miq., Mor.
*C. palmata* Willd., Mor.
*Ceiba pentandra* (L.) Gaertn., Bomb.
*Cenchrus echinatus* L., Gram.
*Centrolobium paraense* Tul., Leg./Pap.
*Cereus hexagonus* (L.) Mill., Cact.
*Chamaecrista rotundifolia* (Pers.) Greene, Leg./Caes.
*Cheiloclinium cognatum* (Miers) A.C. Smith, Hippocr.
*Chomelia barbellata* Standl., Rub.
*Chrysochlamys weberbauerii* Engl., Gutt.
*Clarisia racemosa* R. & P., Mor.
*Clathrotropis macrocarpa* Ducke, Leg./Pap.
*Clusia minor* L., Gutt.

*C. renggerioides* Tr. & Pl., Gutt.
*C.* sp., Gutt.
*Cnidoscolus* sp. aff. *urens* (L.) Arthur, Euph.
*Cochlospermum orinocense* (Kunth) Steud., Cochl.
*Combretum rotundifolium* Rich., Comb.
*Cordia nodosa* Lam., Bor.
*C. sellowiana* Cham., Bor.
*Costus arabicus* L., Cost.
*C. scaber* R. & P., Cost.
*Couepia* sp. 1 (R6240) 'Utirana', Chrys.
*Couratari oblongifolia* Ducke & Knuth, Lecy.
*Coursetia ferruginea* (Kunth) Lavin, Leg./Pap.
*Coussapoa villosa* Poepp. & Endl., Mor.
*Crepidospermum goudotianum* (Tul.) Tr. & Pl., Burs.
*Cupania rubiginosa* (Poir.) Radlk., Sapind.
*C.* cf. *scrobiculata* L.C. Rich., Sapind.
*Cuphea antisyphilitica* Kunth, Lyth.
*Curatella americana* L., Dill.
*Cycolosorus interruptus* (Willd.) H. Ito, Thely.
*Cynometra alexandri* C.H. Wright, Leg./Caes.
*Cyperus simplex* Kunth, Cyp.
*Cyrtopodium poecilum* Reichb. f. & Worm., Orch.
*Dacryodes* cf. *roraimensis* Cuatr., Burs.
*D.* spp., Burs.
*Dalbergia* sp., Leg./Pap.
*Desmoncus polyacanthos* Mart., Palm.
*Dialium guianense* (Aubl.) Sandw., Leg./Caes.
*Didymopanax morototoni* Decne. & Planch., Aral.
*Duguetia cauliflora* R.E. Fries, Annon.
*D. lucida* Urban, Annon.
*D. marcgraviana* R.E. Fries, Annon.
*Duroia eriopila* L.f., Rub.
*Ecclinusa guianensis* Eyma, Sapot.
*Echinodorus scaber* Rataj, Alis.
*Eichhornia azurea* Kunth, Pont.
*E. crassipes* (Mart.) Solms-Laut., Pont.
*Eleocharis acutangula* (Roxb.) Schultes, Cyp.
*E.* spp., Cyp.
*Elizabetha coccinea* Schomb. ex Benth. var. *oxyphylla* (Harms) Cowan, Leg./Caes.
*Emilia coccinea* (Sims) D. Don, Comp.
*E. fosbergii* Nicols, Comp.
*Endlicheria dictifarinosa* Allen, Laur.
*Entada polystachya* (L.) DC., Leg./Mim.
*Enterolobium cyclocarpum* (Jacq.) Griseb., Leg./Mim.

*E. schomburgkii* Benth., Leg./Mim.
*Eriochloa punctata* Ham., Gram.
*Erythroxylum rufum* Cav., Erythr.
*E. suberosum* St. Hil., Eryth.
*Eschweilera albiflora* (A.P. DC.) Miers, Lecy.
*E. pedicellata* (Richard) Mori, Lecy.
*E. subglandulosa* (Steud. ex Berg.) Miers, Lecy.
*Eugenia* sp. (R5868), Myrt.
*E.* sp. (R6265), Myrt.
*Euterpe precatoria* Mart., Palm.
*Faramea crassifolia* Benth., Rub.
*F. sessilifolia* (Kunth) DC., Rub.
*Ficus eximia* Schott., Mor.
*F. gomelleira* Kunth & Bouché, Mor.
*F. guianensis* Desv. & Ham., Mor.
*F.* cf. *pakkensis* Standl., Mor.
*F.* spp., Mor.
*Genipa americana* L. var. *caruto* (Kunth) Steyerm., Rub.
*G. spruceana* Steyerm., Rub.
*Geonoma deversa* (Poit.) Kunth, Palm.
*G. maxima* (Poit.) Kunth, Palm.
*Goupia glabra* Aubl., Celast.
*Guatteria schomburgkiana* Mart., Annon.
*Guazuma ulmifolia* Lam., Sterc.
*Guettarda spruceana* M. Arg., Rub.
*Gustavia augusta* L., Lecy.
*G. hexapetala* (Aubl.) Smith, Lecy.
*Gymnopteris rufa* (L.) Underw., Adiant.
*Heliconia bihai* (L.) L., Mus.
*H. psittacorum* L.f., Mus.
*Helicostylis tomentosa* (P. & E.) Rusby, Mor.
*Hemionitis palmata* L., Adiant.
*Heteropsis* sp., Arac.
*Hevea guianensis* Aubl., Euph.
*Himatanthus articulatus* (Vahl) Woods., Apoc.
*Homalium guianense* (Aubl.) Oken, Flac.
*Humiria balsamifera* Aubl., Hum.
*Hydrolea spinosa* L., Hydroph.
*Hymenaea courbaril* L., Leg./Caes.
*H. parvifolia* Huber, Leg./Caes.
*Hymenolobium petraeum* Ducke, Leg./Pap.
*Inga ingoides* Willd., Leg./Mim.
*I. obidensis* Ducke, Leg./Mim.
*I. splendens* Willd., Leg./Mim.
*Ischnosiphon arouma* (Aubl.) Koern., Marant.
*Isertia parviflora* Vahl, Rub.
*Jacaranda copaia* (Aubl.) D. Don ssp. *spectabilis* (Mart. ex DC.) Gentry, Bign.
*Jessenia bataua* (Mart.) Burret, Palm.
*Justicia polystachya* Lam., Acanth.

*Laetia procera* (Poepp.) Eichl., Flac.
Lauraceae sp. indet. 1 (M649), Laur.
Lauraceae sp. indet. 2 (M673), Laur.
*Lecythis corrugata* Poit. ssp. *rosea* (Spruce ex Berg) Mori, Lecy.
Leguminosae sp. indet. (R6318), Leg.
*Licania apetala* (E. Mey.) Fritsch var. *apetala*, Chrys.
*L. heteromorpha* Benth., Chrys.
*L. kunthiana* Hook. f., Chrys.
*L. sprucei* (Hook. f.) Fritsch, Chrys.
*Licaria chrysophylla* (Meissn.) Kostermans, Laur.
*Lindackeria paludosa* (Benth.) Gilg., Flac.
*Lonchocarpus margaritensis* Pittier vel. aff., Leg./Pap.
*L. sericeus* Kunth, Leg./Pap.
*Ludwigia nervosa* (Poir.) Hara, Onag.
*L. torulosa* (Arnott) Hara, Onag.
*Lueheopsis duckeana* Bussett, Til.
*Luziola subintegra* Swallen, Gram.
*Macrolobium acaciifolium* (Benth.) Benth., Leg./Caes.
*Maieta guianensis* Aubl., Mela.
*Maprounea guianensis* Aubl., Euph.
*Maranta protracta* Miq., Marant.
*Matayba* sp. 1 (M106), Sapind.
*Mauritia flexuosa* L.f., Palm.
*Maximiliana maripa* (Correa) Drude, Palm.
*Mayaca fluviatilis* Aubl., May.
*Maytenus* sp. (M472), Celast.
*Melocactus smithii* (Alexander) Buin., Cact.
*Melochia simplex* St. Hil., Sterc.
*Mezilaurus lindaviana* Schwacke & Mez, Laur.
*M.* sp. 1 (R5661), Laur.
*M.* sp., Laur.
*Miconia lepidota* DC., Mela.
*M.* cf. *punctata* D. Don ex DC., Mela.
*M.* cf. *regelii* Cogn., Mela.
*Micropholis melinoniana* Pierre, Sapot.
*Mimosa pellita* H. & B. ex Willd., Leg./Mim.
*M. pudica* L., Leg./Mim.
*Minquartia guianensis* Aubl., Olac.
*Monotagma plurispicatum* (Koern.) K. Schum., Marant.
*M.* spp., Marant.
*Monstera dubia* Engl. & Krause, Arac.
*Montrichardia linifera* Schott, Arac.
*Morinda tenniflora* (Benth.) Steyerm., Rub.
*Mourera fluviatilis* Aubl., Podos.
*Myrcia* cf. *splendens* (Sw.) DC., Myrt.
*Myrciaria floribunda* (West ex Willd.) Berg, Myrt.
*Neea* sp. (R5508), Nyct.

*Norantea guianensis* Aubl., Marc.
*Nymphaea* cf. *wittiana* Ule, Nymph.
*Oceoclades maculata* Lindl., Orch.
*Ocotea bracteosa* (Meissn.) Mez, Laur.
*O. glomerata* (Nees) Mez, Laur.
*O. sandwithii* Kostermans, Laur.
*O.* sp. aff. *amazonica* (Meissn.) Mez, Laur.
*Oenocarpus bacaba* Mart., Palm.
*Olyra longifolia* Kunth, Gram.
*O.* sp., Gram.
*Ormosia smithii* Rudd., Leg./Pap.
*O.* sp. (R5626), Leg./Pap.
*Osteophloem platyspermum* (A. DC.) Warb., Myrist.
*Parahancornia fasciculata* (Poir.) R. Ben., Apoc.
*Parinari excelsa* Sabine, Chrys.
*Parkia pendula* (Willd.) Benth., Leg./Mim.
*Passiflora longiracemosa* Ducke, Passi.
*Peltogyne gracilipes* Ducke, Leg./Caes.
*P. paniculata* Benth. ssp. *pubescens* (Benth.) M.F. da Silva, Leg./Caes.
*Peperomia alata* R. & P., Pip.
*Phenakospermum guyannense* (L.C. Rich.) Endl. ex Miq., Mus.
*Philodendron melinonii* Brongn. ex Regel, Arac.
*Polygonum acuminatum* Kunth, Polygon.
*Pourouma cucura* Standl. & Cuatrec., Mor.
*Pouteria caimito* (R. & P.) Radlk., Sapot.
*P. hispida* agg. Eyma, Sapot.
*P. surumuensis* Baehni, Sapot.
*P. venosa* (Mart.) Baehni ssp. *amazonica* Penn., Sapot.
*Pradosia surinamensis* (Eyma) Penn., Sapot.
*Protium pedicellatum* Sw., Burs.
*P. polybotryum* (Turcz.) Engl., Burs.
*Pseudolmedia laevis* (R. & P.) Macbr., Mor.
*Psychotria poeppigiana* M. Arg. ssp. *barcellana* (M. Arg.) Steyerm., Rub.
*Quiina* cf. *rhytidopus* Tul., Quiin.
*Renealmia alpinia* (Rottb.) Maas, Zing.
*Rheedia macrophylla* (Mart.) Tr. & Pl., Gutt.
*R.* sp., Gutt.
*Rhipsalis baccifera* (Mill.) Stearn, Cact.
*Rinorea brevipes* (Benth.) Blake, Viol.
*R. pubiflora* (Benth.) Sprague & Sandw., Viol.
*R.* sp., Viol.
*Rollinia exsucca* (Dun.) A. DC., Annon.
*Roupala montana* Aubl., Prot.
*Rudgea crassiloba* (Benth.) Rob., Rub.
*Ryania speciosa* Vahl var. *bicolor* DC., Flac.
*Sagittaria* spp., Alis.
*Sagotia racemosa* Baill., Euph.

*Sapium* sp., Euph.
*Sauvagesia rubiginosa* St Hil., Ochn.
*Scleria secans* (L.) Urb., Cyp.
*S. sprucei* C.B. Cl., Cyp.
*S. stipularis* Nees, Cyp.
*Securidaca diversifolia* (L.) Blake, Polygal.
*Senna alata* (L.) Roxb., Leg./Caes.
*Simaba paraensis* Ducke vel aff., Simar.
*Simarouba amara* Aubl., Simar.
*Sinningia incarnata* (Aubl.) Denham, Gesn.
*Siparuna guianensis* Aubl., Monim.
*Sloanea garckeana* K. Schum., Elaeo.
*Socratea exorrhiza* (Mart.) H.A. Wendl., Palm.
*Spondias mombin* L., Anac.
*Swartzia diphylla*, Leg./Pap.
*S. grandifolia* Bong. ex Benth., Leg./Pap.
*S. laurifolia* Benth., Leg./Pap.
*Tabebuia capitata* (Bur. & K. Schum.) Sandw., Bign.
*T. ulei* (Kranzl.) A. Gentry, Bign.
*Tabernaemontana siphilitica* (L.f.) Leeuwenberg, Apoc.
*Tapirira guianensis* Aubl., Anac.
*Tetragastris altissima* (Aubl.) Sw., Burs.
*T. panamensis* (Engl.) Kuntze, Burs.
*Thalia geniculata* L., Marant.
*T. trichocalyx* Gagnepain, Marant.
*Tocoyena neglecta* N.E. Brown, Rub.
*Toulicia* sp. (R5728), Sapind.
*Trattinnickia* cf. *glaviozii* Sw., Burs.
*T. rhoifolia* Willd., Burs.
*Trichanthera gigantea* (H. & B.) Nees, Acanth.
*Trichilia cipo* (A. Juss.) C. DC., Meli.
*T.* ? *septentrionalis* C. DC., Meli.
*T. surumuensis* C. DC., Meli.
*T.* sp. 'Gitó branco', Meli.
*Triplaris surinamensis* Cham., Polygon.
*Uncaria guianensis* (Aubl.) Gmel, Rub.
*Urera baccifera* (L.) Gaud., Urt.
*U. caracasana* (Jacq.) Griseb., Urt.
*Utricularia foliosa* L., Lent.
*U.* sp., Lent.
*Vanilla* sp., Orch.
*Vernonia brasiliana* (L.) Druce, Comp.
*Virola surinamensis* (Rol.) Warb., Myrist.
*Vismia cayennensis* (Jacq.) Pers., Gutt.
*Vitex schomburgkiana* Schauer, Verb.
*Xiphidium caeruleum* Aubl., Haem.
*Xylopia aromatica* (Lam.) Mart., Annon.
*X. discreta* (L.f.) Prague & Hutch., Annon.
*Xyris laxifolia* Mart., Xyr.
*Zanthoxylum* sp. aff. *rigidum* H. & B. in Willd., Rut.

# Appendix 2 (Chapter 10)
# Key to small mammals known from the Ilha de Maracá

NB Given that the small mammal fauna of the Ilha de Maracá has almost certainly been under-collected, it should be remembered that other species may be encountered which are not included in this key.

1  All teeth similar, small and sharply pointed; diastema* absent; pollex** without nail .................................................. 2
   Teeth differentiated, incisors prominent, molars flat; diastema present; pollex vestigial or with nail .................................. 3
2  Tail prehensile; black ring of fur around eye .............. *Marmosa murina*
   Tail not prehensile; no eye ring...................... *Monodelphis brevicaudata*
3  Tail absent or vestigial† ........................................ 4
   Tail obvious and well developed, or if only a stump not tapered at its end ....... 6
4  Flanks with rows of white spots ........................... *Agouti paca*
   Flanks without rows of white spots ............................ 5
5  Head and neck elongated, slender body form; head and body (adult) >400 mm; tail 10–25 mm .......................... *Dasyprocta agouti*
   Head rounded, neck short, stocky body form; head and body (adult) <350 mm; tail virtually absent .......................... *Cavia aperea*
6  Fur with spines ................................................. 7
   Fur without spines‡ ............................................ 9
7  Spines very long (>3 cm) and tubular, hollow, much longer than 'normal' hairs; a tree porcupine ........................ *Coendou prehensilis*
   Spines short (<3 cm), flat in cross-section, solid, some 'normal' hairs as long as spines; a spiny mouse ............................ 8
8  Spines with slender grey or black tips and paler shafts, forming a 'saddle' on the back ending well before the rump; no spines on flanks; hind foot long; tail bicoloured, with coarse hairs ... *Proechimys guyannensis*
   Spines broad, with yellow tips and dark shafts, covering the entire back and flanks; hind foot short and broad; tail unicoloured, with a terminal tuft ....................................... *Mesomys hispidus*
9  Tail furry or woolly for all or some of its length ...................... 10
   Tail naked or sparsely hairy ..................................... 11
10 Tail densely furry only at base, rest of tail naked .......... *Dactylomys dactylinus*
   Tail bushy, covered with hair for entire length .............. *Sciurus pyrrhonotus*
11 Hind foot webbed ............................................. 12
   Hind foot not webbed ......................................... 13
12 Distal part of underside of foot with scales; scales on tail in irregular annular rings ................................. *Nectomys squamipes*
   Distal part of underside of foot without scales; scales on tail in regular annular rings ................................ *Holochilus brasiliensis*
13 Adults small,§ head and body <85 mm; fur pale sandy brown; hind toes with tufts of pale bristles projecting beyond end of claws ....................................... *Oryzomys delicatus*
   Adults >85 mm; hind toes without tufts of bristles projecting beyond end of claws  14
14 Dorsal fur a rich orangey-red; eye proportionately large; tail unicoloured and longer than head and body; furred at base for 0.5–1.0 cm ..................................... *Rhipidomys mastacalis*

Dorsal fur any of various shades of brown or grey, never
orangey-red; eye not proportionately large; tail lacking basal furring .......... 15

15 Tail longer than head and body, unicoloured; hairs on throat white
with grey bases; tail hairs not projecting beyond tip of tail .... *Oryzomys concolor*
Tail shorter or equal to head and body, never longer ....................... 16

16 Upper parts grey to brown, no mid-dorsal stripe; tail bicoloured,
sparsely haired, almost naked; distal part of underside of foot
with small black spots ......................... *Zygodontomys* cf. *lasiurus*
Upper parts dark buffy brown to yellowish brown, never grey
in adults, darker mid-dorsal stripe present; hairs on throat white
from tip to base; tail with hair cover, tail hairs projecting
beyond tip of tail, tail bicoloured; hind foot without black
spots ........................................... *Oryzomys megacephalus*

\*    Naturally tooth-free gap between incisors and first pre-molars.
\*\*   First (innermost) digit of the forefoot (corresponding to the human thumb).
†    The tails of *Proechimys* and *Mesomys* break easily near the base. Specimens may be found with no or little apparent tail. This is a natural phenomenon, believed to be related to predator avoidance. The tail does not regenerate and the stump will not taper as it would in a naturally short-tailed species.
‡    A rare recessive gene codes for spinelessness in *Proechimys*. Some individuals may be encountered with no dorsal spines. Such specimens may be recognized as echimyids by the presence of a vestigial pollex on the front paws. However, the species is likely to be sufficiently common that such individuals, which resemble normal members of the species in every other way, can probably be identified by comparison with other captured individuals.
§    Juveniles of species that reach a larger size when adult will, when less than 85 mm, be in juvenile pelage. This is normally much greyer than either the normal colour of *Oryzomys delicatus* or the adults of their own species. Juveniles of *O. delicatus* are minute (less than 6 g). Head and feet are disproportionately large in juveniles. Langguth (pers. comm.) reports that in many species of rodents the juveniles have a small ball of cartilage on either side of the digital joints. The resulting protrusions disappear in adults.

# Appendix 3 (Chapter 10)
# Other mammals on the Ilha de Maracá

In the course of this study the following additional mammal species were recorded visually (primates and bats are omitted since they are reported elsewhere):

**ARTIODACTYLA**
  TAYASSUIDAE
    *Tayassu pecari* — Queixada, Porco da mata — White-lipped peccary
    *Tayassu tajacu* — Caititu — Collared peccary
  CERVIDAE
    *Mazama americana* — Veado mateiro — Red brocket deer
    *Odocoileus virginianus* — Veado galeiro — White-tailed deer

**CARNIVORA**
  CANIDAE
    *Cerdocyon thous* — Raposa — Crab-eating fox
  FELIDAE
    *Panthera onca* — Onça pintada — Jaguar
    *Felis concolor* — Suçuarana, Onça vermelha — Puma
  MUSTELIDAE
    *Eira barbara* — Irara, Papa-mel, Furão — Tayra
    *Lutra longicaudis* — Lontra — Otter
    *Pteronura brasiliensis* — Ariranha — Giant otter
  PROCYONIDAE
    *Nasua nasua* — Cuati — Coati

**CETACEA**
  PLATANISTIDAE
    *Inia geoffrensis* — Boto cor-de-rosa — Pink river dolphin
  DELPHINIDAE
    *Sotalia fluviatilis* — Tucuxi — Grey dolphin

**LAGOMORPHA**
  LEPORIDAE
    *Sylvilagus brasiliensis* — Coelho — Rabbit

**PERISSODACTYLA**
  TAPIRIDAE
    *Tapirus terrestris* — Anta — Brazilian tapir

**RODENTIA**
  HYDROCHAERIDAE
    *Hydrochaeris hydrochaeris* — Capivara — Capybara

**XENARTHRA**
  BRADYPODIDAE
    *Bradypus tridactylus* — Preguiça — Pale-throated three-toed sloth
  DASYPODIDAE
    *Dasypus novemcinctus* — Tatú comum, Tatú galinha — Nine-banded armadillo
    *Priodontes maximus* — Tatú canastra — Giant armadillo

MYRMECOPHAGIDAE
*Cyclopodes didactylus*     Tamanduaí     Silky anteater
*Myrmecophaga tridactyla*     Tamanduá bandeira     Giant anteater
*Tamandua tetradactyla*     Tamanduá colete, Mambira, Mixila     Collared anteater

Reports of small spotted cats (probably ocelot, *Felis pardalis*, and margay, *F. wiedii*) were also received. There were also several reports of animals that may have been the hog-nosed skunk, *Conepatus*.

# Appendix 4 (Chapter 11)
# Bird species of the Ilha de Maracá

*Key*:

Record (1st):
- A  Moskovits *et al.* (1985)
- B  José Maria Cardoso da Silva
- C  Douglas F. Stoltz
- D  Andrew Whittaker/Mario Cohn-Haft
- E  Mike Hopkins
- F  Duncan Scott
- G  Jenevora Searight

Diet:
- a  Small insects
- b  Large insects and other invertebrates
- c  Fish
- d  Terrestrial vertebrates
- e  Carrion
- f  Fruits and/or large seeds
- g  Small seeds
- h  Nectar
- i  Leaves

Habitat:
1. *Terra firme* forest
2. Riverine forest
3. *Buritizal* and associated damp forest
4. *Campina* forest
5. *Capoeira*
6. Seasonally wet *campo*
7. *Curatella americana/Byrsonima crassifolia* campo

\*  Migrant

|  | English name | Record/Habitat/Diet |
|---|---|---|
| **TINAMIDAE (7)** | **Tinamous** | |
| *Tinamus tao* | Grey tinamou | A1f |
| *T. major* | Great tinamou | A1f |
| *Crypturellus cinereus* | Cinereous tinamou | A15f |
| *C. soui* | Little tinamou | A125f |
| *C. undulatus* | Undulated tinamou | A25f |
| *C. variegatus* | Variegated tinamou | A1f |
| *C. erythropus* | Red-legged tinamou | A1f |
| **PHALACROCORACIDAE (1)** | **Cormorants** | |
| *Phalacrocorax olivaceus* | Neotropic cormorant | A6c |
| **ANHINGIDAE (1)** | **Darters** | |
| *Anhinga anhinga* | Anhinga | A6c |
| **ARDEIDAE (13)** | **Herons/Egrets/Bitterns** | |
| *Ardea cocoi* | White-necked heron | A6c |
| *Casmerodius albus* | Great (common) egret | A6c |
| *Egretta thula* | Snowy egret | A6c |
| *Florida caerulea* | Little blue heron | A6c |
| *Butorides striatus* | Striated heron | A6c |
| *Bubulcus ibis* | Cattle egret | A67b |
| *Pilherodius pileatus* | Capped heron | A6c |
| *Nycticorax nycticorax* | Black-crowned night-heron | A2c |
| *Tigrisoma lineatum* | Rufescent tiger-heron | A6c |
| *Zebrilus undulatus* | Zigzag heron | A12b |
| *Ixobrychus involucris* | Stripe-backed bittern | A6c |

|  | English name | Record/Habitat/Diet |
|---|---|---|
| *I. exilis* | Least bittern | A6c |
| *Botaurus pinnatus* | Pinnated bittern | A6c |
| **COCHLEARIIDAE (1)** | **Boat-billed herons** | |
| *Cochlearius cochlearius* | Boat-billed heron | A6c |
| **CICONIIDAE (3)** | **Storks** | |
| *Mycteria americana* | Wood-stork | A6c |
| *Euxenura maguari* | Maguari stork | E6c |
| *Jabiru mycteria* | Jabirú | A6c |
| **THRESKIORNITHIDAE (3)** | **Ibises** | |
| *Theristicus caudatus* | Buff-necked ibis | A6cb |
| *Cercibis oxycerca* | Sharp-tailed ibis | A6cb |
| *Mesembrinibis cayennensis* | Green ibis | A23b |
| **ANATIDAE (4)** | **Ducks/Geese** | |
| *Dendrocygna viduata* | White-faced tree-duck | A6fb |
| *D. autumnalis* | Black-bellied tree-duck | A6fb |
| *Amazonetta braziliensis* | Brazilian duck | D6fb |
| *Cairina moschata* | Muscovy duck | A6fb |
| **CATHARTIDAE (5)** | **Vultures** | |
| *Sarcoramphus papa* | King vulture | A1e |
| *Coragyps atratus* | Black vulture | A67e |
| *Cathartes aura* | Turkey vulture | A67e |
| *C. burrovianus* | Lesser yellow-headed vulture | A6e |
| *C. melambrotos* | Greater yellow-headed vulture | A1e |
| **ACCIPITRIDAE (23)** | **Kites/Hawks/Eagles** | |
| *Elanus leucurus* | White-tailed kite | A7bd |
| *Gampsonyx swainsonii* | Pearl kite | A7bd |
| *Elanoides forficatus* | Swallow-tailed kite | A1bd |
| *Leptodon cayanensis* | Grey-headed kite | A1bd |
| *Harpagus bidentatus* | Double-toothed kite | A1bd |
| *Ictinia plumbea* | Plumbeous kite | A12bd |
| *Rostrhamus sociabilis* | Snail kite | A6b |
| *Accipiter bicolor* | Bicoloured hawk | A1bd |
| *A. poliogaster* | Grey-bellied hawk | A1bd |
| *Buteo albicaudatus* | White-tailed hawk | A67bd |
| *B. albonotatus* | Zone-tailed hawk | A7bd |
| *B. magnirostris* | Roadside hawk | A7bd |
| *B. nitidus* | Grey hawk | A7bd |
| *Leucopternis albicollis* | White hawk | A1d |
| *L. melanops* | Black-faced hawk | A1d |
| *Busarellus nigricollis* | Black-collared hawk | A6cd |
| *Heterospizias meridionalis* | Savanna hawk | A7bd |
| *Buteogallus urubitinga* | Great black hawk | A6cb |
| *Morphnus guianensis* | Crested eagle | A1d |
| *Harpia harpyja* | Harpy eagle | A1d |
| *Spizaetus ornatus* | Ornate hawk-eagle | A1d |
| *S. tyrannus* | Black hawk-eagle | A1d |
| *Geranospiza caerulescens* | Crane hawk | B1bd |

# APPENDIX 4

|  | English name | Record/Habitat/Diet |
|---|---|---|
| **PANDIONIDAE (1)** | **Ospreys** | |
| *Pandion haliaetus** | Osprey | A6c |
| **FALCONIDAE (11)** | **Falcons/Caracaras** | |
| *Herpetotheres cachinnans* | Laughing falcon | A12bd |
| *Micrastur semitorquatus* | Collared forest-falcon | A1bd |
| *M. ruficollis* | Barred forest-falcon | A1bd |
| *M. gilvicollis* | Lined forest-falcon | D1bd |
| *Daptrius ater* | Black caracara | A12bf |
| *D. americanus* | Red-throated caracara | A12bf |
| *Milvago chimachima* | Yellow-headed caracara | A23be |
| *Polyburus plancus* | Crested caracara | A23be |
| *Falco rufigularis* | Bat falcon | A1bd |
| *F. femoralis* | Aplomado falcon | A1bd |
| *F. sparverius* | American kestrel | E3bd |
| **CRACIDAE (5)** | **Chachalacas/Guans/Curassows** | |
| *Ortalis motmot* | Little chachalaca | A125f |
| *Penelope marail* | Marail guan | A1fb |
| *Pipile pipile* | Blue-throated piping guan | A1fb |
| *Crax alector* | Black curassow | A1fb |
| *Mitu tomentosa* | Lesser razor-billed curassow | B12fb |
| **PHASIANIDAE (2)** | **Wood-quails/Bobwhites** | |
| *Colinus cristatus* | Crested bobwhite | A7fg |
| *Odontophorus gujanensis* | Marbled wood-quail | A1f |
| **OPISTHOCOMIDAE (1)** | **Hoatzins** | |
| *Opisthocomis hoazin* | Hoatzin | A2i |
| **ARAMIDAE (1)** | **Limpkins** | |
| *Aramus guarauna* | Limpkin | A6b |
| **PSOPHIIDAE (1)** | **Trumpeters** | |
| *Psophia crepitans* | Grey-winged trumpeter | A1bf |
| **RALLIDAE (6)** | **Rails/Coots** | |
| *Aramides cajanea* | Grey-necked wood-rail | A23bf |
| *Porzana albicollis* | Ash-throated crake | A6bf |
| *Laterallus exilis* | Grey-breasted crake | B67b |
| *L. viridis* | Russet-crowned crake | A67bf |
| *Porphyrula martinica* | Purple gallinule | A6bf |
| *P. flavirostris* | Azure gallinule | D2bf |
| **HELIORNITHIDAE (1)** | **Sungrebes** | |
| *Heliornis fulica* | Sungrebe | C6b |
| **EURYPYGIDAE (1)** | **Sunbittern** | |
| *Eurypyga helias* | Sunbittern | A1bc |
| **JACANIDAE (1)** | **Jacanas** | |
| *Jacana jacana* | Wattled jacana | A67bf |
| **CHARADRIIDAE (2)** | **Lapwings/Plovers** | |
| *Vanellus chilensis* | Southern lapwing | A6b |
| *Hoploxypterus cayanus* | Pied lapwing | A6b |

|  | English name | Record/Habitat/Diet |
|---|---|---|
| **SCOLOPACIDAE (6)** | **Snipes/Sandpipers** | |
| *Tringa solitaria** | Solitary sandpiper | A6b |
| *T. flavipes** | Lesser yellowlegs | A6b |
| *T. melanoleuca** | Greater yellowlegs | A6b |
| *Actitis macularia** | Spotted sandpiper | A6b |
| *Gallinago gallinago* | Common snipe | A6b |
| *G. undulata* | Giant snipe | A6b |
| **BURHINIDAE (1)** | **Thick-knees** | |
| *Burhinus bistriatus* | Double-striped thick-knee | A7b |
| **LARIDAE (2)** | **Gulls/Terns** | |
| *Phaetusa simplex* | Large-billed tern | A6c |
| *Sterna superciliaris* | Yellow-billed tern | A6c |
| **RYNCHOPIDAE (1)** | **Skimmers** | |
| *Rynchops nigra* | Black skimmer | F6c |
| **COLUMBIDAE (10)** | **Pigeons/Doves** | |
| *Columba cayennensis* | Pale-vented pigeon | A123f |
| *C. subvinacea* | Ruddy pigeon | A12f |
| *Zenaida auriculata* | Eared dove | A7fg |
| *Columbina passerina* | Common ground-dove | A67g |
| *C. minuta* | Plain-breasted ground-dove | A7fg |
| *C. talpacoti* | Ruddy ground-dove | A7fg |
| *Claravis pretiosa* | Blue ground-dove | A12fg |
| *Leptotila verreauxi* | White-tipped dove | B12fg |
| *L. rufaxilla* | Grey-fronted dove | A1fg |
| *Geotrygon montana* | Ruddy quail-dove | A1fg |
| **PSITTACIDAE (19)** | **Macaws/Parrots/Parakeets** | |
| *Ara ararauna* | Blue-and-yellow macaw | A123f |
| *A. macao* | Scarlet macaw | A1f |
| *A. chloroptera* | Red-and-green macaw | A12f |
| *A. severa* | Chestnut-fronted macaw | A12f |
| *A. manilata* | Red-bellied macaw | A32f |
| *A. nobilis* | Red-shouldered macaw | A32f |
| *Aratinga leucophthalmus* | White-eyed parakeet | A123f |
| *A. solstitialis* | Sun parakeet | A2f |
| *A. pertinax* | Brown-throated parakeet | A42f |
| *Pyrrhura picta* | Painted parakeet | A12f |
| *Forpus passerinus* | Green-rumped parrotlet | A7g |
| *Brotogeris chrysopterus* | Golden-winged parakeet | A12f |
| *Pionotes melanocephala* | Black-headed parrot | A1f |
| *Pionopsitta barrabandi* | Orange-cheeked parrot | D1f |
| *Pionus menstruus* | Blue-headed parrot | A12f |
| *Amazona ochrocephala* | Yellow-headed parrot | A1f |
| *A. amazonica* | Orange-winged parrot | A1f |
| *A. farinosa* | Mealy parrot | A1f |
| *Deroptyus accipitrinus* | Red-fan parrot | D1f |
| **CUCULIDAE (10)** | **Cuckoos/Anis** | |
| *Coccyzus pumilus* | Dwarf cuckoo | D7a |
| *C. americanus* | Yellow-billed cuckoo | A12ab |

# APPENDIX 4

|  | English name | Record/Habitat/Diet |
|---|---|---|
| *C. melacoryphus* | Dark-billed cuckoo | A15ab |
| *Piaya cayana* | Squirrel cuckoo | A25ab |
| *P. melanogaster* | Black-bellied cuckoo | A1ab |
| *P. minuta* | Little cuckoo | A1ab |
| *Crotophaga major* | Greater ani | A2ab |
| *C. ani* | Smooth-billed ani | A67b |
| *Tapera naevia* | Striped cuckoo | A67b |
| *Neomorphus rufipennis* | Rufous-winged ground-cuckoo | A1ab |
| **STRIGIDAE (9)** | **Owls** | |
| *Otus choliba* | Tropical screech-owl | A1bd |
| *O. watsonii* | Tawny-bellied screech-owl | A1bd |
| *Bubo virginianus* | Great horned owl | A1bd |
| *Pulsatrix perspicillata* | Spectacled owl | A1bd |
| *Glaucidium hardyi* | Least pygmy-owl | A1bd |
| *G. brasilianum* | Ferruginous pygmy-owl | A7bd |
| *Athene cunicularia* | Burrowing owl | A7bd |
| *Ciccaba virgata* | Mottled owl | A1bd |
| *Rhinoptynx clamator* | Striped owl | A7bd |
| **NYCTIBIIDAE (3)** | **Potoos** | |
| *Nyctibius grandis* | Great potoo | A1ab |
| *N. aethereus* | Long-tailed potoo | A1ab |
| *N. griseus* | Common potoo | A1ab |
| **CAPRIMULGIDAE (10)** | **Nighthawks/Nightjars** | |
| *Chordeiles pusillus* | Least nighthawk | A67a |
| *C. rupestris* | Sand-coloured nighthawk | A67a |
| *C. acutipennis* | Lesser nighthawk | A6a |
| *C. minor\** | Common nighthawk | A67a |
| *Nyctiprogne leucopyga* | Band-tailed nighthawk | F1a |
| *Podager nacunda* | Nacunda nighthawk | A67a |
| *Nyctidromus albicollis* | Pauraque | A125a |
| *Caprimulgus cayennensis* | White-tailed nightjar | A67a |
| *C. maculicaudus* | Spot-tailed nightjar | B67a |
| *Hydropsalis climacocerca* | Ladder-tailed nightjar | E2a |
| **APODIDAE (5)** | **Swifts** | |
| *Streptoprocne zonaris* | White-collared swift | A1a |
| *Chaetura spinicauda* | Band-rumped swift | C1a |
| *C. andrei\** | Ashy-tailed swift | D1a |
| *C. brachyura* | Short-tailed swift | C1a |
| *Reinarda squamata* | Fork-tailed palm-swift | A3a |
| **TROCHILIDAE (15)** | **Hummingbirds** | |
| *Threnetes leucurus* | Pale-tailed barbthroat | A1ah |
| *Phaethornis superciliosus* | Long-tailed hermit | A12ah |
| *P. squalidus* | Dusky-throated hermit | A15ah |
| *P. ruber* | Reddish hermit | A12ah |
| *Anthracothorax nigricollis* | Black-throated mango | A25ah |
| *Chrysolampis mosquitus* | Ruby-topaz hummingbird | A7ah |
| *Chlorestes notatus* | Blue-chinned sapphire | A15ah |
| *Chlorostilbon mellisugus* | Blue-tailed emerald | A7ah |
| *Thalurania furcata* | Fork-tailed woodnymph | D15ah |

|  | English name | Record/Habitat/Diet |
|---|---|---|
| *Polytmus guainumbi* | White-tailed goldenthroat | A23ah |
| *Amazilia chionopectus* | White-chested emerald | G7ah |
| *A. versicolor* | Versicoloured emerald | A1ah |
| *A. fimbriata* | Glittering-throated emerald | A1ah |
| *Heliothryx aurita* | Black-eared fairy | A1ah |
| *Heliomaster longirostris* | Long-billed starthroat | A1ah |
| **TROGONIDAE (3)** | **Trogons** | |
| *Trogon melanurus* | Black-tailed trogon | A1fb |
| *T. viridis* | White-tailed trogon | A12fb |
| *T. violaceus* | Violaceous trogon | A1fb |
| **ALCEDINIDAE (5)** | **Kingfishers** | |
| *Ceryle torquata* | Ringed kingfisher | A6cb |
| *Chloroceryle amazona* | Amazon kingfisher | A6cb |
| *C. americana* | Green kingfisher | A6cb |
| *C. inda* | Green-and-rufous kingfisher | A12cb |
| *C. aenea* | Pygmy kingfisher | A1a |
| **MOMOTIDAE (1)** | **Motmots** | |
| *Momotus momota* | Blue-crowned motmot | A12bf |
| **GALBULIDAE (3)** | **Jacamars** | |
| *Galbula galbula* | Green-tailed jacamar | A1a |
| *G. dea* | Paradise jacamar | A12a |
| *Jacamerops aurea* | Great jacamar | A12a |
| **BUCCONIDAE (5)** | **Puffbirds** | |
| *Notharchus macrorhynchus* | White-necked puffbird | C1ab |
| *N. tectus* | Pied puffbird | C1ab |
| *Bucco macrodactylus* | Chestnut-capped puffbird | A1ab |
| *Monasa atra* | Black nunbird | A1ab |
| *Chelidoptera tenebrosa* | Swallow-wing | A1ab |
| **CAPITONIDAE (1)** | **Barbets** | |
| *Capito niger* | Black-spotted barbet | A1bf |
| **RAMPHASTIDAE (6)** | **Toucans/Aracaris** | |
| *Pteroglossus pluricinctus* | Many-banded aracari | A125f |
| *P. viridis* | Green aracari | A15f |
| *P. flavirostris* | Ivory-billed aracari | A125f |
| *Ramphastos vitellinus* | Channel-billed toucan | A1fb |
| *R. tucanus* | Red-billed toucan | A12fb |
| *R. toco* | Toco toucan | A73fb |
| **PICIDAE (13)** | **Piculets/Woodpeckers** | |
| *Piculus flavigula* | Yellow-throated woodpecker | A1a |
| *Picumnus spilogaster* | White-bellied piculet | A7a |
| *P. exilis* | Golden-spangled piculet | A1a |
| *Celeus grammicus* | Scale-breasted woodpecker | A1af |
| *C. flavus* | Cream-coloured woodpecker | A12af |
| *C. torquatus* | Ringed woodpecker | A1af |
| *C. jumana* | Chestnut woodpecker | A1af |
| *Dryocopus lineatus* | Lineated woodpecker | A1af |
| *Melanerpes cruentatus* | Yellow-tufted woodpecker | A1af |
| *Veniliornis passerinus* | Little woodpecker | A1a |

|  | English name | Record/Habitat/Diet |
|---|---|---|
| *V. cassini* | Golden-collared woodpecker | B1a |
| *Campephilus melanoleucos* | Crimson-rested woodpecker | A1af |
| *C. rubricollis* | Red-necked woodpecker | A1af |
| **DENDROCOLAPTIDAE (14)** | **Woodcreepers** | |
| *Dendrocincla fuliginosa* | Plain-brown woodcreeper | A12a |
| *D. merula* | White-chinned woodcreeper | A1a |
| *Deconychura longicauda* | Long-tailed woodcreeper | A1a |
| *Sittasomus griseicapillus* | Olivaceous woodcreeper | A12a |
| *Glyphorynchus spirurus* | Wedge-billed woodcreeper | A1a |
| *Dendrexetastes rufigula* | Cinnamon-throated woodcreeper | A2a |
| *Hylexetastes perrotii* | Red-billed woodcreeper | A1a |
| *Xiphocolaptes promeropirhynchus* | Strong-billed woodcreeper | A1a |
| *Dendrocolaptes certhia* | Barred woodcreeper | A12a |
| *D. picumnus* | Black-banded woodcreeper | A1a |
| *Xiphorhynchus picus* | Straight-billed woodcreeper | A2a |
| *X. pardalotus* | Chestnut-rumped woodcreeper | A1a |
| *X. guttatus* | Buff-throated woodcreeper | A1a |
| *Lepidocolaptes albolineatus* | Lineated woodcreeper | A1a |
| **FURNARIIDAE (8)** | **Horneros/Spinetails/Foliage gleaners, etc.** | |
| *Furnarius leucopus* | Pale-legged hornero | A7a |
| *Synallaxis gujanensis* | Pale-crowned spinetail | A6a |
| *S. rutilans* | Ruddy spinetail | A1a |
| *Certhiaxis cinnamomea* | Yellow-throated spinetail | A6a |
| *Automolus rufipileatus* | Chestnut-crowned foliage-gleaner | A12a |
| *A. infuscatus* | Olive-backed foliage-gleaner | D1a |
| *Xenops minutus* | Plain xenops | A1a |
| *Sclerurus caudacutus* | Black-tailed leafscraper | A1a |
| **FORMICARIIDAE (27)** | **Antbirds** | |
| *Taraba major* | Great antshrike | A25a |
| *Sakesphorus canadensis* | Black-crested antshrike | A25a |
| *Thamnophilus doliatus* | Barred antshrike | A25a |
| *T. aethiops* | White-shouldered antshrike | A1a |
| *T. murinus* | Mouse-coloured antshrike | A1a |
| *T. punctatus* | Slaty antshrike | A1a |
| *Pygiptila stellaris* | Spot-winged antshrike | A1a |
| *Thamnomanes ardesiacus* | Dusky-throated antshrike | A1a |
| *Myrmotherula brachyura* | Pygmy antwren | A1a |
| *M. guttata* | Rufous-bellied antwren | A1a |
| *M. haematonota* | Stipple-throated antwren | C1a |
| *M. axillaris* | White-flanked antwren | B1a |
| *M. menetriesii* | Grey antwren | B1a |
| *Herpsilochmus rufimarginatus* | Rufous-winged antwren | A1a |
| *Formicivora grisea* | White-fringed antwren | A7a |
| *Cercomacra cinerascens* | Grey antbird | C1a |
| *C. tyrannina* | Dusky antbird | A1a |
| *Myrmoborus leucophrys* | White-browed antbird | A1a |
| *Hypocnemis cantator* | Warbling antbird | A1a |
| *Hypocnemoides melanopogon* | Black-chinned antbird | D1a |
| *Myrmeciza longipes* | White-bellied antbird | A1a |

|  | English name | Record/Habitat/Diet |
|---|---|---|
| *M. atrothorax* | Black-throated antbird | A125a |
| *Pithys albifrons* | White-plumed antbird | D1a |
| *Gymnopithys rufigula* | Rufous-throated antbird | A1a |
| *Hylophylax poecilinota* | Scale-backed antbird | A1a |
| *Myrmothera campanisoma* | Thrush-like antpitta | A1a |
| *Formicarius colma* | Rufous-capped antthrush | A1a |
| **COTINGIDAE (11)** | **Cotingas/Fruiteaters/Bellbirds/Becards** | |
| *Iodopleura fusca* | Dusky purpletuft | C1a |
| *Lipaugus vociferans* | Screaming piha | A1fb |
| *Xenopsaris albinucha* | White-naped xenopsaris | B6a |
| *Pachyramphus rufus* | Cinereous becard | F1fa |
| *P. polychopterus* | White-winged becard | A15fa |
| *Tityra cayana* | Black-tailed tityra | A12fb |
| *T. inquisitor* | Black-crowned tityra | A1fb |
| *Querula purpurata* | Purple-throated fruitcrow | A1fb |
| *Cephalopterus ornatus* | Amazonian umbrellabird | A2f |
| *Perissocephalus tricolor* | Capuchinbird | A12f |
| *Gymnoderus foetidus* | Bare-necked fruitcrow | A21f |
| **PIPRIDAE (8)** | **Manakins** | |
| *Pipra erythrocephala* | Golden-headed manakin | A1f |
| *P. pipra* | White-crowned manakin | D1f |
| *P. coronata* | Blue-crowned manakin | D1f |
| *P. filicauda* | Wire-tailed manakin | A1f |
| *Chiroxiphia pareola* | Blue-backed manakin | A1f |
| *Manacus manacus* | White-bearded manakin | A25f |
| *Machaeropterus pyrocephalus* | Fiery-capped manakin | A1f |
| *Tyranneutes stolzmanni* | Dwarf tyrant manakin | A1f |
| **TYRANNIDAE (49)** | **Tyrants/Flycatchers** | |
| *Fluvicola pica* | Pied water-tyrant | A6a |
| *Arundinicola leucocephala* | White-headed marsh-tyrant | A6a |
| *Pyrocephalus rubinus* | Vermilion flycatcher | A7a |
| *Ochthornis littoralis* | Drab water-tyrant | A2a |
| *Tyrannus melancholicus* | Tropical kingbird | A7a |
| *T. dominicensis** | Grey kingbird | A7a |
| *T. albogularis* | White-throated kingbird | F7a |
| *T. savana** | Fork-tailed flycatcher | A67a |
| *Tyrannopsis sulphurea* | Sulphury flycatcher | A12af |
| *Empidonomus varius* | Variegated flycatcher | A15af |
| *Legatus leucophaius* | Piratic flycatcher | A25af |
| *Conopias parva* | White-ringed flycatcher | A12af |
| *Megarhynchus pitangua* | Boat-billed flycatcher | A25af |
| *Myiodynastes maculatus* | Streaked flycatcher | A25af |
| *Myiozetetes cayanensis* | Rusty-margined flycatcher | A67af |
| *M. similis* | Social flycatcher | A67af |
| *Pitangus sulphuratus* | Great kiskadee | A67bf |
| *P. lictor* | Lesser kiskadee | A67af |
| *Attila spadiceus* | Bright-rumped attila | A1af |
| *A. cinnamomeus* | Cinnamon attila | A1af |
| *Laniocera hypopyrrha* | Cinereous mourner | A1af |
| *Rhytipterna simplex* | Greyish mourner | A1af |

|  | English name | Record/Habitat/Diet |
|---|---|---|
| *Myiarchus ferox* | Short-crested flycatcher | A1af |
| *M. tuberculifer* | Dusky-capped flycatcher | A1af |
| *Lathotriccus euleri* | Euler's flycatcher | A1a |
| *Terenotriccus erythrurus* | Ruddy-tailed flycatcher | B1a |
| *Myiobius barbatus* | Sulphur-rumped flycatcher | A1a |
| *M. atricaudus* | Black-tailed flycatcher | D1a |
| *Myiophobus fasciatus* | Bran-coloured flycatcher | A7a |
| *Tolmomyias sulphurescens* | Yellow-olive flycatcher | A1af |
| *T. poliocephalus* | Grey-crowned flycatcher | A1af |
| *T. flaviventris* | Yellow-breasted flycatcher | A1af |
| *Ramphotrigon ruficauda* | Rufous-tailed flatbill | A12a |
| *Todirostrum cinereum* | Common tody-flycatcher | B67a |
| *T. maculatum* | Spotted tody-flycatcher | A67a |
| *T. sylvia* | Slate-headed tody-flycatcher | A1a |
| *T. pictum* | Painted tody-flycatcher | A1a |
| *Colopteryx galeatus* | Helmeted pygmy-tyrant | A1a |
| *Myiornis ecaudatus* | Short-tailed pygmy-tyrant | A1a |
| *Inezia subflava* | Pale-tipped tyrannulet | A7a |
| *Elaenia flavogaster* | Yellow-bellied elaenia | A7af |
| *E. chiriquensis* | Lesser elaenia | A7af |
| *E. ruficeps* | Rufous-crowned elaenia | A42af |
| *Myiopagis gaimardii* | Forest elaenia | A1af |
| *Phaeomyias murina* | Mouse-coloured tyrannulet | A7af |
| *Camptostoma obsoletum* | Southern beardless tyrannulet | A7a |
| *Tyrannulus elatus* | Yellow-crowned tyrannulet | A1af |
| *Ornithion inerme* | White-lored tyrannulet | A1a |
| *Mionectes oleagineus* | Ochre-bellied flycatcher | A1af |
| **HIRUNDINIDAE (9)** | **Swallows/Martins** | |
| *Tachycineta albiventer* | White-winged swallow | A6a |
| *Phaeoprogne tapera* | Brown-chested martin | A67a |
| *Progne chalybea* | Grey-breasted martin | A67a |
| *Atticora fasciata* | White-banded swallow | A6a |
| *A. melanoleuca* | Black-collared swallow | F6a |
| *Alopochelidon fucata* | Tawny-headed swallow | E6a |
| *Stelgidopteryx ruficollis* | Rough-winged swallow | A67a |
| *Riparia riparia\** | Bank swallow | A6a |
| *Hirundo rustica\** | Barn swallow | A67a |
| **CORVIDAE (1)** | **Crows/Jays** | |
| *Cyanocorax violaceus* | Violaceous jay | A25fb |
| **TROGLODYTIDAE (6)** | **Wrens** | |
| *Campylorhynchus griseus* | Bicoloured wren | A67a |
| *Thryothorus coraya* | Coraya wren | D1a |
| *T. leucotis* | Buff-breasted wren | A1a |
| *Troglodytes aedon* | House wren | A67a |
| *Microcerculus bambla* | Wing-banded wren | A1a |
| *Cyphorhinus arada* | Musician wren | A1a |
| **MIMIDAE (2)** | **Mockingbirds/Thrashers** | |
| *Mimus gilvus* | Tropical mockingbird | A7af |
| *Donacobius atricapillus* | Black-capped mockingthrush | A67b |

|  | English name | Record/Habitat/Diet |
|---|---|---|
| **TURDIDAE (7)** | **Thrushes/Solitaires** | |
| *Catharus fuscescens** | Veery | B1f |
| *C. minimus** | Grey-cheeked thrush | D17 |
| *Turdus olivater** | Black-hooded thrush | A1af |
| *T. leucomelas* | Pale-breasted thrush | A15af |
| *T. fumigatus* | Cocoa thrush | B1af |
| *T. nudigenis* | Bare-eyed thrush | A1af |
| *T. albicollis* | White-necked thrush | A1af |
| **SYLVIIDAE (3)** | **Gnatwrens/Gnatcatchers** | |
| *Microbates collaris* | Collared gnatwren | D1a |
| *Ramphocaenus melanurus* | Long-billed gnatwren | A1a |
| *Polioptila plumbea* | Tropical gnatcatcher | A67a |
| **MOTACILLIDAE (1)** | **Pipits** | |
| *Anthus lutescens* | Yellowish pipit | A7ag |
| **VIREONIDAE (5)** | **Peppershrikes/Shrike-vireos/ Vireos/Greenlets** | |
| *Cyclarhis gujanensis* | Rufous-browned peppershrike | A25af |
| *Smaragdolanius leucotis* | Slaty-capped shrike-vireo | A1af |
| *Vireo olivaceus* | Red-eyed vireo | A1af |
| *Hylophilus pectoralis* | Ashy-headed greenlet | A1af |
| *H. muscicapinus* | Buff-cheeked greenlet | B1a |
| **ICTERIDAE (14)** | **Oropendolas/American orioles/ Blackbirds, etc.** | |
| *Molothrus bonariensis* | Shiny cowbird | A67ag |
| *Scaphidura oryzivora* | Giant cowbird | D67ag |
| *Psaracolius decumanus* | Crested oropendola | A1af |
| *P. viridis* | Green oropendola | A1af |
| *Cacicus cela* | Yellow-rumped cacique | A1af |
| *C. haemorrhous* | Red-rumped cacique | A1af |
| *C. solitarius* | Solitary black cacique | A1af |
| *Lampropsar tanagrinus* | Velvet-fronted grackle | A6a |
| *Agelaius icterocephalus* | Yellow-hooded blackbird | A6ag |
| *Icterus chrysocephalus* | Moriche oriole | A13af |
| *I. nigrogularis* | Yellow oriole | A6a |
| *Gymnomystax mexicanus* | Oriole blackbird | A6ag |
| *Leistes militaris* | Red-breasted blackbird | A67ag |
| *Sturnella magna* | Eastern meadowdark | A7a |
| **PARULIDAE (5)** | **Wood-warblers** | |
| *Dendroica petechia** | Yellow warbler | A1af |
| *D. striata** | Blackpoll warbler | A1af |
| *Geothlypis aequinoctialis* | Masked yellowthroat | A67a |
| *Setophaga ruticilla** | American redstart | C3ab |
| *Basileuterus rivularis* | Buff-rumped warbler | A12a |
| **COEREBIDAE (6)** | **Honeycreepers/Dacnis** | |
| *Coereba flaveola* | Bananaquit | A25fh |
| *Cyanerpes caeruleus* | Purple honeycreeper | D1fh |
| *C. cyaneus* | Red-legged honeycreeper | A1fh |
| *Chlorophanes spiza* | Green honeycreeper | A1fh |

|  | English name | Record/Habitat/Diet |
|---|---|---|
| *Dacnis cayana* | Blue dacnis | A1af |
| *D. lineata* | Black-faced dacnis | D1af |
| **TERSINIDAE (1)** | **Swallow-tanagers** | |
| *Tersina viridis* | Swallow-tanager | A1af |
| **THRAUPIDAE (22)** | **Tanagers/Euphonias** | |
| *Euphonia xanthogaster* | Orange-bellied euphonia | A1af |
| *E. finschi* | Finsch's euphonia | B7af |
| *E. chlorotica* | Purple-throated euphonia | A1f |
| *E. violacea* | Violaceous euphonia | A1f |
| *E. chrysopasta* | Golden-billed euphonia | D1f |
| *Tangara velia* | Opal-rumped tanager | D1af |
| *T. chilensis* | Paradise tanager | D1af |
| *T. xanthogastra* | Yellow-bellied tanager | D1af |
| *T. mexicana* | Turquoise tanager | A1af |
| *T. cayana* | Burnished-buff tanager | B3af |
| *Thraupis episcopus* | Blue-grey tanager | A25af |
| *T. palmarum* | Palm tanager | A25af |
| *Ramphocelus carbo* | Silver-beaked tanager | A25af |
| *Piranga flava* | Hepatic tanager | A7af |
| *Tachyphonus cristatus* | Flame-crested tanager | D1af |
| *T. surinamus* | Fulvous-crested tanager | D1af |
| *T. luctuosus* | White-shouldered tanager | A1af |
| *Nemosia pileata* | Hooded tanager | F1af |
| *Hemithraupis guira* | Guira tanager | A1af |
| *H. flavicollis* | Yellow-backed tanager | A1af |
| *Cissopis leveriana* | Magpie tanager | F1af |
| *Schistoclamys melanopis* | Black-faced tanager | A7af |
| **FRINGILLIDAE (21)** | **Cardinals/Saltators/American sparrows/Sierra finches** | |
| *Saltator maximus* | Buff-throated saltator | A25af |
| *S. caerulescens* | Greyish saltator | A67af |
| *Pitylus grossus* | Slate-coloured grosbeak | A1af |
| *Paroaria gularis* | Red-capped cardinal | A6ag |
| *Cyanocompsa cyanoides* | Ultramarine grosbeak | A1af |
| *Volatinia jacarina* | Blue-black grassquit | A67g |
| *Sporophila schistacea* | Slate-coloured seedeater | A12g |
| *S. intermedia* | Grey seedeater | B67g |
| *S. plumbea* | Plumbeous seedeater | A67g |
| *S. lineola* | Lined seedeater | A6g |
| *S. bouvronides* | Lesson's seedeater | B6g |
| *S. minuta* | Ruddy-breasted seedeater | A7g |
| *S. castaneiventris* | Chestnut-bellied seedeater | A67g |
| *Oryzoborus crassirostris* | Large-billed seed-finch | A67g |
| *O. angolensis* | Lesser seed-finch | A67g |
| *Sicalis citrina* | Stripe-tailed yellow-finch | A67g |
| *S. luteola* | Grassland yellow-finch | A67g |
| *Arremon taciturnus* | Pectoral sparrow | A1f |
| *Arremonops conirostris* | Black-striped sparrow | A67g |
| *Myospiza humeralis* | Grassland sparrow | A7g |
| *Emberizoides herbicola* | Wedge-tailed grass-finch | A7g |

# Appendix 5 (Chapter 12)
# Reptiles of the Ilha de Maracá

A checklist of known species, incorporating earlier records from Celso Morato de Carvalho and Márcio Martins

NR  Previous record, not recorded during Maracá Rainforest Project
FR  First record for Maracá made during Maracá Rainforest Project

| Scientific name | English name | Portuguese name |
| --- | --- | --- |
| **Turtles** | | |
| CHELIDAE | | |
| *Chelus fimbriatus* (Schneider 1783)[NR1] | Mata-mata | Matá-matá |
| *Platemys platycephala* (Schneider 1792)[FR2] | Side- or twist-neck turtle | Lala, Perema |
| PELOMEDUSIDAE | | |
| *Podocnemis unifilis* Troschel 1848[FR2] | Yellow-necked Amazon side-neck turtle | Tracajá |
| KINOSTERNIDAE | | |
| *Kinosternon s. scorpioides* (Linnaeus 1766)[FR2] | Scorpion mud turtle | Muçuã |
| EMYDIDAE | | |
| *Rhinoclemmys p. punctularia* (Daudin 1801)[FR2] | Spotted-legged terrapin | Jaboti-aperema |
| TESTUDINIDAE | | |
| *Geochelone carbonaria* (Spix 1824) | Red-footed tortoise | Jaboti |
| *Geochelone denticulata* (Linnaeus 1766) | Yellow-footed tortoise | Jaboti |
| **Caiman** | | |
| CROCODYLIDAE | | |
| *Caiman c. crocodilus* (Linnaeus 1758)[3] | Spectacled caiman | Jacarétinga |
| **Worm lizards** | | |
| AMPHISBAENIDAE | | |
| *Amphisbaena alba* Linnaeus 1758 | Red worm lizard | Cobra-de-duas-cabeças |
| *Amphisbaena fuliginosa amazonica* Vanzolini 1951 | Speckled worm lizard | Cobra-de-duas-cabeças |
| *Amphisbaena* sp.[4] | ? | ? |
| **Lizards** | | |
| SCINCIDAE | | |
| *Mabuya carvalhoi* Rebouças-Spieker & Vanzolini 1990[5] | Roraima mabuya | Mabuia |
| *Mabuya nigropunctata* (Spix 1825)[6] | Black-spotted mabuya | Mabuia |

| Scientific name | English name | Portuguese name |
|---|---|---|
| *Mabuya* sp. [7] | ? | ? |
| **GEKKONIDAE** | | |
| *Coleodactylus septentrionalis* Vanzolini 1980 | Maracá leaf-litter gecko | Lagartixa |
| *Gonatodes humeralis* (Guichenot 1855) | Amazonian bent-toed gecko | Lagartixa |
| *Hemidactylus palaichthus* Kluge 1969 | Guianan house gecko | Lagartixa |
| *Thecadactylus rapicauda* (Houttuyn 1782)[NR] | Turnip-tailed gecko | Lagartixa |
| **GYMNOPHTHALMIDAE**[8] | | |
| *Cercosaura o. ocellata* Wagler 1830 | Ocellated microteiid | Lagarto |
| *Gymnophthalmus underwoodi* Grant 1958 | Spectacled microteiid | Lagarto |
| *Leposoma percarinatum* (Müller 1923) | Water microteiid | Lagarto |
| *Neusticurus racenisi* Roze 1958[NR] | Brown water microteiid | Lagarto |
| **TEIIDAE** | | |
| *Ameiva ameiva* (Linnaeus 1758) | Common ameiva | Jacaré-pinima |
| *Cnemidophorus lemniscatus* (Linnaeus 1758)[FR] | Rainbow whiptail | Calango |
| *Kentropyx calcarata* Spix 1825[FR] | Keeled teiid | Lagarto |
| *Kentropyx striata* (Daudin 1802) | Keeled teiid | Lagarto |
| *Tupinambis teguixin* (Linnaeus 1758)[9] | Common black tegu | Teiuaçu, Jacuruaru |
| **IGUANIDAE** | | |
| *Iguana i. iguana* (Linnaeus 1758) | South American green iguana | Camaleão, Sinimbu |
| **POLYCHROTIDAE**[10] | | |
| *Anolis auratus* Daudin 1802 | Grass anole | Papa-vento |
| *Anolis ortonii* Cope 1868[FR][11] | Orton's anole | Papa-vento |
| *Polychrus marmoratus* (Linnaeus 1758)[FR] | Many-coloured bush anole | Camaleão pequeno |
| **TROPIDURIDAE**[12] | | |
| *Plica plica* (Linnaeus 1758) | Harlequin racerunner | ? |
| *Tropidurus hispidus* (Spix 1825)[13] | Spiny lava lizard | Taraguira |
| *Uranoscodon superciliosus* (Linnaeus 1758) | Mop-headed iguanid | Tamacuaré |
| **Snakes** | | |
| **LEPTOTYPHLOPIDAE** | | |
| *Leptotyphlops dimidiatus* (Jan 1861) | Slender blindsnake, thread snake | Cobra cega, Minhocão |
| *Leptotyphlops macrolepis* (Peters 1857)[NR] | Large-scaled slender blindsnake | Cobra cega, Minhocão |
| *Leptotyphlops septemstriatus* (Schneider 1801)[FR] | Seven-striped slender blindsnake | Cobra cega, Fura-terra |

| Scientific name | English name | Portuguese name |
|---|---|---|
| **TYPHLOPIDAE** | | |
| *Typhlops reticulatus* (Linnaeus 1766)[FR] | White-headed blindsnake | Cobra cega, Minhocão |
| **BOIDAE** | | |
| *Boa c. constrictor* Linnaeus 1768[FR] | Common boa constrictor | Jibóia |
| *Corallus h. hortulanus* (Linnaeus 1758)[14] | Amazonian tree boa | Cobra de veado, Suaçuaboia |
| *Epicrates cenchria cenchria* (Linnaeus 1758) | Brazilian rainbow boa | Jibóia vermelha |
| *Epicrates cenchria maurus* Gray 1849[FR15] | Guyanan rainbow boa | ? |
| *Eunectes murinus gigas* (Latreille 1802)[FR] | Northern green anaconda | Sucuri, Sucurijú |
| **COLUBRIDAE** | | |
| *Atractus trilineatus* Wagler 1828 | Three-lined ground snake | Cobra-de-terra, Fura-terra |
| *Chironius carinatus* (Linnaeus 1758)[FR16] | Green treesnake | Cutimboia, Cobra-cipó, Cobra-cutia |
| *Chironius exoletus* (Linnaeus 1758) | Green treesnake | Cutimboia, Cobra-cipó, Cobra-cutia |
| *Dipsas catesbyi* (Sentzen 1796)[NR] | Catesby's slug-eater | Dorminhoca, Cobra-cipó |
| *Drymarchon c. corais* (Boie 1867)[FR] | Yellow-tailed cribo | Papa-ova, Papa-pinto, Caninana |
| *Drymobius rhombifer* (Günther 1860)[FR17] | Rhombic racer | ? |
| *Drymoluber dichrous* (Peters 1863)[NR] | Forest racer | ? |
| *Erythrolamprus a. aesculapii* (Linnaeus 1766) | False coral snake | Cobra coral falsa, Boicorá |
| *Helicops angulatus* (Linnaeus 1758)[FR] | Angulate watersnake | Cobra de água, Jararaca falsa |
| *Leptodeira a. annulata* (Linnaeus 1758) | Cat-eyed snake | Dormideira, Cacaual, Jararaca falsa |
| *Leptophis a. ahaetulla* (Linnaeus 1758)[FR] | Parrot snake | Cobra-de-papagaio, Azulão-boia |
| *Liophis* (*Lygophis*) *l. lineatus* (Linnaeus 1758)[FR] | Savanna lined snake | Jararaca-listrada |
| *Liophis* (*Leimadophis*) *poecilogyrus* (Wied 1825) | Variegated grass snake | Cobra-capim, Cobra-de-lixo |
| *Mastigodryas b. boddaerti* (Sentzen 1796) | Striped whipsnake | Buru-listrada, Jararaca-listrada |
| *Oxybelis aeneus* (Wagler 1824)[NR] | Brown vinesnake | Cobra-cipó, Cobra-flecha, Bicuda |
| *Oxyrhopus petola digitalis* (Reuss 1834) | False coral snake | Cobra-coral falsa |
| *Philodryas olfersii* (Lichtenstein 1823)[FR18] | Lora, Parrot snake | Tucanabóia, Cobra-verde |
| *Philodryas viridissimus* (Linnaeus 1758)[FR] | Lora, Parrot snake | Tucanabóia, Cobra-verde |

| Scientific name | English name | Portuguese name |
| --- | --- | --- |
| *Pseudoboa neuwiedii* (Duméril, Bibron & Duméril 1854)[FR] | Neuwied's ground snake | Cobra-coral falsa, Cobra-de-sangue |
| *Spilotes p. pullatus* (Linnaeus 1758) | Tiger ratsnake | Caninana, Cobra-tigre |
| *Tantilla m. melanocephala* (Linnaeus 1758) | Black-headed snake | Cobra-de-terra |
| *Thamnodynastes* sp.[NR19] | ? | Cobra-da-mata |
| **ELAPIDAE** | | |
| *Micrurus h. hemprichii* (Jan 1858)[FR] | Hemprich's coral snake | Cobra-coral verdadeiro |
| *Micrurus l. lemniscatus* (Linnaeus 1758)[FR20] | South American coral snake | Cobra-coral verdadeiro |
| *Micrurus lemniscatus diutius* Burger 1955 | South American coral snake | Cobra-coral verdadeiro |
| **VIPERIDAE** | | |
| *Bothrops atrox* (Linnaeus 1758)[FR] | Common lancehead | Jararaca-do-norte, Caiçaca |
| *Crotalus durissus ruruima* Hoge 1966[21] | Mt Roraima rattlesnake | Cascavel |

[1] Recorded from a specimen preserved at INPA Roraima.
[2] No previous survey of freshwater turtles on Maracá. Other *Podocnemis* spp. probably occur in the Rio Uraricoera but they are difficult to capture.
[3] *Caiman crocodilus crocodilus* was present in the lagoons below the Ecological Station and also in the Rio Uraricoera. Very large caiman sighted in the river by other project members may be the black caiman, *Melanosuchus niger*, which occurs elsewhere in Amazonian Brazil (Avila-Pires, pers. comm.) and in neighbouring Guyana (O'Shea, pers. obs.), but this species has not been confirmed for Maracá. In addition, small caiman sighted by colleagues in a *buriti* palm swamp in the forest may represent the secretive dwarf caiman of the genus *Palaeosuchus*.
[4] A small white amphisbaenian which may represent a species other than *A. alba*. Not included in statistical analysis.
[5] Listed as *Mabuya* sp. nov. by O'Shea (1989). Since described by Rebouças-Spieker and Vanzolini (1990) and believed to be confined to two localities in the State of Roraima.
[6] Listed as *Mabuya bistriata* by O'Shea (1989). Avila-Pires (1995) confined the name *M. bistriata* (including *M. ficta*) to specimens from open habitats in the river valleys of the Amazon and its major tributaries. The more widely distributed forest specimens are allocated the resurrected name *Mabuya nigropunctata*.
[7] A small *Mabuya* which appears to differ from *M. nigropunctata*. Not included in the statistical analysis.
[8] Formerly a subfamily of the Teiidae. Now considered a separate family (Estes *et al.*, 1988).
[9] Listed as *Tupinambis nigropunctatus* by O'Shea (1989). This taxa should now be known as *T. teguixin* (Avila-Pires, 1995).
[10] Formerly a subfamily of the Iguanidae. Now considered a separate family (Frost and Etheridge, 1989).
[11] Listed in error as *Anolis chrysolepis* by O'Shea (1989).
[12] Formerly a subfamily of the Iguanidae. Now considered a separate family (Frost and Etheridge, 1989).
[13] Formerly *Tropidurus torquatus* but now correctly *T. hispidus*.
[14] Listed as *Corallus enydris enydris* by O'Shea (1989) but corrected to *C. hortulanus hortulanus* by McDiarmid *et al.* (1996).
[15] An aberrant savanna taxon which differs from typical *Epicrates cenchria maurus* but which is recorded as such, pending further investigation.
[16] *Chironius carinatus* and *C. exoletus* are the source of some confusion since they exhibit shared characters. Both species appeared to be present on Maracá.

[17] A single specimen which differed significantly from typical *Drymobius rhombifer* since it lacked the usual rhomboid pattern. This specimen, the first representative of the genus from Brazil, may represent a new species but in the absence of further specimens it was recorded as *D. rhombifer* (O'Shea and Stimson, 1993).

[18] *Philodryas olfersii* and *P. viridissimus* were separated by scale counts and the condition of their apical pits (paired or single).

[19] Two species of the genus *Thamnodynastes* occur in the northern Amazonian/Guianan region (Peters *et al.*, 1986) and it was not known which species, *T. pallidus* (Linnaeus 1758) or *T. strigilis* (Thunberg 1787) had previously been collected from Maracá. Since from the other eight sites only Iquitos possesses *T. pallidus*, the inability to determine the species present on Maracá does not greatly affect the statistical analysis.

[20] A specimen collected in the forest agreed more closely with the forest subspecies *Micrurus l. lemniscatus* than the previously reported savanna subspecies, *M. l. diutius*.

[21] *Crotalus durissus ruruima* is tentatively identified to subspecies since the Maracá and Boa Vista specimens require comparison with Venezuelan *C. d. ruruima* and Guyanan *C. d. trigonicus* (see da Cunha and Nascimento, 1980).

# Appendix 6 (Chapter 16)
# Butterflies collected on the Ilha de Maracá

**Collection dates:**
I  24–31 August 1987, Mielke and Casagrande
II  26 November–2 December 1987, Mielke and Casagrande
III  23–28 February 1988, Mielke and Casagrande
IV  21–26 May 1988, Mielke and Casagrande
V  2–13 May 1987, J.A. Rafael
VI  19–24 July 1987, J.A. Rafael
VII  27 August 1987, J.A. Rafael

**Collection localities:**
E  Track between Ecological Station and Furo de Maracá
S  Savanna (within forest)
F  Forest

|  | Date/locality |
|---|---|
| **PAPILIONIDAE** | |
| **PAPILIONINAE** | |
| PAPILIONINI | |
| *Heraclides anchisiades anchisiades* (Esper, 1788) | 2♂ – I. 1♂ – III. F. |
| *H. androgeus androgeus* (Cramer, 1775) | 1♂ – I. F. |
| *H. astyalus phanias* (Rothschild & Jordan, 1906) | 1♂ – I. F. |
| *H. thoas thoas* (Linnaeus, 1771) | 4♂ – I. F. |
| TRIODINI | |
| *Battus polydamas polydamas* (Linnaeus, 1758) | 1♂ – I. E. |
| *Parides sesostris sesostris* (Cramer, 1779) | 1♀ – I. F. |
| GRAPHIINI | |
| *Eurytides dolicaon dolicaon* (Cramer, 1775) | 7♂ – I. 1♂ – IV. F. |
| *E. protesilaus* (Linnaeus, 1758) | 1♂ – I. F. |
| **PIERIDAE** | |
| **PIERINAE** | |
| *Appias drusilla drusilla* (Cramer, 1777) | 8♂ – I. 2♂ – II. 4♂ – III. 1♂, 1♀ – IV. F.E.S. |
| *Itaballia demophile demophile* (Linnaeus, 1767) | 3♂ – I. 3♂, 1♀ – II. 8♂, 2♀ – III. F. |
| *Melete lycimnia lycimnia* (Cramer, 1777) | 4♂ – I. F. |
| *Perrhybris pyrrha pyrrha* (Fabricius, 1775) | 1♂, 1♀ – I. 1♂ – III. F. |
| **COLIADINAE** | |
| *Anteos clorinde* (Godart, 1824) | 1♂ – I. 1♂ – IV. E.S. |
| *Aphrissa statira statira* (Cramer, 1777) | 5♂, 2♀ – I. 4♂ – IV. S.F. |
| *Eurema (Eurema) albula* (Cramer, 1775) | 3♂, 1♀ – I. 1♂ – II. 2♂ – III. E. |
| *E. (E.) elathea elathea* (Cramer, 1777) | 5♂, 6♀ – I. 1♀ – II. 2♂, 2♀ – III. 1♂, 3♀ – IV. E. |
| *E. (E.) phiale phiale* (Cramer, 1775) | 6♂, 3♀ – I. 1♀ – II. E.S. |
| *E. (Pyrisitia) nise nise* (Cramer, 1775) | 5♂, 7♀ – I. 1♂, 1♀ – II. E. |

|  | Date/locality |
|---|---|
| *Phoebis argante argante* (Fabricius, 1775) | 1♂ – I. S. |
| *P. sennae sennae* (Linnaeus, 1758) | 1♂ – IV. E. |
| **DISMORPHIINAE** | |
| *Moschoneura pinthous pinthous* (Linnaeus, 1758) | 1♂ – I. F. |

**NYMPHALIDAE**
**ITHOMIINAE**

|  |  |
|---|---|
| TITHOREINI | |
| *Tithorea harmonia harmonia* (Cramer, 1777) | 4♂, 5♀ – I. 1♂, 3♀ – II. 2♀ – III. F. |
| MECHANITINI | |
| *Mechanitis polymnia polymnia* (Linnaeus, 1758) | 5♂, 7♀ – I. 2♂ – II. 1♂ – III. F. |
| *Sais rosalia rosalia* (Cramer, 1779) | 13♂, 9♀ – I. 5♂, 3♀ – II. 1♂, 1♀ – III. F. |
| METHONINI | |
| *Methona confusa* Butler, 1873 | 2♂, 3♀ – I. 2♀ – II. 3♂, 1♀ – III. F. |
| NAPEOGENINI | |
| *Hypothyris euclea barii* (Bates, 1862) | 2♂ – I. F. |
| *H. ninonia colophonia* D'Almeida, 1945 | 13♂, 24♀ – I. 2♂, 2♀ – II. 4♂, 1♀ – III. F. |
| OLERIINI | |
| *Aeria elara elara* (Hewitson, 1855) | 16♂, 7♀ – I. 5♂, 2♀ – II. F. |
| DIRCENNINI | |
| *Callithomia lenea lenea* (Cramer, 1779) | 2♂, 1♀ – I. 3♂, 1♀ – II. F. |
| *Ceratinia neso neso* (Huebner, 1806) | 2♂ – I. F. |

**DANAINAE**

|  |  |
|---|---|
| EUPLOEINI | |
| *Lycorea pasinuntia pasinuntia* (Stoll, 1780) | 1♂ – II. F. |
| DANAINI | |
| *Anosia eresimus eresimus* (Cramer, 1777) | 1♂, 2♀ – III. E. |

**SATYRINAE**

|  |  |
|---|---|
| HAETERINI | |
| *Pierella hyalinus hyalinus* (Gmelin, 1790) | 8♂, 8♀ – I. 1♂, 1♀ – II. F. |
| *P. lena lena* (Linnaeus, 1767) | 2♂, 3♀ – I. F. |
| *P. rhea chalybaea* Godman, 1905 | 3♂, 2♀ – I. F. |
| EUPTYCHINI | |
| *Caeruleuptychia brixius* (Godart, 1824) | 2♂, 7♀ – I. 1♂ – II. 1♂ – III. F. |
| *C. coerulea* (Butler, 1869) | 1♀ – I. F. |
| *Cepheuptychia cephus* (Fabricius, 1775) | 1♀ – IV. F. |
| *Chloreuptychia arnaea arnaea* (Fabricius, 1777) | 1♂ – I. F. |
| *C. chloris* (Cramer, 1780) | 3♂, 2♀ – I. 1♂ – III. F. |
| *C. herse* (Cramer, 1775) | 1♂, 2♀ – I. 1♀ – III. F. |
| *C. hewitsonii* (Butler, 1866) | 1♂, 1♀ – I. F. |
| *Cissia ocypete* (Fabricius, 1777) | 4♂, 4♀ – I. 1♂ – II. 1♂, 1♀ – III. F. |
| *C. penelope* (Fabricius, 1775) | 5♂, 3♀ – I. 1♀ – II. 1♂ – IV. F. |
| *C. terrestris* (Butler, 1866) | 3♂, 1♀ – I. 1♀ – II. 1♂ – IV. F. |

|  | Date/locality |
|---|---|
| *Erichthodes erichtho* (Butler, 1866) | 1♂ – I. 1♀ – III. F. |
| *Hermeuptychia hermes* (Fabricius, 1775) | 4♂ – I. 2♂, 1♀ – II. 4♂, 2♀ – III. E. F. |
| *Magneuptychia helle* (Cramer, 1779) | 4♂, 3♀ – I. 1♂ – II. 1♂ – III. F. |
| *M. lea* (Cramer, 1777) | 4♂, 4♀ – I. 1♀ – II. 1♂, 4♀ – III. F. |
| *M. libye* (Linnaeus, 1767) | 1♂, 1♀ – I. 2♂ – III. 1♀ – IV. F. |
| *M. ocnus* (Butler, 1866) | 1♀ – I. 1♀ – III. 1♀ – IV. F. |
| *Megeuptychia antonoe* (Cramer, 1775) | 4♂, 2♀ – I. 1♂ – III. F. |
| *Pareuptychia ocirrhoe ocirrhoe* (Fabricius, 1777) | 4♂, 5♀ – I. 1♀ – II. 1♀ – III. F. |
| *P. o. ocirrhoe* (Fabricius, 1777) *f. lydia* (Cramer, 1777) | 3♂, 2♀ – I. 1♀ – III. F. |
| *Pseudodebis marpessa* (Hewitson, 1862) | 2♂, 3♀ – I. 4♂, 5♀ – III. F. |
| *Splendeuptychia itonis* (Hewitson, 1862) | 1♂ – I. F. |
| *Taygetis celia* (Cramer, 1779) | 3♂, 2♀ – I. 1♂, 1♀ – II. 4♂, 1♀ – III. F. |
| *T. echo echo* (Cramer, 1775) | 7♂, 6♀ – I. 3♂, 3♀ – II. 13♂, 3♀ – III. F. |
| *T. erubescens* Butler, 1868 | 2♂, 2♀ – I. 1♂ – II. 1♂, 4♀ – III. F. |
| *T. kerea* Butler, 1869 | 3♂ – I. 2♀ – II. 1♂ – III. F. |
| *T. laches* (Fabricius, 1793) | 4♂, 2♀ – I. 3♂, 3♀ – II. 3♂, 5♀ – III. F. |
| *T. larua* Felder & Felder, 1867 | 3♂, 1♀ – I. 12♂, 3♀ – II. 11♂, 1♀ – III. 2♀ – IV. F. |
| *T. penelea penelea* (Cramer, 1777) | 8♂, 1♀ – I. 4♂, 4♀ – II. 1♂, 3♀ – III. F. |
| *T. thamyra thamyra* (Cramer, 1779) | 1♂, 2♀ – I. 1♂, 1♀ – II. 2♀ – III. F. |
| *T. virgilia* (Cramer, 1776) | 3♂ – I. 1♂, 1♀ – III. F. |
| *T. xenana xenana* Butler, 1870 | 6♂, 1♀ – I. 5♂, 3♀ – II. 10♂, 3♀ – III. F. |
| *Ypthimoides argyrospila* (Butler, 1867) | 2♂, 3♀ – I. 6♂, 9♀ – III. 1♀ – IV. E. |

PRONOPHILINI
| *Amphidecta calliomma calliomma* (Felder & Felder, 1862) | 2♀ – I. 3♂, 3♀ – III. F. |
|---|---|

BIINI
| *Bia actorion* (Linnaeus, 1767) | 6♂, 3♀ – I. 1♂, 1♀ – II. 3♂, 1♀ – III. F. |
|---|---|

TRIBE UNCERTAIN
| *Manataria maculata* (Hopffer, 1874) | 4♂, 6♀ – IV. F. |
|---|---|

**BRASSOLINAE**
| *Caligo eurilochus eurilochus* (Cramer, 1775) | 3♂ – III. F. |
|---|---|
| *C. idomeneus apollonides* Fruhstorfer, 1912 | 2♂, 1♀ – I. 2♂ – II. 9♂, 1♀ – III. 1♂ – IV. 1♀ – V. F. |
| *C. illioneus polyxenus* Stichel, 1903 | 1♂, 2♀ – I. 1♀ – II. 4♂ – III. F. |
| *C.* sp. | 1♂ – I. 1♀ – V. F. |
| *C. teucer* ssp. | 1♂ – III. F. |
| *Catoblepia berecynthia halli* Bristow, 1981 | 3♂, 2♀ – I. 3♂ – II. 4♂, 1♀ – III. 1♀ – IV. F. |
| *C. soranus* (Westwood, 1851) | 8♂, 7♀ – I. 2♂ – II. 11♂, 10♀ – III. 1♂, 8♀ – IV. F. |
| *Eryphanes polyxena polyxena* (Meerburgh, 1775) | 1♂ – I. 3♂, 1♀ – III. F. |
| *Opsiphanes cassiae* ssp. | 1♂ – II. F. |

|  | Date/locality |
|---|---|
| *O. cassina merianae* Stichel, 1902 | 1♂ – I. F. |
| *O. invirae intermedius* Stichel, 1902 | 11♂, 5♀ – I. 5♂ – II. 14♂ – III. 1♀ – IV. F. |
| *O. quiteria obidonus* Fruhstorfer, 1907 | 1♂ – I. 1♂ – II. F. |

**MORPHINAE**

| | |
|---|---|
| *Morpho helenor helenor* (Cramer, 1775) | 16♂, 15♀ – I. 3♂, 12♀ – II. 11♂, 3♀ – III. F. |
| *M. pseudagamedes* Weber, 1944 | 6♂, 2♀ – I. 2♂, 1♀ – II. 4♂, 1♀ – III. F. |

**NYMPHALINAE**

HELICONIINI

| | |
|---|---|
| *Dryadula phaetusa* (Linnaeus, 1758) | 1♂ – II. E. |
| *Dryas iulia iulia* (Fabricius, 1775) | 4♂ – I. E. |
| *Eueides aliphera aliphera* (Godart, 1819) | 1♂, 1♀ – I. E. |
| *E. vibilia vibilia* (Godart, 1819) | 4♂, 1♀ – I. 1♂, 1♀ – II. F. |
| *Heliconius antiochus antiochus* (Linnaeus, 1767) | 3♂, 2♀ – I. 4♂, 3♀ – II. 4♂, 5♀ – III. F. |
| *H. burneyi catharinae* Standinger, 1885 | 1♂, 1♀ – I. F. |
| *H. doris doris* (Linnaeus, 1771) | 12♂, 1♀ – I. 1♀ – II. 2♂, 1♀ – III. F. |
| *H. erato magnificus* Riffarth, 1900 | 12♂, 10♀ – I. 5♂, 3♀ – II. 4♂, 3♀ – III. F. |
| *H. ethilla hyalinus* Neustetter, 1928 | 1♂ – I. F. |
| *H. luciana* Lichy, 1960 | 4♂, 5♀ – I. 1♀ – II. 6♂, 2♀ – III. F. |
| *H. melpomene pyrforus* Kaye, 1907 | 8♂, 4♀ – I. 1♂ – III. 1♂ – IV. F. |
| *H. metharme* Erichson, 1848 | 1♂ – I. 1♂ – II. F. |
| *H. ricini* (Linnaeus, 1758) | 1♂, 3♀ – I. 2♂ – II. 1♂, 1♀ – III. E. F. |
| *H. sara thamar* (Huebner, 1806) | 1♂ – I. 2♀ – II. 1♂ – III. F. |
| *H. wallacei flavescens* Weymer, 1890 | 4♂, 2♀ – I. 1♂ – II. F. |
| *Philaethria dido* (Linnaeus, 1763) | 5♀ – I. 1♀ – II. 2♂ – III. F. |

MELITAEINI

| | |
|---|---|
| *Eresia eunice eunice* (Huebner, 1807) | 1♂ – I. 1♂, 3♀ – II. 1♂ – III. F. |
| *Ortilia liriope* (Cramer, 1775) | 1♂ – II. E. |
| *Tegosa similis* Higgins, 1981 | 3♂ – I. E. |

ARGYNNINI

| | |
|---|---|
| *Euptoieta hegesia* (Cramer, 1779) | 2♂, 1♀ – I. E. |

VANESSINI

| | |
|---|---|
| *Anartia jatrophae jatrophae* (Linnaeus, 1763) | 2♂, 1♀ – I. 1♂ – III. E. |
| *Junonia genoveva* (Cramer, 1780) | 6♂, 3♀ – I. 1♀ – II. 3♂, 2♀ – III. S. |
| *Victorina stelenes stelenes* (Linnaeus, 1758) | 1♂ – I. F. |

COLOBURINI

| | |
|---|---|
| *Coea achaeronta achaeronta* (Fabricius, 1775) | 1♀ – I. 2♂, 1♀ – IV. F. |
| *Colobura dirce* (Linnaeus, 1758) | 7♂, 1♀ – I. 4♂ – III. 2♂ – IV. F. |
| *Historis odius* (Fabricius, 1758) | 2♂, 1♀ – I. 2♀ – IV. F. |
| *Tigridia acesta* (Linnaeus, 1758) | 1♂ – III. 1♂ – V. F. |

EURYTELINI

| | |
|---|---|
| *Mestra cana cowiana* (Butler, 1902) | 8♂, 7♀ – I. 3♂, 1♀ – II. 2♂ – III. E. |

|  | Date/locality |
|---|---|
| **MARPESINI** | |
| *Marpesia chiron* (Fabricius, 1775) | 2♂ – I. 1♂ – III. 1♀ – IV. F. |
| *M. norica* (Hewitson, 1852) | 2♂ – I. F. |
| *M. petreus* (Cramer, 1776) | 1♂ – I. 1♀ – II. F. |
| **EPICALIINI** | |
| *Catonephele acontius acontius* (Linnaeus, 1771) | 5♂, 1♀ – I. 1♂ – II. 2♂, 2♀ – III. F. |
| *C. numilia numilia* (Cramer, 1775) | 1♀ – I. 1♀ – III. F. |
| *Eunica caresa* (Hewitson, 1857) | 2♂, 2♀ – IV. F. |
| *E. malvina* (Bates, 1863) | 1♂ – I. F. |
| *E. tatila bellaria* Fruhstorfer, 1908 | 1♀ – I. F. |
| *Nessaea obrina obrina* (Linnaeus, 1758) | 4♂ – I. 2♂ – II. 4♂, 1♀ – III. F. |
| *Pyrrhogyra amphiro* ssp. | 1♀ – I. 1♂ – II. F. |
| *Temenis laothoe laothoe* (Cramer, 1777) | 6♂, 7♀ – I. 2♂, 1♀ – II. 1♂, 1♀ – III. F |
| **HAMADRYINI** | |
| *Ectima iona* Doubleday, 1848 | 3♀ – I. 5♂, 2♀ – III. 1♀ – IV. F. |
| *Hamadryas amphinome amphinome* (Linnaeus, 1767) | 1♂, 1♀ – I. 2♂, 1♀ – III. F. |
| *H. februa ferentina* (Godart, 1824) | 2♂ – I. 1♀ – III. F. |
| *H. feronia feronia* (Linnaeus, 1758) | 2♂, 2♀ – I. 1♀ – II. 4♂, 2♀ – III. 1♀ – IV. F. |
| **CALLICORINI** | |
| *Callicore cyllene cyllene* (Doubleday, 1847) | 25♂, 36♀ – I. 1♂, 1♀ – II. 4♂, 1♀ – III. F. |
| *Dynamine arene arene* (Huebner, 1823) | 1♀ – III. F. |
| *D. mylitta* (Cramer, 1779) | 2♂, 3♀ – I. 1♀ – II. 1♂ – III. F. |
| **LIMINITINI** | |
| *Adelpha cocala cocala* (Cramer, 1779) | 1♂ – I. F. |
| *A. cytherea cytherea* (Linnaeus, 1758) | 2♀ – I. 1♂ – III. F. |
| *A. delphicola delphicola* Frukstorfer, 1910 | 1♀ – I. 2♂ – III. F. |
| *A. iphicla iphicla* (Linnaeus, 1758) | 11♂ – I. 2♂ – II. 3♂ – III. F. |
| *A. naxia* (Felder and Felder, 1867) | 1♂ – I. F. |
| *A. nea* Hewitson, 1847 | 1♂ – I. F. |
| *A. plesaure phliassa* (Godart, 1824) | 1♂, 1♀ – I. 1♂ – II. F. |
| *A. thesprotia* (Felder & Felder, 1867) | 1♂, 1♀ – I. F. |
| *A.* sp. | 4♂, 1♀ – I. 3♂ – II. F. |
| **APATURINAE** | |
| *Doxocopa agathina* (Cramer, 1777) | 1♂ – I. F. |
| *D. laure laure* (Drury, 1773) | 4♂ – I. F. |
| **CHARAXINAE** | |
| *Anaea morvus mormus* (Fabricius, 1775) | 5♂, 2♀ – I. F. |
| *A. oenomais* (Boisduval, 1870) | 2♂, 1♀ – I. 1♂ – II. 1♂ – III. F. |
| *A. polycarmes* (Fabricius, 1775) | 1♂, 4♀ – I. 2♂, 1♀ – III. F. |
| *A. ryphea ryphea* (Cramer, 1775) | 7♂, 6♀ – I. 3♂, 2♀ – III. F. |
| *A. vicinia* Staudinger, 1887 | 1♂ – I. F. |
| *Hypna clytemnestra clytemnestra* (Cramer, 1777) | 4♂, 5♀ – I. 2♂ – II. 1♂, 1♀ – III. 1♂ – IV. F. |
| *Prepona demophon demophon* (Linnaeus, 1758) | 1♂, 1♀ – I. 7♂ – III. 1♂, 1♀ – IV. F. |
| *P. demophoon demophoon* (Huebner, 1814) | 1♂, 1♀ – II. 3♂ – III. 1♂ – IV. F. |

|  | Date/locality |
|---|---|
| *P. dexamenes* Hopffer, 1874 | 1♂, 1♀ – I. 1♂, 1♀ – II. F. |
| *P. eugenes eugenes* (Bates, 1865) | 1♂ – II. 2♂, 1♀ – III. F. |
| *P. joiceyi* Le Moult, 1932 | 3♂ – I. 1♂ – II. 3♂, 1♀ – III. 1♂ – IV. F. |
| *P. laertes ikarios* Fruhstorfer, 1904 | 1♂, 1♀ – III. F. |
| *P. meander meander* (Cramer, 1775) | 1♂ – I. 1♂ – III. F. |
| *P. omphale* (Huebner, 1819) | 5♂, 1♀ – I. 2♂ – II. 4♂, 1♀ – III. 2♂ – IV. F. |
| *P. pheridamas pheridamas* (Cramer, 1777) | 1♂, 4♀ – I. 6♂, 2♀ – III. 1♂ – IV. F. |
| *P.* sp. | 1♂ – III. 1♀ – IV. F. |
| *Siderone marthesia* (Cramer, 1777) | 2♂ – I. F. |
| *Zaretis itys itys* (Cramer, 1777) | 9♂, 5♀ – I. 2♂ – II. 5♂, 5♀ – III. F. |

**LIBYTHEIDAE**

| | |
|---|---|
| *Libytheana carinenta carinenta* (Cramer, 1777) | 3♂ – I. E. |

**LYCAENIDAE**

**RIODININAE**

EUSELASIINI

| | |
|---|---|
| *Euselasia arbas arbas* (Stoll, 1782) | 1♂ – I. F. |
| *E. cafusa cafusa* (Bates, 1868) | 4♂ – II. F. |
| *E. eubotes* (Hewitson, 1864) | 4♂ – I. 11♂, 1♀ – II. F. |
| *E. eugeon* (Hewitson, 1856) | 1♂ – II. 1♂ – III. F. |
| *E. gelanor gelanor* (Stoll, 1780) | 6♂ – I. 1♂ – II. F. |
| *E. hygenius eustola* Stichel, 1919 | 3♂ – I. 1♂ – II. F. |
| *E. lysias* (Cramer, 1777) | 1♂ – I. 2♂ – II. F. |
| *E. mys mys* (Hewitson, 1856) | 1♂, 2♀ – I. 1♂ – III. F. |
| *E.* sp. | 3♂, 1♀ – I. 2♂, 2♀ – II. F. |

EURYBIINI

| | |
|---|---|
| *Cremna actoris actoris* (Cramer, 1776) | 2♂ – I. 2♂, 1♀ – II. 1♂ – III. F. |
| *C. eucharila* Bates, 1867 | 1♀ – I. F. |
| *Eurybia halimede halimede* Huebner, 1819 | 1♂ – I. F. |
| *E. lamia lamia* (Cramer, 1777) | 2♂, 1♀ – I. 1♀ – II. 1♀ – III. F. |
| *E. nicaea nicaea* (Fabricius, 1775) | 2♂ – I. F. |
| *Mesophthalma idotea* Westwood, 1851 | 1♂ – I. F. |
| *Mesosemia anthaerice* Hewitson, 1859 | 1♂, 3♀ – I. 5♂, 4♀ – III. F. |
| *M. coea coea* Huebner, 1819 | 2♂, 9♀ – I. 1 ♂, 11♀ – II. 2♀ – III. F. |
| *M. euphyne* (Cramer, 1777) | 2♀ – I. F. |
| *M. ibycus* Hewitson, 1859 | 2♀ – I. 1♂ – II. F. |
| *M. philocles philocles* (Linnaeus, 1758) | 7♂, 2♀ – I. 1♂, 1♀ – II. 1♀ – III. F. |
| *M. steli* Hewitson, 1857 | 29♂, 18♀ – I. 4♂, 3♀ – II. 1♂, 7♀ – III. F. |
| *Perophthalma tullia tullia* (Fabricius, 1782) | 1♀ – I. 1♀ – II. F. |

RIODININI

| | |
|---|---|
| *Adelotypa aristus cretata* (Stichel, 1911) | 2♀ – I. 1♀ – II. 1♂ – III. F. |
| *Anteros formosus formosus* (Cramer, 1777) | 1♀ – II. F. |
| *Audre epulus epulus* (Cramer, 1775) | 1♂ – III. S. |
| *A.* sp. 1 | 3♂, 2♀ – I. 2♂, 2♀ – III. E. |
| *A.* sp. 2 | 2♀ – I. 1♂ – II. 1♂ – III. S. |

|  | Date/locality |
|---|---|
| *Baeotis zonata zonata* Felder & Felder, 1869 | 3♀ – II. 2♂, 2♀ – III. E. |
| *Calephelis* sp. | 12♂, 7♀ – I. 3♂, 2♀ – III. 2♀ – IV. E. |
| *Calospila emylius* (Cramer, 1775) | 1♂ – I. 3♂, 1♀ – II. 1♀ – III. F. |
| *C. lucianas* (Fabricius, 1793) | 2♂ – I. 1♂, 1♀ – II. F. |
| *Calydna candace* Hewitson, 1858 | 1♀ – II. 1♀ – III. F. |
| *Caria castalia castalia* (Ménétries, 1857) | 1♂ – I. F. |
| *Chamaelimnas villagomes villagomes* Hewitson, 1870? | 1♂, 1♀ – I. 1♂ – II. F. |
| *Emesis cerea* (Linnaeus, 1767) | 1♂ – I. F. |
| *E. lucinda lucinda* (Cramer, 1775) | 4♂ – I. 1♂ – II. F. |
| *E. mandana* (Cramer, 1780) | 1♂ – I. E. |
| *E. spreta* Bates, 1868 | 2♂ – I. 1♂ – II. F. |
| *Isapis agyrtus agyrtus* (Cramer, 1777) | 1♂ – II. 1♀ – III. F. |
| *Juditha azan lamiola* Callaghan, 1986 | 1♀ – III. F. |
| *J.* sp. | 1♂ – I. 3♂ – II. 10♂ – III. 1♀ – IV. F. |
| *Lasaia meris* (Cramer, 1781) | 2♂ – I. 1♂ – II. F. |
| *Mesene phareus phareus* (Cramer, 1777) | 3♂ – I. 5♂, 6♀ – II. 1♂ – III. F. |
| *M.* sp. | 2♂ – I. 1♂ – II. F. |
| *Metacharis lucius* (Fabricius, 1793) | 1♂ – I. 1♀ – III. F. |
| *M. regalis regalis* Butler, 1867 | 2♂, 2♀ – I. 2♂, 1♀ – II. F. |
| *Notheme erota erota* (Cramer, 1780) | 1♂ – II. 1♀ – III. F. |
| *Nymphidium aurum* Callaghan, 1985 | 1♂ – III. F. |
| *N. azanoides amazonensis* Callaghan, 1986 | 1♂ – I. 2♂ – II. 2♂, 2♀ – III. F. |
| *N. cachrus* (Fabricius, 1787) | 1♂, 2♀ – I. 5♂ – II. F. |
| *N. caricae caricae* (Linnaeus, 1758) | 1♂, 1♀ – I. 2♂ – II. 2♂, 2♀ – III. F. |
| *N. fulminans fulminans* Bates, 1868 | 1♂ – II. F. |
| *N. mantus* (Cramer, 1775) | 4♂ – II. 5♂ – III. F. |
| *Panara phereclus phereclus* (Linnaeus, 1758) | 3♂, 1♀ – I. F. |
| *Phaenochitonia pyrsodes* (Bates, 1868) | 2♂ – I. 2♂ – II. 1♂ – III. F. |
| *Riodina lysippus lysippus* (Linnaeus, 1758) | 3♂, 3♀ – I. 2♂, 1♀ – II. F. |
| *Symmachia batesi* (Staudinger, 1888) | 1♂ – II. F. |
| *Synargis calyce calyce* (Felder, 1862) | 2♀ – I. 1♀ – II. 1♀ – III. 2♀ – IV. F. |
| *S. gela gela* (Hewitson, 1852) | 1♂ – II. F. |
| *S. tytia tytia* (Cramer, 1777) | 1♀ – II. F. |
| *Themone carveri* (Weeks, 1906) | 1♂ – III. F. |
| *Theope lycaenina* Bates, 1868 | 1♂ – II. 1♂ – III. F. |
| *Thisbe irenea irenea* (Cramer, 1780) | 1♂, 1♀ – I. F. |

STALACHTINI

| | |
|---|---|
| *Stalachtis calliope calliope* (Linnaeus, 1758) | 2♂, 1♀ – I. 3♂, 2♀ – II. 9♂, 7♀ – III. F. |

**LYCAENINAE**

POLYOMMATINI

| | |
|---|---|
| *Hemiargus hanno* (Stoll, 1790) | 4♂, 1♀ – II. 3♂ – IV. E. S. |
| *Leptotes cassius* (Cramer, 1775) | 2♂ – III. E. |

THECLINI

| | |
|---|---|
| '*Thecla*' *adela* Staudingen, 1888 | 2♂ – III. F. |
| '*T.*' *celmus* (Cramer, 1775) | 1♂, 1♀ – III. F. |

|  | Date/locality |
|---|---|
| 'T.' cylarissus (Herbst, 1800) | 1♂ – II. 1♂ – III. F. |
| 'T.' falerina Hewitson, 1867 | 1♂ – I. 1♂ – II. F. |
| 'T.' mecrida Hewitson, 1867 | 1♂ – II. 1♂ – III. |
| 'T.' munditia Druce, 1907 | 1♂, 1♀ – I. E. |
| 'T.' philinna Hewitson, 1868 | 1♂ – II. 1♂ – III. F. |
| 'T.' phoster Druce, 1907 | 1♂ – II. 1♂ – III. F. |
| 'T.' strephon (Fabricius, 1775) | 1♂ – III. F. |
| 'T.' sydera Hewitson, 1867 | 1♂ – III. F. |
| 'T.' tyriam Druce, 1907 | 1♂ – III. F. |
| 'T.' sp. 1 | 1♀ – I. |
| 'T.' sp. 2 | 1♂ – I. 1♀ – III. |
| 'T.' sp. 3 | 1♀ – IV. E. |
| Arawacus aetolus (Sulzer, 1776) | 3♂ – I. 1♂, 1♀ – II. 1♂ – III. F. |
| A. dolylas (Cramer, 1777) | 1♂ – II. F. 1♂ – III. E. |
| Atlides bacis (Godman and Salvin, 1887) | 1♂ – III. F. |
| Calycopis atrius (Herrich-Schaeffer, 1853) | 1♂ – I. |
| C. calus (Godart, 1824) | 1♂ – I. F. |
| C. chacona (Jörgensen, 1932) | 3♂, 2♀ – I. 10♂, 1♀ – II. 8♂ – III. 1♂ – IV. F. |
| Evenus atrox (Butler, 1877) | 3♂ – I. 1♂ – II. |
| E. gabriela (Cramer, 1775) | 1♂, 1♀ – I. F. |
| E. satyroides (Hewitson, 1865) | 2♂ – III. F. |
| 'T.' tagyra (Hewitson, 1865) | 1♀ – III. F. |
| Jaspis verania Hewitson, 1868 | 1♂ – II. 1♂ – III. F. |
| Ministrymon azia (Hewitson, 1873) | 1♀ – III. E. |
| M. una (Hewitson, 1873) | 1♂, 2♀ – I. 1♂ – III. E. |
| Olynthus essus (Herrich-Schaeffer, 1853) | 2♂ – III. F. |
| O. punctum (Herrich-Schaeffer, 1858) | 1♂ – III. F. |
| Panthiades bitias (Cramer, 1777) | 1♂ – III. F. |
| P. phaleros (Linnaeus, 1767) | 2♂ – III. F. |
| Pseudolycaena marsyas marsyas (Linnaeus, 1758) | 1♀ – I. F. |
| Rekoa marius (Lucas, 1857) | 1♂, 1♀ – I. 1♀ – II. E. |
| R. palegon (Cramer, 1780) | 1♀ – I. E. |
| Siderus leucophaeus (Huebner, 1818) | 2♂ – I. 3♂ – II. 5♂ – III. F. |
| Strymon albata (Felder and Felder, 1865) | 1♂ – III. E. |
| S. bazochii (Godart, 1822) | 1♀ – II. S. |
| S. bubastus (Stoll, 1780) | 1♀ – I. 1♂ – III. E. |
| S. mulucha (Hewitson, 1867) | 1♀ – III. E. |
| S. ziba (Hewitson, 1868) | 1♂ – I. F. |
| Theclopsis eryx (Cramer, 1777) | 1♂ – I. F. |
| Theritas mavors (Huebner, 1818) | 1♀ – I. F. |
| Tmolus echion (Linnaeus, 1767) | 1♂ – III. |
| T. mutina Hewitson, 1867 | 1♂ – I. 1♂ – III. |

## HESPERIIDAE

### PYRRHOPYGINAE

| | |
|---|---|
| Jemadia hewitsonii hewitsonii (Mabille, 1878) | 2♂ – I. E. |
| Mysoria barcastus venezuelae (Scudder, 1872) | 2♂ – I. 2♂, 1♀ – II. 3♂ – III. E. |
| Myscelus assaricus assaricus (Cramer, 1779) | 1♂ – I. F. |
| Pyrrhopyge amyclas denticulata Herrich-Schaeffer, 1869 | 1♂ – I. 1♀ – II. 1♂ – III. E. |

|  | Date/locality |
|---|---|

## PYRGINAE

### URBANINI

| | |
|---|---|
| *Aguna albistria leucogramma* (Mabille, 1888) | 1♂, 1♀ – I. 4♂, 1♀ – II. 1♀ – III. E. |
| *A. aurunce* (Hewitson, 1867) | 1♀ – II. 1♂, 1♀ – III. F. |
| *A. coelus* (Stoll, 1781) | 1♀ – III. F. |
| *A. metophis* (Latreille, 1824) | 1♂ – I. 1♂ – III. E. |
| *Astraptes alector hopfferi* (Ploetz, 1882) | 1♂ – I. F. |
| *A. anaphus anaphus* (Cramer, 1777) | 1♂ – I. 1♂ – III. F. |
| *A. fulgerator fulgerator* (Walch, 1775) | 4♂, 3♀ – I. 1♂ – II. 1♂, 1♀ – III. 1♂, 1♀ – IV. F. |
| *A. janeira* (Schaus, 1902) | 1♂ – I. 2♀ – III. F. |
| *Augiades crinisus* (Cramer, 1780) | 1♀ – IV. F. |
| *Autochton longipennis* (Ploetz, 1882) | 1♀ – I. F. |
| *A. neis* (Geyer, 1832) | 1♂ – I. 2♂, 1♀ – II. 1♀ – III. F. |
| *A. zarex* (Huebner, 1818) | 2♂, 2♀ – I. F. |
| *Cephise procerus* (Ploetz, 1880) | 1♂ – I. F. |
| *Chioides catillus catillus* (Cramer, 1780) | 3♂, 3♀ – I. 3♂, 2♀ – II. 1♂ – III. E. S. |
| *Chrysoplectrum perniciosum perniciosum* (Herrich-Schaeffer, 1869) | 1♂ – I. 2♂ – VI. F. |
| *Drephalys alcmon* (Cramer, 1779) | 1♀ – I. F. |
| *D. oriander oriander* (Hewitson, 1867) | I. F. |
| *Entheus* sp. | 1♀ – I. F. |
| *Epargyreus socus sinus* Evans, 1952 | 2♂, 1♀ – I. F. |
| *E. spina spina* Evans, 1952 | 2♂, 1♀ – I. 2♀ – III. 3♂ – V. F. |
| *Hyalothyrus leucomelas* (Geyer, 1832) | 1♀ – I. 2♂ – II. 4♂, 6♀ – III. F. |
| *H. nitocris* (Stoll, 1782) | 1♀ – II. 1♂ – III. F. |
| *Narcosius colossus granadensis* (Moeschler, 1879) | 1♂, 1♀ – III. F. |
| *Nascus paulliniae* (Sepp, 1842) | 1♂ – III. F. |
| *Polythrix auginus* (Hewitson, 1867) | 1♂ – III. F. |
| *P. caunus* (Herrich-Schaeffer, 1869) | 1♂ – I. F. |
| *P. octomaculata octomaculata* (Sepp, 1844) | 1♀ – III. F. |
| *Typhedanus optica optica* Evans, 1952 | 2♂ – I. 1♂ – II. 1♀ – III. E. |
| *Udranomia kikkawai* (Weeks, 1906) | 1♂ – V. F. |
| *Urbanus acawoios* (Williams, 1926) | 1♂, 1♀ – III. 6♂, 1♀ – IV. F. |
| *U. albimargo takuta* Evans, 1952 | 1♀ – I. F. |
| *U. cindra* Evans, 1952 | 1♂, 1♀ – I. 2♂ – II. S. E. |
| *U. dorantes dorantes* (Stoll, 1790) | 2♂, 2♀ – I. 1♂ – II. F. |
| *U. doryssus doryssus* (Swainson, 1831) | 2♂, 1♀ – I. 1♂ – II. F. |
| *U. esma* Evans, 1952 | 1♂ – IV. F. |
| *U. esmeraldus* (Butler, 1877) | 1♀ – I. F. |
| *U. pronta* Evans, 1952 | 1♀ – I. F. |
| *U. proteus proteus* (Linnaeus, 1758) | 3♂ – I. 1♀ – II. F. |
| *U. simplicius* (Stoll, 1790) | 1♂, 1♀ – I. 2♂ – II. E. |
| *U. tanna* Evans, 1952 | 7♂, 4♀ – I. 3♂, 1♀ – II. 2♂, 1♀ – III. E. |

### PYRGINI

| | |
|---|---|
| *Achlyodes mithridates thraso* (Huebner, 1807) | 5♂, 1♀ – I. E. F. |
| *Anastrus obscurus narva* Evans, 1953 | 1♀ – I. 2♀ – II. F. |
| *A. tolimus robigus* (Ploetz, 1884) | 3♂, 1♀ – II. 1♂ – III. F. |
| *Anisochoria pedaliodina polysticta* Mabille, 1876 | 14♂ – I. 5♂, 2♀ – II. 1♂ – V. F. |

|  | Date/locality |
|---|---|
| *Antigonus erosus* (Huebner, 1812) | 11♂, 2♀ – I. 1♀ – II. 1♀ – III. 1♂ – IV. E. F. |
| *A. nearchus* (Latreille, 1817) | 10♂ – I. 1♂ – II. E. F. |
| *Camptopleura auxo* (Moeschler, 1879) | 1♂ – IV. F. |
| *Chiomara asychis asychis* (Stoll, 1780) | 6♂ – I. 2♂ – II. E. |
| *C. basigutta* (Ploetz, 1884) | 2♂, 4♀ – I. S. E. |
| *C. mithrax* (Moeschler, 1879) | 2♂ – I. E. |
| *Clito littera anda* Evans, 1953 | 1♂ – V. |
| *Cogia calchas* (Herrich-Schaeffer, 1869) | 6♂ – I. 5♂, 1♀ – II. E. S. |
| *Cycloglypha thrasibulus thrasibulus* (Fabricius, 1793) | 1♂, 1♀ – I. 2♂ – II. 1♂ – III. F. |
| *Cyclosemia herennius herennius* (Stoll, 1782) | 2♂ – I. 3♂ – II. F. |
| *Ebrietas anacreon anacreon* (Staudinger, 1876) | 1♂ – I. F. |
| *Gorgythion begga pyralina* (Moeschler, 1877) | 5♂, 2♀ – I. 4♂ – II. F. |
| *Helias phalaenoides phalaenoides* Fabricius, 1807 | 1♀ – III. F. |
| *Heliopetes alana* (Reakirt, 1868) | 2♂ – I. 2♂ – II. E. |
| *H. arsalte arsalte* (Linnaeus, 1758) | 3♀ – I. 3♂, 1♀ – II. E.S. |
| *H. macaira nivella* (Mabille, 1883) | 3♂ – I. 1♀ – II. 1♂ – III. E. S. |
| *H. macaira orbigera* (Mabille, 1888) | 4♂ – I. 2♂ – II. E. S. |
| *Milanion hemes hemes* (Cramer, 1777) | 1♂ – II. F. |
| *M. hemestinus* Mabille & Boullet, 1917 (*sensu* Evans, 1953) | 2♂ – II. 2♀ – III. F. |
| *Mylon jason* (Ehrmann, 1907) | 6♂, 1♀ – II. 1♂ – III. E. |
| *M. pelopidas* (Fabricius, 1793) | 1♂ – IV. E. |
| *Nisoniades macarius* (Herrich-Schaeffer, 1870) | 1♂ – I. F. |
| *N. rubescens* (Moeschler, 1877) | 1♂ – I. F. |
| *Onenses kelso* Evans, 1953 | 1♀ – I. F. |
| *Ouleus fridericus fridericus* (Geyer, 1832) | 1♂ – I. F. |
| *Paches loxus loxana* Evans, 1953 | 1♂ – III. F. |
| *Pellicia costimacula costimacula* Herrich-Schaeffer, 1870 | 1♂ – I. E. |
| *P. theon* ssp. | 1♂ – I. E. |
| *Pyrgus oileus orcus* (Stoll, 1780) | 2♂, 2♀ – I. 3♀ – II. 2♂ – IV. E. S. |
| *Pythonides herennius herennius* Geyer, 1838 | 2♂, 1♀ – I. 1♂ – II. F. |
| *P. jovianus jovianus* (Stoll, 1782) | 3♀ – I. F. |
| *P. limaea limaea* (Hewitson, 1868) | 1♂ – I. F. |
| *P. maraca* Mielke & Casagrande, 1991 | 1♀ – I. F. |
| *Quadrus cerialis* (Stoll, 1782) | 3♀ – I. F. |
| *Sostrata bifasciata adamas* (Ploetz, 1884) | 1♂ – I. 1♂ – II. 1♂, 1♀ – III. F. |
| *Spathilepia clonius* (Cramer, 1775) | 1♀ – I. 1♂ – II. F. |
| *Spioniades artemides* (Stoll, 1782) | 1♂, 1♀ – I. F. |
| *Staphylus vulgata sinepunctis* Kaye, 1904 | 1♂ – II. F. |
| *Timochares trifasciata trifasciata* (Hewitson, 1868) | 1♂ – IV. E. |
| *Timochreon satyrus forta* Evans, 1953 | 1♂ – I. 1♂ – III. E. |
| *Viola violella* (Mabille, 1897) | 13♂, 3♀ – I. 7♂, 1♀ – II. 7♂, 5♀ – III. 12♂, 5♀ – IV. 1♀ – V. E. S. |
| *Xenophanes tryxus* (Stoll, 1780) | 4♂ – I. 3♂ – II. 1♂ – III. F. E. |

**HESPERIINAE**

| | |
|---|---|
| *Aides aegita* (Hewitson, 1866) | 1♂ – III. 1♂ – VI. F. |
| *Anthoptus epictetus* (Fabricius, 1793) | 1♂ – I. F. |
| *A. insignis* (Ploetz, 1882) | 1♂, 3♀ – I. F. |

# APPENDIX 6

|  | Date/locality |
|---|---|
| *Arita arita* (Schaus, 1902) | 1♂ – V. F. |
| *Artines aepitus* (Geyer, 1832) | 1♀ – II. F. |
| *A. focus* Evans, 1955 | 2♂ – VII. F. |
| *Callimormus alsimo* (Moeschler, 1882) | 1♀ – II. 1♂ – III. F. |
| *C. saturnus* (Herrich-Schaeffer, 1869) | 1♂ – I. 1♂ – II. F. E. |
| *Calpodes ethlius* (Stoll, 1782) | 1♂ – III. 1♂, 1♀ – IV. E. |
| *Carystoides maroma* (Moeschler, 1877) | 2♂ – II. F. |
| *Cobalus virbius virbius* (Cramer, 1777) | 2♂ – I. 1♂ – II. F. |
| *Copaeodes jean jean* Evans, 1955 | 2♂ – I. 2♂ – II. 1♀ – III. S. |
| *Corticea corticea* (Ploetz, 1883) | 1♀ – I. 1♂, 2♀ – II. F. |
| *C. oblinita* (Mabille, 1891) | 2♀ – II. 1♂ – III. S. |
| *C. sp.* | 1♀ – I. S. |
| *Cymaenes odilia edata* (Ploetz, 1883) | 1♂ – II. F. |
| *Cynea bistrigula* (Herrich-Schaeffer, 1869) | 1♂ – I. 1♂ – II. F. |
| *C. cynea* (Hewitson, 1876) | 1♂ – I. F. |
| *C. diluta* (Herrich-Schaeffer, 1869) | 1♂ – I. F. |
| *C. irma* (Moeschler, 1879) | 1♂ – III. F. |
| *C. megalops* (Godman, 1900) | 2♂ – I. F. |
| *C. popla* Evans, 1955 | 1♂ – I. 1♂ – II. 1♂ – III. F. |
| *C. robba robba* Evans, 1955 | 1♂ – II. F. |
| *Damas clavus* (Herrich-Schaeffer, 1869) | 1♂ – II. 1♂ – III. F. |
| *Enosis blotta* Evans, 1955 | 1♂ – V. F. |
| *E. iccius* Evans, 1955 | 1♀ – II. F. |
| *E. pruinosa agassus* (Mabille, 1891) | 1♂, 1♀ – III. F. |
| *Euphyes kayei* (Bell, 1931) | 1♀ – I. F. |
| *E. peneia* (Godman, 1900) | 1♀ – I. E. |
| *Eutocus mathildae vinda* Evans, 1955 | 1♂ – II. F. |
| *Eutychide subcordata subcordata* (Herrich-Schaeffer, 1869) | 1♂ – I. F. |
| *Flaccilla aecas* (Stoll, 1781) | 1♀ – I. 1♂ – III. F. |
| *Hylephila phyleus phyleus* (Drury, 1773) | 1♀ – I. 3♂, 3♀ – III. 1♀ – V. E. |
| *Joanna boxi* Evans, 1955 | 2♂ – II. 2♂ – III. F. |
| *Justinia justinianus dappa* Evans, 1955 | 1♀ – V. F. |
| *J. maculata* (Bell, 1930) | 1♂ – I. 1♂ – II. F. |
| *Lerema ancillaris ancillaris* (Butler, 1877) | 2♂ – I. 1♂ – II. 1♂ – III. E. |
| *Lerodea eufala eufala* (Edwards, 1869) | 2♂, 1♀ – I. 3♀ – II. 1♂ – III. 1♂ – V. E. S. |
| *Megistias ignarus* Bell, 1931 |  |
| *Mellana clavus* (Erichson, 1848) | 1♂ – I. E. |
| *M. eulogius* (Ploetz, 1883) | 1♀ – II. E. |
| *Methionopsis ina* (Ploetz, 1882) | 3♂, 1♀ – II. 1♂ – III. F. |
| *Metron fasciata* (Moeschler, 1876) | 1♂ – I. 1♂ – II. F. |
| *Mnaseas bicolor inca* Bell, 1930 | 3♂ – I. 1♂ – V. S. |
| *Mnasilus allubita* (Butler, 1877) | 1♂ – I. 1♀ – II. F. |
| *Mnasitheus cephoides* Hayward, 1943 | 3♂ – I. F. |
| *Morys compta compta* (Butler, 1877) | 3♂ – I. 1♀ – V. F. |
| *M. geisa geisa* (Moeschler, 1879) | 3♂, 2♀ – I. 2♂, 3♀ – II. 2♂ – III. F. |
| *M. valerius valerius* (Moeschler, 1879) | 1♀ – II. F. |
| *Nastra chao* (Mabille, 1897) | 6♂ – I. 2♂ – II. 1♂, 1♀ – III. 1♀ – IV. S. |
| *Niconiades xanthaphes* Huebner, 1821 | 1♀ – I. F. |
| *Nyctelius nyctelius nyctelius* (Latreille, 1824) | 2♂ – III. E. |

|  | Date/locality |
|---|---|
| *Orses cynisca* (Swainson, 1821) | 1♂, 1♀ – I. F. |
| *Panoquina fusina fusina* (Hewitson, 1868) | 1♂ – II. E. |
| *P. hecebolus* (Scudder, 1872) | 2♂, 1♀ – I. 1♂, 1♀ – II. 1♂ – III. F. E. |
| *P. ocola* (Edwards, 1863) | 4♂, 3♀ – I. 2♂ – II. 4♂ – III. 1♂ – IV. F. E. |
| *P. sylvicola* (Herrich-Schaeffer, 1865) | 1♂ – I. 6♂, 7♀ – IV. F. E. |
| *P. trix* Evans, 1955 | 3♂ – II. F. |
| *Papias subcostulata subcostulata* (Herrich-Schaeffer, 1870) | 1♂ – I. F. |
| *P. ignarus* (Bell, 1931) | 1♂ – I. F. |
| *P. phainis* Godman, 1900 | 3♂, 2♀ – II. 4♂, 1♀ – III. F. |
| *Parphorus decora* (Herrich-Schaeffer, 1869) | 5♂ – I. 3♂ – II. 1♀ – III. F. |
| *P. prosper* Evans, 1955 | 4♂, 2♀ – I. 1♀ – II. F. |
| *P. storax storax* (Mabille, 1891) | 14♂, 5♀ – I. 3♂, 2♀ – II. 4♀ – III. F. |
| *Peba striata* Mielke, 1968 | 2♂, 1♀ – II. 1♀ – III. F. |
| *Penicula reducta* (Bell, 1930) | 1♂, 1♀ – I. F. |
| *Perichares philetes philetes* (Gmelin, 1790) | 1♀ – II. F. |
| *Phanes aletes* (Geyer, 1832) | 1♀ – I. F. |
| *P. almoda* (Hewitson, 1866) | 1♂ – VII. F. |
| *Phlebodes campo sifax* Evans, 1955 | 1♂, 1♀ – I. F. |
| *P. pertinax* (Stoll, 1781) | 1♂ – III. F. |
| *Polites vibex catilina* (Ploetz, 1886) | 4♂, 2♀ – I. 1♂ – II. 1♀ – III. E. S. |
| *P. vibicoides* Jong, 1983 | 2♂, 3♀ – I. 3♂ – II. S. |
| *Pompeius amblyspila* (Mabille, 1897) | 1♂, 2♀ – I. 5♂, 2♀ – II. 2♂ – III. S. |
| *P. dares* (Ploetz, 1883) | 1♀ – I. S. |
| *Remella remus* (Fabricius, 1798) | 1♂ – I. F. |
| *Saliana nigel* Evans, 1955 | 1♂ – I. F. |
| *S. saladin culta* Evans, 1955 | 1♂ – I. 1♀ – II. F. |
| *S. salius* (Cramer, 1775) | 1♂ – II. 1♂ – III. F. |
| *Saturnus saturnus saturnus* (Fabricius, 1787) | 2♂, 1♀ – I. 4♂, 1♀ – II. 1♂ – III. F. |
| *Talides sinois sinois* Huebner, 1819 | 1♂ – I. F. |
| *Thoon canta* Evans, 1955 | 1♀ – II. F. |
| *Vehilius almoneus almoneus* Schaus, 1902 | 2♂ – I. S. |
| *V. inca* (Scudder, 1872) | 1♀ – V. |
| *V. stictomenes stictomenes* (Butler, 1877) | 4♂ – I. F. |
| *V. vetulus* (Mabille, 1978) | 3♂ – I. F. |
| *Venas evans* (Butler, 1877) | 1♀ – I. F. |
| *Vettius artona* (Hewitson, 1868) | 1♂ – III. F. |
| *V. lafrenaye pica* (Herrich-Schaeffer, 1869) | 1♂ – III. F. |
| *V. marcus marcus* (Fabricius, 1787) | 1♂ – I. F. |
| *Vidius anna* (Mabille, 1897) | 2♀ – I. 1♀ – III. 1♀ – V. S |
| *Virga virginius* (Moeschler, 1882) | 1♂ – I. 4♂, 1♀ – II. 3♂ – III. E. |
| *Wallengrenia otho curassavica* (Snellen, 1887) | 1♂, 1♀ – II. E. |
| *W. premnas* (Wallengren, 1860) | 2♂ – I. 4♂, 3♀ – II. E. S. |
| Gen. and sp. indet. | 1♀ – I. F. |

# Appendix 7 (Chapter 18)
# Preliminary list of spiders collected on the Ilha de Maracá*

Order ARANEAE
  Suborder MYGALOMORPHAE
    **Actinopodidae**
      1 morphotype
    **Ctenizidae**
      *Celaetychius* sp.
    **Dipluridae**
      1 morphotype
    **Theraphosidae**
      1 morphotype
  Suborder ARANEOMORPHAE
    **Amaurobiidae**
      *Retiro* sp.
    **Anyphaenidae**
      *Anyphaena* sp.
      *Hibana melloleitoi*
      *Patrera* sp.
      *Teudis* sp.
      *Wulfila modesta*
      *Wulfila* spp. (3)
      Gen. indet., 3 morphotypes
    **Aphantochilidae**
      *Aphantochilus inermipes*
      *Aphantochilus rogersi*
      *Bucranium taurifrons*
    **Araneidae**
      *Aculepeira travassosi*
      *Alpaida carminea*
      *Alpaida truncata*
      *Argiope argentata*
      *Amazonepeira masaka*
      *Araneus guttatus*
      *Araneus venatrix*
      *Cyclosa* spp. (3)
      *Enacrosoma anomalum*
      *Eriophora* sp.
      *Eustala fuscovittata*
      *Eustala* spp. (8)
      *Hypognatha deplanata*
      *Hypognatha scutata*
      *Madrepeira amazonica*
      *Mangora* spp. (3)
      *Mecynogea* sp.
      *Metazygia* sp.
      *Micrepeira hoeferi*
      *Micrepeira pachitea*
      *Parawixia hypocrita*
      *Parawixia kochi*
      *Parawixia tomba*
      *Spilasma duodecimguttata*
      *Taczanowskia* sp.
      *Verrucosa* spp. (2)
      *Wagneriana neglecta*
      Gen. indet., 13 morphotypes
    **Caponiidae**
      1 morphotype
    **Clubionidae**
      *Castianeira* sp.
      *Clubionoides valvula*
    **Corinnidae**
      *Apochinoma* sp.
      *Corinna* sp.
      *Myrmecium* sp.
      *Myrmecotipus olympus*
      *Myrmecotipus* spp. (2)
      *Trachela* spp. (2)
      *Trachelopachis* sp.
      Gen. indet., 2 morphotypes
    **Ctenidae**
      *Ctenus* sp.
      *Notroctenus marshii*
    **Deinopidae**
      *Deinopis* sp.
    **Dictynidae**
      1 morphotype
    **Gnaphosidae**
      *Camillina* sp.
      *Zimiromus* sp.
    **Hahniidae**
      *Hahnia* sp.
    **Heteropodidae**
      8 morphotypes
    **Linyphiidae**
      1 morphotype
    **Mimetidae**
      *Ero* sp. (2)
      *Gelanor heraldicus*
      *Gelanor* spp. (2)
      *Mimetus* sp.
    **Mysmenidae**
      *Mysmenopsis ischnamigo*
    **Miturgidae**
      *Eutichurus* sp.
    **Ochyroceratidae**
      1 morphotype
    **Oonopidae**
      *Oonops* sp.
      *Hytanes oblonga*

**Oxyopidae**
   *Hamataliwa* sp.
   *Oxyopes salticus*
   *Peucetia* sp.
   *Schaenicoscelis* sp.
   *Papinillus longipes*
**Palpimanidae**
   *Otiothops oblongus*
   *Otiothops hoeferi*
**Philodromidae**
   *Gephyrellula* sp.
**Pholcidae**
   1 morphotype
**Pisauridae**
   *Architis nitidopilosa*
   *Ancylometes* sp.
   Pisaurina sp.
   Gen. indet., 11 morphotypes
**Salticidae**
   *Acragas miniaceus*
   *Acragas* sp.
   *Asaracus fimbriatus*
   *Breda* spp (2)
   *Chloridusa* sp.
   *Cotinusa dimidiata*
   *Dryphias maccuni*
   *Encolpius* sp.
   *Freya perelegans*
   *Hypaeus flavipes*
   *Hypaeus miles*
   *Lyssomanes unicolor*
   *Lyssomanes bitaeniatus*
   *Mago acutidens*
   *Noegus fuscimanus*
   *Noegus pallidus*
   *Noegus* spp. (2)
   Phiale crocea
   *Sarinda silvatica*
   *Scopocira* sp.
   *Synemosina* spp. (2)
  Sect. **Unidentatae**
   20 morphotypes
  Sect. **Fisidentatae**
   3 morphotypes
  Sect. **Pluridentatae**
   10 morphotypes
**Scytodidae**
   *Scytodes* sp.
**Senoculidae**
   *Senoculus (Labdacus)* sp.
   *Senoculus (Stenoctenus)* sp.
**Synotaxidae**
   *Synotaxus monocerus*
**Tetragnathidae**
   *Leucauge* spp. (5)
   *Nephila clavipes*
**Theridiidae**
   *Achaearanea dalana*
   *Achaearanea inops*
   *Achaearanea maraca*
   *Achaearanea pydanieli*
   *Achaearanea taenia*
   *Achaearanea nigrovittata*
   *Achaearanea trapezoidalis*
   *Anelosimus chikeringi*
   *Anelosimus eximius*
   *Argyrodes altus*
   *Argyrodes attenuatus*
   *Argyrodes caudatus*
   *Argyrodes dracus*
   *Argyrodes metaltissimus*
   *Chryssso* spp. (2)
   *Dipoena alta*
   *Dipoena atlantica*
   *Dipoena cordiformis*
   *Dipoena hortoni (cf.)*
   *Dipoena inca*
   *Dipoena kuyuwini*
   *Dipoena tingo*
   *Dipoena tiro*
   *Dipoena woytkowskii*
   *Dipoena* spp. (2)
   *Echinotheridion utibili*
   *Episinus bruneoviridis*
   *Episinus crysus*
   *Episinus erythrophtalmus*
   *Euryopis taczanowskii*
   *Helvibis* sp.
   *Latrodectus geometricus*
   *Phoroncidia moyobamba*
   *Tekellina bella*
   *Theridion artum*
   *Theridion crispulum*
   *Thymoites cravilus*
   *Thymoites* sp.
   *Thwaitesia bracteata*
   *Tidarren* sp.
   Gen indet., 1 morphotype
**Theridiosomatidae**
   *Cthonos* sp.
   *Wendilgarda* sp.
**Thomisidae**
Subfam. **Strophiinae**
   *Strophius nigricans*
Subfam. **Stephanopinae**
   *Epicadus heterogaster*
   *Epicadus* sp.
   *Onocolus trifolius*
   *Onocolus* sp.
   *Tobias* spp. (2)

Subfam. **Thomisinae**
   *Acentroscelus guyanensis*
   *Misumenoides magnus*
   *Misumenoides* sp.
   *Misumenops pallens*
   *Misumenops silvarum*
   *Synema* sp.
   *Titidius quinquenotatus*
   *Tmarus* spp. (3)
**Uloboridae**
   *Miagrammopes* spp. (2)
   *Philoponella* spp. (2)
   *Uloborus penicellatus*
   *Uloborus* spp. (4)
**Zodariidae**
   *Leprolochus spinifrons*

---

\* Including material collected by Erica Helema Buckup and Alexandre Bragio Bonaldo (Museu de Ciências – Fundação Zoobotânica do Rio Grande do Sul). Material identified by Arno A. Lise, Erica H. Buckup, Alexandre B. Bonaldo, Antonio D. Brescovit, Maria Aparecida de L. Marques and Augusto Braul.

# Appendix 8 (Chapter 20)
# Species of rotifers collected on the Ilha de Maracá

Relative abundance of each species in the four samples (SP = seasonal pond sample; Sav 1–3 = seasonally flooded savanna samples) is represented as follows: vr = very rare (1–5); r = rare (6–10); c = common (11–20); ab = abundant (>20). Notes on habits and distributions are abbreviated as follows: li = littoral; pl = planktonic; s = sessile; trop = tropical; end = endemic of South America; cosm = cosmopolitan; indet. = undetermined.

| Species | SP | Sav 1 | Sav 2 | Sav 3 | Notes |
|---|---|---|---|---|---|
| MONOGONONTA | | | | | |
| *Asplanchna girodi* | – | – | – | r | pl, cosm |
| *Beauchampia crucigera* | r | r | – | – | li, s, cosm |
| *Brachionus calyciflorus* (f. *amphiceros*) | vr | – | – | – | pl, cosm |
| *B. falcatus falcatus* | – | r | – | – | pl, trop |
| *B. patulus macracanthus* | r | – | – | r | li, trop |
| *B. patulus patulus* | c | c | – | c | li, cosm |
| *B. quadridentatus melhemi* | – | r | – | c | pl, trop |
| *Cephalodella forficula* | – | vr | vr | – | li, cosm |
| *C. gibba* | r | r | vr | vr | li, cosm |
| *C. giganthea* | vr | vr | – | – | li, cosm |
| *C. mucronata* | r | vr | – | r | li, trop |
| *C. nana* | – | vr | – | – | li, cosm |
| *C. panarista* | r | r | r | – | li, trop |
| *C. tenuiseta* | – | vr | – | – | li, cosm |
| *C.* sp. indet. | r | r | r | r | li, ? |
| *Collotheca campanulata* | vr | vr | – | – | li, cosm, s |
| *C. edentata* | – | vr | – | – | li, cosm, s |
| *C. ornata* | vr | r | – | – | li, cosm, s |
| *C. tenuilobata* | c | c | r | r | li, trop, s |
| *C.* sp. indet. | r | r | r | r | li, ?, s |
| *Colurella uncinata* | vr | vr | r | r | li, cosm |
| *Conochilus unicornis* | – | r | – | – | pl, cosm |
| *Cupelopagis vorax* | vr | – | – | – | li, trop, s |
| *Dicranophorus caudatus* | r | r | r | – | li, cosm |
| *D. forcipatus* | r | r | r | r | li, cosm |
| *D. luetkeni* | vr | – | – | – | li, trop, s |
| *D. sebastus* | c | – | – | – | li, ? |
| *Enteroplea lacustris* | vr | – | vr | – | li, cosm |
| *Euchlanis incisa* | vr | r | vr | vr | li, cosm |
| *E. meneta* | – | – | vr | – | li, cosm |
| *Filinia longiseta* | – | – | ab | – | pl, cosm |
| *Floscularia conifera* | vr | r | r | r | li, cosm, s |
| *F. decora* | vr | vr | vr | vr | li, end, s |
| *F. janus* | – | r | – | vr | li, cosm, s |
| *F. melicerta* | c | – | – | – | li, cosm, s |
| *F. ringens* | ab | c | c | c | li, cosm, s |
| *Itura aurita* | r | – | – | – | li, cosm |
| *Keratella americana* | r | – | – | – | pl, end, trop |
| *K. lenzi* | r | – | – | ab | pl, trop |

| Species | SP | Sav 1 | Sav 2 | Sav 3 | Notes |
|---|---|---|---|---|---|
| K. procurva | – | – | c | – | pl, trop |
| Lacinularia flosculosa | – | c | – | – | pl, cosm |
| Lecane acus | – | – | r | r | li, cosm |
| L. amazonica | – | r | vr | vr | li, end, trop |
| L. arcuata | – | – | r | – | li, cosm |
| L. astia | r | – | – | – | li, end |
| L. bulla | ab | ab | c | c | li, cosm |
| L. clara | ab | vr | vr | ab | li, cosm |
| L. closterocerca | – | – | r | – | li, cosm |
| L. cornuta | c | c | r | c | li, cosm |
| L. crenata | c | ab | ab | c | li, trop |
| L. crepida | – | r | r | r | li, trop |
| L. curvicornis | ab | ab | c | c | li, cosm, trop |
| L. curvicornis nitida | r | r | r | – | li, trop |
| L. doryssa | c | r | r | r | li, trop |
| L. elongata | vr | – | r | c | li, trop |
| L. elsa | vr | – | – | – | li, cosm |
| L. flexilis | – | – | vr | – | li, cosm |
| L. haliclysta | – | vr | vr | vr | li, trop |
| L. hamata | – | – | c | vr | li, cosm |
| L. hornemanni | vr | vr | vr | vr | li, cosm |
| L. kutikova | – | vr | – | – | li, trop, end |
| L. leontina | c | c | c | c | li, trop |
| L. ludwigi ercodes | – | – | vr | – | li, trop |
| L. ludwigi ludwigi | r | vr | c | c | li, trop |
| L. methoria | – | vr | – | – | li, trop |
| L. monostyla | r | r | vr | r | li, trop |
| L. murrayi | vr | – | – | – | li, trop |
| L. obtusa | vr | vr | vr | vr | li, cosm |
| L. ohioensis | – | – | vr | – | li, cosm |
| L. papuana | – | – | vr | vr | li, trop |
| L. pertica | r | vr | vr | – | li, trop |
| L. ploenensis | r | – | vr | – | li, trop |
| L. quadridentata | r | r | r | r | li, cosm |
| L. rhytida | – | – | vr | – | li, trop |
| L. scutata | – | – | – | vr | li, cosm |
| L. signifera | vr | r | r | r | li, trop |
| L. stichaea amazonica | – | – | vr | – | li, trop, end |
| L. styrax | – | – | vr | vr | li, trop |
| L. subulata | r | – | vr | vr | li, cosm |
| L. sympoda | – | vr | vr | – | li, trop |
| L. tenuiseta | – | – | – | vr | li, cosm |
| L. ungulata | r | vr | vr | vr | li, cosm |
| L. wulferti | r | r | r | r | li, trop, end |
| Lepadella benjamini | r | vr | r | vr | li, trop |
| L. christinei nov. spec. | – | – | r | vr | li, trop |
| L. cristata | c | c | c | c | li, trop |
| L. donneri | r | r | r | vr | li, trop |
| L. latusinus | r | – | c | – | li, trop |
| L. monodactyla | r | r | vr | – | li, trop |
| L. patella | – | – | r | – | li, cosm |
| L. quadricarinata | r | – | r | vr | li, cosm |

| Species | SP | Sav 1 | Sav 2 | Sav 3 | Notes |
|---|---|---|---|---|---|
| L. quinquecosta | vr | – | vr | vr | li, cosm |
| L. rhomboides | r | r | r | – | li, cosm |
| L. tricostata nov. spec. | – | – | – | r | li, end? |
| L. triptera | – | vr | vr | – | li, cosm |
| Limnias ceratophylli | r | – | r | – | li, cosm, s |
| L. melicerta | r | r | r | – | li, cosm, s |
| Lindia truncata | – | – | vr | – | li, cosm |
| Macrochaetus collinsi | r | ab | c | – | li, cosm |
| Manfredium eudactylum | r | – | vr | – | li, cosm |
| Monommata grandis | r | c | c | c | li, cosm |
| M. maculata | c | c | c | r | li, trop |
| M. sp. indet. | r | vr | vr | vr | li, trop? |
| Mytilina bisulcata | – | – | vr | – | li, cosm |
| M. macrocera | r | – | – | – | li, trop |
| M. trigona | vr | – | – | vr | li, cosm |
| M. ventralis macracantha | r | r | r | c | li, cosm |
| M. ventralis michelangellii | vr | – | – | vr | li, trop, end |
| M. ventralis ventralis | r | – | r | r | li, cosm |
| Notommata copeus | r | r | r | r | li, cosm |
| N. glyphura | r | r | r | – | li, cosm |
| N. pachyura | – | vr | – | – | li, cosm |
| N. tripus | r | – | – | – | li, cosm |
| N. sp. indet. | r | – | – | – | li |
| Octotrocha speciosa | vr | – | – | vr | li, trop, s |
| Platyias leloupi latiscapularis | vr | r | vr | c | li, trop, end |
| P. leloupi leloupi | r | r | vr | – | li, trop |
| Polyarthra sp. indet. | r | – | c | – | pl, cosm |
| Ptygura barbata | r | – | – | – | li, trop, s |
| P. linguata | c | r | r | r | li, trop, s |
| P. longicornis | r | vr | – | – | li, cosm, s |
| P. mucicola | r | – | – | – | li, cosm, s |
| P. tacita | r | r | vr | – | li, trop, s |
| Resticula melandocous | r | r | r | r | li, cosm |
| Scaridium longicaudum | vr | – | vr | vr | li, cosm |
| Sinantherina spinosa | r | – | – | c | pl, trop |
| Taphrocampa selenura | vr | – | vr | r | li, cosm |
| Testudinella ahlstromi | c | – | c | – | li, trop |
| T. mucronata | vr | vr | vr | vr | li, cosm |
| T. ohlei | c | – | – | – | li, trop, end |
| T. parva | vr | c | c | vr | li, cosm |
| T. patina | c | c | r | c | li, cosm |
| T. robertsoni nov. spec. | c | – | – | – | li, trop |
| Tetrasiphon hydrocora | c | – | vr | – | li, cosm |
| Trichocerca bidens | vr | – | – | vr | li, cosm |
| T. bicristata | vr | r | vr | r | li, cosm |
| T. brasiliensis | vr | vr | vr | vr | li, trop |
| T. collaris | c | – | c | vr | li, cosm |
| T. elongata | – | – | – | vr | li, cosm |
| T. flagellata | – | vr | r | vr | li, trop |
| T. iernis | vr | – | vr | – | li, cosm |
| T. insignis | r | vr | c | – | li, cosm |
| T. rosea | c | – | c | vr | li, cosm |

# APPENDIX 8

| Species | SP | Sav 1 | Sav 2 | Sav 3 | Notes |
|---|---|---|---|---|---|
| *T. scipio* | – | vr | vr | vr | li, cosm |
| *T. similis* | r | r | r | r | pl, cosm |
| *T. tenuior* | vr | – | vr | – | li, cosm |
| *T. tigris* | – | r | vr | vr | li, cosm |
| *Trichotria tetractis* | r | – | c | – | li, cosm |
| **BDELLOIDEA** | | | | | |
| *Dissotrocha aculeata* | r | – | – | – | li, cosm |
| *D. macrostyla* | – | – | – | r | li, cosm |
| *Habrotrocha angusticollis* | c | – | c | c | li, cosm |
| *H.* sp. indet. | r | – | – | – | li, cosm |
| *Macrotrachela* sp. indet. | r | – | vr | – | li, cosm |
| *Mniobia* sp. indet. | vr | – | – | – | li, cosm |
| *Philodina* sp. indet. | r | r | r | r | li, cosm |
| *Rotaria neptunia* | c | – | – | – | li, cosm |
| *R. rotatoria* | r | r | r | r | li, cosm |
| Bdelloidea sp. indet. | c | r | c | r | li, cosm |
| **Totals**: | | | | | |
| Monogononta | 106 | 82 | 102 | 79 | 136 li, 13 pl |
| Bdelloidea | 9 | 3 | 5 | 5 | 10 li, 0 pl |
| | 115 | 85 | 107 | 84 | 146 li, 13 pl |

# Appendix 9 (Chapter 22)
## Soil profile descriptions from representative vegetation types at the eastern end of the Ilha de Maracá

| Horizon | Profile A1 | | | Profile A2 | | | | Profile A3 | | |
|---|---|---|---|---|---|---|---|---|---|---|
| | $A_1$ | $B_1$ | $B_{2/3}$ | $A_1$ | $A_2$ | $B_1$ | $B_{2/3}$ | $A_{1/2}$ | $Bg$ | $IIB/Cg$ |
| Depth (cm) | 0–17 | 18–60 | 61–110+ | 0–17 | 18–32 | 33–105 | 106–130 | 0–19 | 20–61 | 62–125 |
| Moisture loss (%) | 0.12 | 0.08 | 1.80 | 0.02 | 0.06 | 0.02 | 0.06 | 0.10 | 0.10 | 0.02 |
| Loss on ignition (%) | 3.5 | 2.8 | 6.8 | 11.0 | 7.7 | 7.8 | 6.3 | 7.5 | 6.1 | 6.1 |
| Colour | 10 yr 4/2 | 2.5 yr 7/4 | 2.5 yr 4/8 | 7.5 yr 5/4 | 2.5 yr 4/8 | 2.5 yr 4/8 | 10 yr 7/3 | 10 yr 5/2 | 10 yr 5/1 | 10 yr 5/1–6/1 |
| Organic C% | 2.25 | 0.70 | 0.30 | 1.85 | 0.85 | 0.20 | 0.10 | 1.40 | 0.10 | 0.10 |
| Total N% | 0.29 | 0.18 | 0.19 | tr | 0.12 | 0.06 | 0.00 | 0.18 | 0.01 | tr |
| Ex. Ca (cmol/kg) | 3.52 | 0.21 | 0.01 | 2.48 | 0.04 | 0.01 | 0.02 | 4.96 | 1.68 | 0.16 |
| Ex. Mg (cmol/kg) | 0.63 | 0.16 | 0.70 | 3.28 | 1.97 | 0.82 | 2.54 | 3.21 | 6.41 | 0.41 |
| Ex. K (cmol/kg) | 0.18 | 0.07 | 0.15 | 0.93 | 0.56 | 0.20 | 0.19 | 0.33 | 0.10 | 0.01 |
| Ex. Na (cmol/kg) | 0.16 | 0.15 | 0.14 | 0.21 | 0.15 | 0.13 | 0.28 | 0.26 | 1.45 | 0.18 |
| Sum (cmol/kg) | 4.49 | 0.59 | 1.00 | 6.9 | 2.7 | 1.2 | 3.0 | 8.76 | 9.64 | 0.76 |
| Fe (%) | 0.54 | 1.05 | 0.01 | 1.20 | 1.50 | 1.40 | 1.50 | 0.07 | 0.58 | 0.02 |
| $Fe_2O_3$ (%) | 0.77 | 1.50 | 0.02 | 1.71 | 1.99 | 1.99 | 2.13 | 0.10 | 0.82 | 0.03 |
| Ex. H (cmol/kg) | 0.10 | 0.26 | 0.22 | 0.12 | 0.28 | 0.40 | 0.16 | 0.16 | 0.08 | 0.20 |
| Ex. Al (cmol/kg) | 0.02 | 0.34 | 0.48 | 0.04 | 0.44 | 0.76 | 0.08 | 0.04 | 0.02 | 0.22 |
| Ex. acidity (cmol/kg) | 0.12 | 0.60 | 0.70 | 0.16 | 0.72 | 1.16 | 0.24 | 0.20 | 0.10 | 0.42 |
| pH ($H_2O$) | 6.0 | 4.9 | 5.1 | 5.6 | 5.1 | 5.4 | 5.8 | 6.3 | 7.8 | 7.0 |
| pH ($CaCl_2$) | 5.4 | 4.3 | 4.4 | 5.1 | 4.5 | 4.5 | 5.1 | 5.8 | 6.7 | 6.9 |
| Base saturation (%) | 92 | 50 | 59 | 97 | 80 | 49 | 94 | 97 | 99 | 62 |
| Av. P (mg/100g) | 0.18 | 0.12 | 0.11 | 0.24 | 0.16 | 0.13 | 0.05 | 0.26 | 0.10 | 0.15 |
| Sand (%) | 93 | 92 | 63 | nd | nd | nd | nd | 51 | 42 | 27 |
| Silt + clay (%) | 7 | 8 | 37 | nd | nd | nd | nd | 49 | 58 | 73 |

|  | Profile B | | | | Profile C | | | | Profile D | | | |
|---|---|---|---|---|---|---|---|---|---|---|---|---|
| Horizon | $A_1$ | $B_1$ | $B_2$ | II | $A_1$ | $A_2$ | $B_{1g}$ | $B_{2g}$ | $A_1$ | $A_2$ | $B_{1g}$ | $B_{2g}$ |
| Depth (cm) | 0–20 | 21–60 | 61–107 | 108–120 | 0–16 | 17–30 | 31–93 | 94–135 | 0–13 | 14–68 | 69–80 | 81–120+ |
| Moisture loss (%) | 1.50 | 1.00 | 0.40 | 0.60 | 0.04 | 0.02 | 0.04 | 0.02 | 0.02 | 0.02 | 0.12 | 0.14 |
| Loss on ignition (%) | 8.0 | 4.3 | 4.1 | 3.9 | 6.1 | 11.6 | 4.6 | 3.3 | 1.5 | 1.0 | 1.7 | 5.1 |
| Colour | 10 yr 3/2 | 10 yr 5/2 | 10 yr 7/4 | 2.5 yr 6/0 | 10 yr 5/3 | 10 yr 5/4 | 10 yr 5/6 | 5 yr 4/2 | 10 yr 6/2 | 10 yr 7/3 | 10 yr 7/4 | 5 yr 7/4 |
| Organic C% | 5.3 | tr | tr | tr | 1.00 | 0.70 | 0.10 | 0.10 | 1.00 | 0.70 | 0.70 | 0.20 |
| Total N% | tr | tr | 0.10 | 0.20 | 0.06 | 0.04 | 0.02 | 0.17 | 0.06 | 0.10 | 0.07 | 0.11 |
| Ex. Ca (cmol/kg) | 0.10 | 0.06 | 0.06 | 0.08 | 0.54 | 0.02 | 0.08 | 0.18 | 0.28 | 0.03 | 0.02 | 0.05 |
| Ex. Mg (cmol/kg) | 0.80 | 0.02 | 0.77 | 0.77 | 1.80 | 1.64 | 6.32 | 7.39 | 0.16 | 0.14 | 0.80 | 9.80 |
| Ex. K (cmol/kg) | 0.19 | tr | tr | tr | 0.41 | 0.51 | 0.13 | 0.15 | 0.14 | 0.04 | 0.06 | 0.13 |
| Ex. Na (cmol/kg) | 0.43 | 0.36 | 0.36 | 0.43 | 0.18 | 0.23 | 0.37 | 0.75 | 0.15 | 0.14 | 0.18 | 0.37 |
| Sum (cmol/kg) | 1.52 | 0.44 | 1.19 | 1.28 | 2.93 | 2.40 | 6.90 | 8.47 | 0.73 | 0.35 | 1.06 | 10.45 |
| Fe (%) | 0.05 | tr | tr | 0.02 | 1.10 | 1.05 | 1.50 | 0.58 | 0.02 | 0.02 | 0.08 | 0.85 |
| $Fe_2O_3$ (%) | 0.07 | tr | tr | 0.02 | 1.56 | 1.49 | 2.13 | 0.82 | 0.02 | 0.03 | 0.11 | 1.20 |
| Ex. H (cmol/kg) | 0.60 | 0.12 | 0.12 | 0.12 | 0.16 | 0.22 | 0.60 | 0.06 | 0.20 | 0.16 | 0.20 | 0.80 |
| Ex. Al (cmol/kg) | 0.24 | 0.13 | 0.04 | 0.12 | 0.10 | 0.40 | 0.02 | 0.02 | 0.12 | 0.38 | 1.40 | 3.00 |
| Ex. acidity (cmol/kg) | 1.02 | 0.16 | 0.16 | 0.14 | 0.26 | 0.62 | 0.08 | 0.08 | 0.32 | 0.54 | 1.60 | 3.80 |
| pH ($H_2O$) | 4.2 | 5.4 | 5.8 | 5.8 | 5.4 | 5.4 | 6.3 | 8.0 | 5.3 | 5.3 | 5.4 | 5.4 |
| pH ($CaCl_2$) | 3.6 | 4.4 | 5.9 | 4.8 | 4.8 | 4.7 | 6.1 | 7.0 | 4.9 | 4.7 | 4.5 | 4.5 |
| Base saturation (%) | 44 | 85 | 73 | 90 | 92 | 79 | 98 | 98 | 70 | 39 | 40 | 73 |
| Av. P (mg/100g) | 0.24 | 0.13 | 0.04 | 0.12 | 0.24 | 0.13 | 0.04 | 0.12 | 0.59 | 0.33 | 0.34 | 0.43 |
| Sand (%) | 84 | 94 | 99 | 97 | nd | nd | nd | nd | 90 | 96 | 58 | nd |
| Silt + clay (%) | 16 | 6 | 1 | 3 | nd | nd | nd | nd | 10 | 4 | 42 | nd |

*continues overleaf*

Appendix 9 (continued)

| Horizon | Profile E | | | | | Profile F | | | |
|---|---|---|---|---|---|---|---|---|---|
| | $A_1$ | $A_2$ | $B_{21}$ | $B_{22}$ | $B_3$ | $A_1$ | $A_2$ | $B_{1g}$ | $B_{2g}$ |
| Depth (cm) | 0–7 | 8–24 | 25–60 | 61–80 | 81–120+ | 0–10 | 11–57 | 58–68 | 69–130 |
| Moisture loss (%) | 4.50 | 3.40 | 7.20 | 4.10 | 3.10 | 0.04 | 0.02 | 0.04 | 0.02 |
| Loss on ignition (%) | 15.4 | 1.8 | 7.2 | 6.5 | 4.8 | 1.20 | 0.60 | 2.90 | 1.20 |
| Colour | 10 yr 4/2 | 10 yr 4/2 | 10 yr 5/2 | 10 yr 5/2 | 10 yr 5/2 | 10 yr 7/3 | 10 yr 6/2 | varied | 2.5 yr 5/2 |
| Organic C% | 14.80 | 2.50 | 0.80 | tr | tr | 1.40 | tr | tr | tr |
| Total N% | 1.34 | tr | 0.29 | tr | 0.11 | tr | 0.10 | 0.10 | tr |
| Ex. Ca (cmol/kg) | 9.85 | 4.58 | 5.73 | 9.10 | 8.98 | 0.06 | 0.05 | 0.20 | 0.24 |
| Ex. Mg (cmol/kg) | 4.31 | 2.05 | 3.59 | 5.34 | 6.50 | 0.06 | 0.05 | 0.15 | 0.16 |
| Ex. K (cmol/kg) | 1.59 | 2.55 | 0.19 | 0.06 | 0.06 | 0.03 | 0.03 | 0.03 | 0.03 |
| Ex. Na (cmol/kg) | 0.71 | 0.43 | 0.71 | 0.86 | 1.01 | 0.45 | 0.42 | 0.48 | 0.59 |
| Sum (cmol/kg) | 16.50 | 9.60 | 10.20 | 15.60 | 16.60 | 0.60 | 0.60 | 0.90 | 0.90 |
| Fe (%) | 0.17 | 0.75 | 1.02 | 0.20 | 0.18 | 0.04 | 0.02 | 0.09 | 0.17 |
| $Fe_2O_3$ (%) | 0.24 | 1.06 | 1.44 | 0.28 | 0.25 | 0.05 | 0.02 | 0.12 | 0.22 |
| Ex. H (cmol/kg) | 0.30 | 0.55 | 0.60 | 0.18 | 0.10 | 0.22 | 0.08 | 0.18 | 0.18 |
| Ex. Al (cmol/kg) | 0.46 | 0.48 | 0.34 | 0.02 | 0.00 | 0.18 | 0.18 | 0.13 | 0.18 |
| Ex. acidity (cmol/kg) | 0.76 | 1.03 | 0.94 | 0.20 | 0.10 | 0.27 | 0.10 | 0.30 | 0.32 |
| pH ($H_2O$) | 4.7 | 4.9 | 4.8 | 6.5 | 7.5 | 4.9 | 5.5 | 5.5 | 5.7 |
| pH ($CaCl_2$) | 4.4 | 4.3 | 4.3 | 5.3 | 6.4 | 4.0 | 4.3 | 4.4 | 4.7 |
| Base saturation (%) | 96 | 90 | 92 | 99 | 99 | 60 | 67 | 74 | 72 |
| Av. P (mg/100g) | 0.60 | tr | tr | 0.00 | 0.00 | 0.93 | 0.93 | tr | 0.79 |
| Sand (%) | nd | 70 | 43 | 41 | 57 | 91 | 99 | 96 | 96 |
| Silt + clay (%) | nd | 30 | 55 | 59 | 43 | 9 | 1 | 4 | 4 |

Profile A1 – Dense forest (Ecological Station)
Profile A2 – Dense forest, mesotrophic site (Angico)
Profile A3 – Dense forest, poorly drained (Ecological Station, *buriti* site)
Profile B – Predominantly low or open forest (*campina*) at edge of *buritizal*
Profile C – Forest edge/transition (Angico transect)
Profile D – Tree savanna (Santa Rosa)
Profile E – *Vazante* vegetation (Preguiça trail)
Profile F – Herbaceous grassland, *campo* (Preguiça trail)

# Index

*Abuta grandiflora*, 81; *imene*, 93
*Accipiter poliogaster*, 225
*Achaearanea*, 380; *trapezoidalis*, 380
Acidity, soil, 54, 421–6, 427; water, 399–400, 405–6, 412
*Acosmium tomentellum*, 97
*Acrocinus longimanus*, 385
*Acromyrmex subterraneus*, 335
Acta Amazonica, special issue, xviii
*Actitis macularia*, 221
*Adenomera*, 297
Adiantaceae, 120–1
*Adiantopsis radiata*, 115, 120
*Adiantum latifolium*, 121; *lucidum*, 121; *petiolatum*, 121; *pulverulentum*, 91, 121; *serratodentatum*, 121; *terminatum*, 121
Aerial photographs, 14, 27, 31, 34, 35, 44
*Agave*, 104
*Agelaius icterocephalus*, 225
Aggregations, avifaunal, 220–1
*Agouti paca*, 157, 159, 202, 204
Agoutidae, 202
Agoutis, see *Dasyprocta agouti*
*Akodon urichi*, 204
*Alchornea schomburgkii*, 99
*Alexa canaracunensis*, 89
Alfisols, 63
Algae, see Phytoplankton
*Allamanda nobilis*, 275, 281, 282
Alluvium, 8, 39, 49, 417, 425, 427; Quaternary, 5; Tertiary, 5
*Alouatta seniculus*, 143, 146
*Alseis longifolia*, 85
Alto Alegre, 3, 364
Aluminium, soil content, 56, 421, 424, 425, 427
Amajari, Rio, 31
*Amazona amazonica*, 219
Amazonas, forest diversity, 106; mosquitoes, 374
Amazonia, fern diversity, 113, 114, 116; frogs, 288; primates, 143; pseudoscorpions, 381; reptile fauna, 243–50; termites, 348–9
*Ameiva*, 258; *ameiva*, 234, 236, 237, 243

*Americhernes bethaniae*, 384; *incertus*, 384; *longimanus*, 384–5
Amphibole, 5
Amphibolite, 54, 56
*Amphisbaena fuliginosa*, 234
Amphisbaenids, 234, 255–6
Ampullaridae, 410
*Anacardium occidentale*, 108, 316
Anacondas, 238, 239, 258, 259
*Anadenanthera peregrina*, 84; relationship with soils, 422–3
*Ananthocorus angustifolius*, 121
Andes, 218
Andesite, 5
*Andira surinamensis*, 96, 98
*Andropogon angustatus*, 9
*Anemia oblongifolia*, 119
*Anetium citrifolium*, 114, 122
Angico forest, 84–5, 422–3
*Angiopolybia pallens*, 309, 318
*Aningal*, 438
*Anolis*, 247; *auratus*, 235, 246; *fuscoauratus*, 243; *ortonii*, 235, 243
*Anopheles*, 369–75; *albitarsis*, 372–3, 374; *argyritarsis*, 372–3, 374; *braziliensis*, 372; *darlingi*, 372–3, 374; *mediopunctatus*, 372; *nuneztovari*, 374; *oswaldoi*, 372, 373, 374; *triannulatus*, 372–3, 374
*Anopheles*, as vectors, 370, 372–3, 374; biting activity, 372; collecting techniques, 370; density, 374; diversity, 372, 373; habitat preferences, 372; seasonality, 372, 373
*Anosia*, 357
*Anthurium gracile*, 98
Anteaters, 10
Ants, army, 268, 313, 335–6, 340–4; bait removal, 195; eaten by reptiles, 257, 258; guards, 335–6, 340–3; in *Cecropia* & *Triplaris*, 97, 101; leaf-cutter, 339–40, 345–7; on *Passiflora longiracemosa*, 334, 335–6
Anurans, see Frogs
Anyphaenidae, 380
*Aotus trivirgatus*, 143, 146

*Apeiba schomburgkii*, 85, 91, 101, 105, 437
Apiaú, 144, 147, 148, 213, 219, 221
Apicotermitinae, 349
*Apinagia tenuifolia*, 103
*Apocephalus*, 341, 342; *hispidus*, 341; *luteihalteratus*, 340
*Apoica flavissima*, 309, 314, 318; *pallens*, 313; *pallida*, 309, 312, 314, 318; *thoracica*, 309, 314, 318–19
*Apolpium*, 383; *ecuadorense*, 383; *minutum*, 383
*Apterophora*, 343–4
Aquatic habitats, biological indicators, 403–14
Aquatic vegetation, 102–3
*Ara ararauna*, 219; *manilata*, 219; *nobilis*, 219; *severa*, 219
Aracaris, 224–5
Arachnids, 377–89
*Aramides*, 267
Araneidae, 379
*Aratinga pertinax*, 220; *solstitialis*, 219
Archaeology, 434–6
Ariquemes, Rondônia, 374
Armadillos, 273
Army ants, *see* Ants
*Arremonops conirostris*, 225
*Arribalzagia*, 370
Artefacts, archaeological, 434, 436
*Artibeus*, 172, 173, 183–4; *anderseni*, 183; *cinereus*, 183; *concolor*, 172, 183; *fuliginosus*, 183; *jamaicensis*, 183; *lituratus*, 183; *planirostris*, 183
*Aspidosperma nitidum*, 106
Aspleniaceae, 128
*Asplenium*, 117; *auritum*, 128; *serratum*, 124
Assassin bugs, 366–8
*Astrocaryum*, 314; *aculeatum*, 81, 132, 435; *gynacanthum*, 89, 310, 315, 320; *jauari*, 98; *murumuru*, 159; *standleyanum*, 159
*Astronium lecointei*, 80
Atelabinae, 345–7
*Ateles belzebuth*, 143, 146, 148, 154; *paniscus*, 143, 147
Atemnidae, 384
*Athene cunicularia*, 225
*Atractus trilineatus*, 235, 246, 257, 259
*Atta cephalotes*, 339–40, 343–4
*Atta*, frogs in nest, 300
*Attalea*, 367
Audre, 357
*Automolus ochrolaemus*, 219; *rubiginosus*, 219

Autotomy, 201
Avifauna, 211–29
Axes, stone, 434, 435, 436
Axial plane foliation, 16, 18
*Azteca*, 319, 342–3

B horizon, in soil classification, 61, 62, 68
*Bacopa reflexa*, 102
*Bactris maraja*, 95, 96, 97, 101
*Bagassa guianensis*, 275
*Baixadas*, 95, *see also* Buritizal
Balbina, 373
Bamboo, 96, 196
Bananal, Ilha de, 107
Barcelos, 3, 4
Barro Colorado, 160; reptiles, 240, 243–50, 255
Basal area, forest, 85, 107, 437; secondary vegetation, 132
Base column, army ants, 336, 341
Base saturation, soils, 61, 62, 66, 422, 425, 426
Base status, soils, 54–6, 420, 421–6, 427
*Basileuterus mesoleuca*, 216; *rivularis*, 214, 216
Bats, 165–87; capture rates, 173–4; collecting techniques, 166–8; diversity, 168–71; ectoparasites, 167; habitat preferences, 171–3; in buildings, 186, 366–7; pollination, 92; sex ratio, 178; vampire, 184
*Bauhinia outimouta*, 84; *ungulata*, 104
Bdelloidea, 400
Bearded pig, 152
Beating umbrella, 377
Bees, on *Passiflora*, 332–5
Beetles, cerambycid, 381, 382, 384–5; eaten by lizards, 257; weevils, 345–7
Beetles, passalid, *see* Passalidae
Belém, forest diversity, 106; reptiles, 240, 241, 243–50
Berlese funnels, 382
Bessa, Hollanda, 432, 434, 439
Bignoniaceae, 104
Biogeography, birds, 224–5; ferns, 114–15; frogs, 286; reptiles, 240–2, 243–50; tortoises, 269
Biomass, forest, 105–6
Birds, 211–29; annotated species list, 451–61; diversity, 218, 219; eaten by snakes, 259; endemism, 215, 224–5, 226; feeding preferences, 217; fish-eating, 220–1; geographical distributions, 214–16, 224–5; habitat preferences, 216–17; insectivorous, 221; migration, 221–4; nests, 366–7; of

*buritizal*, 219; of *campina*, 219–20; of forest–savanna boundary, 220; of riverine forest, 219; of savanna, 220; of secondary vegetation, 219; of *terra firme* forest, 218–19; of the understorey, 218–19; predation of earthworms, 391; predation of tortoises, 267–8; seed dispersal; seed-eating, 220
Bivouac, army ants, 336, 341, 342
*Bixa orellana*, 96
Blechnaceae, 124
*Blechnum serrulatum*, 103, 124
Blue-green algae, 140
Bluffs, fringing, 34–5
*Boa*, 247; *constrictor*, 238, 243, 256, 258
Boa Esperança, Fazenda, 439
Boa Vista, xv, 2, 3, 26, 41, 147, 148, 362; frogs, 287, 303; ponds, 412; population, 3; savanna, 108; soils, 416
Boa Vista Formation, 5
Bolivia, Hybridization zone, 214
*Bombacopsis quinata*, 80
Boqueirão, 39
Borrow pits (lagoons), 102, 104, 370, 373, 374, 405, Plate 4; frogs, 290; reptiles, 237
*Bothrops atrox*, 236, 243, 256, 258
*Bowdichia virgilioides*, 108
*Brachygastra lecheguana*, 313; *smithii*, 309, 320
Branco (Rio), xv, 1, 31, 41, 42, 143, 148, 410; birds, 212, 226; soils, 418
*Brazilatemnus browni*, 381
Brazilian Classification, soils, 52, 56, 57, 61, 62
Brazilian Shield, 218
Bromeliaceae, 93, 98, 257, 273, 380
*Brosimum lactescens*, 80, 85
*Brotogeris chrysopterus*, 213, 214, 224; *cyanoptera*, 213, 214
Bryophytes, 91
*Buchenavia tetraphylla*, 106
*Bufo*, 290–2; *granulosus*, 290–1; *guttatus*, 291; *marinus*, 291–2; *typhonius*, 292
Bufonidae, 290–2
*Burhinus bistriatus*, 225
Buriti, *see Mauritia flexuosa*
Buriti bridge, 95, 405
Buritizal (*Mauritia* swamp), 94–6, 102–3, 391; birds, 219; reptiles, 234–5; underlying soils, 423–4, 427; use by peccaries, 154, 156–7, 159, 160; use by tortoises, 269
Burning, of forest, 131; of savanna, 10, 100, 136, 140, 141, 439–40
Bushmaster, *see Lachesis*

*Buteo albicaudatus*, 225
*Butorides striatus*, 220
Butterflies, 226, 355–8; annotated species list, 467–78; collecting methods, 356; diversity, 356–8; habitat preferences, 358
Buttress roots, 80, 90
*Byrsonima coccolobifolia*, 108; *crassifolia*, 9, 74, 92, 93, 94, 98–100, 102, 104, 108, 109, 137, 140, 425; *schomburgkiana*, 94, 98; *verbascifolia*, 108

Caapoeira, *see* Secondary vegetation
Caatinga, 3, 10, 92, 108, 219, 224; small mammals, 190, 194
*Cabomba piauhiensis*, 102
Cacti, 104
Caiman, 234, 237; *crocodilus*, 237, 243, 250, 465
*Calathea*, 84, 91
Calazar, 364
Calcium, food content, 282; soil content, 56, 84–5, 422, 423, 428, 435; water content, 406–7, 410
*Callicebus torquatus*, 143, 147, 148
*Callicora*, 358
*Calophyllum lucidum*, 97
*Caluromys philander*, 204
Cambisols, 65
Cameroon, small mammals, 193
Campina, 92–3, 107–8, 424, 428; birds, 219–20; underlying soils, 424, 426–7
Campinarana, 92
Campo, *see* Savanna
Campo sujo, 99
*Camponotus*, 341–2
*Camptopleura auxo*, 358
*Campyloneurum latum*, 114, 125; *phyllitidis*, 125; *repens*, 125
*Campylorhynchus griseus*, 225
*Canna glauca*, 101
Cannibalism, snakes, 258
Canopy, access, 232–3; fauna, 357–8; fogging, 378–9, 382; height, 80, 89; wasps, 314
Capão, 100
Capoeira, *see* Secondary vegetation
Capture rates, bats, 173–4; small mammals, 193–5, 203, 205
Capuchin monkeys, 146
Caracaraí, 2, 3, 148, 226, 362
Carajás, reptiles, 240, 241, 243–50
Carbon, soil content, 421, 424, 425, 426
*Carollia*, 175, 182; *brevicauda*, 182; *castanea*, 182; *perspicillata*, 172, 182
Carrasco, 93–4, 424

*Casearia sylvestris*, 99, 105
*Casmerodius albus*, 220
Cataclasis, 14, 18, 20, 21
*Catharus fuscescens*, 223; *minimus*, 223
Catrimani, Rio, 14, 147, 239
Cattle, 438, 440; in Roraima, 3
*Cattleya violacea*, 98
Causeway, 100, 102, 166, 171, 237, Plate 4a; vegetation, 104–5
*Cavia aperea*, 196, 203, 204
Caya-Caya, 432
CDC traps, 362
*Cebus*, 206; *apella*, 143, 146; *nigrivittatus*, 143, 146
*Cecropia*, 95, 101, 104, 132, 219, 435; *latiloba*, 97, 105; *palmata*, 97, 105
Cedar, Brazilian, 438
*Cedrela odorata*, 438
*Ceiba pentandra*, 80, 96, 435
*Celeus elegans*, 215; *flavus*, 219; *grammicus*, 215; *jumana*, 214, 215; *undatus*, 214, 215
*Cenchrus echinatus*, 104
Census techniques, 144, 153, 265
Central Highlands (Maracá), 48, 50, 52, 65
*Centrolobium paraense*, 80
*Cephalodella*, 400
*Cephalopterus ornatus*, 219
*Cepheuptychia cephis*, 358
Cerambycid beetles, 381, 382, 384–5
*Cercomacra carbonaria*, 226
*Cercosaura ocellata*, 234, 256
*Cerdocyon*, 268
*Cereus hexagonus*, 104
Cerrado, 9, 99, 108, 224, 269; butterflies, 358; root depth, 140; small mammals, 190, 193, 199; soils, 425–6, 428
*Chaetura andrei*, 223
Chagas' disease, 366, 367–8
*Chamaecrista rotundifolia*, 104
Channel patterns, 28, 30, 31
*Chara*, 374
Charcoal, 437
*Chartergus chartarius*, 309, 310, 320–1
*Cheiloclinium cognatum*, 81
Cheiridiidae, 383–4
Chelonians, 234, 239
*Chelus fimbriatus*, 234
Chemistry, soil, 52–6, 411, 420–7; water, 373, 405–8, 410–13
Chernetidae, 384–7
Chewits, 168, 175–6
*Chiomara*, 358
*Chiroderma*, 184; *trinitatum*, 184; *villosum*, 172, 184

*Chironectes minimus*, 204, 206
*Chironius carinatus*, 238, 239, 465; *exoletus*, 238, 239, 250, 465
*Chiropotes satanas*, 143, 147, 148
Chiroptera, 165–87
Chlorophyta, 77
*Choeroniscus*, 181–2; *minor*, 172, 181–2
*Chordeiles minor*, 223; *rupestris*, 225
Chrysophyta, 77
Chrysopidae, 386
Chthoniidae, 382
Clay, 39, 418; activity, 61, 62; alluvial & colluvial, 49
Clearings, *see* Forest clearings
*Clelia clelia*, 236
Climate, 4, 288; of Roraima, 3–4
Climate change, Pleistocene, 37, 42–4
Clubionidae, 380
*Clusia*, 89; *renggerioides*, 95, 99, 424
Cluster analysis, soils, 54
*Cnemidophorus*, 249, 256, 258; *lemniscatus*, 237, 245, 247
*Cnidoscolus urens*, 104
*Coccyzus pumilus*, 213, 214
Cocha Cashu (Peru), 146, 148, 221; mammal fauna, 205; reptiles, 240, 243–50
*Cochlospermum orinocense*, 84, 92, 95, 96, 97, 104, 106, 275, 281, Plate 1b
Coefficient of biogeographical resemblance, 242–6
*Cogia*, 357
*Colaenis*, 357
*Coleodactylus septentrionalis*, 233, 234, 235, 239, 246
*Colinus cristatus*, 225
Collecting techniques, *Anopheles*, 369; bats, 166–8, 171, 173–4; butterflies, 356; frogs, 286–7; Passalidae, 324; pseudoscorpions, 382; reptiles, 232–3; sandflies, 362; small mammals, 192; spiders, 377–8; termites, 349; wasps, 309
*Collotheca*, 400
Colluvium, 37, 39
Colonization, xx, 3, 10, 148–9, 364
Colonizing plants, *see* Secondary vegetation
*Colostethus*, 303
Colour morphs, wasps, 313–14
Columbite, 22
Column raid, army ants, 340
*Combretum rotundifolium*, 96
Communication, peccaries, 155–6
Community structure, frogs, 289–90
*Comolia*, 140
Concretions, iron, 39
Conductivity, water, 405–7

INDEX 493

Conference, on Maracá, xviii
Conglomerate, 49
Conservation, xviii–xx, 212
Conservation status, birds, 226; frogs, 290; peccaries, 161; primates, 148
*Copaeodes*, 357
Coppice regeneration, 92–3, 105, 132, 435
Coral snakes, 236, 237–8
*Corallus enydris*, 185; *hortulanus*, 235, 238, 243, 250, 258, 259
*Cordia nodosa*, 95, 343
Corduroy bridge, *see Buriti* bridge
*Cordylochernes*, 385; *scorpioides*, 381
*Cormura*, 177; *brevirostris*, 172, 177
*Cornitermes*, 348
*Corticea*, 357
*Corvomeyenia thumi*, 411
*Costus arabicus*, 95; *scaber*, 91, 95, 101
*Couendou prehensilis*, 190, 202, 204
*Couratari oblongifolia*, 80, 345
*Coursetia ferruginea*, 105
*Coussapoa villosa*, 98
*Crax*, 267
*Crematogaster*, 343
*Cremersia*, 340, 342; *setitarsus*, 342; *spinicosta*, 342
*Crepidospermum goudotianum*, 105
Cretaceous, 7, 27
Cricetidae, 196–9
*Crotalus*, 249, 256; *durissus*, 236, 238, 245, 247, 250, 258, 465
*Crotophaga major*, 219
Ctenidae, 379, 380
*Ctenitis refulgens*, 116, 123
Cultivation, shifting, 101, 105, 131, 132, 312, 435, 438, 439, 440
*Cuphea antisyphilitica*, 99
*Curatella americana*, 9, 74, 92, 93, 94, 98–100, 102, 104, 108, 109, 137, 140, 425, Plate 1b
Cutoff morphology, 35
Cyanophyta, 77, 140
*Cyathea*, 117; *microdonta*, 128; *oblonga*, 122
Cyatheaceae, 122, 128
*Cyclosa*, 379
*Cyclosorus interruptus*, 103, 125
*Cymbilaymus lineatus*, 219
*Cynometra alexandri*, 107
*Cyrtopodium poecilum*, 104

*Daceton armigerum*, 335
Dacite, 5
*Dactylomys dactylinus*, 200, 204, 206

*Dalbergia*, 91
*Danaus*, 357
*Dasyprocta agouti*, 157, 159–60, 201, 204, 273, 274
Dasyproctidae, 201
*Dasypus*, 273, 274
Dead wood, 437
Deciduous trees, 10, 84, 90, 91
*Deconychura longicauda*, 225
Deer, brocket, *see Mazama americana*
Deformation (geological), 18, 21, 22; ductile, 21
Demini, Rio, 146; Serra de, 5
*Dendrexetastes rufigula*, 225
*Dendrobates leucomelas*, 235, 292–3
Dendrobatidae, 292–3
*Dendrocincla fuliginosa*, 225; *merula*, 225
*Dendrocolaptes certhia*, 225; *picumnus*, 225
Dendrocolaptidae, 217, 225
*Dendroica petechia*, 223; *striata*, 223
*Dendrophryniscus*, 292
Dennstaedtiaceae, 123
Desmids, 411
*Desmodus rotundus*, 184
*Desmoncus polyacanthos*, 101
Development, xx, 3
Diamonds, 3, 22, *see also* Mineral prospecting
Diatoms, 407
*Dicranophorus sebastus*, 401
*Dicranopteris pectinata*, 128
Didelphidae, 195–6
*Didelphis*, 268; *marsupialis*, 204
*Didymopanax morototoni*, 84, 95, 106–7
Diel activity, reptiles, 255–7
Diets, *see* Feeding preferences
*Diocophora appretiata*, 341; *disparifrons*, 341
Diorite, 5
*Dipsas*, 247, 249, 256; *catesbyi*, 236, 243, 250, 257, 259
Diptera, 361–5
Discharge, river, 28
Disease, xix; Chagas', 366, 367–8; leishmaniasis, 364; malaria, 268, 373, 374–5
Dispersal, seeds, 219
Dissection (terrain), 26, 31, 33, 34, 36–7, 41
Distributaries, 28
Distributions, *see* Geographical distributions
Disturbance indicators, 435, 437
Disturbance, human, xvi, xix–xx, 10, 107, 131, 136, 219, 226, 432–42

Diversity, *Anopheles*, 372, 373; bats, 168–71; birds, 218, 219; butterflies, 357–8; earthworms, 391; ferns, 113–16; forest trees, 85, 89, 91, 92, 106–7, 421; frogs, 288; passalid beetles, 328–9; plants, 77; pseudoscorpions, 387; reptiles, 242–3; sandflies, 362, 364; savanna plants, 108; secondary vegetation, 132; small mammals, 204–5; spiders, 378; wasps, 317
DNPM, 14, 22
Dogs, 364
*Dohrniphora ecitophila*, 343
Dolerite, 14, 20–1
*Dorymyrmex*, 343
*Doryopteris*, 117; *collina*, 114, 116
Drainage, 5, 14, 23, 26, 27–31, 42, 49, 95, 99, 101
Dredging, gold, xix, 22
Drift, 48–50, 51, 67–8
Drift fences, 192, 193, 232, 234, 255
Drought, 307, 412
Dry forest, 107
*Drymarchon corais*, 238, 243, 247, 250, 257; *rhombifer*, 236, 465
*Drymoluber dichrous*, 235, 243, 250
*Dryocopus lineatus*, 220
Dryopteridaceae, 123–4
*Duguetia*, 126, 274
Dunes, of the Tacutu, 42–3, 44
Dykes, dolerite, 16, 20–1
Dystrophic soils, 63, 65, 68, 416, 427

Earthworms, 391–7; diversity, 391; eaten by snakes, 257; habitat preferences, 392–3; identification key, 393; new species description, 394–7
Eastern Shield (Maracá), 48, 421
*Ecclinusa guianensis*, 85, 92, 132
ECEC, soils, 54–6, 420, 421–6, 427
Echimyidae, 200–1
*Echinodorus*, 372; *scaber*, 102
*Echinotheridion urarum*, 379
*Eciton*, 268; *burchelli*, 313, 319, 336, 340–4; *hamatum*, 340
*Ecitophora collegiana*, 343
*Ecituncula tarsalis*, 343
Ecological heterogeneity, 114, 205, 218, 226, 288, 391
Ecological Station, xv–xvi, 166; assassin bugs, 366–7, 372, 373; geomorphological situation, 48, 49; reptiles, 238, 259; small mammals, 199, 204; underlying soils, 421–2; vegetation, 79, 105
Economic geology, 22

*Ectatomma edentatum*, 335; *tuberculatum*, 335, 336
Ectoparasites, on bats, 167; on small mammals, 194, 200–1
Edge habitats, use by tortoises, 269
Education pack, Maracá, xviii
*Egretta thula*, 220
*Eichhornia*, 294, 373; *azurea*, 102, 405, 410, Plate 4b; *crassipes*, 103
*Eira*, 268
*Elachistochleis ovalis*, 301–2
*Elaenia ruficeps*, 220
*Elaphoglossum*, 117
*Electron platyrhynchum*, 213, 214
*Eleocharis*, 102, 373, 405, Plate 4b; *acutangula*, 102
*Eleutherodactylus*, 303
ELISA test, 373
*Elizabetha coccinea*, 97
Emballonuridae, 171, 173, 176–7
Emergents, 10, 80, 84, 92
*Emilia coccinea*, 104; *fosbergii*, 104
Endemism, 204; birds, 215, 224–5, 226; centres of, 42; ferns, 113–14; frogs, 286, 288–9, 290; reptiles, 247; rotifers, 401
*Endlicheria dictifarinosa*, 95, 99
Endoparasitism, *see* Parasitism
*Entada polystachya*, 96
*Enterolobium cyclocarpum*, 96, 434, 435; *schomburgkii*, 80, 106
Entisols, 417, 418
*Eperua*, 107; *leucantha*, 10
*Epicrates cenchria*, 236, 238, 243, 250, 256
Epiphytes, plants, 84, 98, 203; spiders, 380
*Eptesicus*, 185
*Eratyrus mucronatus*, 368
Erethizontidae, 202
*Eriochloa punctata*, 101
*Erythrolampus aesculapii*, 236, 243, 257, 259
*Erythroxylum suberosum*, 99
*Eschweilera pedicellata*, 95
Essequibo, river, 143
Etchplanation, 7, 39, 41
*Eugenia*, 91, 92, 94, 424
Euglenophyta, 77
*Eunectes murinus*, 238, 239, 257, 259
*Eunica caresa*, 358
*Euphonia finschi*, 225; *laniirostris*, 213, 216; *violacea*, 213, 216, 219
*Euptychia*, 357
*Eurytides*, 358
*Euterpe precatoria*, 97, 98

Eutrophic soils, 7–8, 63, 65, 68, 74, 417
Evergreen seasonal forest, 84
Exchange capacity, *see* ECEC

*Falco sparverius*, 225
False coral snakes, 236
*Faramea crassifolia*, 98; *sessilifolia*, 94
Farms, xx, 184, 372; on the Ilha de Maracá, 437–8, 439
Faulting, geological, 5, 14, 16, 18, 23, 26, 27, 31
*Fazendas, see* Farms
Feeding preferences, birds, 217; peccaries, 152, 156–61; reptiles, 257–60; termites, 349–50, 352–3; tortoises, 274–7, 281–2
Feldspar, 51, 52
Fermentation, microbial, 200, 282
Ferns, 113–29; Andean elements, 116; biogeography, 114–15; diversity, 89, 113–16; endemism, 113–14; homosporous, 116; of western Maracá, 116–17; relationship with small mammal density, 196
Fertility, soil, 7, 52
*Ficus*, 95; *guianensis*, 89; *pakkensis*, 89
Filhote trail, soils, 50, 52, 54, 57, 63
Fire, 10, 100, 131, 136, 140, 141, 439–40
Fish, eaten by bats, 177; eaten by reptiles, 257; in savanna, 140
Fish-eating birds, 220–1
FITOPAC, 76
Flocks, mixed, 220–1
Flooding, *see* Inundation
Floodplains, 6, 8, 33, 34–5; soils, 418
Florida, tortoises, 281
*Florida caerulea*, 220
*Floscularia decora*, 401
Fogging, canopy, 378–9, 382
Folding, geological, 16
Foliation, 16, 18, 19, 21, 22
Foraging efficiency, 220
Forest, 3, 8, 10, 79–98; Amazonian inventories, 106–8; *angico*, 84–5, 422–3; basal area, 85, 107, 437; birds, 218–20; *buritizal*, 94–6, 102, 391; burning, 131; buttress roots, 80, 90, 91; *campina*, 92–3, 107–8, 424, 428; canopy height, 80, 89, 90; *carrasco*, 93–4, 424; composition, 432; deciduous trees, 10, 84, 90, 91; diversity, 85, 89, 91, 92, 106–7; dry, 107; emergents, 10, 80, 84, 92; expansion, 100, 107, 109, 117, 218, 269, 413; fringe Amazonian, 106–7; frogs, 289–90; ground layer, 81, 84, 89, 91, 95; low, 92–4; management, 109, 440; *manguezal*, 95–6, 424; maturation, 109; monodominant, 107–8; of western Maracá, 86–9, 107; *Peltogyne*, 56, 78, 89–92, 107–8, 422–3, 439, Plate 2a, Plate 3b; regeneration studies, 131; relationship with soils, 65–7, 420–4; reptile fauna, 233–7; riverine, 92, 435, Plate 2; slope profiles, 37; structure, 85, 89, 92, 171; survey techniques, 74–6; *terra firme*, 79–89, 420–2, Plate 3a; understorey, 81, 89, 91, 95; variation in, 85–6, 87, 89, 91–2, 107; *várzea*, 97
Forest clearings, 218; termites in, 349–51; wasps in, 309, 310–12, 314
Forest–savanna boundary, 10, 42, 44, 92, 94, 100, 107, 273, 279, 416, 440, Plate 1; birds, 220; phenology, 135–42; underlying soils, 424–5, 427
Founder-controlled communities, 313
Fourier series, 144
French Guiana, frogs, 288; tortoises, 269
Fringillidae, 217
Frogs, 285–309; biogeography, 286; collecting methods, 286–7; conservation status, 290; distribution patterns, 288–9; diversity, 288; endemism, 286, 288–9, 290; habitat preferences, 289–90; in ants' nests, 303; of Roraima, 288–9; preyed on by snakes, 234; spawn eaten by reptiles, 257–8; species on Maracá, 290–306
Fruit trees, 136, 435, 438
Fumaça rapids, 16, 20, 38
Fumaça trail, soils, 49, 50, 56, 57, 63; vegetation, 80, 90, 94, 101, 105
Fungi, 77, 274, 277
*Furnarius leucopus*, 225
Furo de Maracá, 92, 97–8, 176, 371–3, Plate 2; geology, 13, 14, 16; geomorphology, 27, 30–1, 35; human occupation, 435, 438, 439
Furo de Santa Rosa, 5, 98, 371–3, 374; geology, 13, 14, 16, 19, 20, 21, 22, 23; geomorphology, 26, 27, 29, 30–1, 35; human occupation, 432, 434, 435, 438

Galleries, Passalidae, 324, 328, 382–5
*Garimpeiros, see* Mineral prospecting
Gastropods, eaten by snakes, 257
*Genipa americana*, 105, 273, 275, 437
*Genipa spruceana*, 98
*Geochelone carbonaria, see* Tortoises
*Geochelone denticulata, see* Tortoises
Geographical distribution, frogs, 288–9; reptiles, 250; sandflies, 362–4; birds, 214–16, 224–5; wasps, 308, 309–10
Geological mapping, 13, 14–15

Geology, 5, 13–23, 67; economic, 22; of Roraima, 4–5
Geomorphology, 6, 25–46, 47–50; Amazonian, 41, 42, 43; of Roraima, 5–7, 26; pre-Quaternary, 41–2; Quaternary, 42–4; regional, 41–4
*Geonoma*, 89; *baculifera*, 147; *deversa*, 89; *maxima*, 89
Gibbsite, 39
Gleicheniaceae, 128
Gleying, 7, 418, 423–4, 425, 427
*Glossodrilus*, 392; *arapaco*, 395; *itajo*, 395; *motu*, 394–7; *oliveirae*, 395, 397; *ortonae*, 395
*Glossophaga*, 181; *longirostris*, 181; *soricina*, 174, 181
Gneiss, 5, 14, 29; mylonitized tonalitic, 14, 21–2; quartz-feldspar, 5, 16, 26; tonalitic, 14, 16–18
Goethite, 51, 52
Gold prospecting, *see* Mineral prospecting
*Gonatodes humeralis*, 235, 243, 256
Gondwana, herpetofauna, 247
*Goniopteris abrupta*, 125
*Gopherus polyphemus*, 281
Gran Sabana, 204
Granite, 5, 14, 23, 29, 97–8; outcrops, 103–4, 114–15; ridges, 48, 49, 50, 65; tonalitic, 5, 16, 26; vegetation on, 103
Granodiorite, hornblende, 14, 20
Gravels, 37, 43, 44, 425
Guayana, *see* Guiana
*Guazuma ulmifolia*, 105, 437
Guiana Complex, 5, 26, 27, 364
Guiana Highland, 311; ferns, 116
Guiana Shield, xiii, 3, 4–5, 6, 14, 26, 31, 41, 218, 411; small mammals, 195, 205
Guiana, avifauna, 224–5; ferns, 114; frogs, 289, 290; herpetofauna, 246, 247, 256, 260, 286
*Gustavia augusta*, 97, 101
Guyana, 2, 3, 9, 14, 27, 41, 42, 57, 98, 107, 108, 143, 440; ferns, 116
*Gymnomystax mexicanus*, 225
*Gymnophthalmus*, 258; *underwoodi*, 233, 246
*Gymnopteris rufa*, 114, 120

Habitat disturbance, 148
Habitat preferences, *Anopheles*, 372; bats, 171–3; birds, 216–17; butterflies, 358; earthworms, 392–3; ferns, 116–17; frogs, 289–90; pseudoscorpions, 387–8; reptiles, 250–5; small mammals, 203; spiders, 378, 379–80; tortoises, 264–5, 269–73; wasps, 308, 309, 310–14

Haematite, 51, 52
Halloysite, 51, 52
Hamilton-Rice, Alexander, 434, 439
Harp traps, 167, 171–4
*Harpia harpyja*, 226
Heath forest, Amazonian, 437
*Hecistopteris pumila*, 115
*Heliconia psittacorum*, 101
*Helicops angulatus*, 236, 243, 257
*Heliopetes*, 358
Heliophily, in snakes, 258; in spiders, 378
*Hemidactylus brooki*, 247; *palaichthus*, 235, 238, 245, 246, 247
*Hemidictyum marginatum*, 116, 123
*Hemionitis palmata*, 114, 120
*Hemithraupis guira*, 221
Herons, 220–1
Herpetofauna, 231–62, 263–84, 285–309; Guianan, 246, 247, 256, 260, 286, 288
*Herpsilochmus*, 221; *dorsimaculatus*, 221; *rufimarginatus*, 221
Hesperiidae, 356, 357
Heterogeneity, ecological, 114, 205, 218, 226, 288, 391
*Heteromeyenia stepanowi*, 409
Heteropodidae, 379
*Heteropsis*, 89
*Hevea*, 81
*Himatanthus articulatus*, 95, 108, 132
Hirundinidae, 217
*Hirundo rustica*, 223
Historical faunal assemblages, reptiles, 247–50
*Holochilus brasiliensis*, 196, 204
*Homalium guianense*, 97, 101
Homosporous ferns, 116
Hornblende granodiorite, 14, 20
Hornblende-biotite tonalite, 16, 19–20
Howler monkeys, 146
Human disturbance, xvi, xix–xx, 10, 107, 131, 136, 219, 226, 432–42
Human influence, reptiles, 239
Human occupation, 105, 432–42
Humane research, 174–6
Humidity, 274, 378; soil, 421, 424
*Humiria balsamifera*, 10, 94, 99
Hummingbirds, hermit, 332–5
Hummocking, 49, 91, 94, 95, 99
Hunting, 439; monkeys, 149; on Maracá, xix; peccaries, 153, 158, 160; tortoises, 268
*Huperzia dichotoma*, 118
Hybridization zones, birds, 214
*Hydrolea spinosa*, 102
Hydromorphic soils, 7–8, 65, 66, 68, 418

*Hyla*, 287, 293–6; *boans*, 297; *boesmani*, 293; *crepitans*, 293–4; *exigua*, 303; *fuscomarginata*, 294; *geographica*, 294; *marmorata*, 296; *microcephala*, 294–5; *minuta*, 303; *nebulosa*, 303; *raniceps*, 303; *rubra*, 285, 296; *wavrini*, 295–6
Hylaea, 77, 109
*Hylephila*, 357
*Hylexetastes perrotii*, 225
Hylidae, 293–7
*Hylophilus muscicapinus*, 221; *pectoralis*, 221
*Hylophylax punctulata*, 219
*Hymenaea courbaril*, 160
Hymenophyllaceae, 122
*Hymenophyllum*, 117
Hyperseasonal savanna, *see* Savanna
*Hypocnemoides melanopogon*, 219
*Hypognatha*, 379

IBAMA, xv
Ice ages, 224
*Icterus chrysocephalus*, 219
*Igapó* forest, 411
*Igarapés, see* Streams
*Iguana*, 239, 247, 257, 258; *iguana*, 239
Illite, 39
*Imantodes*, 247, 249; *cenchoa*, 243
Imeri, area of endemism, 224–5
Inajá, *see Maximiliana maripa*
Inajá, Igarapé, 438
Inceptisols, 417, 418
Indians, *see* Indigenous peoples
Indigenous peoples, xix, 3, 10, 107, 148, 432, 434, 439–40; population, 3
*Inezia subflava*, 225
*Inga*, 91, 95, 101
INPA, xv
Insectivores, bats, 171; birds, 221
Inselbergs, 7, 41
Interfluves, 33
Introduced species, reptiles, 239; small mammals, 204
Intrusions (geological), 16, 18, 23, 54; metamorphic, 65
Inundation, seasonal, 10, 48, 66, 97, 100–1, 115, 416, 417, 425, 427, 399–400, Plate 4
Iquitos, forest diversity, 106; reptiles of, 240, 241, 243–50
Ireng, Rio, 31, 41–2
*Iriartea deltoidea*, 159
Iron, concretions, 39; enrichment of alluvia, 49; soil content, 56, 417, 425; water content, 406–7
Ironstone, 50

*Ischnosiphon*, 324; *arouma*, 95, 310, 314
*Isertia parviflora*, 95, 137
Islands, riverine, xv; of the Furo de Maracá, 92, 97–8, Plate 2b; savanna, 10, 93, 424–6, 427
Itacoatiara, soils, 54

*Jacaranda copaia*, 104, 107, 134, 435, 437
Jaguar, 268
*Jessenia bataua*, 89
*Junonia*, 357
*Justicia polystachya*, 95

Kalotermitidae, 348
Kaolinite, 51, 52
*Kentropyx calcarata*, 236; *striata*, 237, 245, 256, 258
*Keratella americana*, 401
King's method, 144, 153
Kink-bands, 18, 19, 22
*Kinosternon scorpioides*, 234, 243
Kjeldahl digest, 56
Koch-Grünberg, Theodor, 432, 434

La Selva, Costa Rica, reptiles, 240, 243–50
*Labidus coecus*, 344
*Lachesis muta*, 237, 243
Lagoons, *see* Borrow pits
Lakes, 439; Central Amazonian, 400; frogs, 290; seasonal, 102–3, 269, 399–400, 403–14; *see also* Borrow pits
*Lamprochernes*, 384
LANDSAT, *see* Satellite imagery
Landscape rejuvenation, 5
*Lasius*, 343
Laterites, hydromorphic, 416
Lateritic gravels, 37, 44
Laterization, 39, 56
Latosols, 7–8, 56, 63, 65, 68, 416
Leaf cigars, 345–7
Leaf litter, *see* Litter
Leaf-cutter ants, *see* Ants
*Lecane amazonica*, 401; *astia*, 401; *clara*, 401; *stichaea*, 401; *wulferti*, 401
*Lechytia chthoniiformis*, 382, 387
*Lecythis corrugata*, 92
Leguminosae, 74
*Leipomeles*, 309, 310, 317; *dorsata*, 317; *nana*, 317
Leishmaniasis, 364; visceral, 364
*Lepadella*, 400; *christinei*, 401; *donneri*, 401; *rhomboides*, 401; *tricostata*, 401
*Lepidocolaptes albolineatus*, 225
Lepidoptera, 355–8
*Leposoma*, 258; *percarinatum*, 233

*Leprolocus spinifrons*, 380
Leptodactylidae, 297–301
*Leptodactylus*, 297–300; *amazonicus*, 299; *bolivianus*, 298–9; *fuscus*, 298; *knudseni*, 298; *longirostris*, 303; *macrosternum*, 299; *mystaceus*, 299; *ocellatus*, 299; *pentadactylus*, 298; *podicipinus*, 300; *wagneri*, 234, 300
*Leptodeira*, 247, 256; *annulata*, 236, 238, 243, 257
*Leptophis ahaetulla*, 236, 243
*Leptotyphlops dimidiatus*, 234, 245; *macrolepis*, 234; *septemstriatus*, 234
*Lerodea*, 357
*Leucage*, 379
Levees, 96, Plate 4b
Lianas, 84, 89, 91, 93, 96, 101, 105
Libytheidae, 356
*Licania kunthiana*, 92, 107
Lightbox, 324
*Lindackeria paludosa*, 437
*Lindsaea*, 117; *portoricensis*, 123
*Liophis lineatus*, 238, 245; *poecilogyrus*, 234
Literature, on Maracá, xvi–xviii
*Lithodytes lineatus*, 300
Lithological units, 14–16
Lithosols, 7–8, 417
Lithotypes, of Maracá, 16–22
Litter, 89, 91, 289, 421, 422, 424; consumption, 348–53; production, 352
Liverworts, 91
Livramento, 432
Lizards, 231–62; arboreal, 235; *see also* Reptiles
*Llanos*, 9, 42, 146, 288–9; small mammals, 190, 193
Load, river, 28
Logging, on Maracá, 438
*Lomariopsis japurensis*, 124
*Lonchorhina*, 181; *aurita*, 172, 181; *marinkellei*, 181; *orinocensis*, 181
Longworth traps, 192
*Lophopodomiya*, 370
Loranthaceae, 104
Lowland terrain, 33–4
Lua, Serra da, 5
*Ludwigia nervosa*, 103; *torulosa*, 103
*Lueheopsis duckeana*, 10
*Lustrochernes*, 385–6; *intermedius*, 381, 385; *reimoseri*, 385
*Lutreolina crassicaudata*, 204
*Lutzomyia amazonensis*, 364; *anduzei*, 364; *antunesi*, 364; *aragoi*, 362; *ayrozai*, 364; *baityi*, 362; *carrerai*, 362; *chagasi*, 362; *choti*, 362; *claustrei*, 362; *dasypodogeton*, 362; *davisi*, 362, 364; *driesbachi*, 362; *evandroi*, 362; *flaviscutellata*, 364; *hirsuta*, 364; *longipalpis*, 362, 364; *longispina*, 362; *monstruosa*, 362; *olmeca*, 362; *panamensis*, 362; *paraensis*, 364; *rorotaensis*, 362; *ruii*, 362; *scaffi*, 362; *sherlocki*, 362; *spathotrichia*, 362; *spinosa*, 362; *squamiventris*, 362; *umbratilis*, 364
*Luziola subintegra*, 103
Lycaenidae, 356, 357
Lycopodiaceae, 118
Lycosidae, 380
*Lygodium venustum*, 119; *volubile*, 119
*Lysapsus limellus*, 302

*Mabuya*, 239, 258, 465; *bistriata*, 243, 465; *carvalhoi*, 239, 246, 250, 256, 258; *nigropunctata*, 235, 237, 243, 465
Macaws, 219
*Macrolobium acaciifolium*, 96, 97, 98, Plate 2b
Macrophytes, 404–5, 409
Macuxi (Makuxi) Indians, 3, 153, 432
Madeira, Rio, 357
Magnesium, soil content, 56, 66, 84–5, 108, 422, 423, 425, 428; water content, 407
Maiongong Indians, 239, 397
*Makalata armatus*, 204
Malacofauna, *see* Molluscs
Malaise traps, 362, 382
Malaria, 268, 371, 372–3, 374
Malaysia, detrivory, 352
Mammals, 449–50; *see also* Bats, Peccaries, Primates, Small mammals
*Manacus manacus*, 219
Management, forest, 109, 440
*Manataria maculata*, 358
Manaus, 3, 4; birds, 221; *campina* forest, 92; diversity of Passalidae, 329; fern diversity, 114; forest, 437; litter consumption, 352; mammal fauna, 205; reptiles, 240, 243–50; soils, 74
*Manguezal*, 95–6; underlying soils, 424
*Manicaria saccifera*, 159
Manu (Peru), reptiles, 240, 243–50, 280–1
Mapping, geological, 13, 14–15; morphostructural, 26; soils, 62–4, 68, 416–17; vegetation, 77–8
Maracá Rainforest Project, xv, xix, 73–4, 213
*Marmosa emilae*, 204; *murina*, 195, 204; *robinsoni*, 195
Marsupials, *see* Small mammals
Martin, grey-breasted, 366–7
*Mastigodryas boddaerti*, 234, 250
*Matayba*, 91

INDEX

Materials, landscape, 38–41
Mato Grosso, forest diversity, 106–7
*Mauritia flexuosa*, 9, 80, 94–6, 98, 99, 102, 119, 391, 404, 411, 412, 424, 426, 428, Plate 1; eaten by peccaries, 152, 154, 156–7, 158–60; eaten by tortoises, 157, 264, 275, 277, 281; ferns on, 124, 126; phenology, 157, 158–9, 264, 275; *see also* Buritizal
*Maximiliana maripa*, 81, 84, 85, 92, 96, 97, 107, 122, 236, 264, 421, Plate 4a; eaten by peccaries, 152, 156–7, 158–60; Triatomine bugs and, 366, 367; as disturbance indicator, 435, 437; phenology, 157, 158–9; wasps' nests on, 318, 321
*Mayaca*, 372; *fluviatilis*, 102
*Mazama americana*, 157
Medical entomology, 361–75
Medicinal plants, 435
*Megeuptychia antonoe*, 359
Melastomataceae, 140
*Melanosuchus niger*, 465
*Melocactus smithii*, 104
*Melochia simplex*, 101
*Mesomys hispidus*, 200, 204
*Mesophylla macconnelli*, 172, 184
Mesotrophic soils, 422–3, 425, 427
*Metachirus nudicaudatus*, 204
Metamorphism, secondary, 21, 22
*Metania spinata*, 411
*Metapolybia cingulata*, 321; *unilineata*, 310, 314, 321
*Metaxya rostrata*, 123
Metaxyaceae, 123
Mexico, passalid beetle diversity, 328–9
*Miagrammopes*, 380
Mica, *see* Schist
*Micrastur*, 268
*Micrathena*, 379; *schreibersi*, 379
*Microbates collaris*, 221
*Microgramma percussa*, 128; *persicariifolia*, 126
Microhylidae, 301–2
*Micronycteris*, 179; *megalotis*, 172, 179; *minuta*, 172, 179; *nicefori*, 172, 179
*Micrurus*, 257, 259; *lemniscatus*, 237, 243, 465
Middle American Unit, herpetofauna, 247–50, 256
Migmatitic features, 16, 23
Migration, birds, 221–4; peccaries, 152, 154, 160–1
*Mimon*, 180; *bennetti*, 180; *crenulatum*, 172, 180
*Mimosa pellita*, 101; *pudica*, 104
Mineral nutrients, richness, 218

Mineral prospecting, xviii–xix, 3, 22, 148–9, 158, 160, 268, 364, 374, 432, 438–9
Mineralogy, soils, 50–2
Minimum collection policy, bats, 167, 175; reptiles, 233
Miocene, 26, 269
*Mischocyttarus alboniger*, 309, 313, 315–16; *carbonarius*, 309, 315; *granadaensis*, 309, 313, 316; *injucundus*, 309, 313, 316; *labiatus*, 309, 315; *maracaensis*, 309, 314, 316, 317; *prominulus*, 309, 316–17; *surinamensis*, 309, 317
Mist-nets, 167, 171–4
Mixed bands, monkeys, 146
Mixed flocks, 220–1
*Mnaseas*, 357
Mocidade, Serra de, 8
Molluscs, 409–13
Molossidae, 171, 173, 185–6
*Molossus*, 186; *ater*, 186; *molossus*, 186, 238, 366–7
*Momotus*, 268
Monkeys, *see* Primates
*Monodelphis brevicaudata*, 195–6, 204
Monodominant forests, 107–8
*Monotagma*, 84; *plurispicatum*, 95
*Monstera dubia*, 84, Plate 3a
*Montrichardia linifera*, 96, 98, 102, 197, 404, 412
*Mora*, 107
Mora forest, 107–8
Morato de Carvalho, Celso, xvi, 232
Moreno de Oliveira, Guttemberg, 437
Mormoopidae, 171, 173, 178–9
*Morphnus guianensis*, 226
Mosquitoes, 268; *see also* Anopheles
Mosses, *see* Bryophytes
Moths, sphingid, 384
*Motommata*, 400
Mounds, *see* Hummocking
Mountains, 8
*Mourera fluviatilis*, 103
Mucajaí, granites, 16; Rio, 3, 7, 31, 41, 147; Serra de, 8
Mud, rotten, 400
Multiple stems, *see* Coppice regeneration
*Mus*, 204
Museu Integrado de Roraima, xvi, 73
Mushrooms, *see* Fungi
Mussels, 410
Mygalomorphae, 380
*Myiobius barbatus*, 221; *gaimardii*, 221
*Mylon pelopidas*, 358
Mylonite, 21, 23
Mylonitized tonalitic gneiss, 14, 21–2

*Myotis*, 185; *albescens*, 185; *nigricans*, 185, 366-7
*Myrapetra*, 313
*Myriaspora*, 275
*Myrmeciza atrothorax*, 220
Myrmecophily, 97, 338
*Myrmoborus leucophrys*, 219; *myotherinus*, 219
*Myrmornis torquata*, 219
*Myrmotherula axillaris*, 221; *longipennis*, 219; *menetriesii*, 221
*Mytilina*, 400

Names, of places, xvi
Napo, area of endemism, 224-5
Nassasseira (Nassazeira) catchment (Maracá), 48
*Nastra*, 357
*Nasutitermes*, 349-54; *guayanae*, 350, 353
Natural classification, soils, 56, 57-61
*Neacomys guianae*, 204, 206
Nearctic, rotifers, 401
Nectar thieves, on *Passiflora*, 335
Nectaries, *Passiflora longiracemosa*, 335-6
*Nectomys squamipes*, 196-7, 204
Negro (Rio), 2, 10, 147; *caatingas*, 3, 10, 92, 108
*Neivamyrmex*, 342
*Neocheiridium corticum*, 383-4, 387; *triangulare*, 384
*Neodohrniphora*, 338, 340; *curvinervis*, 340
*Neomorphus*, 267
*Neotoma*, 367
*Nephrolepis biserrata*, 115, 124
Nests, birds, 366-7; termites, 349, 351, 352, 391; wasps, 308, 310, 313, 314, 315-21, 342-3
Neuroptera, 386
*Neusticurus*, 257, 258; *racenisi*, 233, 246
Night monkeys, 146
*Niphidium crassifolium*, 129
Nipper traps, 192
Nitrogen, food content, 279, 282; soil content, 421, 424, 425, 426; water content, 406, 408
*Noctilio*, 177-8; *albiventris*, 177-8; *leporinus*, 172, 173, 177-8
Noctilionidae, 177-8
Nocturnal activity, Phoridae, 341; wasps, 314
Nogueira Neto, Paulo, xiii
*Nomamyrmex*, 342
Non-passerines, 217
*Norantea guianensis*, 98
*Norops*, 247
Nova Olinda, Fazenda, 39
Nutrient composition, tortoise food items, 267, 279, 280, 282
Nutrients, recycling, 348; released by savanna burning, 141; turnover, 423
*Nymphaea*, 103, 140, 372, 404, 412; *wittiana*, 102
Nymphalidae, 356, 357
*Nymphoides*, 373
*Nyssorhynchus*, 370

Occupation, human, 105, 432-42
*Ochthornis littoralis*, 219
*Oenocarpus*, 159; *bacaba*, 81, 85, 89
Old Northern Unit, herpetofauna, 247-50, 256
Oleandraceae, 124
Oligocene, 6, 41
Oligochaetes, *see* Earthworms
Oligotrophic conditions, 424, 426, 427
*Ololygon*, 287
Olpiidae, 383
*Olyra*, 101; *longifolia*, 95, 119
Onça, falls, 98
*Oncosclera*, 409
Ophioglossaceae, 119
*Ophioglossum ellipticum*, 119
*Opisthocomis hoazin*, 219
*Ormosia smithii*, 96, 97
*Oryzomys bicolor*, 197, 204; *capito*, 203, 206; *concolor*, 197, 198, 204; *delicatus*, 197-8, 204; *fulvescens*, 197-8; *goeldi*, 198; *macconnelli*, 204; *megacephalus*, 198, 204
Oscines, 217
Outcrops, *see* Rock outcrops
Oxisols, 63, 416
*Oxybelis aeneus*, 236, 250
Oxygen, water content, 406-7, 411-12
*Oxyrhopus petola*, 237, 250

Pacaraima, frogs, 287, 306; Serra, 2, 144, 147, 148
Pacas, *see* Agouti paca
*Pachycondyla commutata*, 351
*Pachyolpium*, 383; *irmgardae*, 383
*Palaeosuchus palpebrosus*, 234; *trigonatus*, 234
Palm tanager, 366-7
Palms, abundance, 89; interaction with peccaries, 152, 156-61; phenology, 154, 157
Pampel solution, 324
*Pandion haliaetus*, 221, 223
*Panthera onca*, 268
*Papilio*, 358
Papilionidae, 356

*Parachernes*, 386–7; *adisi*, 386; *albomaculatus*, 386; *plumosus*, 386–7
*Paramo*, 193
Parasitism, 339–43
*Paratemnoides minor*, 384
*Parawithius gracilimanus*, 387
*Parawixia*, 379
Parent material, soils, 8, 47–50, 56, 67–8, 418, 423, 427
*Parides sesostris*, 358
Parima, metamorphic suite, 16; Serra, 2, 147
*Parinari excelsa*, 97
Parthenogenesis, reptiles, 233
Passalidae, 324–9; associated with pseudoscorpions, 382, 383, 385, 387; collecting, 324; diversity, 328–9; galleries, 324, 328, 382–5; shared colonies, 327–8
*Passalus convexus*, 329; *glaberrimus*, 324, 325, 327; *interruptus*, 324, 327; *interstitialis*, 324, 327, 329, 383; *latifrons*, 329
Passerines, 217
*Passiflora longiracemosa*, 84, 331–6; ant guards, 335–6; flowers, 332–3; nectar thieves, 335; pollination, 332–5
*Passiflora vitifolia*, 334
*Paxillus leachi*, 324, 327, 329
Peat, 412–13
Peccaries, 151–63, 184; 268; behaviour, 154–7; collared, 152; communication, 155–6; conservation status, 161; feeding ecology, 152, 156–61; foraging, 156–7; group size, 156; hunting, 153, 158, 160; migration, 152, 154, 160–1; population densities, 153, 154, 158; preying on earthworms, 391; range size, 152; seeds dropped by squirrels, 203
*Pecluma plumula*, 126; *recurvata*, 115, 116, 126, 127
Pediplanation, 39
Pediplanes, 6–7, 26–7; *see also* Surfaces
Pedogenic activity, 418
Pedra Sentada catchment (Maracá), 48, 49
*Peltogyne* forest, 56, 78, 89–92, 107–8, 439; underlying soils, 65, 66, 67, 422–3, Plate 2a, Plate 3b
*Peltogyne gracilipes*, 10, 78, 79, 80, 81, 84, 85, 86, 89–92, 94, 96, 97, 98, 105, 107, 438–9, 422, Plate 2a, Plate 3b
*Penelope*, 267
Pennsylvania BLB traps, 382
Peripheric Province, 411
Peripheric regions of Amazonia, 218
*Peropteryx*, 177
Petersen method, 265

*Petiveria alliacea*, 435
pH, *see* Acidity
*Phaethornis squalidus*, 332; *superciliosus*, 332
*Pheidole*, 344
*Phenakospermum guyannense*, 81, 95, 96, 98, 99
Phenology, 277, 278; forest–savanna boundary, 135–42; palms, 154, 157, 158–9
*Philodendron melinonii*, 89
*Philodryas*, 256; *olfersii*, 234, 245, 465; *viridissimus*, 234, 465
*Philoponella*, 380
*Philydor ruficaudatus*, 219
*Phoebis sennae*, 358
Phoretic behaviour, 381–9
Phoridae, 338–43; airlifting, 343; endoparasitism, 338–42; nocturnal activity, 341; non-parasitic, 343–4; ovipositor, 340; sex ratio, 343
Phosphate, water content, 406–8
Phosphorus, food content; soil content, 421, 424, 425, 426
Photographs, aerial, 14, 27, 31, 34, 35, 44
*Phrynohyas venulosa*, 296
*Phyllomedusa bicolor*, 303; *hypocondrialis*, 296–7
Phyllostomidae, 179–84
*Phyllostomus*, 180; *elongatus*, 180; *hastatus*, 180
*Physalaemus enesefae*, 301; *ephippifer*, 234, 300–1
*Phytelephas*, 159
Phytoplankton, 77, 407, 408
*Piaya cayana*, 219; *melanogaster*, 219; *minuta*, 220
*Picumnus spilogaster*, 225
Pieridae, 356, 357, 358
*Pilherodius pileatus*, 220
Pioneer formations, 10
Pipe traps, 233
*Pipra coronata*, 213, 215; *exquisita*, 215; *iris*, 215; *nattereri*, 215; *pipra*, 219; *serena*, 215; *vilasboasi*, 215
*Piranga flava*, 224
Pisauridae, 379, 380
Pisoliths, 39
Pitfall traps, 192, 233, 378
*Pithecia pithecia*, 143, 147
*Pithys albifrons*, 219
*Pityrogramma calomelanos*, 115, 120
Place names, xvi
Planar surfaces, *see* Surfaces
Plankton, 399–401

Plantations, 105, 435, 438, 439, 440
Plants, species cited in vegetation account, 443–6
*Plasmodium*, 373, 374; *falciparum*, 373
*Platemys platycephala*, 234, 243
*Platyias leloupi*, 401
Pleistocene, 37, 41–2, 109, 226; climate change, 42–4
*Pleopeltis macrocarpa*, 126
*Pleurodema brachyops*, 303
*Plica plica*, 235, 257, 258
Plinthite, 8, 39, 43
Plio-Pleistocene, 6, 26
Pliocene, 6, 26
Pneumatophores, 95, 102
*Podocnemis*, 465; *unifilis*, 239
Podzolics, 7, 63, 65, 68
Podzols, 7–8, 54, 416
*Poeciluus kollari*, 226
Point-centred quarter transects, 74–6
*Polistes billardieri*, 313; *brevifissus*, 309, 314–15; *canadensis*, 313; *versicolor*, 309, 315
*Polites*, 357; *vibicioides*, 358
Pollen, balls, 328; robbers on *Passiflora*, 332–4
Pollination, bat, 92; of *Passiflora*, 332–5
*Polyarthra*, 400
*Polybia belemensis*, 313; *dimidiata*, 309, 310, 312, 319; *dimorpha*, 309, 319; *ignobilis*, 309, 320; *liliacea*, 309, 312, 319, 342; *occidentalis*, 309, 310, 313, 314, 319–20; *rejecta*, 309, 319; *roraimae*, 309, 310, 313, 320; *sericea*, 309, 314, 320
*Polybotrya osmundacea*, 114, 124
*Polychrus marmoratus*, 235
Polycyclic development, 5, 26, 41
*Polygonum acuminatum*, 102
Polypodiaceae, 125–6, 128–9
*Polypodium polypodioides*, 126; *triseriale*, 129
*Polytmus guainumbi*, 225
*Pomaea*, 410
*Pompeius*, 357
Ponds, *see* Lakes, Borrow pits
*Pontoscolex*, 392
Population densities, earthworms, 391; peccaries, 153, 154, 158; primates, 148; spiders, 379; tortoises, 267, 280–1; wasps, 310–12
Population, of Roraima, xix, 3, 148–9
Porcupines, 202; attacked by snake, 238
Pot sherds, 434, 435, 436, 438
*Pouteria*, 132; *hispida*, 85, 126, 275; *surumuensis*, 275

*Pradosia surinamensis*, 79, 85, 91, 92, 275, 421
Pre-Cambrian, 5, 23, 26, 48
Pre-Colombian occupation, 3, 440
Pre-Quaternary geomorphology, 41–2
Predation risk, 220
Preguiça trail, soils, 49, 54, 56, 57, 63, 424, 426; vegetation, 79, 90, 93, 94, 99, 100, 101, 105
*Prepona*, 359
Primates, 143–50; conservation status, 148; densities, 145, 148; of Amazonia, 143; of Roraima, 143
*Proechimys*, 190, 197; *cuvieri*, 201; *guyannensis*, 190, 200–1, 204; *semispinosus*, 201
Profile classification, soils, 56–62
Profile descriptions, soils, 485–7
*Progne chalybea*, 366–7
Projeto Flora Amazônica, xvi, 73
Projeto Maracá, *see* Maracá Rainforest Project
Proterozoic, 5, 26
Proto-Berbice, 7, 41, 42
*Protopolybia exigua*, 309, 310, 319
Prumama, *see* Purumame
*Pseudacteon*, 342–3
Pseudidae, 302–3
*Pseudis paradoxa*, 302–3
*Pseudoboa neuwiedii*, 236, 245
*Pseudochthonius*, 382
*Pseudopaludicola boliviana*, 301; *pusillus*, 239, 301
*Pseudopolybia vespiceps*, 309, 313, 314, 317
Pseudoscorpions, 381–9; collecting techniques, 382; diversity, 387; habitat preferences, 387–8; of Amazonia, 381
*Psophia crepitans*, 267, 268
Psychodidae, *see* Sandflies
*Psychotria poeppigiana*, 95
Pteridaceae, 120
Pteridophytes, *see* Ferns
*Pteris pearcei*, 120; *pungens*, 116, 120, 123
*Pteroglossus aracari*, 213, 214–15, 224; *flavirostris*, 224; *pluricinctus*, 213, 214–15, 224
*Pteronotus*, 178–9; *gymnonotus*, 178–9; *parnelli*, 178; *personatus*, 171, 178–9
*Ptygura linguata*, 401
Publications, on Maracá, xvi–xviii, 74
Purumame falls, 13, 16, 21, 22, 48
Purumame trail, soils, 50, 54
*Pygiptila stellaris*, 221
*Pyrgus*, 357

Pyrrhophyta, 77
*Pythonides maraca*, 356, 358

Quartz, 51, 52, 417
Quartz-biotite schist, 14, 16, 18–19, 26
Quartz-feldspathic gneiss, 16, 26
Quartzite, 29
Quaternary, 269; alluvium, 5; geomorphology, 42–4

Rabies, 174
RADAMBRASIL, xvi, 26, 62–3, 68, 416, 418
Radar imagery, 36–7, 41, 44
*Radiospongilla amazonensis*, 409
Rain, mineral content, 410
Rainfall, 3–4, 141, 264–5, 288, 378
*Ramphastos tucanus*, 224; *vitellinus*, 224
*Ramphocelus carbo*, 219
*Ramphotrigon ruficauda*, 219
*Rana palmipes*, 303
Ranching, 3, *see also* Farming
Range size, peccaries, 152
Rapids, 27–31, 97
Raptors, threatened species, 226
Rattlesnakes, 236, 238
*Rattus*, 204
Records, new for Brazil, 214, 313, 387; new for Roraima, 114, 115, 312, 324
Refuge theory, 224–5, 226
Refugia, 204
Regosols, 54
*Reinarda squamata*, 219
Relief, patterns, 31–7; relationship to soils, 68
Remote sensing, *see* Satellite imagery
*Renealmia alpinia*, 95
Reptiles, 231–62; biogeography, 240–2, 243–50; *buritizal*, 234–5; collecting techniques, 232–3; diel activity, 255–7; distribution, 250; diversity, 242–3; endemism, 247; feeding preferences, 257–60; forest, 233–7; habitat preferences, 250–5; historical faunal assemblages, 246–50; introduced species, 239; river, 239–40; rock outcrops, 235; savanna, 237–9, 247; savanna relicts, 246, 251; species composition, 242–3; species on Maracá, 462–6
Research, previous & past (on Maracá), xvi, 73–4, 190, 212–13, 232
Residua, 48–50, 51, 67–8
Resource partitioning, reptiles, 250–60
*Rheedia*, 101
*Rhinoclemmys punctularia*, 234, 250

*Rhinophylla*, 175, 182; *pumilio*, 172, 182; *fischerae*, 182
Rhinotermitidae, 348
*Rhipidomys mastacalis*, 198–9, 204
*Rhipsalis baccifera*, 98
*Rhodnius prolixus*, 368
Rhodophyta, 77
*Rhynchonycteris*, 176–7; *naso*, 172, 173, 176–7
Rhyolite, 5
Ridges, gneiss, 21; granite, 48, 49, 50, 65
*Rinorea brevipes*, 91, 92; *pubiflora*, 81
*Riparia riparia*, 221
River capture, 27, 31, 41, 42
Riverine forest, 92, 435, Plate 2; birds, 219
Rivers, *see* Uraricoera
Rock outcrops, 16, 18, 19, 20, 38; ferns, 114–15; frogs, 290; of Uraricoera, 104, Plate 2; reptiles, 235; vegetation, 103–4
Rodents, *see* Small mammals
*Rollinia exsucca*, 104
Roots, breathing, 95; buttress, 80, 90; depth, 140; feeding, 89; stilt, 95, 96
Roraima, Mt, 2, 5, 190
Roraima State, xv, 1–11; avifauna, 217, 225; climate, 3–4; fauna, 190; fern diversity, 115–16; frogs, 288–9; geology, 4–5; geomorphology, 5–7, 26; indigenous population, 3; population, xix, 3, 148–9; primates, 143; sandflies, 362, 363; small mammals, 204; soils, 7–8, 416–18; vegetation, 8–10
Roraima Formation, 5, 14, 26, 49
Roraima Sandstone, 3, 5, 8, 42, 49, 116, 358
Rotifera, *see* Rotifers
Rotifers, 399–401; annotated species list, 481–4
Rotten mud, 400
*Roupala montana*, 108
Royal Geographical Society, xv
*Ruptitermes*, 349–53; *reconditus*, 349–50; *silvestrii*, 349–50
Rupununi, 9, 41, 77, 98; highlands, 108
Rutile, 417

*Saccoloma inaequale*, 123
*Saccopteryx*, 177; *bilineata*, 177; *canescens*, 177; *leptura*, 172, 177
*Sagittaria*, 102
*Saguinus midas*, 143, 147
*Saimiri*, 206; *sciureus*, 143, 144
*Sakesphorus canadensis*, 219

Sakis, 147
*Saltator maximus*, 219
Salticidae, 379, 380
Sand, deposits, 49; quartzose, 63, 65, 416, 426; white, 65, 93, 425, 426
Sandbanks, 28, 97–8
Sandflies, 361–5; as vectors, 364; collecting techniques, 362; diversity, 362, 364; geographical distribution, 362–4; of Roraima, 362
Sandstone, Roraima, 3, 5, 8, 42, 49, 116, 358
Santa Cecilia (Ecuador), reptiles, 240, 243–50
Santa Rosa, Casa de, 166, 238, 369, 372, 373; cultivation, 107; human occupation, 434; savanna, 92, 98–100, 135–42, 238, 424, 425, 426; soils, 52
Sapará Indians, 432
Sapling density, 91, Plate 3b
Sapotaceae, 421
Saprolite, 7, 38–9
Satellite imagery, 26, 42, 76, 77, 90, 99, 415–16, 418–19
Satyrinae, 358
*Sauvagesia rubiginosa*, 95, 99
Savanna, 3, 8, 9–10, 93, 98–102, 440, Plate 1, Plate 4; birds, 220; burning, 10, 100, 136, 140, 141, 439–40; *Curatella/Byrsonima*, 9, 98–100, 108, 425–6, Plate 1b; diversity, 107–8; drainage patterns, 37; frogs, 290; islands, 10, 93, 424–6, 427; phenology, 135–42; Pleistocene, 42; relicts, 246, 251; reptile fauna, 237–9, 247; Roraima-Rupununi, 77; Santa Rosa, 92, 98–100, 135–42, 424–6; seasonally flooded, 96, 100–1, 264, 399–400, Plate 4; slope profiles, 36; soils, 52, 65–6, 425–6, 427; steppe, 3, 8, 10; *vazante*, 98, 101–2, Plate 4a
Savanna–forest boundary, *see* Forest–savanna boundary
Scarps, 5
*Scheelea*, 367
Schist, 29; mica, 5, 38, 39, 421, 423, 424, 425; quartz-biotite, 5, 14, 16, 18–19, 26
Schizaeaceae, 119
Schomburgk, Robert, 432
*Scirpus*, 373
Sciuridae, 202–3
*Sciurus aestuans*, 204; *granatensis*, 202; *pusillus*, 204; *pyrrhonotus*, 202–3, 204
*Scleria*, 264, 373; *secans*, 93, 94; *sprucei*, 101; *stipularis*, 95
*Sclerurus mexicanus*, 219

Seasonal lakes, 102–3, 269
Seasonality, 4, 9, 102–3, 140–1, 159, 194–5, 203, 224, 274, 281, 288, 307, 358; of *Anopheles*, 372, 373
Secondary vegetation, 97, 104–5, 130–4, 136, 435, 439, 440; birds, 219; diversity, 132; structure, 132
*Securidaca diversifolia*, 104
Seed dispersal, 219; birds, 140; tortoises, 281
Seed eating birds, 220
Seed predation, 157, 158–60
*Selaginella*, 117; *brevifolia*, 119; *erythropus*, 118; *seemanii*, 119; *umbrosa*, 115, 118
Selaginellaceae, 118–19
SEMA, xiii, xviii
Semi-evergreen seasonal forest, 8, 84, 91
*Senna alata*, 101
Serritermitidae, 348
*Setophaga ruticilla*, 223
Sex ratio, bats, 178; Phoridae, 345
Shannon traps, 382
Shannon-Weaver index, 168
Shared colonies, Passalidae, 327–8
Shear zones, 23
Sheetwash, 43, 44
Sherds, ceramic, 434, 435, 436, 438
Shifting cultivation, 101, 105, 131, 132, 312, 435, 438, 439, 440
*Sigmodon alstoni*, 204
Silica, water content, 406–7, 411–12
Sillimanite, 19
*Simarouba amara*, 104
*Sinningia incarnata*, 104
*Siparuna guianensis*, 95, 99
*Sittasomus griseicapillus*, 221, 225
Slash-and-burn, *see* Shifting cultivation
Slope, profiles, 37–8; undercutting, 37
Small mammals, 189–210; arboreal, 203; collecting techniques, 192; community composition, 205–6; diversity, 204–5; eaten by reptiles, 257, 258, 259; ectoparasites, 194, 200–1; habitat preferences, 203; identification key, 447–8; in Ecological Station, 367; of Roraima, 204; trapping success, 193–5, 203, 205
Snails, 435
Snakes, 231–62; anuran-eating, 259; arboreal, 236, 256; cannibalism, 258; mammal-eating, 257, 258, 259; of western Maracá, 236–7; saurophagous, 234, 258, 259; terrestrial nature, 255; *see also* Reptiles
Social wasps, *see* Wasps
*Socratea exorrhiza*, 80, 89, 96, 98, 159

Soil, 7–8, 47–69, 415–29; acidity, 54, 421–6, 427; alluvial, 8, 39, 49, 417, 425, 427; aluminium content, 56, 421, 424, 425, 427; analytical techniques, 419–20; base saturation, 61, 62, 66, 422, 425, 426; base status & ECEC, 54–6, 420, 421–6, 427; Brazilian Classification, 52, 56, 57, 61, 62; calcium content, 56, 84–5, 422, 423, 428, 435; chemical properties, 52–6, 411, 420–6; contemporary development, 49, 50; dystrophic, 63, 65, 68, 416, 427; eutrophic, 7–8, 63, 65, 68, 74, 417; fertility, 52; gleying, 7, 418, 423–4, 425, 427; hummocking, 91, 94, 95, 99; hydromorphic, 7–8, 65, 66, 68, 418; iron content, 56, 417, 425; laterization complex, 56; magnesium content, 84–5, 108; mapping, 62–4, 68, 416–17; mesotrophic, 422–3, 425, 427; mineralogy, 50–2; nitrogen content, 421, 424, 425, 426; of Brazil, 68; of Roraima, 7–8, 416–18; parent material, 8, 47–50, 56, 67–8, 418, 423, 427; profile classification, 56–62; profile descriptions, 485–7; relationship with vegetation, 65, 415–16, 420–6; spiders in, 380; suitability for tortoises, 273, 280; texture, 8, 39–40, 421–6, 427; under *angico* forest, 422–3; under *campina* forest, 424, 426–7; under *Curatella/Byrsonima* savanna, 425–6, 427; under forest–savanna transition, 424–5, 427; under *manguezal*, 424; under *Peltogyne* forest, 422–3; under savanna, 52, 65–6, 425–7
*Solenopsis*, 335, 343
Solimões Formation, 5
South America, frogs, 288
South American Unit, herpetofauna, 247–50, 256
South Central Highlands (Maracá), 48
Southwestern Highlands (Maracá), 48, 49
Speciation, Pleistocene, 42
Species composition, reptiles, 242–3
Species, new, 239, 313, 356, 358, 379, 394–7, 401
*Sphagnum*, 401
Spider monkeys, 146, 148, 154
Spiders, 377–80; collecting techniques, 377–8; diversity, 378; epiphytic, 380; habitat preferences, 378, 379–80; population densities, 379; soil-dwelling, 380; species found on Maracá, 479–80; webs, 379–80
*Spilotes pullatus*, 238, 239, 257
*Spondias mombin*, 92, 96, 106, 274, 275

Sponges, 409, 411–13; species associations, 409
*Sporophila intermedia*, 220, 225; *lineola*, 223–4; *minuta*, 220; *plumbea*, 220, 225
Squirrel monkeys, 144
Squirrels, 202–3, 204
*Stelopolybia angulata*, 309, 314, 317; *cajennensis*, 309, 317–18; *fulvofasciata*, 309, 310, 318; *multipicta*, 309, 318; *ornata*, 309, 314, 318; *pallipes*, 313; *testacea*, 309, 318
*Stenodontes spinibarbis*, 382, 384, 385
Stenotherms, 401
Steppe savanna, 3, 8, 10
Stilt roots, 95, 96
Stone axes, 434, 435, 436
Streams, 403–14
Stripped profiles, 43
Structural control, 23
*Sturnira*, 182–3; *lilium*, 172, 182–3; *tildae*, 172, 182–3
Sub-oscines, 217
Summits, 27, 31, 33, 36, 37, 38, 39, 43
Superspecies, birds, 214–16
Surfaces, planar, 5–7, 8, 14, 23, 26–7, 31–4, 36–8, 41, 44
Surinam, 269, 440; fern diversity, 116; tortoises, 269
Surucucu granite, 5
Surumu Formation, 5
Surumu, frogs, 287, 306; Rio, 3, 10
Survey techniques, forest, 74–6
*Sus barbatus*, 152
Swamp forest, 94–6
Swarm raid, army ants, 268, 336, 340–3
*Swartzia diphylla*, 108; *grandifolia*, 108; *laurifolia*, 105, 108
Sweep-net, 378
*Synoeca surinama*, 309, 314, 321; *virginea*, 309, 314, 321
*Syntermes*, 348–53; *calvus*, 352; *chaquimayensis*, 349–50; *molestus*, 349–50, 352; *parallelus*, 349–50; *solidus*, 352
Syrinx, 217

*Tabebuia uleana*, 10
*Tachyphonus luctuosus*, 221
Tacutu, fossil dunes, 42–3, 44; Rio, 7, 31, 41
*Tadarida*, 185–6; *laticaudata*, 172, 173, 185–6
Tamarins, 147
*Tantilla melanocephala*, 235, 257
*Tapirira guianensis*, 95
Tapirs, *see Tapirus terrestris*
*Tapirus terrestris*, 157, 184

*Taraba major*, 220
Taulipáng, *see* Taurepang
Taurepang Indians, 433
*Tayassu pecari*, 151–63; *tajacu*, 152; *see also* Peccaries
*Tectaria incisa*, 118, 124; *plantaginea*, 115, 124
Tectonic activity, 5, 19, 21, 23, 26
TEMA mill, 50
Temperature, 3, 264, 274, 314; water, 399–400
Tepequém, 5, 14, 22, 48, 114, 116; frogs, 287, 303
*Tepuis*, 48
*Terenura*, 221; *humeralis*, 221; *spodioptila*, 221
Termites, 348–53; activity, 391; collecting methods, 349; feeding preferences, 349–50, 352–3; nests, 349, 351, 352, 391; predation of, 351
Termitidae, 349
*Terra firme* forest, 79–89, Plate 3; birds, 218–19; ferns, 115; underlying soils, 65, 66, 67, 420–3, 426; *see also* Forest
*Terra roxa*, 8, 417
Terracettes, 37
Terrain, classification, 31–7, 44; dissected, 26, 33–4; lowland, 33–4; upland, 31–3
Tertiary, 6, 26, 41; alluvium, 5
*Testudinella*, 400; *ohlei*, 401; *robertsoni*, 401
*Tetragastris panamensis*, 85, 421
Tetragnathidae, 379
*Tetragona dorsalis*, 332–5; *handlirschii*, 332–5
Texture, soil, 8, 39–40, 421–6, 427
*Thalia*, 404, 412; *geniculata*, 101, 102; *trichocalyx*, 102
*Thamnodynastes*, 235, 465
*Thamnomanes ardesiacus*, 221; *caesius*, 219, 221
*Thamnophilus doliatus*, 220; *punctatus*, 221
*Thecadactylus rapicauda*, 235, 243, 250
*Thecla*, 357
Thelypteridaceae, 125
*Thelypteris tetragona*, 115, 125
Theraphosidae, 380
Theridiidae, 379, 380
Thiaridae, 410
Thicket vegetation, *see Carrasco*
Thomisidae, 379
*Thraupis episcopus*, 219; *palmarum*, 219, 367
*Tibouchina aspera*, 140
Ticks, 268–9

Tilting, regional, 31, 35
*Timochares trifasciata*, 358
Tipurema rapids, 19, 28, 31, 98
Titis, 147
Tocantins, forest diversity, 106
*Tolmomyias flaviventris*, 221; *sulphurescens*, 221
Tonalite, 14, 18; hornblende-biotite, 16, 19–20
Tonalitic gneiss, 14, 16–18
Tonalitic granite, 5, 16, 26
*Tonatia*, 180; *bidens*, 172, 180; *brasiliense*, 172, 180
Toototobi, Rio, 146, 147
Topé rapids, 16
Topography, 27
Tors, 38
Tortoises (*Geochelone*), 219, 234, 237, 242, 243, 245, 250, 263–84; biogeography, 269; census methods, 265; feeding behaviour, 277–9, 281–2; feeding preferences, 274–7, 281–2; habitat preferences, 264–5, 269–73; mating aggregations, 273; *Mauritia flexuosa* in diet, 157, 264, 275, 277, 281; nesting sites, 273; nutrient composition of food, 267, 279, 280, 282; population densities, 267, 280–1; predation & mortality, 267–9; shelters, 273–4; species ratio, 279–80; tracking, 266
Toucans, 224–5
*Toulicia*, 99
Tourmaline, 417
*Trachops*, 181; *cirrhosus*, 181
*Trachypogon spicatus*, 9, 10
Tracking, tortoises, 266
Traditional medicine, 435
Trails, on Maracá, xvi, xvii, 73, 144, 436
Trairao, colonization, xx, 149; Rio, 22, 23
Transects, forest, 74–6, 79, 87, 90
Transitional vegetation, *see* Forest–savanna boundary
Trapliner hummingbirds, 334–5
Trapping success, small mammals, 193–5, 203, 205
*Trattinickia rhoifolia*, 132
Tree trunks, Passalidae in, 325–9
Treefalls, 264
*Triatoma infestans*, 368; *maculata*, 366–8; *pseudomaculata*, 368
Triatomine bugs, 366–8
*Trichanthera gigantea*, 97
*Trichomanes* 117; *diversifrons*, 122; *elegans*, 122; *pinnatum*, 118; *tanaicum*, 115; *ankersii*, 115

Tridenchthoniidae, 383
*Tridenchthonius mexicanus*, 383
*Trigona williana*, 335
*Tringa flavipes*, 221; *melanoleuca*, 221; *solitaria*, 221
*Tripanurgos compressus*, 243
*Triplaris surinamensis*, 80, 96–7, 101
*Triplophyllum funestum*, 115
*Trochospongilla variabilis*, 409
Trophi mounts, 400
*Tropidurus*, 258; *hispidus*, 239, 245, 256
Trumpeters, grey-winged, 267, 268
*Trypanosoma cruzi*, 366, 367–8
*Tupinambis*, 258; *teguixin*, 237, 243, 256, 267, 465
*Turdus fumigatus*, 219; *leucomelas*, 219; *olivater*, 224
Turtles, see Chelonians
*Typhlops reticulatus*, 234, 247; *virescens*, 215–16; *stolzmanni*, 214, 215–16, 221
Tyrannidae, 217
*Tyrannus dominicensis*, 214, 216, 223; savana, 223
*Tyto alba*, 198

Uafaranda, Serra, 5
Uloboridae, 380
*Uloborus*, 380
Ultisols, 63, 68, 416
Umbrella, beating, 377
*Uncaria guianensis*, 96, 104, 310, 320
Undercutting, slope, 37
Understorey, birds, 218–19; use by tortoises, 269
Undulatory extinction, 18, 19, 20
União, Fazenda, 104
Upland terrain, 31–3
*Uranoscodon superciliosus*, 235, 256
Uraricaá, Rio, xix, 23, 28, 31, 39
Uraricoera (Rio), 1–2, 3, 7, 13, 22, 103, 212, 364, Plate 2; alluvial soils, 56; ecology, 403–14; geomorphology, 28, 31, 41, 42; metamorphic suite, 16; reptile fauna, 239–40; rock outcrops, 104; Upper, xix; see also Furo de Maracá, Furo de Santa Rosa
*Urbanus esma*, 358
*Uroderma*, 183; *bilobatum*, 183; *magnirostrum*, 183
USDA Soil Taxonomy, 57, 61, 62
*Utricularia*, 102, 373, 404, 411, 412, 401; *foliosa*, 103

*Vampyrops*, 172, 184
*Várzea*, 97; soils, 418

*Vazante*, 98, 101–2, Plate 4a; underlying soils, 63, 66, 67, 426, 427
Vectors, Chagas' disease, 366, 367–8; leishmaniasis, 364; malaria, 372–3, 374–5
Vegetation, 3, 8, 10, 71–112, 420–6; aquatic, 102–3, 290, Plate 4b; classification, 77–8, 418–19; forest, 8, 10, 79–98, Plate 2, Plate 3; heliophilous, 96; mapping, 77–8; of rock outcrops, 103–4; of Roraima, 8–10; physiognomy, 78–9, 80; Pleistocene, 42, 44; relationship with soils, 65–7, 415–16, 420–6; savanna, 9–10, 93, 98–102, Plate 1, Plate 4; secondary, 97, 104–5, 130–4, 136, 435, 439, 440; *vazante*, 98, 101–2, Plate 4a
*Vehilius*, 358
Venezuela, 2, 14, 42, 98, 143, 437; assassin bugs, 367; frogs, 288; small mammals, 193; tortoises, 269; see also Llanos
*Verres furcilabris*, 324, 325, 327, 328, 329, 382, 385
*Verrucosa*, 379
Vespertilionidae, 171, 173, 184–5
Vespidae, 307–21
*Veturius*, 385
*Vidius*, 357
Vine forest, 106
Vines, see Lianas
*Vireo olivaceus*, 221
*Virola surinamensis*, 95, 97
*Vismia cayenensis*, 95
*Vitex schomburgkiana*, 101
*Vittaria lineata*, 122
Vittariaceae, 121–2
*Volatinia jacarina*, 220

Wallaba forest, 107–8
*Wallengrenia*, 357
Wapixana Indians, 3
Ward's method, 50
Wash features, 37
*Wasmania auropunctata*, 338
Wasps, 307–21; canopy species, 314; collecting, 308; colour morphs, 313–14; diversity, 313; geographical distributions, 308, 309–10; habitat preferences, 308, 309, 310–14; nests, 308, 310, 313, 314, 315–21, 344–5; nocturnal activity, 314; use of forest clearings, 308, 309, 310–12, 314
Water table, 91, 108, 137, 140, 418, 428, 440
Water, acidity, 399–400, 405–6, 412; analytical techniques, 405–8; chemistry, 373, 405–8, 410–13; temperature, 399–400
Waterlogging, soil, 61
Wayumara Indians, 432

Weathering, 7, 28, 41, 44, 50, 56, 104, 407, 416, 417
Webb method, 153
Webs, spiders', 379–80
Weeds, 105
Weevils, *see* Beetles
White sand forest, *see* Campina
Withiidae, 387
Wolfram, 22
Wood, dead, 437
Woodcreepers, 225
Woodsiaceae, 123

X-ray diffraction, 50
*Xenodon severus*, 243
Xenoliths, 16, 19
*Xenops minutus*, 221

*Xenopsaris albinucha*, 225
Xingu, 437; forest diversity, 106
*Xiphorhynchus guttatus*, 225; *pardalotus*, 219; *picus*, 220
Xylophagy, 348
*Xylopia aromatica*, 105, 108; *discreta*, 97
*Xyris laxifolia*, 102

Yanomami (Yanomama) Indians, xix, 3
Young Northern Unit, herpetofauna, 247–50, 256

Zircon, 20, 417
Zodariidae, 380
*Zollernia grandifolia*, Plate 3a
Zoochlorellas, 411
*Zygodontomys lasiurus*, 199, 204